SENSORS AND SIGNAL CONDITIONING

SENSORS AND SIGNAL CONDITIONING

Second Edition

RAMON PALLÀS-ARENY
Universitat Politècnica de Catalunya

JOHN G. WEBSTER
University of Wisconsin—Madison

A Wiley-Interscience Publication
JOHN WILEY & SONS, INC.
New York · Chichester · Weinheim · Brisbane · Singapore · Toronto

Copyright © 2001 by John Wiley & Sons. All rights reserved.

Published simultaneously in Canada.

For ordering and customer service, call 1-800-CALL-WILEY.

Library of Congress Cataloging-in-Publication Data:
Pallàs-Areny, Ramon.
   Sensors and signal conditioning / Ramon Pallàs-Areny, John G. Webster.—2nd ed.
      p.   cm.
   "A Wiley-Interscience publication."
   Includes bibliographical references.
   ISBN 0-471-33232-1 (cloth : alk. paper)
   1. Transducers. 2. Detectors. 3. Interface circuits. I. Webster, John G., 1932– II. Title.
TK7872.T6 P25   2000
621.3815—dc21   00-028293

Printed in the United States of America.

10 9 8 7 6

CONTENTS

PREFACE

Sensors have been traditionally used for industrial process control, measurement, and automation, often involving temperature, pressure, flow, and level measurement. Nowadays, sensors enable a myriad of applications fostered by developments in digital electronics and involving the measurement of several physical and chemical quantities in automobiles, aircraft, medical products, office machines, personal computers, consumer electronics, home appliances, and pollution control.

Many of the new application areas for sensors do not pose any severe working conditions and are high-volume consumers. This makes those applications a target for semiconductor-based sensors, particularly sensors built by microfabrication techniques (microsensors), which can be manufactured in large scale. Annual sales of accelerometers and pressure sensors in the automotive industry, along with the annual sales of blood pressure sensors in the medical industry, amount to tens of millions units. Gas sensors, rate sensors, CMOS image sensors, and biosensors can similarly boom.

Classical sensors (or macrosensors) have not been superseded by the new microsensors. Many conventional sensors are still required for specialized applications, so there is no replacement for them in the foreseeable future. Nevertheless, the performance of several integrated circuits commonly used in signal conditioning has improved and allows the design of simpler circuits. Also, there are specific integrated circuits intended for conditioning the signals of common sensors such as thermocouples, RTDs, capacitive sensors, and LVDTs, and microcontrollers have become an inexpensive resource for low-cost, low-resolution analog-to-digital interfacing. Furthermore, the low cost of digital computing has moved part of the calculations and compensations closer to the sensor. The communication with a central controller is increasingly digital, and intelligent (or smart) sensors are being installed in new factories.

This second edition responds to this new scenario from the same point of view of the first edition: that of electronic engineering students or professionals interested in designing measurement systems using available sensors and integrated circuits. For each sensor we describe the working principle, advantages, limitations, types, equivalent circuit, and relevant applications. To clarify sensor types and materials, there is a new section on sensor materials and another on microsensor technology. Microsensors available for different applications are mentioned in the corresponding sections. Sensors are grouped depending on whether (a) they are variable resistors, inductors, capacitors, (b) they generate voltage, charge, or current, or (c) they are digital, semiconductor-junction based, or use some form of radiation. This approach simplifies the study of signal conditioners, which are instrumental in embedding sensors in any electronic system. Basic measurement methods and primary sensors for common physical quantities are described in an expanded section. Further information can be found in J. G. Webster (ed.), *The Measurement, Instrumentation, and Sensors Handbook*, CRC Press, 1999.

Some new sensors covered are giant magnetoresistive sensors, resistive gas sensors, liquid conductivity sensors, magnetostrictive sensors, SQUIDs, flux-gate magnetometers, Wiegand and pulse-wire sensors, position-sensitive detectors (PSDs), semiconductor-junction nuclear radiation detectors, CMOS image sensors, and biosensors. Several of these have moved from the research stage to the commercialization stage since the publication of the first edition. Velocity sensors, fiber-optic sensors, and chemical sensors, in general, receive expanded coverage because of their wider use.

Signal conditioners use new ICs with improved parameters, which often enable novel approaches to circuit design. Some new topics are error analysis of single-ended amplifiers, current feedback amplifiers, composite amplifiers, and IC current integrators. The section on noise now includes noise fundamentals, noise analysis of transimpedance and charge amplifiers, and noise and drift in resistors. Chapter 8, on digital and intelligent sensors, has been expanded by adding sections on variable oscillators including a sensor, direct microcomputer interfacing, sensor communications, and intelligent sensors.

Because the selection of the sensor influences the sensitivity, accuracy, and stability of the measurement system, we describe a broad range of sensors and list the actual specifications of several commercial sensors in tables elsewhere in the book. We have summarized several relevant specifications of common integrated circuits for signal conditioning in tables. New sections deal with basic statistical analysis of measurement results, and reliability. We give 68 worked-out examples and include a total of 103 end-of-chapter problems, many from actual design cases. The annotated solution to the problems is in an appendix at the end of the book. End-of-chapter references have been updated. For ease of reference, figures for examples or problems are respectively preceded by an E or a P. Line crossings in figures are not a connection, unless indicated by a dot.

In the study of any field, the knowledge of important dates adds perspective. Hence, this book names the discoverer and approximate date of the discovery of different physical laws applied in sensors. This may also help in preventing professionals from thinking that sensors are subsequent to the transistor (1947), the operational amplifier (1963), or the microprocessor (1971). Some sensors existed long before all of them. It is the work of electronic engineers to apply all the capabilities of integrated circuits in order that the information provided by sensors results in more economical, reliable, and efficient systems for the benefit of the humans, who certainly have limited perception but who have unmatched intelligence and creativity.

RAMON PALLÀS-ARENY
elerpa@eel.upc.es

JOHN G. WEBSTER
webster@engr.wisc.edu

Barcelona, Spain
Madison, Wisconsin
August, 2000

SENSORS AND SIGNAL CONDITIONING

1

INTRODUCTION TO SENSOR-BASED MEASUREMENT SYSTEMS

Measurements pervade our life. Industry, commerce, medicine, and science rely on measurements. Sensors enable measurements because they yield electric signals with embedded information about the measurand. Electronic circuits process those signals in order to extract that information. Hence, sensors are the basis of measurement systems. This chapter describes the basics of sensors, their static and dynamic characteristics, primary sensors for common quantities, and sensor materials and technology.

1.1 GENERAL CONCEPTS AND TERMINOLOGY

1.1.1 Measurement Systems

A system is a combination of two or more elements, subsystems, and parts necessary to carry out one or more functions. The function of a measurement system is the objective and empirical assignment of a number to a property or quality of an object or event in order to describe it. That is, the result of a measurement must be independent of the observer (objective) and experimentally based (empirical). Numerical quantities must fulfill the same relations fulfilled by the described properties. For example, if a given object has a property larger than the same property in another object, the numerical result when measuring the first object must exceed that when measuring the second object.

One objective of a measurement can be process monitoring: for example, ambient temperature measurement, gas and water volume measurement, and clinical monitoring. Another objective can be process control: for example, for temperature or level control in a tank. Another objective could be to assist

1

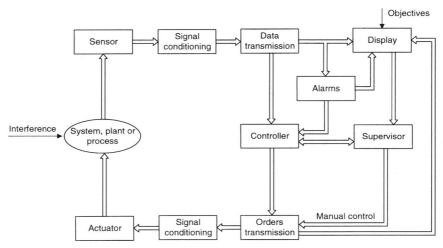

Figure 1.1 Functions and data flow in a measurement and control system. Sensors and actuators are transducers at the physical interface between electronic systems and processes or experiments.

experimental engineering: for example, to study temperature distribution inside an irregularly shaped object or to determine force distribution on a dummy driver in a car crash. Because of the nature of the desired information and its quantity, computer-aided design (CAD) does not yield complete data for these experiments. Thus measurements in prototypes are also necessary to verify the results of computer simulations.

Figure 1.1 shows the functions and data flow of a measurement and control system. In general, in addition to the acquisition of information carried out by a sensor, a measurement requires the processing of that information and the presentation of the result in order to make it perceptible to human senses. Any of these functions can be local or remote, but remote functions require information transmission. Modern measurement systems are not physically arranged according to the data flow in Figure 1.1 but are instead arranged according to their connection to the digital bus communicating different subsystems (Sections 8.6 and 8.7).

1.1.2 Transducers, Sensors, and Actuators

A *transducer* is a device that converts a signal from one physical form to a corresponding signal having a different physical form. Therefore, it is an energy converter. This means that the input signal always has energy or power; that is, signals consist of two component quantities whose product has energy or power dimension. But in measurement systems, one of the two components of the measured signal is usually so small that it is negligible, and thus only the remaining component is measured.

When measuring a force, for example, we assume that the displacement in the transducer is insignificant. That is, that there is no "loading" effect. Otherwise it might happen that the measured force is unable to deliver the needed energy to allow the movement. But there is always some power taken by the transducer, so we must ensure that the measured system is not perturbed by the measuring action.

Since there are six different kinds of signals—mechanical, thermal, magnetic, electric, chemical, and radiation (corpuscular and electromagnetic, including light)—any device converting signals of one kind to signals of a different kind is a transducer. The resulting signals can be of any useful physical form. Devices offering an electric output are called *sensors*. Most measurement systems use electric signals, and hence rely on sensors. Electronic measurement systems provide the following advantages:

1. Sensors can be designed for any nonelectric quantity, by selecting an appropriate material. Any variation in a nonelectric parameter implies a variation in an electric parameter because of the electronic structure of matter.

2. Energy does not need to be drained from the process being measured because sensor output signals can be amplified. Electronic amplifiers yield (low) power gains exceeding 10^{10} in a single stage. The energy of the amplifier output comes from its power supply. The amplifier input signal only controls (modulates) that energy.

3. There is a variety of integrated circuits available for electric signal conditioning or modification. Some sensors integrate these conditioners in a single package.

4. Many options exist for information display or recording by electronic means. These permit us to handle numerical data and text, graphics, and diagrams.

5. Signal transmission is more versatile for electric signals. Mechanical, hydraulic, or pneumatic signals may be appropriate in some circumstances, such as in environments where ionizing radiation or explosive atmospheres are present, but electric signals prevail.

Sensor and transducer are sometimes used as synonymous terms. However, sensor suggests the extension of our capacity to acquire information about physical quantities not perceived by human senses because of their subliminal nature or minuteness. Transducer implies that input and output quantities are not the same. A sensor may not be a transducer. The word *modifier* has been proposed for instances where input and output quantities are the same, but it has not been widely accepted.

The distinction between input-transducer (physical signal/electric signal) and output-transducer (electric signal/display or actuation) is seldom used at present. Nowadays, input transducers are termed *sensors*, or *detectors* for radiation, and output transducers are termed *actuators* or *effectors*. Sensors are intended to acquire information. Actuators are designed mainly for power conversion.

Sometimes, particularly when measuring mechanical quantities, a *primary sensor* converts the measurand into a measuring signal. Then a sensor would convert that signal into an electric signal. For example, a diaphragm is a primary sensor that stresses when subject to a pressure difference, and strain gages (Section 1.7.2 and Section 2.2) sense that stress. In this book we will designate as sensor the whole device, including the package and leads. We must realize, however, that we cannot directly perceive signals emerging from sensors unless they are further processed.

1.1.3   Signal Conditioning and Display

Signal conditioners are measuring system elements that start with an electric sensor output signal and then yield a signal suitable for transmission, display, or recording, or that better meet the requirements of a subsequent standard equipment or device. They normally consist of electronic circuits performing any of the following functions: amplification, level shifting, filtering, impedance matching, modulation, and demodulation. Some standards call the sensor plus signal conditioner subsystem a *transmitter*.

One of the stages of measuring systems is usually digital and the sensor output is analog. Analog-to-digital converters (ADCs) yield a digital code from an analog signal. ADCs have relatively low input impedance, and they require their input signal to be dc or slowly varying, with amplitude within specified margins, usually less than ± 10 V. Therefore, sensor output signals, which may have an amplitude in the millivolt range, must be conditioned before they can be applied to the ADC.

The display of measured results can be in an analog (optical, acoustic, or tactile) or in a digital (optical) form. The recording can be magnetic, electronic, or on paper, but the information to be recorded should always be in electrical form.

1.1.4   Interfaces, Data Domains, and Conversion

In measurement systems, the functions of signal sensing, conditioning, processing, and display are not always divided into physically distinct elements. Furthermore, the border between signal conditioning and processing may be indistinct. But generally there is a need for some signal processing of the sensor output signal before its end use. Some authors use the term *interface* to refer to signal-modifying elements that operate in the electrical domain, even when changing from one data domain to another, such as an ADC.

A *data domain* is the name of a quantity used to represent or transmit information. The concept of data domains and conversion between domains helps in describing sensors and electronic circuits associated with them [1]. Figure 1.2 shows some possible domains, most of which are electrical.

In the *analog domain* the information is carried by signal amplitude (i.e.,

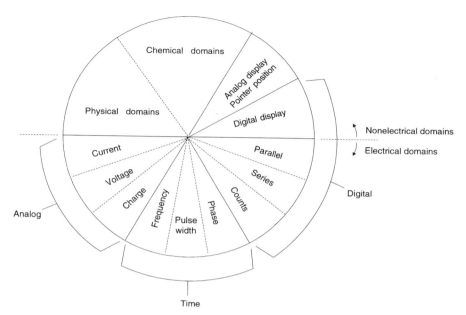

Figure 1.2 Data domains are quantities used to represent or transmit information [1]. (From H. V. Malmstadt, C. G. Enke, and S. R. Crouch, *Electronics and Instrumentation for Scientists*, copyright 1981. Reprinted by permission of Benjamin/Cummings, Menlo Park, CA.)

charge, voltage, current, or power). In the *time domain* the information is not carried by amplitude but by time relations (period or frequency, pulse width, or phase). In the *digital domain*, signals have only two values. The information can be carried by the number of pulses or by a coded serial or parallel word.

The analog domain is the most prone to electrical interference (Section 1.3.1). In the time domain, the coded variable cannot be measured—that is, converted to the numerical domain—in a continuous way. Rather, a cycle or pulse duration must elapse. In the digital domain, numbers are easily displayed.

The structure of a measurement system can be described then in terms of domain conversions and changes, depending on the direct or indirect nature of the measurement method.

Direct physical measurements yield quantitative information about a physical object or action by direct comparison with a reference quantity. This comparison is sometimes simply mechanical, as in a weighing scale.

In *indirect physical measurements* the quantity of interest is calculated by applying an equation that describes the law relating other quantities measured with a device, usually an electric one. For example, one measures the mechanical power transmitted by a shaft by multiplying the measured torque and speed of rotation, the electric resistance by dividing dc voltage by current, or the traveled distance by integrating the speed. Many measurements are indirect.

1.2 SENSOR CLASSIFICATION

A great number of sensors are available for different physical quantities. In order to study them, it is advisable first to classify sensors according to some criterion. White [10] provides additional criteria to those used here.

In considering the need for a power supply, sensors are classified as modulating or self-generating. In modulating (or active) sensors, most of the output signal power comes from an auxiliary power source. The input only controls the output. Conversely, in self-generating (or passive) sensors, output power comes from the input.

Modulating sensors usually require more wires than self-generating sensors, because wires different from the signal wires supply power. Moreover, the presence of an auxiliary power source can increase the danger of explosion in explosive atmospheres. Modulating sensors have the advantage that the power supply voltage can modify their overall sensitivity. Some authors use the terms *active* for self-generating and *passive* for modulating. To avoid confusion, we will not use these terms.

In considering output signals, we classify sensors as analog or digital. In *analog sensors* the output changes in a continuous way at a macroscopic level. The information is usually obtained from the amplitude, although sensors with output in the time domain are usually considered as analog. Sensors whose output is a variable frequency are called *quasi-digital* because it is very easy to obtain a digital output from them (by counting for a time).

The output of *digital sensors* takes the form of discrete steps or states. Digital sensors do not require an ADC, and their output is easier to transmit than that of analog sensors. Digital output is also more repeatable and reliable and often more accurate. But regrettably, digital sensors cannot measure many physical quantities.

In considering the operating mode, sensors are classified in terms of their function in a deflection or a null mode. In *deflection sensors* the measured quantity produces a physical effect that generates in some part of the instrument a similar but opposing effect that is related to some useful variable. For example, a dynamometer to measure force is a sensor where the force to be measured deflects a spring to the point where the force it exerts, proportional to its deformation, balances the applied force.

Null-type sensors attempt to prevent deflection from the null point by applying a known effect that opposes that produced by the quantity being measured. There is an imbalance detector and some means to restore balance. In a weighing scale, for example, the placement of a mass on a pan produces an imbalance indicated by a pointer. The user has to place one or more calibrated weights on the other pan until a balance is reached, which can be observed from the pointer's position.

Null measurements are usually more accurate because the opposing known effect can be calibrated against a high-precision standard or a reference quantity. The imbalance detector only measures near zero; therefore it can be very

TABLE 1.1   Sensor Classifications According to Different Exhaustive Criteria

Criterion	Classes	Examples
Power supply	Modulating	Thermistor
	Self-generating	Thermocouple
Output signal	Analog	Potentiometer
	Digital	Position encoder
Operation mode	Deflection	Deflection accelerometer
	Null	Servo-accelerometer

sensitive and does not require any calibration. Nevertheless, null measurements are slow; and despite attempts at automation using a servomechanism, their response time is usually not as short as that of deflection systems.

In considering the input–output relationship, sensors can be classified as zero, first, second, or higher order (Section 1.5). The order is related to the number of independent energy-storing elements present in the sensor, and this affects its accuracy and speed. Such classification is important when the sensor is part of a closed-loop control system because excessive delay may lead to oscillation [6].

Table 1.1 compares the classification criteria above and gives examples for each type in different measurement situations. In order to study these myriad devices, it is customary to classify them according to the measurand. Consequently we speak of sensors for temperature, pressure, flow, level, humidity and moisture, pH, chemical composition, odor, position, velocity, acceleration, force, torque, density, and so forth. This classification, however, can hardly be exhaustive because of the seemingly unlimited number of measurable quantities. Consider, for example, the variety of pollutants in the air or the number of different proteins inside the human body whose detection is of interest.

Electronic engineers prefer to classify sensors according to the variable electrical quantity—resistance, capacity, inductance—and then to add sensors generating voltage, charge, or current, and other sensors not included in the preceding groups, mainly p–n junctions and radiation-based sensors. This approach reduces the number of groups and enables the direct study of the associated signal conditioners. Table 1.2 summarizes the usual sensors and sensing methods for common quantities.

1.3   GENERAL INPUT–OUTPUT CONFIGURATION

1.3.1   Interfering and Modifying Inputs

In a measurement system the sensor is chosen to gather information about the measured quantity and to convert it to an electric signal. A priori it would be unreasonable to expect the sensor to be sensitive to only the quantity of interest

TABLE 1.2 Usual Sensors and Sensing Methods for Common Quantities

Sensor type	Acceleration Vibration	Flow Rate Point velocity	Force	Humidity Moisture
Resistive	Mass–spring + strain gage	Anemometer	Strain gage	Humistor
		Thermistor		
		Target + strain gage		
Capacitive	Mass–spring + variable capacitor		Capacitive strain gage	Dielectric-variation capacitor
Inductive and electro-magnetic	Mass–spring + LVDT	Faraday's law	Load cell + LVDT	
		Rotameter + LVDT	Magnetostriction	
Self-generating	Mass–spring + piezo-electric sensor	Thermal transport + thermocouple	Piezoelectric sensor	
Digital		Impeller, turbine		SAW sensor
		Positive displacement		
		Vortex shedding		
PN junction				
Optic, fiber optic		Laser anemometry		Chilled mirror
Ultrasound		Doppler effect		
		Travel time		
		Vortex		
Other		Differential pressure		
		Variable area + level sensor (open channel)		
		Variable area + displacement		
		Coriolis effect + force		

Quantity				
Level	Position Distance Displacement	Pressure	Temperature	Velocity Speed
Float + potentiometer	Magnetoresistor	Bourdon tube + potentiometer	RTD	
LDR	Potentiometer	Diaphragm + strain gage	Thermistor	
Thermistor	Strain gage			
Variable capacitor	Differential capacitor	Diaphragm + variable capacitor		
Magnetostriction	Eddy currents	Diaphragm + LVDT		Eddy currents
Magnetoresistive	Hall effect	Diaphragm + variable reluctance		Hall effect
Float + LVDT	Inductosyn			Faraday's law
Eddy currents	LVDT			LVT
	Resolver, synchro			
	Magnetostriction			
		Piezoelectric sensor	Pyroelectric sensor	
			Thermocouple	
Vibrating rod	Position encoder	Bourdon tube + encoder	Quartz oscillator	Incremental encoder
Float + pulley		Bourdon tube or bellows + quartz resonator		
		Diaphragm + vibrating wire		
Photoelectric	Photoelectric sensor		Diode	
			Bipolar transistor	
			T/I converter	
		Diaphragm + light reflection		
Absorption	Travel time			Doppler effect
Travel time				
Differential pressure		Liquid-based manometer + level sensor		
Microwave radar				
Nuclear radiation				

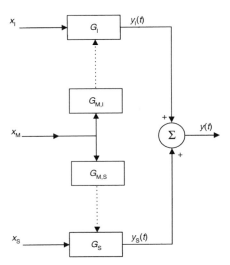

Figure 1.3 Effect of internal and external perturbations on measurement systems. x_S is the signal of interest. $y(t)$ is the system output. x_I is an interference or external perturbation. x_M is a modifying input. (From E. O. Doebelin, *Measurement Systems Application and Design*, 4th ed., copyright 1990. Reprinted by permission of McGraw-Hill, New York.)

and also to expect the output signal to be entirely due to the input signal. No measurement is ever obtained under ideal circumstances; therefore we must address real situations. We follow here the method proposed by Doebelin [2]. Figure 1.3 shows a general block diagram for classifying desired signal gains and interfering input gain for instruments. The desired signal x_S passes through the gain block G_S to the output y. Interfering inputs x_I represent quantities to which the instrument is unintentionally sensitive. These pass through the gain block G_I to the output y. Modifying inputs x_M are the quantities that through $G_{M,S}$ cause a change in G_S for the desired signal and through $G_{M,I}$ cause a change in G_I for interfering inputs. The gains G can be linear, nonlinear, varying, or random.

For example, to measure a force, it is common to use strain gages (Section 2.2). Strain gages operate on the basis of variation in the electric resistance of a conductor or semiconductor when stressed. Because temperature change also yields a resistance variation, we can regard any temperature variation as an interference or external disturbance x_I with gain G_I. At the same time, to measure resistance changes as a result of the stress, an electronic amplifier is required. Since any temperature change x_M through $G_{M,S}$ affects the amplifier gain G_S and therefore the output, it turns out that a temperature variation also acts as a modifying input x_M. If the same force is measured with a capacitive gage (Section 4.1), a temperature variation does not interfere but can still modify the amplifier gain.

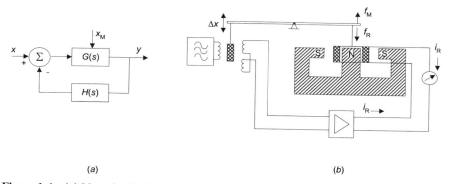

(a)   (b)

Figure 1.4 (*a*) Negative feedback method to reduce the effect of internal perturbations. Block *H* may be insensitive to those perturbations because it handles lower power than block *G*. (*b*) Force-to-current converter that relies on negative feedback and a balance sensor.

1.3.2 Compensation Techniques

The effects of interfering and modifying inputs can be reduced by changing the system design or by adding new elements to it. The best approach is to design systems insensitive to interference and that respond only to the desired signals. In the preceding example, it would have been best to use strain gages with a low temperature coefficient ($G_I = 0$). Thin, narrow, long magnetic sensors are only sensitive to magnetic fields parallel to their long dimension. In designing sensors for vector mechanical quantities, it would be best to obtain a unidirectional sensitivity and a low transverse sensitivity—that is, in directions perpendicular to the desired direction. In electronic circuits, low-drift components such as metal-film resistors and NP0 capacitors are less sensitive to temperature. Nevertheless, this method is not always possible for obvious practical reasons.

Negative feedback is a common method to reduce the effect of modifying inputs, and it is the method used in null measurement systems. Figure 1.4*a* shows the working principle. It assumes that the measurement system and the feedback are linear and can be described by their respective transfer functions $G(s)$ and $H(s)$. The input–output relation is

$$\frac{Y(s)}{X(s)} = \frac{G(s)}{1 + G(s)H(s)} \cong \frac{1}{H(s)} \tag{1.1}$$

where the approximation is valid when $G(s)H(s) \gg 1$. If the negative feedback is insensitive to the modifying input, and it has been designed so that the system remains stable, then the output signal is not affected by the modifying input.

The advantage of such a solution stems from the different physical characteristics of the elements described by $G(s)$ and $H(s)$. The probable insensitivity of H to a modifying input is a consequence of its lower power-handling capacity than G. This also results in higher accuracy and linearity for H. Moreover, negative feedback results in less energy extracted from the measured system because G is designed very large. The force-to-current converter in Figure 1.4b relies on negative feedback. The force to be measured, f_M, is compared with a restoring force f_R, generated by an internal moving-coil system. f_R is proportional to the current i_R in the coil, and i_R is proportional to the output voltage from the displacement sensor—here an LVDT (Section 4.2.3)—that senses the balance between f_M and f_R. If the amplifier gain is high enough, a very small input voltage from the sensor yields a current high enough to produce a force f_R able to balance f_M. Because i_R is proportional to f_R and $f_R \approx f_M$, we can determine f_M from i_R, regardless, for example, of the sensor linearity.

Filtering is a common method for interference reduction. A filter is any device that separates signals according to their frequency or another criterion. Filters are very effective when frequency spectra of signals and interference do not overlap. Filters can be placed at the input or at any intermediate stage. They can be electric, mechanical (e.g., to reduce vibrations), pneumatic, thermal (e.g., a high mass covering to reduce turbulence effects when measuring the average temperature of a flowing fluid), or electromagnetic. Filters placed at intermediate stages are usually electric.

Another common compensation technique for interfering and modifying inputs is the use of opposing inputs, often applied to compensate for temperature variations. If, for example, a gain that depends on a resistor having a positive temperature coefficient changes due to a temperature change, another resistor can be placed in series with the affected resistor. If the added resistor has a negative temperature coefficient, it is possible to keep the gain constant in spite of temperature changes. This method is also used for temperature compensation in strain gages, sensor-bridge supply, catalytic gas sensors, resistive gas sensors, and copper-wire coils (e.g., in electromagnetic relays, galvanometers, and tachometers), as well as to compensate vibration in piezoelectric sensors.

Finally, when the mathematical relationship between the interference and sensor output is known, interference can be compensated by digital calculation after measuring the magnitude of the interfering variable—for example, temperature in a pressure sensor. This method is common in smart sensors.

1.4   STATIC CHARACTERISTICS OF MEASUREMENT SYSTEMS

Because the sensor influences the characteristics of the whole measurement system, it is important to describe its behavior in a meaningful way. In most measurement systems the quantity to be measured changes so slowly that it is only necessary to know the static characteristics of sensors.

Nevertheless, the static characteristics influence also the dynamic behavior of the sensor—that is, its behavior when the measured quantity changes with time. However, the mathematical description of the joint consideration of static and dynamic characteristics is complex. As a result, static and dynamic behavior are studied separately. The concepts used to describe static characteristics are not exclusive to sensors. They are common to all measurement instruments.

1.4.1 Accuracy, Precision, and Sensitivity

Accuracy is the quality that characterizes the capacity of a measuring instrument for giving results close to the true value of the measured quantity. The "true," "exact," or "ideal" value is the value that would be obtained by a perfect measurement. It follows that true values are, by nature, indeterminate. The conventional true value of a quantity is "the value attributed to a particular quantity and accepted, sometimes by convention, as having an uncertainty appropriate for a given purpose" [3].

Sensor accuracy is determined through static calibration. It consists of keeping constant all sensor inputs, except the one to be studied. This input is changed very slowly, thus taking successive constant values along the measurement range. The successive sensor output results are then recorded. Their plot against input values forms the calibration curve. Obviously each value of the input quantity must be known. Measurement standards are such known quantities. Their values should be at least ten times more accurate than that of the sensor being calibrated.

Any discrepancy between the true value for the measured quantity and the instrument reading is called an *error*. The difference between measurement result and the true value is called *absolute error*. Sometimes it is given as a percentage of the maximal value that can be measured with the instrument (full-scale output, FSO) or with respect to the difference between the maximal and the minimal measurable values—that is, the measurement range or *span*. Therefore we have

$$\text{Absolute error} = \text{Result} - \text{True value}$$

The common practice, however, is to specify the error as a quotient between the absolute error and the true value for the measured quantity. This quotient is called the *relative error*. Relative error usually consists of two parts: one given as a percentage of the reading and another that is constant (see Problem 1.1). The constant part can be expressed as a percentage of the FSO, a threshold value, a number of counts in digital instruments, or a combination of these. Then,

$$\text{Relative error} = \frac{\text{Absolute error}}{\text{True value}}$$

Because true values are indeterminate, error calculations use a conventional true value.

Some sensors have an error (uncertainty) specified as a percentage of the FSO. If the measurement range includes small values, the full-scale specification implies that for them the measurement error is very large. Some sensors have a relative error specified as a percentage of the reading. If the measurement range includes small values, the percent-of-reading specification implies unbelievable low errors for small quantities.

The *Accuracy Class* concept facilitates the comparison of several sensors with respect to their accuracy. All the sensors belonging to the same class have the same measurement error when the applied input does not exceed their nominal range and work under some specified measurement conditions. That error value is called the *index of class*. It is defined as the percent measurement error, referred to a conventional value that is the measurement range or the FSO. For example, a class 0.2 displacement sensor whose end-of-scale displacement is 10 mm, in the specified reference conditions, has an error lower than 20 μm when measuring any displacement inside its measuring range.

The measured value and its error must be expressed with consistent numerical values. That is, the numerical result of the measurement must not have more figures than those that can be deemed reliable by considering the uncertainty of the result. For example, when measuring ambient temperature, $20\,°C \pm 1\,°C$ is a result correctly expressed, while $20\,°C \pm 0.1\,°C$, $20.5\,°C \pm 1\,°C$ and $20.5\,°C \pm 10\%$ are incorrect expressions because the measured value and the error have different uncertainty (see Problem 1.2).

Care must be taken also when converting units to avoid false gains of accuracy. For example, a 19.0 inch length (1 inch = 25.4 mm) should not be directly expressed as 482.6 mm, since the original figure suggests an uncertainty of tenths of an inch while the converted figure indicates an uncertainty of tenths of a millimeter. That is, the original result indicates that the length is between 485 mm and 480 mm, while the converted result would suggest that it is between 482.5 mm and 482.7 mm.

Precision is the quality that characterizes the capability of a measuring instrument of giving the same reading when repetitively measuring the same quantity under the same prescribed conditions (environmental, operator, etc.), without regard for the coincidence or discrepancy between the result and the true value. Precision implies an agreement between successive readings and a high number of significant figures in the result. Therefore, it is a necessary but not sufficient condition for accuracy. Figure 1.5 shows different possible situations.

The *repeatability* is the closeness of agreement between successive results obtained with the same method under the same conditions and in a short time interval. Quantitatively, the repeatability is the minimum value that exceeds, with a specified probability, the absolute value of the difference between two successive readings obtained under the specified conditions. If not stated, it is assumed that the probability level is 95%.

The *reproducibility* is also related to the degree of coincidence between successive readings when the same quantity is measured with a given method, but in this case with a long-term set of measurements or with measurements carried out by different people or performed with different instruments or in different laboratories. Quantitatively, the reproducibility is the minimal value that

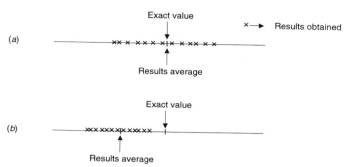

Figure 1.5 Measurement situations illustrating the difference between accuracy and precision. In case (*a*) there is a high accuracy and a low repeatability. In case (*b*) the repeatability is higher but there is a low accuracy.

exceeds, with a given probability, the absolute value of the difference between two single measurement results obtained under the above-mentioned conditions. If not stated, it is assumed that the probability level is 95%.

When a sensor output changes with time (for a constant input), it is sometimes said that there are instabilities and that the sensor drifts. In particular, some sensors have zero and scale factor drifts specified. The *zero drift* describes output variations when the input is zero. *Scale factor drift* describes sensitivity changes.

The *sensitivity* or scale factor is the slope of the calibration curve, whether it is constant or not along the measurement range. For a sensor in which output y is related to the input x by the equation $y = f(x)$, the sensitivity $S(x_a)$, at point x_a, is

$$S(x_a) = \frac{dy}{dx}\bigg|_{x=x_a} \tag{1.2}$$

It is desirable in sensors to have a high and, if possible, constant sensitivity. For a sensor with response $y = kx + b$ the sensitivity is $S = k$ for the entire range of values for x where it applies. For a sensor with response $y = k^2 x + b$ the sensitivity is $S = 2kx$, and it changes from one point to another over the measurement range.

1.4.2 Other Characteristics: Linearity and Resolution

Accuracy, precision, and sensitivity are the characteristics that sufficiently describe the static behavior of a sensor. But sometimes others are added or substituted when it is necessary to describe alternative behavior or behavior that is of particular interest for a given case; likewise, characteristics can be added that are complementary to describe the suitability of a measurement system for a specific application.

The *linearity* describes the closeness between the calibration curve and a specified straight line. Depending on which straight line is considered, several definitions apply.

Independent Linearity. The straight line is defined by the least squares criterion. With this system the maximal positive error and the minimal negative error are equal. This is the method that usually gives the "best" quality.

Zero-Based Linearity. The straight line is also defined by the least squares criterion but with the additional restriction of passing through zero.

Terminal-Based Linearity. The straight line is defined by the output corresponding to the lower input and the theoretical output when the higher input is applied.

End-Points Linearity. The straight line is defined by the real output when the input is the minimum of the measurement range and the output when the input is the maximum (FSO).

Theoretical Linearity. The straight line is defined by the theoretical predictions when designing the sensor.

Figure 1.6 shows these different straight lines for a sensor with a given calibration curve. In sum, the linearity of the calibration curve indicates to what

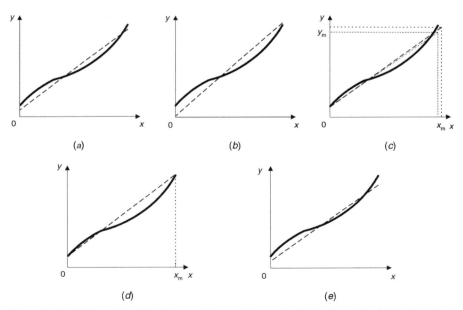

Figure 1.6   Different straight lines used as a reference to define linearity: (*a*) independent linearity (least squares method); (*b*) zero-based linearity (least squares adjusted to zero); (*c*) terminal-based linearity; (*d*) end-points-defined linearity; (*e*) theoretical linearity.

extent a sensor's sensitivity is constant. Nevertheless, for a sensor to be acceptable, it does not need to have a high linearity. The interest of linearity is that when sensitivity is constant we only need to divide the reading by a constant value (the sensitivity) in order to determine the input. In linear instruments the nonlinearity equals the inaccuracy.

Current measurement systems incorporate microprocessors so that there is more interest in repeatability than in linearity, because we can produce a look-up table giving input values corresponding to measured values. By using interpolation, it is possible to reduce the size of that table to a reasonable dimension.

The main factors that influence linearity are resolution, threshold, and hysteresis. The *resolution* (or discrimination) is the minimal change of the input necessary to produce a detectable change at the output. When the input increment is from zero, then it is called the *threshold*. When the input signal can display fast changes, the *noise floor* of the sensor determines the resolution. Noise is a random fluctuation of the sensor output unrelated to the measured quantity.

The *hysteresis* refers to the difference between two output values that correspond to the same input, depending on the direction (increasing or decreasing) of successive input values. That is, similarly to the magnetization in ferromagnetic materials (Section 1.8.2), it can happen that the output corresponding to a given input depends on whether the previous input was higher or lower than the present one.

1.4.3 Systematic Errors

The static calibration of a sensor allows us to detect and correct the so-called systematic errors. An error is said to be *systematic* when in the course of measuring the same value of a given quantity under the same conditions, it remains constant in absolute value and sign or varies according to a definite law when measurement conditions change. Because time is also a measurement condition, the measurements must be made in a short time interval. Systematic errors yield measurement bias.

Such errors are caused not only by the instrument, but also by the method, the user (in some cases), and a series of factors (climatic, mechanical, electrical, etc.) that never are ideal—that is, constant and known.

The presence of systematic errors can therefore be discovered by measuring the same quantity with two different devices, by using two different methods, by using the readings of two different operators, or by changing measurement conditions in a controlled way and observing their influence on results. To determine the consistency of the different results it is necessary to use statistical methods [4]. In any case, even in high-accuracy measurements, there is always some risk that a systematic error may remain undetected. The goal therefore is to have a very low risk for large errors to remain undetected.

Errors in indirect measurements propagate from each measured quantity to the estimated quantity, so that indirect measurements are usually less accurate than direct measurements (see Problem 1.3).

Example 1.1   In order to measure the drop in voltage across a resistor, we consider two alternative methods: (1) Use a voltmeter, whose accuracy is about 0.1 % of the reading. (2) Use an ammeter, whose accuracy is also about 0.1 % of the reading and apply Ohm's law. If the resistor has 0.1 % tolerance, which method is more accurate?

We first differentiate Ohm's law to obtain

$$dV = RdI + IdR$$

Dividing each term by V yields

$$\frac{dV}{V} = \frac{RdI + IdR}{V} = \frac{RdI + IdR}{IR} = \frac{dI}{I} + \frac{dR}{R}$$

For small variations, we can approximate differentials by increments to obtain

$$\frac{\Delta V}{V} = \frac{\Delta I}{I} + \frac{\Delta R}{R}$$

The relative uncertainty for the current and resistance add together. Therefore, the uncertainty in the voltage when measuring current is

$$\frac{\Delta V}{V} = \frac{0.1}{100} + \frac{0.1}{100} = 0.2\%$$

The uncertainty when measuring voltage directly is 0.1 %, hence lower.

1.4.4   Random Errors

Random errors are those that remain after eliminating the causes of systematic errors. They appear when the same value of the same quantity is measured repeatedly, using the same instrument and the same method. They have the following properties:

1. Positive and negative random errors with the same absolute value have the same occurrence probability.
2. Random errors are less probable as the absolute value increases.
3. When the number of measurements increases, the arithmetic mean of random errors in a sample (set of measurements) approaches zero.
4. For a given measurement method, random errors do not exceed a fixed value. Readings exceeding that value should be repeated and, if necessary, studied separately.

Random errors are also called *accidental* (or fortuitous) errors, thus meaning that they may be unavoidable. The absence of changes from one reading to another when measuring the same value of the same quantity several times does not necessarily imply an absence of random errors. It may happen, for example, that the instrument does not have a high enough resolution—that is, that its ability to detect small changes in the measured quantity is rather limited and therefore the user does not perceive them.

The presence of random errors implies that the result of measuring n times a measurand x is a set of values $\{x_1, x_2, \ldots, x_n\}$. If there is no systematic error, the best estimate of the actual value of the measurand is the average of the results:

$$\hat{x}_n = \frac{x_1 + x_2 + \cdots + x_n}{n} = \frac{\sum\limits_{i=1}^{n} x_i}{n} \tag{1.3}$$

Were n infinite, (1.3) would yield a conventional true value for x. When n is finite, however, each set of n measurements yields different x_i and a different average. These averages follow a Gaussian distribution whose variance is σ^2/n, where σ^2 is the variance of x. Then,

$$\text{Prob}\left[-k \le \frac{\hat{x}_n - x}{\sigma/\sqrt{n}} \le +k\right] = 1 - \alpha \tag{1.4}$$

where k and α can be obtained from tables for the unit normal (Gaussian) distribution. From (1.4), we obtain

$$\text{Prob}\left[\hat{x}_n - k\frac{\sigma}{\sqrt{n}} \le x \le \hat{x}_n + k\frac{\sigma}{\sqrt{n}}\right] = 1 - \alpha \tag{1.5}$$

which yields the (*confidence*) *interval* with a probability $1 - \alpha$ of including the true value x. The term $\pm k\sigma/\sqrt{n}$ is also called the *uncertainty* (see Problems 1.4 and 1.5).

Example 1.2   Determine the confidence interval that has 50% probability of including the true value of a quantity when the average from n measurements is $\hat{x}_n$ and the variance is σ^2.

For $k = 0.67$ the tail area of unit normal distribution is 0.2514, and for $k = 0.68$ the tail area is 0.2483. We need the value for a tail area of $(1 - 0.5)/2 = 0.25$ because we are looking for a two-sided interval. By interpolating, we obtain

$$k = 0.67 + \frac{0.68 - 0.67}{0.2483 - 0.2514}(0.25 - 0.2514) = 0.67 + 0.0045 = 0.6745$$

Hence, the interval $\hat{x}_n \pm 0.6745\sigma/\sqrt{n}$ has 50% probability of including the true value. $0.6745\sigma/\sqrt{n}$ is sometimes termed *probable error*; but it is not an "error," nor is it "probable."

Often, however, the variance of the population of (infinite) possible results for the measurand is unknown. If we estimate that variance from a sample of n results by

$$s_n^2 = \frac{\sum_{i=1}^{n} (x_i - \hat{x}_n)^2}{n - 1} = \frac{\sum_{i=1}^{n} x_i^2 - \frac{\left(\sum_{i=1}^{n} x_i\right)^2}{n}}{n - 1} \qquad (1.6)$$

it is not possible to directly substitute s_n for σ in (1.4). Nevertheless, when the distribution of possible results is Gaussian and $n > 31$, (1.4) holds true even if s_n replaces σ. For $n < 31$, $(\hat{x}_n - x)/(s_n/\sqrt{n})$ follows a Student t distribution instead. Therefore,

$$\text{Prob}\left[-t_{1-\alpha/2}(n - 1) \le \frac{\hat{x}_n - x}{s_n/\sqrt{n}} \le +t_{1-\alpha/2}(n - 1)\right] = 1 - \alpha \qquad (1.7a)$$

where $t_{1-\alpha/2}(n - 1)$ is the probability point of the t distribution with $n - 1$ degrees of freedom, corresponding to a tail area probability α. The confidence interval follows from

$$\text{Prob}\left[\hat{x}_n - t_{1-\alpha/2}(n - 1)\frac{s_n}{\sqrt{n}} \le x \le \hat{x}_n + t_{1-\alpha/2}(n - 1)\frac{s_n}{\sqrt{n}}\right] = 1 - \alpha \qquad (1.7b)$$

Example 1.3 Determine the confidence interval that has a 99% probability of including the true value of a quantity when the average from 10 measurements is $\hat{x}_n$ and the sample variance is s_n^2. Compare the result with that when the population variance σ^2 is known.

For $10 - 1 = 9$ degrees of freedom, the t value for a $(1 - 0.99)/2 = 0.005$ tail area probability is $t_{0.995}(9) = 3.250$. The corresponding confidence interval is $\hat{x}_n \pm 3.25 s_n/\sqrt{10} = \hat{x}_n \pm 1.028 s_n$. Had we known σ, for a tail area of $(1 - 0.99)/2 = 0.005$ the normal distribution yields $k = 2.576$. Hence, the confidence interval would be $\hat{x}_n \pm 2.576\sigma/\sqrt{10} = \hat{x}_n \pm 0.815\sigma$, which is narrower than that when σ is unknown.

If s_n has been calculated from a sample of n results, perhaps from previous experiments, it is still possible to determine the confidence interval for a set of m data points by replacing $s_n/\sqrt{m}$ for $s_n/\sqrt{n}$ in (1.7b).

If there are systematic errors in addition to random errors, when calculating the mean of several readings, random errors cancel and only systematic errors remain. Because systematic errors are reproducible, they can be determined for some specified measurement conditions, and then the reading can be corrected when measuring under the same conditions. This calculation of the difference between the true value and the measured value is performed during the calibration process under some specified conditions. Furthermore, during that process the instrument is usually adjusted to eliminate that error. When making a single measurement, under the same conditions, only the random component of error remains.

In practice, however, during the calibration process, only systematic errors for some very specific conditions can be eliminated. Therefore, under different measurement conditions some systematic errors even greater than the random ones may be present. Product data sheets state these errors, usually through the range b_x having a given probability $1 - \alpha$ of enclosing the true value. The overall uncertainty can be then calculated by [5]

$$u_x = \pm t_{1-\alpha/2}\sqrt{\left(\frac{b_x}{2}\right)^2 + \left(\frac{s_n}{\sqrt{n}}\right)^2}\tag{1.8}$$

The usual confidence level in engineering is 95%, so that for $n > 31$, $t_{97.5} = 1.96$.

1.5 DYNAMIC CHARACTERISTICS

The sensor response to variable input signals differs from that exhibited when input signals are constant, which is described by static characteristics. The reason is the presence of energy-storing elements, such as inertial elements (mass, inductance, etc.) and capacitance (electric, thermal, fluid, etc.). The dynamic characteristics are the dynamic error and speed of response (time constant, delay). They describe the behavior of a sensor with applied variable input signals.

The *dynamic error* is the difference between the indicated value and the true value for the measured quantity, when the static error is zero. It describes the difference between a sensor's response to the same input magnitude, depending on whether the input is constant or variable with time.

The *speed of response* indicates how fast the measurement system reacts to changes in the input variable. A delay between the applied input and the corresponding output is irrelevant from the measurement point of view. But if the sensor is part of a control system, that delay may result in oscillations.

To determine the dynamic characteristics of a sensor, we must apply a variable quantity to its input. This input can take many different forms, but it is usual to study the response to transient inputs (impulse, step, ramp), periodic inputs (sinusoidal), or random inputs (white noise). In linear systems, where superposition holds, any one of these responses is enough to fully characterize the system. The selection of one input or another depends on the kind of sensor. For example, it is difficult to produce a temperature with sinusoidal variations, but it is easy to cause a sudden temperature change such as a step. On the other hand, it is easier to cause an impulse than to cause a step of acceleration.

To mathematically describe the behavior of a sensor, we assume that its input and output are related through a constant-coefficient linear differential equation, and that therefore we are dealing with a linear time-invariant system. Then the relation between sensor output and input can be expressed in a simple form, as a quotient, by taking the Laplace transform of each signal and the transfer function of the sensor [2]. Recall that the transfer function gives a general relation between output and input, but not between their instantaneous values. Sensor dynamic characteristics can then be studied for each applied input by classifying the sensors according to the order of their transfer function. It is generally not necessary to use models higher than second-order functions.

1.5.1   Zero-Order Measurement Systems

The output of a zero-order sensor is related to its input through an equation of the type

$$y(t) = k \cdot x(t) \tag{1.9}$$

Its behavior is characterized by its static sensitivity k and remains constant regardless of input frequency. Hence, its dynamic error and its delay are both zero.

An input–output relationship such as that in (1.9) requires that the sensor does not include any energy-storing element. This is, for example, the case of potentiometers applied to the measurement of linear and rotary displacements (Section 2.1). Using the notation of Figure 1.7, we have

$$y = V_r \frac{x}{x_m} \tag{1.10}$$

where $0 \le x \le x_m$ and V_r is a reference voltage. In this case $k = V_r/x_m$.

Models such as the previous one are always a mathematical abstraction because we cannot avoid the presence of imperfections that restrict the applicability of the model. For example, for the potentiometer, it is not possible to apply it to fast-varying movements because of the friction of the wiper.

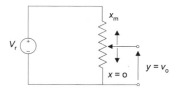

Figure 1.7   A linear potentiometer used as a position sensor is a zero-order sensor.

TABLE 1.3   Output Signal of a First-Order Measurement System for Different Common Test Inputs

Input	Output
Step $u(t)$	$k(1 - e^{-t/\tau})$
Ramp Rt	$Rkt - Rk\tau u(t) + Rk\tau e^{-t/\tau}$
Sinusoid A, ω	$\dfrac{kA\tau\omega e^{-t/\tau}}{1 + \omega^2\tau^2} + \dfrac{kA}{\sqrt{1 + \omega^2\tau^2}}\sin(\omega t + \phi)$
	$\phi = \arctan(-w\tau)$

1.5.2   First-Order Measurement Systems

In a first-order sensor there is an element that stores energy and another one that dissipates it. The relationship between the input $x(t)$ and the output $y(t)$ is described by a differential equation with the form

$$a_1 \frac{dy(t)}{dt} + a_0 y(t) = x(t) \tag{1.11}$$

The corresponding transfer function is

$$\frac{Y(s)}{X(s)} = \frac{k}{\tau s + 1} \tag{1.12}$$

where $k = 1/a_0$ is the *static sensitivity* and $\tau = a_1/a_0$ is the system's *time constant*. The system's *corner (angular) frequency* is $\omega_c = 1/\tau$. Therefore, to characterize the system two parameters are necessary: k for the static response and ω_c or τ for the dynamic response.

Table 1.3 shows the expression of the output signal for each of the most common test inputs: step, ramp, and sinusoid. The derivation of the complete expressions can be found in most books on control theory or in reference 2. For the sinusoid the transient part of the output has been included. This is important when the reading is taken shortly after applying the input.

The dynamic error and delay of a first-order sensor depend on the input waveform. Table 1.4 shows the dynamic error and delay corresponding to the inputs considered in Table 1.3. The two values for the dynamic error for an input ramp correspond, respectively, to two different definitions:

$$e_d = y(t) - x(t) \tag{1.13}$$

$$e_d = y(t) - kx(t) \tag{1.14}$$

For step and sinusoidal inputs, only (1.14) has been used.

The availability of analytical expressions for the dynamic error may suggest

TABLE 1.4   Dynamic Error and Delay for a First-Order Measurement System for Different Common Test Inputs

Input	Dynamic Error	Delay
Step $u(t)$	0	τ
Ramp Rt	$R[t + k(\tau - t)]$ or $R\tau$	τ
Sinusoid A, ω	$1 - \dfrac{1}{\sqrt{1 + \omega^2 \tau^2}}$	$\dfrac{\arctan \omega\tau}{\omega}$

that it can easily be corrected. In practice, however, the real input will seldom be as simple as the ones considered, and therefore it will not be possible to compensate for the dynamic error. Figure 1.8 shows the response to each of these input waveforms (see Problem 1.6).

An example of a first-order sensor is a thermometer based on a mass M with specific heat c (J/kg·K), heat transmission area A, and (convection) heat trans-

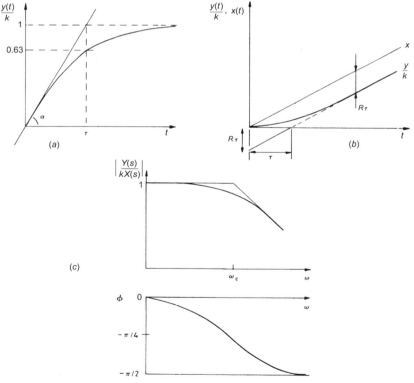

Figure 1.8   First-order system response to (a) a unit step input, (b) a ramp input, and (c) a sinusoidal input (amplitude modulus and phase).

fer coefficient h (W/m^2·K). In steady state, energy balance yields

$$(\text{Heat in}) - (\text{Heat out}) = \text{Energy stored}$$

If we assume that the sensor does not lose any heat—for example, through its leads—and that its mass does not change (negligible expansion), if we call T_i its internal temperature when the external temperature is T_e, we have

$$hA(T_e - T_i)\,dt - 0 = Mc\,dT_i \tag{1.15}$$

$$\frac{dT_i}{dt} = \frac{hA}{Mc}(T_e - T_i) \tag{1.16}$$

By taking the Laplace transform and introducing $\tau^{-1} = hA/Mc$, we obtain

$$\frac{T_i(s)}{T_e(s)} = \frac{1}{1 + \tau s} \tag{1.17}$$

Therefore, the resistance to heat transfer, along with the mass and thermal capacity, will determine the time constant and delay the sensor's temperature change. Nevertheless, once the sensor reaches a given temperature, its response is immediate. There is not any noticeable delay in sensing. The delay is in the sensor achieving the final temperature.

Example 1.4   The approximate time constant of a thermometer is determined by immersing it in a bath and noting the time it takes to reach 63% of the final reading. If the result is 28 s, determine the delay when measuring the temperature of a bath that is periodically changing 2 times per minute.

From the step response we have $\tau = 28$ s. From the last row in Table 1.4, the delay when measuring a cyclic variation will be

$$t_d = \frac{\arctan(\omega\tau)}{\omega}$$

The angular (radian) frequency of the temperature to measure is

$$\omega = 2\pi\frac{2\text{ cycles}}{60\text{ s}} = 0.209\text{ rad/s}$$

The delay will be

$$t_d = \frac{\arctan\left(\dfrac{0.209\text{ rad}}{1\text{ s}} \times 28\text{ s}\right)}{0.209\text{ rad/s}} = 6.7\text{ s}$$

1.5.3 Second-Order Measurement Systems

A second-order sensor contains two energy-storing elements and one energy-dissipating element. Its input $x(t)$ and output $y(t)$ are related by a second-order linear differential equation of the form

$$a_2 \frac{d^2 y(t)}{dt^2} + a_1 \frac{dy(t)}{dt} + a_0 y(t) = x(t) \tag{1.18}$$

The corresponding transfer function is

$$\frac{Y(s)}{X(s)} = \frac{k\omega_n^2}{s^2 + 2\zeta\omega_n s + \omega_n^2} \tag{1.19}$$

where k is the *static sensitivity*, ζ is the *damping ratio*, and ω_n is the *natural undamped angular frequency* for the sensor ($\omega_n = 2\pi f_n$). Two coefficients determine the dynamic behavior, while a single one determines the static behavior. Their expressions for the general second-order system modeled by (1.18) are

$$k = \frac{1}{a_0} \tag{1.20}$$

$$\omega_n^2 = \frac{a_0}{a_2} \tag{1.21}$$

$$\zeta = \frac{a_1}{2\sqrt{a_0 a_2}} \tag{1.22}$$

Notice that these three parameters are related and that a modification in one of them may change another one. Only a_0, a_1, and a_2 are independent.

Doebelin [2] details the procedure to obtain the output as a function of simple test input waveforms. Table 1.5 shows some results. Figure 1.9 shows their graphical characteristics. Note that the system behavior differs for $0 < \zeta < 1$ (underdamped case), $\zeta = 1$ (critically damped case), or $\zeta > 1$ (overdamped case). The initial transient has been omitted for the sinusoidal input.

Example 1.5 In a measurement system, a first-order sensor is replaced by a second-order sensor with the same natural (corner) frequency. Calculate the damping ratio to achieve the same -3 dB attenuation at that frequency.

A -3 dB attenuation means

$$3 = 20 \lg a$$

$$a = 10^{-3/20} = 0.707$$

From the last row in Table 1.5, the relative magnitude for a second-order response is

TABLE 1.5 Outputs of a Second-Order Measuring System for Different Common Test Inputs

Input	Output	

Unit step $u(t)$

$0 < \zeta < 1$

$$1 - \frac{e^{-\delta t}}{\sqrt{1-\zeta^2}}\sin(\omega_d t + \phi)$$

$\delta = \zeta\omega_n$
$\omega_d = \omega_n\sqrt{1-\zeta^2}$
$\phi = \arcsin\dfrac{\omega_d}{\omega_n}$

$\zeta = 1$

$$1 - e^{-\delta t}(1 + \omega_n t)$$

$\zeta > 1$

$$1 + \frac{\omega_n}{2\sqrt{\zeta^2-1}}\left(\frac{e^{-at}}{a} - \frac{e^{-bt}}{b}\right)$$

$a = \omega_n(\zeta + \sqrt{\zeta^2-1})$
$b = \omega_n(\zeta - \sqrt{\zeta^2-1})$

Ramp Rt

$0 < \zeta < 1$

$$R\left\{t - \frac{2\zeta}{\omega_n}\left[1 - \frac{e^{-\zeta\omega_n t}}{2\zeta\sqrt{1-\zeta^2}}\sin\left(\sqrt{1-\zeta^2}\,\omega_n t + \phi\right)\right]\right\}$$

$\phi = \arctan\left(\dfrac{2\zeta\sqrt{1-\zeta^2}}{2\zeta^2-1}\right)$

$\zeta = 1$

$$R\left\{t - \frac{2\zeta}{\omega_n}\left[1 - \left(1 + \frac{\omega_n\tau}{2}\right)e^{-\omega_n t}\right]\right\}$$

$\zeta > 1$

$$R\left\{t - \frac{2\zeta}{\omega_n}\left[1 + \frac{2\zeta(-\zeta-\sqrt{\zeta^2-1}+1)}{4\zeta\sqrt{\zeta^2-1}}e^{-at} + \frac{2\zeta(-\zeta-\sqrt{\zeta^2-1}-1)}{4\zeta\sqrt{\zeta^2-1}}e^{-bt}\right]\right\}$$

Sinusoid A, ω

$$\frac{kA}{\sqrt{\left(1 - \frac{\omega^2}{\omega_n^2}\right)^2 + \left(\frac{2\zeta\omega}{\omega_n}\right)^2}}\sin(\omega t - \phi)$$

$\phi = \arctan\dfrac{2\zeta\omega/\omega_n}{1 - \left(\dfrac{\omega}{\omega_n}\right)^2}$

27

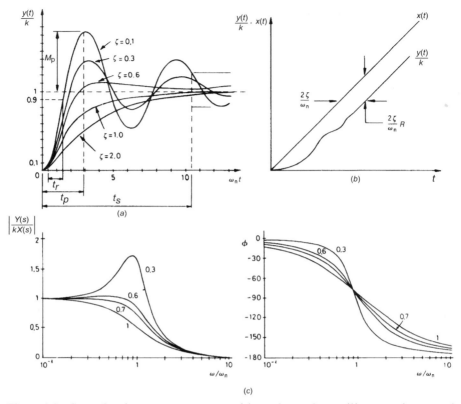

Figure 1.9   Second-order system response to (*a*) a unit step input, (*b*) a ramp input, and (*c*) a sinusoidal input (amplitude modulus and phase), for different damping ratios.

$$\frac{1}{\sqrt{\left(1 - \dfrac{\omega^2}{\omega_n^2}\right)^2 + \left(\dfrac{2\zeta\omega}{\omega_n}\right)^2}}$$

which reduces to $0.5/\zeta$ at ω_n. The condition to fulfill is $0.5/\zeta = 0.707$. This yields $\zeta = 0.707$.

The dynamic error and delay in a second-order system depend not only on the input waveform but also on ω_n and ζ. Their expressions are much more involved than in first-order systems, and to analyze them several factors related to ω_n and ζ are defined.

When the input is a unit step, if the system is overdamped ($\zeta > 1$) or is critically damped ($\zeta = 1$), there is neither overshoot nor steady-state dynamic error in the response.

In an underdamped system, $\zeta < 1$, the steady-state dynamic error is zero, but the speed and the overshoot in the transient response are related (Figure 1.9a). In general, the faster the speed, the larger the overshoot. The *rise time* t_r is the time spent to rise from 10% to 90% of the final output value, and it is given by

$$t_r = \frac{\arctan(-\omega_d/\delta)}{\omega_d} \tag{1.23}$$

where $\delta = \zeta\omega_n$ is the so-called *attenuation*, and $\omega_d = \omega_n\sqrt{1 - \zeta^2}$ is the *natural damped angular frequency*.

The time elapsed to the first peak t_p is

$$t_p = \frac{\pi}{\omega_d} \tag{1.24}$$

and the maximum overshoot M_p is

$$M_p = e^{-(\delta/\omega_d)\pi} \tag{1.25}$$

The time for the output to settle within a defined band around the final value t_s, or *settling time*, depends on the width of that band. For $0 < \zeta < 0.9$, for a $\pm 2\%$ band, $t_s \cong 4/\delta$, and it is minimal when $\zeta = 0.76$; for a $\pm 5\%$ band, $t_s \cong 3/\delta$, and it is minimal when $\zeta = 0.68$. The speed of response is optimal for $0.5 < \zeta < 0.8$ [6].

Figure 1.9a may suggest that underdamped sensors are useless because of their large overshoot. But in practice the input will never be a perfect step, and the sensor behavior may be acceptable. That is the case with piezoelectric sensors (Section 6.2), for example. Nevertheless, a large overshoot can saturate an amplifier output (see Problem 1.8).

When the input is a ramp with slope R, the steady-state dynamic error is

$$e_d = \frac{2\zeta R}{\omega_n} \tag{1.26}$$

and the delay is $2\zeta/\omega_n$ (Figure 1.9b).

To describe the frequency response of a second-order system where $0 < \zeta < 0.707$, we note that the frequency of resonance is not the same as the damped natural frequency

$$\omega_d = \omega_n\sqrt{1 - 2\zeta^2} \tag{1.27}$$

and the amplitude of that resonance at $\omega = \omega_d$ is M_r:

$$M_r = \frac{1}{2\zeta\sqrt{1 - \zeta^2}} \tag{1.28}$$

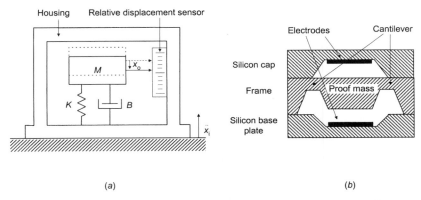

(a)   (b)

Figure 1.10 Second-order, underdamped sensors based on a mass–spring system. (a) The acceleration applied to the housing displaces the proof mass because of the force transmitted through its mechanical links. The spring constant K and viscous friction B do not necessarily belong to separate physical elements. (b) In this micromachined capacitive silicon accelerometer the applied acceleration flexes the cantilevers because of the force exerted by the proof mass, changing the capacitance between that mass surface and the fixed electrodes.

A simple example of a second-order sensor described by (1.19) is a thermometer covered for protection. In this case, we must add the heat capacity and thermal resistance of the covering to the heat capacity of the sensing element and heat conduction resistance from the medium where it is placed. The system has $\zeta > 1$. Liquid in glass manometers also have overdamped response (Problem 1.11).

Examples of underdamped systems are the mass–spring systems used to measure displacement, velocity, and acceleration in vibratory movements or in long-range missiles. They are also the heart of seismographs and micromachined accelerometers for airbag deployment in cars. Using the notation of Figure 1.10a, if we measure the displacement x_o of the mass M with respect to the armature fixed to the element undergoing an acceleration $\ddot{x}_i$, then the force on the mass (Newton's second law) is communicated through the spring deflection (Hooke's law) and the internal viscous friction. The force equation of the system is

$$M(\ddot{x}_i - \ddot{x}_o) = Kx_o + B\dot{x}_o \qquad (1.29)$$

where K is the spring constant or stiffness and B is the viscous frictional coefficient. K and B represent different physical actions, but they are not necessarily separate elements. The Laplace transform of $\ddot{x}_i$ is $s^2 X_i(s)$, from which we obtain

$$Ms^2 X_i(s) = X_o(s)[K + Bs + Ms^2] \qquad (1.30)$$

The transfer function is

$$\frac{X_o(s)}{\ddot{X}_i(s)} = \frac{X_o(s)}{s^2 X_i(s)} = \frac{M}{K} \frac{K/M}{s^2 + s(B/M) + K/M} \tag{1.31}$$

Therefore, $k = M/K$, $\zeta = B/(2\sqrt{KM})$, and $\omega_n = \sqrt{K/M}$. A large mass increases the sensitivity but reduces the natural frequency and the damping ratio. Stiffness increases the natural frequency but reduces the sensitivity and the damping ratio. Viscosity increases the damping ratio without affecting the sensitivity or the natural frequency. Micromachined accelerometers are very stiff and have small mass and friction. Therefore they have a large natural frequency but small sensitivity and damping ratio.

A potentiometer, a capacitive or inductive sensor, or a photodetector (with an ancillary light source and a shutter) can measure the displacement x_o of the proof mass. Alternatively, we can sense the stress of a flexing element holding the mass—for example, by using strain gages or a piezoelectric element. Figure 1.10b shows a capacitive micromachined silicon accelerometer based on a mass–spring system.

To consider also the acceleration of gravity when the axis of the accelerometer forms an angle θ with respect to the horizontal, the term $Mg \sin \theta$ has to be included in the right-hand member of (1.29). Then the output $y(t)$ would be defined as $x_o + (Mg \sin \theta)/K$, and its Laplace transform would be given by (1.31) with $Y(s)$ replacing $X_o(s)$.

If instead of the input acceleration we want to sense the displacement, we would multiply both sides of (1.31) by s^2 to obtain

$$\frac{X_o(s)}{X_i(s)} = \frac{M}{K} \frac{(K/M)s^2}{s^2 + s(B/M) + K/M} \tag{1.32}$$

From (1.31), the response for acceleration measurements is low-pass and ω_n must be higher than the maximal frequency variation of the acceleration to be measured. But for the measurement of vibration displacement—high-pass response (1.32)—ω_n must be lower than the frequency of the displacement and there is no dc response (see Problem 1.13).

1.6   OTHER SENSOR CHARACTERISTICS

Static and dynamic characteristics do not completely describe the behavior of a sensor. Table 1.6 lists other characteristics to consider in sensor selection, relative to the sensor and to the quantity to sense. In addition to those sensor characteristics, the measurement method must always be appropriate for the application. For example, there will be an error if, in measuring a flow, the insertion of the flowmeter significantly obstructs the conduit section.

TABLE 1.6 Characteristics to Consider in Sensor Selection

Quantity to Measure[a]	Output Characteristics	Supply Characteristics	Environmental Characteristics	Other Characteristics
Span	Sensitivity	Voltage	Ambient temperature	Reliability
Target accuracy	Noise floor	Current	Thermal shock	Operating life
Resolution	Signal: voltage, current, frequency	Available power	Temperature cycling	Overload protection
			Atmospheric pressure	
Stability	Signal type: single ended, differential, floating	Frequency (ac supply)	Humidity	Acquisition cost
Bandwidth	Impedance	Stability	Vibration	Weight, size
Response time	Code, if digital		Shock	Availability
Output impedance			Chemical agents	Cabling requirements
Extreme values			Explosion risks	Connector type
Interfering quantities			Dirt, dust	Mounting requirements
Modifying quantities			Immersion	Installation time
			Electromagnetic environment	State when failing
			Electrostatic discharges	Calibration and testing cost
			Ionizing radiation	Maintenance cost
				Replacement cost

[a] Sensor static and dynamic characteristics must be compatible with the requirements of the quantity to measure.

1.6.1   Input Characteristics: Impedance

The output impedance of the quantity to sense determines the input impedance needed for the sensor. Two examples illustrate this connection. To prevent the wiper in a potentiometer (Section 2.1) from losing contact with the resistive element, it is necessary for the wiper to exert a force on it. What would it happen if we desired to measure the movement of an element unable to overcome the friction between the wiper and the resistive element? This effect is not modeled by (1.9).

When we use a thermometer having a considerable mass to measure the temperature reached by a transistor, upon contact, wouldn't the thermometer cool the transistor and give a lower reading than the initial transistor temperature? Equation (1.17) would not describe that effect.

Neither the static nor the dynamic characteristics of sensors that we have defined allow us to describe the real behavior of the combined sensor-measured system. The description of a sensor or a measurement system through block diagrams ignores the fact that the sensor extracts some power from the measured system. When this power extraction modifies the value of the measured variable, we say that there is a *loading error*. Block diagrams are only appropriate when there is no energy interaction between blocks. The concept of input impedance allows us to determine when there will be a loading error.

When measuring a quantity x_1 there is always another quantity x_2 involved, such that the product x_1x_2 has the dimensions of power. For example, when measuring a force there is always a velocity; when measuring flow there is a drop in pressure; when measuring temperature there is a heat flow; when measuring an electric current there is a drop in voltage, and so on.

Nonmechanical variables are designed as *effort variables* if they are measured between two points or regions in the space (voltage, pressure, temperature), and they are designed as *flow variables* if they are measured at a point or region in the space (electric current, volume flow, heat flow). For mechanical variables the converse definitions are used, with effort variables measured at a point (force, torque) and flow variables measured between two points (linear velocity, angular velocity).

For an element that can be described through linear relations, the input impedance, $Z(s)$, is defined as the quotient between the Laplace transforms of an input effort variable and the associated flow variable [7]. The input admittance, $Y(s)$, is defined as the reciprocal of $Z(s)$. $Z(s)$ and $Y(s)$ usually change with frequency. When very low frequencies are considered, *stiffness* and *compliance* are used instead of impedance and admittance.

To have a minimal loading error, it is necessary for the input impedance to be very high when measuring an effort variable. If x_1 is an effort variable, then we obtain

$$Z(s) = \frac{X_1(s)}{X_2(s)} \tag{1.33}$$

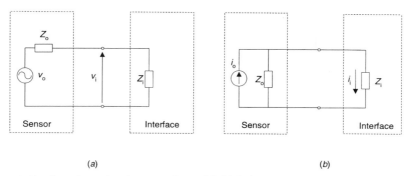

(a)   (b)

Figure 1.11 Interface circuits must have (a) high input impedance for sensors with voltage output and (b) low input impedance for sensors with current output.

The power drained from the measured system will be $P = x_1 x_2$; and if it is to be kept at minimum, x_2 must be as small as possible. Therefore the input impedance must be high.

To keep P very small when measuring a flow variable, it is necessary for x_1 to be very small, and that calls for a low input impedance (i.e., a high input admittance).

To obtain high-valued input impedances, it may be necessary to modify the value of components or to redesign the system and use active elements. For active elements, most of the power comes from an auxiliary power supply, and not from the measured system. Another option is to measure by using a balancing method because there is only a significant power drain when the input variable changes its value.

Sensor output impedance determines the input impedance needed for the interface circuit. A voltage output (Figure 1.11a) demands high input impedance in order for the sensed voltage

$$V_i = V_o \frac{Z_i}{Z_i + Z_o} \tag{1.34}$$

to be close to the sensor output voltage. Conversely, a current output (Figure 1.11b) demands low input impedance in order for the input current

$$I_i = I_o \frac{Z_o}{Z_i + Z_o} \tag{1.35}$$

to be close to the sensor output current.

1.6.2 Reliability

A sensor is reliable when it works without failure under specified conditions for a stated period. Reliability is described statistically: A high reliability means a probability close to 1 of performing as desired (i.e., units of that sensor seldom fail during the period considered). The *failure rate* λ is the number of failures of an item per unit measure of life (time, cycles), normalized to the number of surviving units. If in a time interval dt, $N_f(t)$ units from a batch of N fail and $N_s(t)$ survive, and life is measured in time units, the failure rate is

$$\lambda(t) = \frac{1}{N_s(t)} \frac{dN_f}{dt} \qquad (1.36)$$

The reliability at any time t as a probability is

$$R(t) = \lim_{N \to \infty} \frac{N_s(t)}{N} \qquad (1.37)$$

N will always be finite in practice. Hence, $R(t)$ can only be estimated. Since at any interval between $t = 0$ and any time later t, units either survive or fail,

$$N = N_s(t) + N_f(t) \qquad (1.38)$$

Substituting into (1.37), differentiating, and applying (1.36) yields

$$\frac{dR(t)}{dt} = -\frac{1}{N} \frac{dN_f(t)}{dt} = -\frac{\lambda(t) N_s(t)}{N} = -\lambda(t) R(t) \qquad (1.39)$$

Solving for $R(t)$, we obtain

$$R(t) = e^{-\int \lambda(t)\, dt} \qquad (1.40)$$

Therefore, the reliability can be calculated from the failure rate, which is calculated from experiments that determine its reciprocal, the *mean time between failures* (MTBF):

$$\text{MTBF} = m = \frac{1}{\lambda} \qquad (1.41)$$

Example 1.6   We test 50 units of a given accelerometer for 1000 h. If the failure rate is assumed constant and 2 units fail, determine the failure rate and MTBF.
   From (1.36),

$$\lambda = \frac{1}{50} \frac{2}{1000\ \text{h}} = 40 \frac{\text{failures}}{\text{milion hours}}$$

From (1.41),

$$\text{MTBF} = \frac{10^6\ \text{h}}{40} = 25000\ \text{h}$$

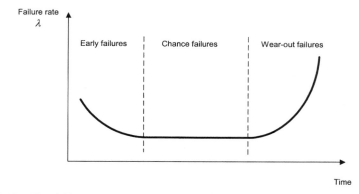

Figure 1.12   The failure rate of many devices follows a bathtub curve that determines three stages in a product life (infant mortality, useful life, and wear out stage) with different failure causes.

Experimental studies of many devices, including sensors, show that their failure rate is not constant but follows the trend in Figure 1.12 after obvious failures have been discarded. Some units of the initial population fail shortly after power up because of *early failures* or *break-in failures*, leading to the so-called infant mortality. Early failure result from microscopic defects in materials and from incorrect adjustments or positioning that went undetected during quality control. Electrical, mechanical, chemical, and thermal stresses during operation sometimes exceed those during product test, and they are withstood by normal units but not by inferior units. Early failures are excluded from MTBF calculations.

The flat segment in Figure 1.12 corresponds to the device useful life. λ is almost constant and is due to *chance failures* (intrinsic or stress-related failures) that result from randomly occurring stresses, the random distribution of material properties and random environmental conditions. Chance failures are present from the beginning, but early failures predominate at that stage.

Some time after placing different units of a device in service, they start to fail at an increasing rate. This is the wear-out stage, wherein parts fail because of the deterioration caused by thermal cycles, wear, fatigue, or any other condition that causes weakening under normal use. In this stage, wear-out failures predominate over chance failures.

Reliability is very important in sensors because they provide information for the control of control systems. Kumamoto and Henzel [8] analyze the reliability of systems that include sensors, alarms and feedback loops. Reference 9 analyzes reliability in depth.

1.7 PRIMARY SENSORS

Primary sensors convert measurands from physical quantities to other forms. We classify primary sensors here according to the measurand. Devices that

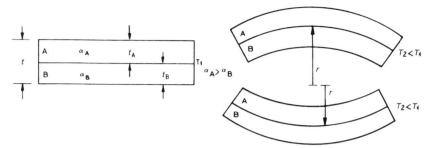

Figure 1.13 A bimetal consists of two metals with dissimilar thermal expansion coefficients, which deforms when temperature changes. Dimensions and curvature have been exaggerated to better illustrate the working principle. (From E. O. Doebelin, *Measurement Systems Application and Design*, 4th ed., copyright 1990. Reprinted by permission of McGraw-Hill, New York.)

have direct electric output are plain sensors and are discussed in Chapters 2, 4, and 6. Radiation-based measurement methods are described in Chapter 9. Khazan [11] and Fraden [12] describe additional primary sensors.

1.7.1 Temperature Sensors: Bimetals

A bimetal consists of two welded metal strips having different thermal expansion coefficients that are exposed to the same temperature. As temperature changes, the strip warps according to a uniform circular arc (Figure 1.13). If the metals have similar moduli of elasticity and thicknesses, the radius of curvature r, when changing from temperature T_1 to T_2, is [2]

$$r \cong \frac{2t}{3(\alpha_A - \alpha_B)(T_2 - T_1)} \qquad (1.42)$$

where t is the total thickness of the piece and where α_A and α_B are the respective thermal expansion coefficients. Therefore the radius of curvature is inversely proportional to the temperature difference. A position or displacement sensor would yield a corresponding electric signal. Alternatively, the force exerted by a total or partially bonded or clamped element can be measured.

The thickness of common bimetal strips ranges from 10 μm to 3 mm. A metal having $\alpha_B < 0$ would yield a small r, hence high sensitivity. Because useful metals have $\alpha_B > 0$, bimetal strips combine a high-coefficient metal (proprietary iron–nickel–chrome alloys) with invar (steel and nickel alloy) that shows $\alpha = 1.7 \times 10^{-6}/°C$. Micromachined actuators (microvalves) use silicon and aluminum.

Bimetal strips are used in the range from $-75\,°C$ to $+540\,°C$, and mostly from $0\,°C$ to $+300\,°C$. They are manufactured in the form of cantilever, spiral, helix, diaphragm, and so on, normally with a pointer fastened to one end of the strip, which indicates temperature on a dial. Bimetal strips are also used as actuators to directly open or close contacts (thermostats, on–off controls, starters for fluorescent lamps) and for overcurrent protection in electric circuits: The current along the bimetal heats it by Joule effect until reaching a temperature high enough to exert a mechanical force on a trigger device that opens the circuit and interrupts the current.

Other nonmeasurement applications of bimetal strips are the thermal compensation of temperature-sensitive devices and fire alarms. Their response is slow because of their large mass. Each October issue of *Measurements & Control* lists the manufacturers and types of bimetallic thermometers.

1.7.2   Pressure Sensors

Pressure measurement in liquids or gases is common, particularly in process control and in electronic engine control. Blood pressure measurement is very common for patient diagnosis and monitoring. Pressure is defined as the force per unit area. *Differential pressure* is the difference in pressure between two measurement points. *Gage pressure* is measured relative to atmospheric pressure. *Absolute pressure* is measured relative to a perfect vacuum. To measure a pressure, it is either compared with a known force or its effect on an elastic element is measured (deflection measurement). Table 1.7 shows some sensing

---

TABLE 1.7   Some Common Methods to Measure Fluid Pressure in Its Normal Range

1. Liquid column + level detection
2. Elastic element
   2.1. Bourdon tube + displacement measurement: Potentiometer
   　　　　　　　　　　　　　　　　　　　　　　LVDT
   　　　　　　　　　　　　　　　　　　　　　　Inductive sensor
   　　　　　　　　　　　　　　　　　　　　　　Digital encoder
   2.2. Diaphragm + deformation measurement
   　　　2.2.1. Central deformation[a]: Potentiometer
   　　　　　　　　　　　　　　　　　　LVDT
   　　　　　　　　　　　　　　　　　　Inductive sensor
   　　　　　　　　　　　　　　　　　　Unbonded strain gages
   　　　　　　　　　　　　　　　　　　Cantilever and strain gages
   　　　　　　　　　　　　　　　　　　Vibrating wire
   　　　2.2.2. Global deformation:　Variable reluctance
   　　　　　　　　　　　　　　　　　　Capacitive sensor
   　　　　　　　　　　　　　　　　　　Optical sensor
   　　　　　　　　　　　　　　　　　　Piezoelectric sensor
   　　　2.2.3. Local deformation: strain gages: Bonded foil
   　　　　　　　　　　　　　　　　　　　　　　Bonded semiconductor
   　　　　　　　　　　　　　　　　　　　　　　Deposited
   　　　　　　　　　　　　　　　　　　　　　　Sputtered (thin film)
   　　　　　　　　　　　　　　　　　　　　　　Diffused/implanted semiconductor

[a] Capsules and bellows yield larger displacements than diaphragms but suit only static pressures.

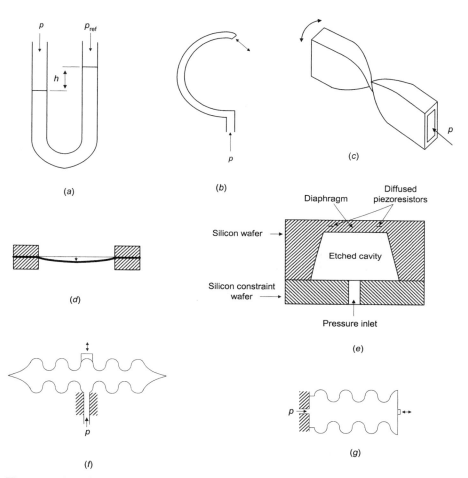

Figure 1.14 Primary pressure sensors. (*a*) Liquid-column U-tube manometer. The liquid must be compatible with the fluid for which pressure is to be measured, and the tube must withstand the mechanical stress. (*b*) C-shaped Bourdon tube. (*c*) Twisted Bourdon tube. (*d*) Membrane diaphragm. (*e*) Micromachined diaphragm. (*f*) Capsule. (*g*) Bellows. The area of the diaphragm in (*e*) is less than 1 mm². All other devices can measure up to several centimeters.

methods. Each issue of *Measurements & Control* lists the manufacturers of different pressure sensors: potentiometric (January); strain-gage and piezo-resistive (April); capacitive (June); digital and reluctive (September); piezoelectric and liquid-column (October); and bellows, Bourdon tube, and diaphragm (December).

A liquid-column manometer such as the U-tube in Figure 1.14*a* compares the pressure to be measured with a reference pressure and yields a difference h of liquid level. When second order effects are disregarded, the result is

$$h = \frac{p - p_{\text{ref}}}{\rho g} \tag{1.43}$$

where ρ is the density of the liquid and g is the acceleration of gravity. A level sensor (photoelectric, float, etc.) yields an electric output signal.

Elastic elements deform under pressure until the internal stress balances the applied pressure. The material and its geometry determine the amplitude of the resulting displacement or deformation, hence the appropriate sensor (Table 1.7). Usual pressure sensors use the Bourdon tube, diaphragms, capsules, and bellows.

The Bourdon tube—patented by Eugene Bourdon in 1849—is a curved (Figure 1.14b) or twisted (Figure 1.14c), flattened metallic tube with one closed end. The tube is obtained by deforming a tube having a circular cross section. When pressure is applied through the open end, the tube tends to straighten. The displacement of the free end indicates the pressure applied. This displacement is not linear along its entire range, but is linear enough in short ranges. Displacement sensors yield an electric output signal. Tube configurations with greater displacements (spiral, helical) have large compliance and length that result in a small-frequency passband. The tube metal (brass, monel, steel) is selected to be compatible with the medium.

A diaphragm is a flexible circular plate consisting of a taut membrane or a clamped sheet that strains under the action of the pressure difference to be measured (Figure 1.14d). The sensor detects the deflection of the center of the diaphragm, its global deformation, or the local strain (by strain gages, Section 2.2). Some metals used are beryllium–copper, stainless steel, and nickel–copper alloys. A micromachined diaphragm is an etched silicon wafer with diffused or implanted gages that sense local strain (Figure 1.14e). Cars and hospitals use silicon pressure sensors by the millions. The diaphragm and elements bonded on it must be compatible with the medium and withstand the required temperature. Stainless steel diaphragms can protect sensing diaphragms from corrosive media, but in order to couple both diaphragms we need to interpose a fluid, which increases the sensor compliance and thermal sensitivity. Ceramic (96% Al_2O_3, 4% SiO_2) and sapphire (Al_2O_3) are highly immune to corrosive attack; but because they are very expensive, their use is restricted to the more demanding applications involving aggressive media, high temperature, or both.

For a thin plate with thickness t and radius R experiencing a pressure difference Δp across it, if the center deflection is $z < t/3$, we have [2]

$$z \cong \frac{3(1 - v^2)R^4}{16Et^3} \Delta p \tag{1.44}$$

where E is Young's modulus and v the Poisson's ratio for the plate material. Large, flexible diaphragms undergo large deflection but have large compliance. Thin plates yield large deflections but are fragile. An alternative to sense the central deflection is to use a rod to transmit force to a cantilever beam with bonded strain gages, away from media temperature. Ceramic and some silicon pressure sensors rely on the capacitance change between an electrode applied on the diaphragm and one fixed electrode.

Piezoresistive sensors distributed on the diaphragm can sense radial and tangential strain. They are connected in a measurement bridge to add their signal and compensate temperature interference (Section 3.4.4).

Capsules and bellows yield larger displacements than diaphragms. A capsule (Figure 1.14*f*) consists of twin corrugated diaphragms joined by their external border and placed on opposite sides of the same chamber. A bellows (Figure 1.14*g*) is a flexible chamber with axial elongation that undergoes deflections larger than capsules, up to 10% of its length. Capsules and bellows are vibration- and acceleration-sensitive, do not withstand high overpressures, and have high compliance, hence poor dynamic response. Their displacement, however, can be sensed by an inexpensive potentiometer.

Pressure between contacting surfaces can be measured by a thin plastic film (Fuji Prescale Film) whose color increases for increasing pressure.

1.7.3 Flow Velocity and Flow-Rate Sensors

Flow is the movement of a fluid in a channel or in open or closed conduits. The flow rate is the quantity of matter, in volume or weight, that flows in a unit time. Flow rate is measured in all energy and mass transport processes to control or monitor those processes and for metering purposes—for example, water, gas, gasoline, diesel, and crude oil. Table 1.8 lists some measurement principles used in flowmeters. Chapters 28 and 29 in reference 13 discuss them. Each issue of *Measurements & Control* lists the manufacturers of different flowmeters: turbine (February); electromagnetic (April); anemometers and vortex (June); differential pressure, rotameters, and mass (September); positive displacement and ultrasonic (October); and open-channel, target, and flowmeters based on laminar flow elements (December).

A viscous or laminar flow is that of a fluid flowing along a straight smooth-walled and uniform transverse section conduit, where all particles have a trajectory parallel to the conduit walls and move in the same direction, each following a streamline. In turbulent flow, in contrast, some of the fluid particles have longitudinal and transverse velocity components—thus resulting in whirls—and only the average velocity is parallel to the axis of the conduit. In laminar flow, the fluid velocity profile across the conduit is parabolic. In turbulent flow, the fluid velocity profile is flatter.

The commonest flowmeters measure the drop in pressure across an obstruction inserted in the pressurized pipe in which we wish to measure the flow rate. Bernoulli's theorem relates fluid pressure, velocity, and height. It applies to an incompressible fluid experiencing only gravity as internal force (i.e., without friction) flowing in stationary movement and with no heat entering or leaving it. Any change in velocity produces an opposite change in pressure that equals the change of kinetic energy per unit of volume added to the change due to any

TABLE 1.8   Measurement Principles Used in Flowmeters

Input Quantity	Measurement Principle	Output Signal
Fluid velocity: local	Pitot probe	Differential pressure
	Thermal (hot wire anemometry)	Temperature
	Laser anemometry	Frequency shift
Fluid velocity: average	Electromagnetic	Voltage
	Ultrasound: transit time	Time
	Ultrasound: Doppler	Frequency
Volume flow rate[a]	Orifice plate	Differential pressure
	Venturi tube	Differential pressure
	Pitot probe	Differential pressure
	Flow nozzle and tube	Differential pressure
	Elbow	Differential pressure
	Laminar flow element	Differential pressure
	Impeller (paddlewheel)	Cycles, revolutions
	Positive displacement	Cycles, revolutions
	Target (drag force)	Force
	Turbine	Cycles, revolutions
	Variable area (rotameter)	Float displacement
	Variable area (weir, flume)	Level
	Vortex shedding	Frequency shift
Mass flow rate	Coriolis effect	Force
	Thermal transport	Temperature

[a] Volume flow rate can also be calculated by multiplying the average fluid velocity by the pipe cross section.

difference in level. That is, along a flow streamline we have

$$p + \rho g h + \frac{\rho v^2}{2} = \text{constant} \tag{1.45}$$

where p is the static pressure, ρ is the fluid density (incompressible), g is the acceleration of gravity, h is the height with respect to a reference level, and v is the fluid velocity at the point considered. When studying actual fluid flows, (1.45) is corrected by experimental coefficients.

The primary sensor in obstruction flowmeters is a restriction having constant cross section that obstructs the flow. For example, if we introduce in a pipe a plate having a hole, the fluid vein contracts, thereby changing from a cross section A_1 (that of the pipe) to a cross section A_2 (that of the hole) (Figure 1.15). Because of the principle of mass conservation, a cross-section change results in a corresponding change of velocity,

$$Q = A_1 v_1 = A_2 v_2 \tag{1.46}$$

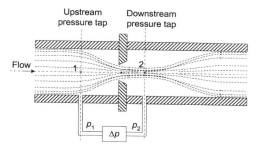

Figure 1.15   An orifice plate inserted in a pipe produces a drop in pressure related to the flow rate.

At the same time, from (1.45) we have

$$p_1 + \rho g h_1 + \frac{\rho v_1^2}{2} = p_2 + \rho g h_2 + \frac{\rho v_2^2}{2} \qquad (1.47)$$

If $h_1 = h_2$, these two equations yield

$$v_2 = \sqrt{\frac{2(p_1 - p_2)}{\rho \left[1 - \left(\dfrac{A_1}{A_2} \right)^2 \right]}} \qquad (1.48)$$

Therefore, we can calculate the velocity from the drop in pressure across the plate, and we can determine the theoretical volumetric flow rate from $Q = A_2 v_2$. The real flow rate is somewhat lower and it is determined by experimentally calculating a correction coefficient, called a discharge factor, C_d, that depends on A_1, A_2, and other parameters. Then, $Q_r = C_d Q$. Tables in standards (ASME, ISO) give C_d for different pipe diameter and hole position and size, flow regimes, and pressure ports placement. For orifice plates we have $C_d \approx 0.6$.

Orifice meters produce a loss in pressure and cannot easily measure fluctuating flows unless the differential pressure sensor is fast enough, including the effects of the hydraulic connections. Flow nozzles and Venturi tubes (Figure 1.16) are based on the same principles but their internal shapes are not so blunt, thus reducing the loss of pressure (C_d can reach 0.97).

Variable-area flowmeters are primary sensors that apply Bernoulli's theorem and the principle of mass conservation in a way reciprocal to that described. They make the fluid pass section variable and keep the difference in pressure between both sides of the obstruction constant. The measured flow rate is then related to the area of the pass section.

The rotameter in Figure 1.17 applies this method. It consists of a uniform conic section tube and a grooved float inside it that is dragged by the fluid to a

(a)

(b)

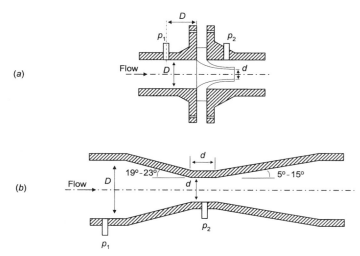

Figure 1.16   Flow nozzles (*a*) and Venturi tubes (*b*) inserted in pipes yield a lower drop in pressure than orifice plates, hence saving energy.

height determined by its weight and the flow. The fluid—gas or liquid—flows upward. When the flow increases, the float rises, thus allowing an increased annular pass section and keeping the pressure difference between both ends constant. The displacement of the float indicates the fluid flow rate. For pressures lower than 3.5 kPa and nonopaque liquids, the tube can be of glass and include the scale to read the float position. For higher pressures and flows the tube must be of metal, and the position of the float is detected magnetically. There are also inexpensive plastic tubes for low-pressure, high flow rates. Adding a solenoid outside the tube enables us to apply the null-measurement method. A photoelectric detector measures the float position. The flow is determined from the amplitude of the current supplied to the solenoid in order to reposition the float at zero.

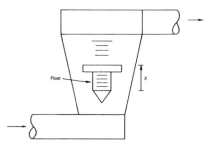

Figure 1.17   A rotameter is a variable area flowmeter in which the position of a float indicates the flow rate.

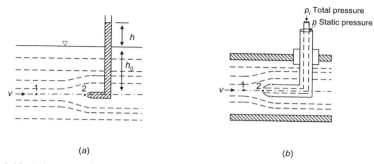

Figure 1.18 Pitot tube for point velocity flow measurement. (*a*) In an open conduit the velocity is indicated by the emerging fluid height. (*b*) In a closed conduit the velocity is calculated from the difference between total pressure and static pressure.

The Pitot tube used to measure the velocity of a fluid at a point also relies on Bernoulli's principle. If a bent open tube is introduced into an open conduit where an incompressible and frictionless fluid flows in a given known direction, and its open end is oriented against the flow (Figure 1.18*a*), the liquid enters into the tube and rises until the pressure exerted by the fluid column balances the force produced by the impacting velocity on the open end. Because in front of the opening the velocity is zero, flow lines distribute around the end, thereby creating a stagnation point. It holds therefore that

$$\frac{v^2}{2g} + \frac{p_1}{\rho g} = \frac{p_2}{\rho g} = h_0 + h \tag{1.49}$$

Also the static pressure in an open conduit comes from the weight of the fluid column, $p_1 = \rho g h_0$. Therefore

$$v = \sqrt{2gh} \tag{1.50}$$

We can thus infer the fluid velocity at the measurement point from the height of the column emerging above the surface.

If the Pitot tube is placed in a pressurized pipe, from (1.45) we obtain

$$v = \sqrt{\frac{2(p_t - p)}{\rho}} \tag{1.51}$$

Therefore, in order to determine the velocity we need to measure the difference between the total or stagnation pressure p_t and the static pressure p, which can be obtained from a port which faces perpendicular to the flow—for example, through a coaxial tube (Figure 1.18*b*). Pitot tubes are very common in laboratories and also for air speed measurement in avionics—in this last case using a modified version of (1.51) that includes temperature and specific heat because air is compressible.

Laminar flowmeters, also called laminar resistance flowmeters, rely on the Poiseuille's law. Jean M. Poiseuille—a physician—established in 1840 that for laminar flow in a tube much longer than wide, the volumetric flow rate is a linear function of the pressure drop according to

$$\Delta p = Q \frac{8\eta L}{\pi r^4} \tag{1.52}$$

where η is the fluid viscosity, L is the tube length, and r is its radius. Laminar flowmeters consist of a bundle or a matrix of capillary tubes, or one or more fine mesh screens and two pressure connections. They are used for leak testing, for calibration work, and in respiratory pneumotachometers.

Target flowmeters sense the fluid force on a target or drag-disk suspended in the flow stream by a sensing tube. The force exerted on the target is measured by strain gages (Section 2.2) placed on the tube, outside the pipe. Target flow-meters can be applied to dirty or corrosive liquids.

Turbine flowmeters consist of a bladed rotor suspended in a moving (clean) fluid that makes it turn at a speed proportional to the volumetric flow rate when it is high enough. The turning velocity is detected by a variable reluctance pickup. Vane flowmeters rely on the same principle.

Positive displacement flowmeters continuously separate the liquid stream into known volumes based on the physical dimensions of the meter, and register flow by counting cycles or revolutions. Figure 1.19 shows two different flow-segmentation methods. In the nutating disk meter, as the liquid attempts to flow through the meter, the pressure drop from inlet to outlet causes the disk to wobble. The sliding vane flowmeter has retractile vanes that seal a volume of liquid between the rotor and the casing and transport it from the inlet to the outlet, where it is discharged.

Weirs and flumes are calibrated restrictions used in open channel flows and

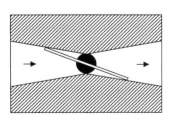
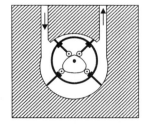

(a)   (b)

Figure 1.19 Two flow-segmentation methods used in positive displacement flowmeters: (a) Nutating disk and (b) sliding vane.

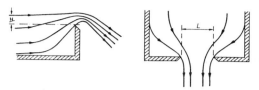

Figure 1.20   A weir is a channel restriction that raises the fluid to a height that depends on the flow rate.

in nonfilled conduits. A *weir* is a dam with a gorge in the top, built perpendicular to the flow direction. The liquid rises to a certain height, and then it flows through the gorge. This device converts part of the kinetic energy of the fluid into potential energy, and the fluid rises to a height relative to the lower point of the gorge that depends on the flow rate. If the gorge is rectangular, as in Figure 1.20, then we obtain

$$Q = kL\sqrt[3]{H^2} \tag{1.53}$$

where Q is the volumetric flow rate, H the height raised by the fluid, L is the weir width, and k is a constant. H can be measured using an upstream fluid level sensor. A *flume* is a channel restriction in area, slope, or both, based in the same principle as weirs.

Mass flow rate can be indirectly measured from volumetric flow rate and density. However, density depends on pressure and temperature, and any error in their measurement will propagate into the calculated flow rate. Thermal and Coriolis flowmeters (Section 8.2.5) yield better accuracy. There are two thermal flowmeters: hot-wire probes and heat transfer flow meters.

Hot wire probes (Figure 1.21*a*) measure the rate of heat loss to the flowing

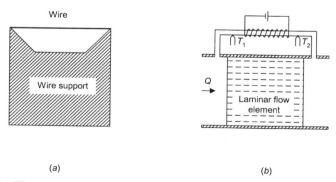

Figure 1.21   Thermal mass flowmeters convert the flow into a temperature change. In hot-wire anemometers (*a*), the rate of heat loss from a heated wire to the fluid depends on the local velocity. In heat-transfer flowmeters (*b*), the temperature rise on the downstream sensor depends on the mass flow.

fluid from a hot body—a resistive wire (Section 2.3), a thermistor (Section 2.4) or a thermopile (Section 6.1)—held perpendicular to the fluid flow. The heat flow rate from the wire to the fluid is proportional to the heat interchanging area A, to the difference in temperature between the wire and the fluid, and to the film coefficient of heat transfer h. The power dissipated by Joule effect is I^2R, and therefore, in equilibrium, we have

$$I^2R = khA(T_w - T_f) \tag{1.54}$$

where k is a unit-conversion constant. The coefficient of heat transfer depends on fluid velocity according to

$$h = c_0 + c_1\sqrt{v} \tag{1.55}$$

where c_0 and c_1 are factors that include the dependence on the dimensions of the wire and on the density, viscosity, specific heat, and thermal conductivity of the fluid. A large mass flow cools the wire to a lower temperature. If the current supply to the wire is constant, the resistance of the wire—or the generated voltage in a thermopile—indicate the mass flow. Alternatively, we can measure the current necessary to keep the wire at constant temperature.

Heat transfer flowmeters measure the rise in temperature of the fluid after a known amount of heat has been added to it. The primary sensor is a capillary tube with a wound heater and two temperature sensors symmetrically mounted upstream and downstream of the heater on the tube surface (Figure 1.21b). When there is no flow, both sensors have the same temperature. As flow increases, the incoming fluid removes heat from the tube and cools the upstream end while it heats the downstream end when passing through it. For low flows, the difference in temperature between sensors is proportional to the mass flow rate. Large flows remove heat even from the hottest point in the tube, and the proportionality is lost. To measure large flows, a laminar flow element in the main pipe causes a drop in pressure proportional to the volumetric flow rate—equation (1.52)—that forces through the capillary a small fraction of the flow. There are micromachined silicon flowmeters that use diffused resistors as heaters and resistor bridges and that use thermodiodes or thermocouples as temperature sensors. They consume low power, their response time is less than 3 ms, and their mass is about 10 g.

1.7.4   Level Sensors

Dipsticks are simple level sensors, but cannot easily provide an electric signal. Floats, based on Archimedes' buoyancy principle, convert liquid level to force or displacement (Figures 1.22a and 1.22b). In sealed or high-pressure containers, the position of the float can be detected magnetically. Build-up and deposits on the float surface limit performance.

The pressure of liquid or solid is proportional to level (Figure 1.22c),

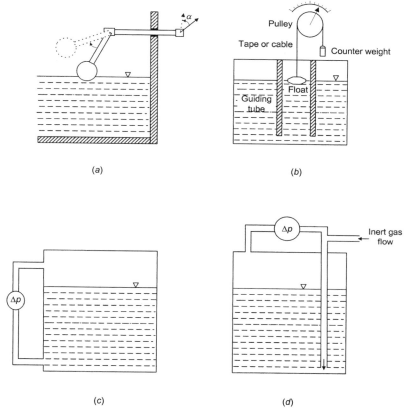

Figure 1.22 Primary level sensors. (*a*) and (*b*) Based on a float. (*c*) and (*d*) Based on differential pressure measurement.

according to

$$h = \frac{\Delta p}{\rho g} \tag{1.56}$$

where ρ is density and g is the acceleration of gravity. This method is suitable for both pressurized and open containers. Temperature interferes because it varies density.

The bubble tube in Figure 1.22*d* overcomes the need for a pressure port near the container bottom, which is a potential leak source. The dip tube has an open end close to the bottom of the tank. An inert gas flows through the dip tube and when gas bubbles escape from the open end, the gas pressure in the tube equals the hydraulic pressure from the liquid. The level can be calculated from (1.56).

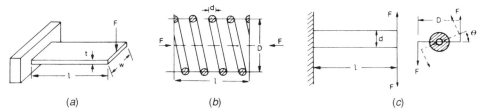

Figure 1.23   (*a*) A cantilever, (*b*) a helical spring, and (*c*) a torsion bar deflect in response to an applied force or torque.

Each April issue of *Measurements & Control* lists manufacturers and types of level measurement and control devices.

1.7.5   Force and Torque Sensors

A method to measure force (or torque) is to compare it with a well-known force, as is done on scales. Another method measures the effect of the force on an elastic element, called a *load cell*. In electric load cells, that effect is a deformation or a displacement. In hydraulic and pneumatic load cells it is an increase in the pressure of, respectively, a liquid or a gas. Each October issue of *Measurements & Control* lists the manufacturers and types of mass/force sensors and load cells.

When a mechanical force is applied to a fixed elastic element, it strains until the strain-generated stresses balance those due to the applied force. The result is a change in the dimensions of the element that is proportional to the applied force, if the shape is appropriate.

Figure 1.23 shows three suitable arrangements. Table 1.9 lists the corre-

TABLE 1.9   **Deflection x or θ and Maximal Stress s_M or τ_M for the Elastic Elements Shown in Figure 1.23**

Element	Deflection	Maximal Stress
Cantilever	$x = \dfrac{4Fl^3}{Ewt^3} = \dfrac{2\sigma l^2}{3Et}$	$s_M = \dfrac{6Fl}{wt^2} = \dfrac{3Etx}{2l^2}$
Helical spring	$x = \dfrac{8FnD^3}{Gd^4} = \dfrac{\pi nD^2\tau}{Gdk_1}$	$\tau_M = \dfrac{8k_1DF}{\pi d^3} = \dfrac{Gdxk_1}{\pi nD^2}$
Torsion bar	$\theta = \dfrac{32FDl}{\pi d^4 G} = \dfrac{2\tau l}{dG}$	$\tau_M = \dfrac{16FD}{\pi d^3} = \dfrac{dG\theta}{2l}$

Source: From H. K. P. Neubert, *Instrument transducers*, copyright 1975. Reprinted by permission of Oxford University Press, Fair Law, NJ.

Note: All quantities are in SI units (lengths in meters, forces in newtons, angles in radians). E = longitudinal modulus of elasticity (Young's modulus), G = modulus of rigidity (torsion elasticity modulus), k_1 = stress factor (function of D/d, valued from 1.1 to 1.6), n = number of turns.

sponding equations. Neubert [14] gives additional shapes and their corresponding equations. Most load cells are underdamped second-order systems (Section 1.5.3), which limits the maximal frequency of dynamic forces that can be accurately measured to a frequency range well below the load cell's natural frequency.

1.7.6   Acceleration and Inclination Sensors

The primary sensor for acceleration is the seismic mass–spring system (Figure 1.10). The output signal is displacement, strain, or capacitance change. Acceleration is measured for structural model verification, engine vibration level measurement in aircraft, machine monitoring, and inertial measuring units (to guide ammunition to a target). It is also used in experimental modal analysis, which is the empirical characterization of structures in terms of their damping, resonant frequencies, and vibration mode shapes; the larger the structure, the lower the frequency of the first vibrating mode. Micromachined accelerometers have found their way in automotive air bags, automotive suspension systems, stabilization systems for video equipment, transportation shock recorders, and activity responsive pacemakers. Each December issue of *Measurements & Control* lists the manufacturers and types of accelerometers and vibration sensors.

Inclinometers measure the attitude of orientation with respect to a reference axis. If the reference axis is defined by gravity (vertical axis), accelerometers work as inclinometers because they sense the acceleration applied along their sensitive axes. Alternatively, the liquid bubble inclinometer works the same as the level vial used by carpenters. In tilt sensors there is a curved tube with a trapped bubble that displaces when the tube tilts (Figure 1.24*a*). Resistive (Section 2.1) or capacitive (Section 4.1) sensors can sense the bubble position. The suspended pendulum (Figure 1.24*b*) is a weight attached to a ball bearing that can rotate. If the case rotates, the mass remains vertical, so that it under-

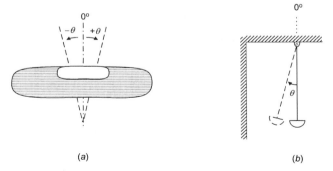

(a)   (b)

Figure 1.24   Inclination sensors. (*a*) The bubble inside a partially filled vial displaces when the vial tilts. (*b*) A mass suspended within a case rotates when the case rotates.

goes an angular displacement relative to the case, equal to the case rotation angle. Horizontal accelerations interfere with both sensors.

A compass senses the inclination with respect to a reference axis defined by a magnetic field. A potentiometer (Section 2.1), reluctive sensor (Section 4.2.1), or variable transformer (Section 4.2.4) can yield an electric signal corresponding to the rotation of the needle.

The spinning wheel of a gyroscope (Section 1.7.7) also defines a reference axis. If the frame in which the wheel rotates is fixed to a vehicle, the change of attitude of the vehicle results in a change in angle between the frame and the axis of rotation of the wheel. Each September issue of *Measurements & Control* lists the manufacturers and types of inclinometers.

1.7.7   Velocity Sensors

Linear velocity can be measured by integrating acceleration or differentiating displacement. Linear velocity can also be converted into rotational velocity by attaching a rack to the moving object and coupling it to a pinion gear that drives a rotor—as in car speedometers.

The seismic sensor in Figure 1.10 can be applied to linear velocity sensing without any link between the moving object and the reference respect to which the velocity is sensed. Integrating the mass displacement, which according to (1.31) is proportional to the input acceleration, yields the input velocity. Alternatively, if we sense the velocity of the mass relative to its housing, by manipulating (1.31) we obtain

$$\frac{\dot{X}_o(s)}{\dot{X}_i(s)} = \frac{sX_o(s)}{sX_i(s)} = \frac{s^2X_o(s)}{s^2X_i(s)} = \frac{M}{K}\frac{s^2(K/M)}{s^2 + s(B/M) + K/M} \tag{1.57}$$

Therefore, at frequencies above the natural frequency of the mass–spring system, the output of the internal velocity sensor is proportional to the input speed $\dot{x}_i$—relative to an inertial reference.

Absolute angular velocity measurement often relies on *gyroscopes* (or *gyros*). In a classic single-axis mechanical gyro, a motor-driven spinning mass (disk or wheel) is supported within a gimbal, held by bearings attached to a case (Figure 1.25a). In a two-axis gyro, the bearings supporting the inner gimbal are attached to an outer gimbal able to rotate with respect to the case.

A rate gyro is a single-axis gyro having an elastic restraint of the spin axis about the output axis (Figure 1.25b). When the gyroscope is rotated around the axis (y-axis) perpendicular to the spinning mass (x-axis), an angular momentum is developed around the z-axis, perpendicular to the x- and y-axes. That momentum is proportional to the angular speed around the y-axis and can be sensed by torque or force sensors [2].

Micromachined gyros have no rotating parts, and thus no bearings. They sense rotation from the Coriolis effect on vibrating mechanical elements [15].

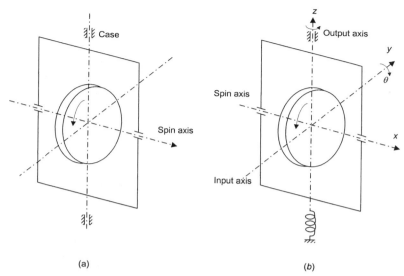

(a)   (b)

Figure 1.25 Single-axis mechanical gyroscope. (*a*) A spinning wheel defines the *x*-axis. (*b*) A rotation around the *y*-axis, perpendicular to the *x*-axis, yields a torque around the *z*-axis, perpendicular to both *x*- and *y*-axes.

The Coriolis effect is an apparent acceleration that arises in a moving element in a rotating body. Consider a traveling particle with velocity v (Figure 1.26) and an observer placed on the *x*-axis watching the particle. If the coordinate system (including the observer) rotates around the *z*-axis with angular velocity Ω, the observer thinks that the particle is moving toward the *x*-axis with acceleration

$$a_{\text{Cor}} = 2\Omega \times v \tag{1.58}$$

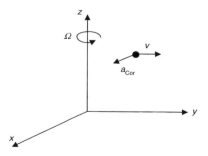

Figure 1.26 The Coriolis acceleration appears on a traveling particle when the coordinate system rotates with angular velocity Ω.

Therefore, when a mechanical element (tuning fork, disk, plate, etc.) is made to oscillate by the application of an alternating force, and this oscillating body is placed in a rotating reference frame, the Coriolis force produces a secondary oscillation perpendicular to the primary oscillation motion. The vibrating structure can be driven by electrostatic, electromagnetic, or piezoelectric force. Capacitive, piezoresistive, or piezoelectric sensors can detect the Coriolis-induced vibrations.

Fiber-optic and laser gyros do not use angular momentum but use the optical heterodyning of counterrotating optical or laser beams produced by Sagnac's effect (Section 9.4).

1.8   MATERIALS FOR SENSORS

Sensors rely on physical or chemical phenomena and materials where those phenomena appear usefully—that is, with high sensitivity, repeatability and specificity. Those phenomena may concern the material itself or its geometry, and most of them have been known for a long time. Major changes in sensors come from new materials, new fabrication techniques, or both.

Solids, liquids, and gases consist of atoms, molecules, or ions—atoms or group of atoms that have lost or gained one or more electrons. Atoms consist of a positive nucleus and electrons orbiting around it in shells. If the outer electron shell is not full, atoms try to gain extra electrons and become bonded in the process, forming molecules or agglomerates. There are four main bond types: ionic, metallic, covalent, and van der Waals [16]. Ionic bonds result from the electrostatic attraction between ions of different polarity. Ionic bonds form crystals—solids whose atoms are arranged in a long-range three-dimensional pattern in a way that reduces the overall energy and maintains electrical neutrality. Ionic crystals, such as NaCl and CsCl, have low electrical conductivity (because there are no free charges), relatively high fusion temperature, and good mechanical resistance, all resulting from the strong cohesion between ions.

The metallic bond also arises from electrostatic forces. But unlike the ionic bond, those forces are not between charges occupying a fixed position but between fixed positive charges and a cloud of electrons moving around the fixed positive ions. Mobile electrons in metals come from the outermost electron shell (valence electrons) of their atoms. Hence, metals have a regular structure (i.e., form crystals), but there is no need for a particular atom arrangement in those crystals to ensure electric neutrality. The swarming electron cloud (or *electron gas*) maintains electroneutrality. The crystal structure is then determined by the packing capability of atoms. Smaller atoms can diffuse through the lattice of higher-radius atoms, such as copper in germanium. Free electrons confer to metals their high electrical and thermal conductivity. The ubiquity of electrostatic forces along the lattice makes metals highly ductile and malleable.

Covalent bonds come from atoms sharing electrons with nearby atoms, so that they "believe" their respective outer electron shell is full. This bond may keep together atoms in a molecule (e.g., chlorine) or in a crystal [e.g., diamond (carbon), silicon, and germanium]. Shared electrons cannot move from their positions, and therefore they are not available to conduct electricity. Hence, materials with covalent bonds have low electrical conductivity.

Van der Waals bonds appear between molecules with intramolecular covalent bonds that have a small dipolar moment because of the lack of coincidence between the centers of positive and negative charge as a result of continuous electron movement. Van der Waals bonds keep together organic molecules to form crystals with low cohesion energy and whose structure depends on how well the molecules can pack together. Because of the low cohesion, materials with van der Waals bonds have low melting and boiling points.

Electrons in atoms can occupy only defined states, or energy levels, even when excited. The gap between the energy level corresponding to a nonexcited state and that corresponding to an excited state equals the amount of energy needed for one electron to jump from the base to the excited state. In a mass of atoms there are many energy levels. Close energy levels form an energy band. We distinguish three energy bands: the saturated or valence band, the conduction or excited band, and the forbidden band between them. Valence electrons cannot leave their positions. Excited electrons are nearly free to move around inside the material.

The relative separation between energy bands determines the electrical conductivity of materials, which is a useful property for sensors. Figure 1.27 shows that valence and conduction bands overlap in conductors, so that there are always free electrons and the electrical conductivity is high. Insulators have

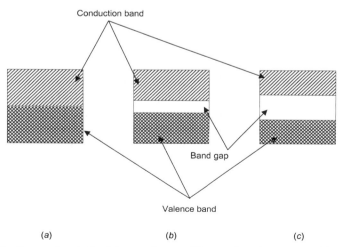

Figure 1.27   Energy bands for (*a*) a conductor, (*b*) a semiconductor, and (*c*) an electrical insulator.

TABLE 1.10   Band Gap Width (in electron-volts) for Various Intrinsic Semiconductors

	ZnS	CdS	CdSe	CdTe	Si	Ge	PbS	InAs	Te	PbTe	PbSe	InSb
Band gap, eV[a]:	3.60	2.40	1.80	1.50	1.12	0.67	0.37	0.35	0.33	0.30	0.27	0.18

[a] 1 eV = 0.16 aJ.

distant valence and conduction bands, and errant vibrations of electrons around their positions (e.g., because of thermal agitation) are unable to furnish enough carriers to significantly conduct electricity. Semiconductors have a narrower forbidden band than do insulators, and electrons excited by thermal, electric, optical or other energy form can jump across that band and contribute to conduct electricity. Each energy form has different efficacy on liberating electrons. Also, impurities—foreign atoms in the lattice—can introduce intermediate energy levels helping in electron transition from the valence band to the conduction band. Group V impurities—elements from column five in the periodic table of elements (antimony, arsenic, phosphorous)—bring an extra electron to silicon or germanium, leading to an *n*-type semiconductor that has a donor level close to the conduction band. Group III impurities (boron, aluminum, gallium, indium) leave a covalent bond—with silicon or germanium— with a missing electron (or "hole"), leading to a *p*-type semiconductor that has an acceptor level above the valence band. Other impurities will behave as either donors or acceptors. Electron jumps to the conduction band are the basis of many sensors. Table 1.10 lists the band gap for various intrinsic (i.e., non-doped) semiconductors.

According to their structure, solids can be single crystal, polycrystalline, amorphous, and glasses [17]. Crystals can be considered the periodic repetition of a "unit cell" totally filling space. Unit cells can be described by a space lattice of points (Figure 1.28a). Atoms can occupy not only the vertices but also the center of the cell, the center of two or more faces, or combinations of them. Directions and lattice planes are referred to by the so-called *Miller indices*. The direction of a vector is specified by the magnitude of its three components along the three axes, usually by writing them side by side enclosed in square brackets. A minus sign above a figure denotes a negative number (Figure 1.28b). Planes are specified by similarly writing inside parentheses the reciprocals of the intercepts of the plane on the axes of the unit cell (Figures 1.28c and 1.28d).

Polycrystalline materials such as metals and ceramics consist of an aggregate of a large number of randomly oriented crystals—called grains—joined via grain boundaries. When the grains are small enough, the physical properties of a polycrystalline material, such as elastic modulus, electrical conductivity, and thermal expansion, are isotropic in spite of the possible anisotropy of constitutive crystals.

Amorphous solids such as resins do not have ordered atoms. They are solidified liquids whose viscosity increases when cooling, thus impeding crystal

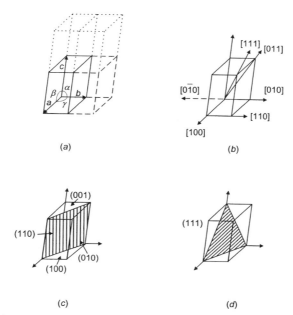

Figure 1.28 (*a*) Space lattice generated from a unit cell defined by vectors *a*, *b*, and *c* and angles α, β, and γ between them. Miller indices describe the direction of a vector by giving the magnitudes of its three components along the three direction axes (*b*), and they describe planes by giving the reciprocals of their intercepts on the axes of the unit cell (*c*) and (*d*).

formation and growing. Glasses have short-range order but lack long-range order. Amorphous-crystalline materials, such as most polymers, are amorphous but partially crystallized.

1.8.1 Conductors, Semiconductors, and Dielectrics

There are two type of conductors: electronic conductors (metals and their alloys) and ionic conductors or electrolytes (acid, base, or salt solutions). A potential difference V applied between the two ends of a solid of length l creates an internal electric field $E = V/l$, which accelerates electrons at

$$a = \frac{qE}{m} = \frac{qV}{ml} \qquad (1.59)$$

where m is the electron mass and q its charge. Hence, electrons moving at random because of thermal agitation have in addition a velocity component in the direction of the applied field. However, collisions deviate electrons from the direction set by E. It can be shown that the average or drift velocity is $v_d = a\tau$, where τ is the average time between collisions, mean free time, relaxation time, or collision time [16]. From (1.59), we obtain

$$v_d = \frac{qE}{m}\tau = \frac{q\tau}{m}E \tag{1.60}$$

where $q\tau/m$ is termed the electron mobility μ_e. The current density crossing a unit area will be

$$J = N_e q v_d = \frac{N_e q^2 \tau}{m}E \tag{1.61}$$

where N_e is the density of electrons (number of electrons per unit volume). This is Ohm's law and

$$\sigma = \frac{N_e q^2 \tau}{m} = \mu_e(N_e q) \tag{1.62}$$

is the electrical conductivity. A high conductivity can result from a high mobility or a high density of electrons. Because of random atom vibration, electron mobility in metals is relatively small. They are good conductors because of the abundance of free electrons. Metals and their alloys are used in sensors because of (a) their thermoelectric properties and (b) the dependence of their electrical conductivity with temperature and stress; they are also used to form electric circuits in which a measurand produces significant changes. Primary sensors such as bimetals and elastic elements (e.g., diaphragms and load cells) also rely on metals and their alloys. Still other metals are used because of their magnetic properties (Section 1.8.2), or they are used as electrodes or to catalyze. Electrolytes are primarily used in chemical sensors (Section 2.9).

Semiconductors are extensively used in sensors [18]. Semiconductors have covalent bonds, hence low electrical conductivity. Because both electrons and holes contribute to the total current, (1.62) transforms into

$$\sigma = q(N_n \mu_n + N_p \mu_p) = q(n \mu_n + p \mu_p) \tag{1.63}$$

where $n = N_n$ and $p = N_p$ are the respective concentrations of free electrons and holes. This conductivity depends on temperature, stress, electric and magnetic fields, corpuscular and electromagnetic radiation (including light), and the absorption of different substances. Adding impurities (dopants) to form an extrinsic semiconductor controls that dependence. Some semiconductors used in sensors are those in Table 1.10 and different oxides. Silicon in particular is a very convenient sensor material because of the deep knowledge of its properties gained in electronic devices, its excellent mechanical properties (higher tensile strength than steel, harder than iron, but brittle), the possibility of integrating signal conditioning circuits in the same chip, and the capability of batch-processing silicon-based devices. Sensors based on semiconductor films use interdigitated metal electrode grids in order to reduce the device resistance.

Dielectric materials have covalent bonds; hence they are electrical insulators. A dielectric is characterized by its dielectric constant or permittivity ϵ, which is the ratio between the electric flux density and the electric field, $\epsilon = D/E$. A vacuum has $\epsilon_0 = 8.85$ pF/m. Dielectrics have $\epsilon_r = \epsilon/\epsilon_0 \gg 1$. They are used as electrical insulators and also for sensing—for example, in variable capacitors. Ceramics, organic polymers, and quartz are also dielectrics used in sensors.

Ceramics resist corrosion, abrasion, and high temperature. They have been the materials of choice for supporting other sensing materials in common sensors and also in thick- and thin-film microsensors. Ceramics are also used as sensors themselves—for example, because of changes in crystal properties (NTC thermistors, oxygen sensors), granularity and grain-dissociation properties (switching PTC thermistors, piezo- and pyroelectric ceramics, ferrites), and surface properties [alumina (Al_2O_3) in humidity sensors, zirconia (ZrO_2) in oxygen sensors, and SnO_2 in gas sensors].

Organic polymers are macromolecules formed when many equal molecules called monomers bind together by covalent bonds. Bonded molecules can form linear or tridimensional structures. Linear arrangements yield flexible, elastic, soft, and thermoplastic materials—that is, materials that are increasingly viscous as temperature increases. Some thermoplastics such as nylon, polyethylene, and polypropylene are crystalline. Polystyrene, polycarbonate, and polyvinyl chloride are amorphous. Thermosetting materials have tridimensional structure. They are stiff, brittle, and scarcely soluble, and they undergo irreversible changes when heated. Silicone, melamine, polyester, and epoxy rosins are common thermosetting materials. Elastomers (neoprene, SBR, urethane) are a third kind of polymers that have rubber-like properties.

Plastics result from adding filler to a polymer—for example, to improve their mechanical properties. Plastics are superb electrical insulators but some have also been used for sensing humidity, force-pressure, and temperature. Some elastomers, for example, encompass a change of electrical conductivity when stretched. Polymers can become conductive by adding to them relatively good conductors such as powered silver or carbon, as well as by adding different counterions during polymer growing. Polymers are also used as membranes in ion-selective sensors and biosensors.

1.8.2 Magnetic Materials

The magnetic flux in vacuum is proportional to the applied magnetic field,

$$B = \mu_0 H \tag{1.64}$$

where $\mu_0 = 4\pi \times 10^{-7}$ H/m is the permeability of vacuum. All materials modify the magnetic flux to some extent, so that

$$B = \mu_0(H + M) = \mu_0\mu_r H \qquad (1.65)$$

where M is the magnetic dipole moment per unit volume, or *magnetization*, and μ_r is the *relative permeability*.

The magnetic properties of solids are related to the properties of the electrons in their atoms. Paramagnetic materials ($\mu_r > 1$) have atoms or ions with incomplete electron shells. Unpaired individual electron spins yield a magnetic moment, but the random orientation of individual dipoles makes the net dipole moment negligible. Nevertheless, upon application of an external field, individual dipoles assume the direction of minimal energy, setting themselves parallel to the applied field. Hence, applied magnetic fields attract paramagnetic materials. Thermal agitation opposes that alignment, so that only at absolute zero (0 K) would the alignment be perfect (Curie–Weiss law).

Diamagnetic materials ($\mu_r < 1$) have atoms or ions with complete electron shells; hence they lack magnetic moment. However, an applied magnetic field imparts an additional rotation to electrons (Larmor precession). This electron movement yields a net magnetic moment in a direction opposed to the field. Hence, applied magnetic fields repel diamagnetic materials. That alignment is not influenced by temperature.

The magnetic induction in paramagnetic and diamagnetic materials is only slightly different from that in vacuum, and its amplitude is independent on that of the applied field. Ferromagnetic and ferrimagnetic materials undergo strong magnetization ($\mu_r \gg 1$), varying with the applied field. Ferromagnetic materials are considered to consist of many elementary volumes called domains, each magnetized in a given direction. When these directions are randomly oriented, the material is not magnetized. When these directions are aligned to some extent, the material is magnetized.

Elementary magnetic dipoles in ferromagnetic materials arise from elementary currents. For iron and metals in its group (cobalt, nickel), those currents come from unpaired electron spins. In rare earth elements, such as gadolinium, orbital unbalance also contributes elementary currents.

A ferromagnetic material magnetizes because of two different processes: displacement and orientation. Displacement refers to the volume change of some domains at the expense of their neighbors. Orientation refers to the alignment of the magnetic domains in the direction of the applied magnetic field. Figure 1.29 describes the magnetization process for an increasing applied field H. When H is small (Figure 1.29a), domains parallel to it or with close direction grow, and the material becomes slightly magnetized in that direction. But upon reducing H the domains resume their initial sizes and the magnetization disappears. Hence, for weak external fields, magnetization is reversible.

As H increases, the induced magnetization increases almost proportionally (Figure 1.29b) because of the reorientation of elementary domains. Nevertheless, after reducing H, some domains shrink and other rotate. As a result, the *remanence*—remaining magnetization—is somewhat smaller than that under the applied field. A strong enough H aligns all magnetic domains (Figure 1.29c)

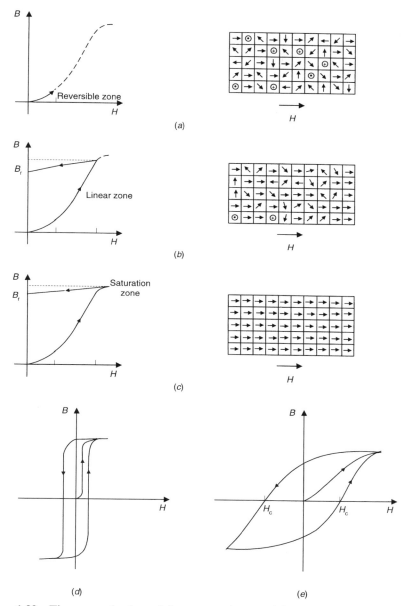

Figure 1.29 The magnetization of ferromagnetic materials results from domain displacement and orientation. (*a*) The magnetization under weak magnetic force is reversible. (*b*) The magnetization for stronger magnetic force is almost proportional to it, but the remanence is smaller than the induction under the applied field. (*c*) An intense external field saturates the material. (*d*) Soft magnetic materials have a narrow hysteresis cycle. (*e*) Hard magnetic materials have a wide hysteresis cycle.

so that further field increases do not cause a larger magnetization. The material becomes *saturated*. If the external field decreases to zero, some domains forced beyond their stable state rotate and the remanence decreases below that corresponding to saturation. The opposing magnetic intensity that should be applied to remove the residual magnetism is termed *coercive force* or *field*.

As the magnetic force cycles from positive to negative values, the material describes a specific B–H curve different for increasing and decreasing H. Magnetic materials are classified according to the relative size of the hysteresis curve. Soft magnetic materials (Figure 1.29d) have a narrow hysteresis cycle with large μ_r (usually larger than 1000) and have a coercive force smaller than 100 µT. They suit ac applications because their magnetization can be easily reversed. Examples are silicon-iron, ferroxcube, permalloy, and mu-metal. Hard magnetic materials follow a wide hysteresis cycle (Figure 1.29e) with relatively small magnetic permeability and large coercive force (more than 100 mT in general), which makes them attractive for permanent magnets. They are also mechanically hard, hence difficult to work. Examples are carbon steel, alnico V (Al–Ni–Co) and hycomax (Al–Ni–Co–Cu).

The magnetic permeability of ferromagnetic materials depends on temperature. It increases with temperature until reaching the Curie point, which is different for each material: 730 °C for iron, 1131 °C for cobalt, 358 °C for nickel, and 16 °C for gadolinium. Beyond the Curie point, μ_r steeply decreases because of thermal vibrations and the material becomes paramagnetic.

Ferrimagnetic materials are crystalline materials with ions whose dipolar moments have antiparallel orientation. Nevertheless, one orientation slightly predominates, so that an external field magnetizes the material. They also have domains and a Curie point, but their temperature dependence is complex, they saturate before ferromagnetic materials, and they display higher electric resistance. Ferrimagnetic materials consist of ferric oxide combined with the oxides of one or more metals (as manganese, nickel, or zinc) and are collectively called ferrites.

Magnetic materials are used as structural elements to convey magnetic flux toward or away from a defined volume. They are also used to sense magnetic quantities because these modify other physical properties, such as electrical conductivity in magnetoresistors (Section 2.5), and to sense physical quantities such as temperature and mechanical stress, able to modify magnetic properties (Section 4.2).

1.9 MICROSENSOR TECHNOLOGY

Microsensor materials are prepared according to their nature, the desired sensing principle, and the intended application. There is an increasing interest in applying integrated circuit (IC) technology and micromachining, because they yield small, reliable sensors produced in large amounts leading to low cost.

1.9.1 Thick-Film Technology

Thick-film technology uses pastes or "inks" with fine particles (5 μm in average diameter) of common or noble metals dispersed in an organic vehicle, along with a glass frit that binds them. Depending on the dispersed particles, the paste can be conductive, resistive, or dielectric. Those pastes are screen-printed on a substrate according to a predefined pattern [19] involving width lines from 10 μm to 200 μm. The printed film is dried by heating at about 150 °C to remove the organic solvent that provided the low viscosity needed for the paste to squeeze through the open areas in the screen. The substrate with the deposited film is then fired on a conveyor belt furnace, usually in air atmosphere, so that the metal powder sinters and the glass frit melts, thereby bonding the film to the substrate. The printing, drying, and firing sequence is repeated for each paste used according to predetermined thermal cycles. The result is a 10 μm to 25 μm thick film, impermeable to many substances but relatively porous for specific chemical or biological agents. Thick-film components have a printed tolerance from $\pm 10\%$ to $\pm 20\%$, but they can be later trimmed to within $\pm 0.2\%$ to $\pm 0.5\%$ through selective abrasion or laser vaporization.

Depending on the firing temperature, there are three basic thick-film circuit types. Low-temperature pastes melt below 250 °C and are deposited on plastic materials, including those for printed circuits (glass fiber with epoxy), or anodized aluminum. There are thermoplastic and thermoset pastes. Thermoplastic pastes are mostly used in membrane switches. Thermoset pastes are epoxy materials with carbon and silver. High-temperature pastes melt at 800 °C to 1000 °C and use alumina, sapphire, or beryl (a silicate of beryllium and aluminum). Conductive pastes embed palladium, ruthenium, gold, and silver. Dielectric pastes use borosilicate glass. Medium-temperature pastes are similar to high-temperature pastes, melt at about 500 °C to 650 °C, and are deposited on low-carbon steel with porcelain enamel.

Thick-film technology finds at least three different uses in sensors. It has been used for years to fabricate hybrid circuits (multichip modules) offering improved performance compared to monolithic integrated circuits for signal conditioning and processing. Thick-film circuits and some sensors can be integrated in the same package, which improves reliability (strong connections), permits functional trimming, and reduces cost. It is also used to create support structures onto which a sensing material is deposited.

Some thick-film pastes directly respond to physical and chemical quantities. F. H. Nicoll and B. Kazan described the fabrication of a screen-printable photoconductor in 1955. There are pastes—some developed for sensing applications—with high temperature coefficient of resistance useful for temperature sensing (Sections 2.3 and 2.4), piezoresistive pastes (Section 2.2), magnetoresistive pastes (Section 2.5), photoconductive pastes (Section 2.6), piezoelectric pastes (Section 6.2), and pastes with high Seebeck coefficient (Section 6.1), among others. Pastes based on organic polymers and metal oxides such as SnO_2 can detect humidity (Section 2.7) and gases (Section 2.8) because of adsorption and absorption. Using thick-film technology, it is straightforward to define the interdigitated structures required for those sensors. Thick-film sensors with ceramic substrate withstand high temperatures, can be driven with relatively large voltages and currents, can integrate heaters, and can resist corrosion. Because the paste is fired into the ceramic, thick-film sensors are compact and sturdy. The printing process is quite inexpensive, which permits competitive low volume fabrication. References 19 and 20 review sensors based on thick-film pastes, as well other applications of thick-film technology in sensors and signal conditioning.

1.9.2   Thin-Film Technology

Thin films are obtained by vacuum deposition on a substrate of polished, high-purity (99.6%) alumina or low-alkalinity glass. Sensor and circuit patterns are defined by masks and transferred by photolitography, similarly to monolithic IC fabrication. Even though their names may suggest that the only difference between thick-film and thin-film technology is in film thickness, they are quite different technologies. In fact, metallized thin films may become thicker than some "thick" films.

Common materials in thin-film circuits are nichrome for resistors, gold for conductors, and silicon dioxide for dielectrics. Many thin-film sensors are resistive [21]. Piezoresistors use nichrome and polycrystalline silicon (Section 2.2), temperature and electrodes for conductivity sensors use platinum (Sections 2.3 and 2.9), anisotropic magnetoresistors use nickel, cobalt, and iron alloys (Section 2.5), gas sensors use zinc oxide (Section 2.8), and conductivity sensors use platinum.

Thin films are deposited by the same techniques used in IC fabrication: spin casting, evaporation, sputtering, reactive growth, chemical vapor deposition, and plasma deposition [18]. In spin casting, the thin-film material is dissolved in a volatile solvent and poured on the fast rotating substrate. Upon rotation, the liquid spreads and the solvent evaporates, thereby leaving a solid, uniform layer $0.1\ \mu m$ to $50\ \mu m$ thick. Thin films can also form by evaporating the material in a vacuum chamber in the presence of the substrate. The thin-film source is held in a heated crucible, and its evaporated atoms condense on the substrate. Sputtering or cathodic deposition also uses a vacuum chamber but the film material is placed on an anode, evaporated by bombardment with plasma of an inert gas inside the chamber, and deposited on a substrate placed on a cathode. Materials able to react with the substrate can be deposited by allowing their reaction. This is termed reactive growth, extensively used to grow silicon dioxide from a silicon surface held at high temperature. In chemical vapor deposition (CVD) there is a heated chamber with gas inlets and outlets. The high temperature breaks down incoming gases (pyrolysis) that contain the atomic components of the film, and the resulting components impinge on the substrate where they nucleate forming a film. The outlet carries away reaction and exhaust gases. CVD films can conform to the substrate. Epitaxial growth is a special CVD process that yields monocrystalline films on crystalline substrates. This is the preferred process to manufacture diaphragms for pressure sensors. CVD can be enhanced by the presence of plasma (PECVD) that induces the decomposition of gaseous compounds into reactive species at lower temperature.

Langmuir–Blodget films are ultrathin films named in honor of Irving Langmuir and Katherine Blodgett, who developed the technique in the 1920s [22]. These are monolayer films of usually insulating materials whose molecules have hydrophilic "head" and hydrophobic "tail." Upon dispersion on the surface of a trough of water (like oil in water), the head orients toward the water while the tail orients out of the water. If hydrophilic and hydrophobic forces are balanced, the result is a monolayer. These films can be transferred from the water by controlled dipping and deposited to form membranes for enzyme immobilization or to attract gas molecules. Monolayers can be transferred one by one and stacked. Langmuir–Blodgett films have been used in ISFETs (Section 9.2) and SAW sensors (Section 8.2.2).

1.9.3 Micromachining Technologies

Micromachining refers to processes to obtain three-dimensional devices whose feature dimensions and spacing between parts are in the range of 1 μm or less. Because it is a batch process (i.e., a process performed on an entire wafer 200 mm in diameter yielding hundreds of devices) and because materials and processes are borrowed from proven IC technology, micromachining has dramatically improved the performance-to-cost ratio over that of conventionally machined sensors. Reduced size and mass have broadened the dynamic range for some mechanical measurands. Placing integral electronics in the sensor housing increases reliability, but at the cost of reduced operating temperature. Micromachined sensors and other semiconductor-based sensors are termed microsensors. The development of microsensors is cost effective for applications requiring several million sensors per year, such as those in the automotive, home appliances, and biomedical industries.

Three-dimensional devices are manufactured in planar technology by using layers of coordinated two-dimensional patterns. The basic planar fabrication processes are deposition, lithography, and etching [18]. Thin-film deposition has been described above. Photolithography is the process of defining a pattern on a layer by first covering that layer with a thin film of photoresist, then exposing it to a light through a quartz mask, followed by chemical development (Figure 1.30). The result is a pattern that leaves parts of the original layer exposed to undergo chemical etching, ion implantation, or other processing. Removing the photoresist leaves the layer with the transferred pattern. Photolithography is used to either (a) define the geometry of overlying thin films such as silicon dioxide, polycrystalline silicon, and silicon nitride, deposited on the substrate or (b) directly modify the properties of the substrate (silicon or other).

Chemical etching relies on the oxidation of silicon to form compounds that can be removed from the substrate. Chemical etchants can be liquid, vapor,

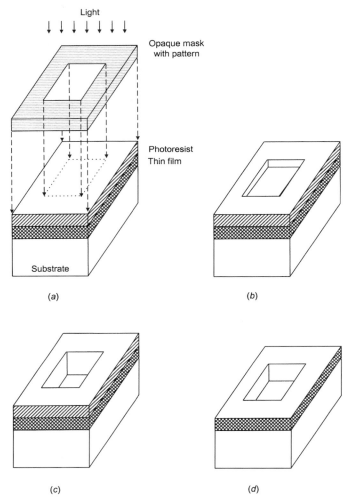

Figure 1.30 Planar photolithography processes. (*a*) The pattern is projected from a quartz mask onto photoresist. (*b*) Chemical development removes unsensitized photoresist. (*c*) Chemical etching removes the film uncovered by photoresist. (*d*) Photoresist removal leaves the desired pattern on film.

or plasma. Wet etching uses aqueous solutions of strong acids or bases that remove silicon from exposed areas of the uppermost layer from an immersed sample with a patterned photoresist. Some wet etches are isotropic; that is, the etching rate is the same for all directions. This creates undercuts that make the patterned structure somewhat smaller than the resist mask (Figure 1.31*a*). Other wet etches are anisotropic because certain silicon planes have different chemical reactivity (Figure 1.31*b*)—reactions are fast in the [100] and [110] di-

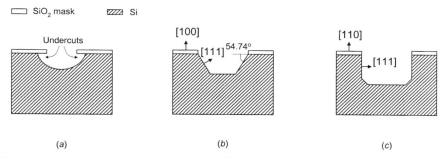

Figure 1.31   Etching can be (*a*) isotropic or (*b*, *c*) anisotropic, depending on the etchants and the etching method.

rections and slow in the [111] direction. Quartz resists etching along planes parallel to its *z*-axis. Etching rate and direction can be controlled by adding dopants to the exposed areas: *n*-type silicon is removed 50 times faster than *p*-type silicon. Vapor-phase etchants are halogen molecules or compounds that dissociate into reactive halogen species upon adsorption by silicon, thus forming volatile silicon compounds. Vapor phase etches are isotropic. Plasma phase etching relies on high reactivity of free halogen radicals created in ionized plasma. Bombarding the ions perpendicularly to the silicon surface enhances the reaction and yields anisotropic etching (Figure 1.31*c*).

Microstructures for sensors are mostly formed by bulk [23] and surface micromachining [24], the former being more common. Bulk micromachining removes significant amounts of material from a relatively thick substrate, usually a single crystal of silicon, but also amorphous glass, quartz (crystalline glass), and gallium arsenide. The etching process can be isotropic (Figure 1.31*a*) or anisotropic (Figures 1.31*b* and 1.31*c*). Some sensors use two or more wafers that are bonded together (Figures 1.10*b* and 1.14*e*). Bulk micromachining is extensively used in pressure sensors and has also been applied in cantilever-beam accelerometers.

Surface micromachining builds three-dimensional structures from stacked and patterned thin films such as polysilicon, silicon dioxide, and silicon nitride. Figure 1.32 illustrates some process steps. First the silicon substrate is thermally oxidized to give a SiO_2 layer, onto which a silicon nitride mask (for protection or electrical insulation) is deposited. Then a sacrificial (or space) layer such as 2 µm thick phosphosilicate glass (PSG) is sputtered and patterned, followed by deposition of the structural layer (often polysilicon 1 µm to 4 µm thick). Upon removal of the sacrificial layer by selective etching, the microstructure stands free and electric contacts are added. Surface micromachining yields smaller devices than bulk micromachining. Some commercial accelerometers are polysilicon surface microstructures with integrated MOS electronics. References 25 and 26 describe the basics of the respective technology processes used to produce different microsensors.

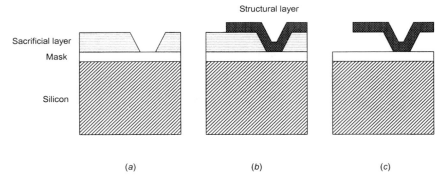

Figure 1.32 Surface micromachining processes include depositing (*a*) a sacrificial layer followed by (*b*) a body or structural layer, which becomes free upon removal of the sacrificial layer by (*c*) selective etching.

1.10 PROBLEMS

1.1 A given sensor has a specified linearity error of 1% of the reading plus 0.1% of the full-scale output (FSO). A second sensor having the same measurement range has a specified error of 0.5% of the reading plus 0.2% FSO. For what range of values is the first sensor more accurate than the second one? If the second sensor had a measurement range twice that of the first one, for what range of values would it be the more accurate?

1.2 Which of the following numerical results of a measurement are incorrectly expressed: $100\,°C \pm 0.1\,°C$, $100\,°C \pm 1\,°C$, $100\,°C \pm 1\%$, $100\,°C \pm 0.1\%$?

1.3 A Pitot tube in a pipe determines flow velocity from a differential pressure measurement. If the manometer has 2% uncertainty, what is the uncertainty in the flow velocity? If the total pressure is about five times the static pressure, what would be the uncertainty in a worst-case condition if two relative manometers were used instead and their readings subtracted?

1.4 Determine what probability has the confidence interval $\hat{x}_n \pm \sigma/\sqrt{n}$ of including the true value x.

1.5 Determine the confidence interval that has 99% probability of including the true value of a quantity when the average from n measurements is $\hat{x}_n$ and the variance is σ^2.

1.6 A bare temperature sensor (first-order dynamic response) is used to measure a turbulent flow with fluctuations of up to 100 Hz. If the dynamic error is to be held smaller than 5%, what time constant should the sensor have?

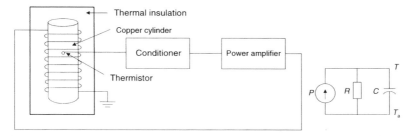

Figure P1.7 System to control the temperature of an oven for a crystal oscillator.

1.7 Figure P1.7 shows the control circuit for an oven that includes a crystal oscillator and its equivalent analog circuit. The aim is to keep the oven at 70 °C for an ambient temperature of 30 °C. The heater is a 50 g coil wounded around a copper cylinder (specific heat $c_p = 390$ J/kg·K) that applies 3 W in average. Heat is lost by conduction through the thermal insulation. Determine the transfer function for the system and its time constant. Assume that copper has negligible thermal resistance.

1.8 The Model 3145 accelerometer (Eurosensor) has a damping ratio of 0.7 typical and 0.4 minimal, along with a typical resonant frequency of 1200 Hz. Calculate the maximal overshoot when applying a 10 g step acceleration for both the typical and minimal damping ratio, and determine the time elapsed to the overshoot for the minimal damping ratio.

1.9 A given load cell has a second-order dynamic response with damping ratio of 0.7. Calculate the amplitude error when measuring a dynamic force at 70% of the load cell's natural frequency.

1.10 A given mass–spring system can be modeled by an underdamped second-order, low-pass transfer function. Determine the damping factor needed to ensure that the maximal frequency response is less than 5% higher than that at low frequency.

1.11 A pressure p is applied to a liquid column manometer like that in Figure 1.14a, where the reference pressure is considered to be constant. If the tube cross section A is uniform, the fluid fills a length L along the tube, its friction coefficient with the walls is R [(N/m^2)/(m/s)], and its density is ρ, what is the transfer function relating the fluid height to the applied pressure?

1.12 We wish to replace a first-order sensor by a second-order sensor with improved frequency response. In order to characterize the first-order sensor, we apply an input step and measure 25 ms to 90% of the final value. The second-order system should have a natural frequency equal to the corner frequency of the first-order sensor, and its relative dynamic error should be less than 10%. Determine the natural frequency, damping ratio, resonant frequency, and dynamic error at resonance.

1.13 Calculate the low-frequency limit of the input velocity applied to a mass–spring system in order for the amplitude error to be smaller than 5% of the high-speed response when the natural period is 10 s and the damping ratio is 0.7.

1.14 Calculate the deflection and maximal stress of a stainless steel cantilever ($l = 3$ cm, $w = 0.5$ cm, $t = 2$ mm, $E = 210$ GPa) when applying 10 kg at its free end.

1.15 To calibrate an accelerometer we have a vibrating table, a frequency meter, a moving-coil linear velocity sensor (Section 4.3.1), and an optical distance measurement system. Discuss which of the following methods is the best in order to determine the applied acceleration depending on the accuracy of each of these instruments: (1) Measure the vibrating frequency and the linear velocity; (2) measure the vibrating frequency and the displacement.

1.16 To calibrate a linear accelerometer, it is placed on a horizontal centrifuge table with radius R, turning at an adjustable speed n, which is indicated on a four-digit panel in revolutions per minute (r/min). The total error of the speed measurement system is ± 1 in the least significant bit (LSB).

   a. If the error in the placement of the accelerometer is insignificant, what is the relative error in the calculated acceleration when the system turns at 5000 r/min?

   b. If the position of the accelerometer is now measured by a digital system having an error of ± 1 LSB, how many bits must it have in order to result in an error in acceleration (due to position uncertainty) lower than the one in part (a)?

   c. To determine the transverse sensitivity of the accelerometer, it is placed with its sensing axis along the tangential direction; a signal 1.7% of that obtained when the axis is along the radial direction results. What must be the accuracy of the angular positioning system so that, in calculating the longitudinal sensitivity the error due to the misalignment between the sensing axis and the radius will be smaller than 0.1%?

REFERENCES

[1] H. V. Malmstadt, C. G. Enke, and S. R. Crouch. *Electronics and Instrumentation for Scientists*. Menlo Park, CA: Benjamin-Cummings, 1981.

[2] E. O. Doebelin. *Measurement Systems: Application and Design*, 4th ed. New York: McGraw-Hill, 1990.

[3] International Organization for Standardization. *Guide to the Expression of Uncertainty in Measurement*. Geneva (Switzerland): ISO, 1993.

[4] C. F. Dietrich. *Uncertainty, Calibration, and Probability*, 2nd ed. Philadelphia: Adam Hilger, 1991.

[5] R. H. Dieck, Measurement accuracy. Chapter 4 in: J. G. Webster (ed.), *The Measurement, Instrumentation, and Sensors Handbook*. Boca Raton, FL: CRC Press, 1999.

[6] K. Ogata. *Modern Control Engineering*. Upper Saddle River, NJ: Prentice-Hall, 1996.

[7] A. C. Bell, Input and output characteristics, Chapter 5 in: C. L. Nachtigal (ed.), *Instrumentation and Control Fundamentals and Applications*. New York: John Wiley & Sons, 1990.

[8] H. Kumamoto and E. J. Henley. *Probabilistic Risk Assessment and Management for Engineers and Scientists*, 2nd ed. New York: IEEE Press, 1996.

[9] W. G. Ireson and C. F. Coombs (eds.). *Handbook of Reliability Engineering and Management*, 2nd ed. New York: McGraw-Hill, 1996.

[10] R. M. White. A sensor classification scheme. *IEEE Trans. Ultrason. Ferroelectr. Freq. Control*, **34**, 1987, 124–126.

[11] A. D. Khazan. *Transducers and Their Elements*. Englewood Cliffs, NJ: Prentice-Hall, 1994.

[12] J. Fraden. *Handbook of Modern Sensors, Physics, Design, and Applications*, 2nd ed. Woodbury, NY: American Institute of Physics, 1997.

[13] J. G. Webster (ed.). *The Measurement, Instrumentation, and Sensors Handbook*. Boca Raton, FL: CRC Press, 1999.

[14] H. K. P. Neubert. *Instrument Transducers*, 2nd ed. New York: Oxford University Press, 1975.

[15] N. Yazdi, F. Ayazi, and K. Nafaji. Micromachined inertial sensors. *Proc. IEEE*, **56**, 1998, 1640–1659.

[16] L. Solymar and D. Walsh. *Electrical Properties of Materials*. New York: Oxford University Press, 1998.

[17] P. T. Moseley and A. J. Crocker. *Sensor Materials*. Philadelphia: IOP Publishing, 1996.

[18] S. M. Sze (ed.). *Semiconductor Sensors*. New York: John Wiley & Sons, 1994.

[19] N. M. White and J. D. Turner. Thick-film sensors: past, present and future. *Meas. Sci. Technol.* **8**, 1997, 1–20.

[20] M. Prudenziati (ed.). *Thick Film Sensors*. Amsterdam: Elsevier, 1994.

[21] P. Ciureanu and S. Middelhoek (eds.). *Thin Film Resistive Sensors*. Philadelphia: IOP Publishing, 1992.

[22] S. S. Chang and W. H. Ko. Thin and thick films. Chapter 6 in: T. Grandke and W. H. Ko (eds.), *Fundamentals and General Aspects*, Vol. 1 of *Sensors, A Comprehensive Survey*, W. Göpel, J. Hesse, and J, N. Zemel (eds.). New York: VCH Publishers (John Wiley & Sons), 1989.

[23] G. T. A. Kovacs, N. I. Maluf, and K. E. Petersen. Bulk micromachining of silicon. *Proc. IEEE*, **86**, 1998, 1536–1551.

[24] J. M. Bustillo, R. T. Howe, and R. S. Muller. Surface micromachining for microelectromechanical systems. *Proc. IEEE*, **86**, 1998, 1552–1574.

[25] L. Ristic (ed.). *Sensor Technology and Devices*. Norwood, MA: Artech House, 1994.

[26] J. W. Gardner. *Microsensors: Principles and Applications*. New York: John Wiley & Sons, 1994.

2

RESISTIVE SENSORS

Sensors based on the variation of the electric resistance of a device are very common. That is because many physical quantities affect the electric resistance of a material. Thus resistive sensors are used to solve many measurement problems. Temperature-dependent resistors can also compensate for thermal interference in systems measuring other quantities.

This chapter discusses sensors based on a variation in resistance. It describes their fundamentals (sensing principle, dynamic model, limitations, advantages), technology, equivalent electric circuit, and applications. Some applications use primary sensors, models, and definitions given in Chapter 1. Chapter 3 describes the circuits that yield a useful electric signal.

The different resistive sensors are classified by the physical quantity being measured as mechanical, thermal, magnetic, optical, and chemical variables.

2.1 POTENTIOMETERS

A potentiometer is a resistive device with a linear or rotary sliding contact (Figure 2.1). The resistance between that contact and the bottom terminal is

$$R = \frac{\rho}{A} x = \frac{\rho l}{A} \alpha \tag{2.1}$$

where ρ is the resistivity, A is the cross section, l is the length, x is the distance traveled from the bottom terminal, and α is the corresponding length fraction. Variable resistance devices were already instrumental in electricity studies when W. Ohm presented his law in 1827. G. Little patented a variable resistance with

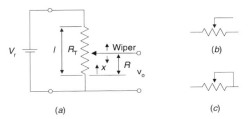

Figure 2.1 (*a*) Ideal (linear) potentiometer connected as a voltage divider and (*b*) its symbol; the arrow indicates that the resistance variation responds to a mechanical action. (*c*) The wiper can also be connected to an end terminal to obtain a variable resistor (rheostat).

insulated metal wire and a slider in 1871. In 1907, H. P. MacLagan was awarded a patent for a rotary rheostat. Arnold O. Beckman patented the first commercially successful 10-turn precision potentiometer in 1945.

A potentiometer is a zero-order system, although it can be itself a component of a nonzero-order sensor—for example, a mass–spring system.

Equation (2.1) means that the resistance is proportional to the travel of the wiper. This implies the acceptance of several simplifications that may not necessarily be true. First, we assume that the resistance is uniform along the length *l*. But the resistance is not perfectly uniform, which limits the linearity of the potentiometer. The agreement between the actual and the theoretical transfer characteristic (here a straight line) is termed *conformity* (here *linearity*). Second, we assume that the sliding contact gives a smooth resistance variation, not a stepped one, and therefore that the resolution is infinite. But that is not true for wound resistive elements or for conductive-plastic potentiometers [1]. Furthermore, the mechanical travel is usually larger than the electrical travel.

For (2.1) to be valid, if the potentiometer is supplied by an alternating voltage, its inductance and capacitance should be insignificant. For low values of the total resistance R_T, the inductance may be significant, particularly in models with wound resistive elements. For high values of R_T, the parasitic capacitance may be important.

Furthermore, resistors drift with temperature. Therefore (2.1) is valid only if resistance changes due to temperature are uniform. Temperature changes can arise not only from fluctuations in ambient temperature but also from self-heating due to the finite power that the potentiometer dissipates. If the power rating is *P*, the maximal rms value of the applied voltage V_r must be

$$V_r < \sqrt{PR_T} \qquad (2.2)$$

Measurement circuits with relatively low input impedance have an electrical loading effect on the potentiometer, and some parts of the potentiometer may heat beyond their rated power (see Problem 2.1).

Friction and inertia of the wiper also limit the model validity because they add a mechanical load to the system being measured. These should be insignificant but at the same time ensure a good contact. As a compromise the force required to displace the wiper is from 3 g to 15 g. For variable movements the starting torque is approximately twice the dynamic torque, and this is reduced by lubrication. For rapid movements there is a risk of losing contact during vibration. Thus some units have two wipers of different arm lengths, and therefore different resonant frequencies. Alternatively, some models have elastomer-damped wipers. Maximal travel speed is limited to about 10 m/s. The axis of rotary potentiometers must be concentric to the axis whose angular displacement is to be measured.

Finally, noise associated with the wiper contact limits resolution. When contact resistance changes with movement from one position to another, current circulating through changes the output voltage, and these fluctuations may be significant for the attached measuring device. Noise can increase because of dust, humidity, oxidation, and wear.

Most of these limitations are outweighed by the advantages of this device. It is simple and robust and yields a high-level voltage with high accuracy relative to its cost.

Models available accept linear and rotary movements (one or more turns in helical units). In some models the output is deliberately nonlinear with respect to the displacement [2]. For example, the output can be a trigonometric function—sine, cosine, tangent—of the angle turned by the sliding contact. A nonlinear relationship can also be obtained by using a nonuniform spacing for the wire or by varying its size along its length. When the measuring circuit loads the potentiometer, the result is also a nonlinear characteristic (Section 3.2.1). Reference 3 describes a computation method to generate resistor geometry with a prescribed potential drop along the wiper path.

Dual potentiometers operated by a single control stick (joysticks) move in four quadrants to locate a point in a plane. Movement along the x-axis controls R_x, and movement along the y-axis controls R_y (Figure 2.2). If both potentiometers are supplied by the same voltage, the output voltages are

$$v_x = V_r(1 - 2\alpha)$$
$$v_y = V_r(1 - 2\beta) \tag{2.3}$$

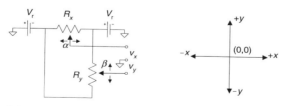

Figure 2.2 Joystick based on a dual potentiometer operated by a single control stick able to move around in four quadrants.

TABLE 2.1   Specifications for Linear and Rotary Potentiometers

Parameter	Linear	Rotary
Input range	2 mm to 8 m	$10°$ to 60 turns
Linearity	0.002% FSO to 0.1% FSO	
Resolution	50 µm	$2°$ to $0.2°$
Maximal frequency	3 Hz	
Power rating	0.1 W to 50 W	
Total resistance	20 Ω to 220 kΩ	
Temperature coefficient	$20 \times 10^{-6}/°C$ to $1000 \times 10^{-6}/°C$	
Life	Up to 10^8 cycles	

where $0 \leq \alpha, \beta \leq 1$. At the center of the plane we have $\alpha = \beta = 0.5$, and the voltages obtained are (0 V, 0 V).

Potentiometers comprise a resistive element, a wiper, an actuating or driving rod, bearings, and housing. The resistive element is not a single wire. Even if the wire were very thin (while retaining enough strength), it would be impossible to obtain a high-enough resistance value as compared to connecting wires. One common configuration is a wire wound around a (ceramic) insulating form. Some of the wire materials used are nickel–chrome, nickel–copper, and precious metal alloys. But then the inductance is high and the resolution is low. Advantages are a low temperature coefficient and a high power rating.

Potentiometers based on a carbon film, sometimes mixed with plastic, deposited on a form, and a noble metal alloy wiper with multiple contacts yield high resolution and long life at a moderate cost. But they have a high temperature coefficient. Other models use a thick-film configuration consisting of silver-filled polymeric ink. Precious metal contacts yield the best electrical and life performance, at a higher cost. For high power dissipation and high resolution, the resistive element of cermet models is based on particles of precious metals fused in a ceramic base and deposited by thick film techniques. Hybrid potentiometers use a wire-wound core coated by a conductive plastic. The conductive plastic confers smoothness but limits the power rating. Table 2.1 gives the range of some specifications of commercially available models.

Electrolytic or liquid potentiometers have a particular arrangement intended for tilt measurement (Section 1.7.6). There is a hermetic, curved vial partially filled with a conductive liquid and three metal electrodes contacting the fluid, a large central electrode, and two shorter electrodes, one at each side (Figure 2.3a). A 0.5 V to 12.5 V ac voltage—20 Hz to 20 kHz—is applied to the outer electrodes (a dc voltage would electrolyze the liquid). When the tube is horizontal, the electric resistance from the center electrode to each outer electrode is equal and the voltage at the central electrode is half the applied voltage (i.e., 0 V). As the tube tilts, the air pocket shifts, the current path between outer electrodes changes, and so does the resistance between the central electrode and each outer electrode. The output voltage is proportional to the tilt angle. The measurement span available in different models ranges from $\pm 0.5°$ to $\pm 60°$,

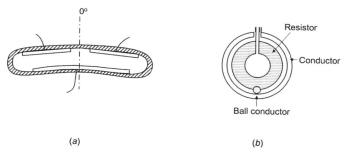

(a)   (b)

Figure 2.3   Tilt sensors based on potentiometers. (*a*) Liquid potentiometer. (*b*) A rolling ball acts as wiper in a pendulum, changing the resistance between the resistor ends and a conductor supported by a vertical plate.

and the resolution is up to 10^{-8} rad. Response time ranges from 100 ms to several minutes. Temperature interferes because it changes the resistivity of the electrolyte. Tilt switches use two close electrodes and mercury that closes the circuit when the vial rotates by a given angle.

Figure 2.3*b* shows another potentiometer intended for inclination measurement. A vertical planar substrate supports a concentric conductor and resistor. A conductive ball is pressed against the resistor by a channeled housing ring (not shown in the figure) and acts as wiper. When the housing tilts, the ball rolls to the lowest point and changes the resistance between the conductor and the resistor terminals. The range is about 360° and the resolution is 0.1°.

The Thévenin equivalent circuit for a potentiometer shows that its output impedance depends on the wiper position. In a linear potentiometer supplied by a direct voltage, the output resistance R_o is the parallel combination of $R_T(1 - \alpha)$ and $R_T\alpha$,

$$R_o = \frac{R_T\alpha R_T(1 - \alpha)}{R_T\alpha + R_T(1 - \alpha)} = R_T\alpha(1 - \alpha) \tag{2.4}$$

The open circuit output voltage is

$$v_o = V_r \frac{R_T\alpha}{R_T} = V_r\alpha \tag{2.5}$$

where $0 \le \alpha \le 1$. The output voltage depends on both the supply voltage and the wiper position. The ratio between the output voltage and the supply voltage depends only on the wiper position. The output voltage is independent of R_T, but the output impedance increases with R_T. Potentiometers can be often directly connected to analog-to-digital converters without any interfacing amplifier (Section 3.2.1).

Potentiometers suit the measurement of linear or rotary displacements exceeding full-scale values of 1 cm or 10°. Displacements of this magnitude can be

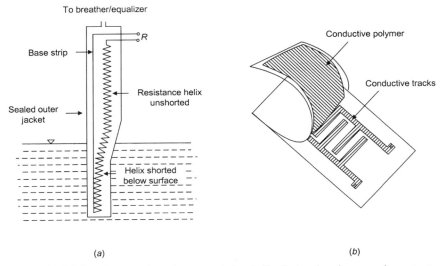

(a)   (b)

Figure 2.4   (*a*) Level sensor based on a resistive helix that, when immersed, contacts a conductive base because of hydrostatic pressure (from Metritape). (*b*) Force-sensitive resistor (FSR^{TM}, Interlink) based on shorting interdigitated electrodes by a conductive polymer.

found in position feedback systems and also in some primary sensors—for example, in pressure sensors based on Bourdon tubes, bellows, or capsules (Section 1.7.2) and in float-based level sensors (Section 1.7.4). Typically, the housing is mounted on a fixed reference frame and the wiper is coupled to the moving element. Belts and pulleys, lead screws, cabled drums, gears, and other mechanisms [4] extend the capabilities of potentiometers.

Figure 2.4*a* shows a level sensor with response αR ($0 \leq \alpha \leq 1$) but with no wiper. There is a stainless steel base strip with a gold contact stripe on one side. A gold-plated nichrome wire is precisely wound to form a resistance helix. When the tape is immersed in liquid, hydrostatic pressure makes the helix wire contact the gold contact stripe and reduces the output resistance. The resistance gradient is about 10 Ω/cm.

Force-sensitive resistors are conductive-polymer film sensors whose conductance is roughly proportional to the applied force and the application point behaves as virtual pressure sensitive wiper. Figure 2.4*b* shows the FSR^{TM} sensor (Interlink) that consists of two polymer films. One film is doped by covalent agents that render its surface conductive. The other film has printed interdigitated electrodes facing the conductive surface of the adjacent film. Contact of the two films makes the conductive surface short electrode fingers, thus reducing the electric resistance between end terminals. Flexiforce^{TM} (Tekscan) sensors use a similar principle but with a conductive material (silver) applied on

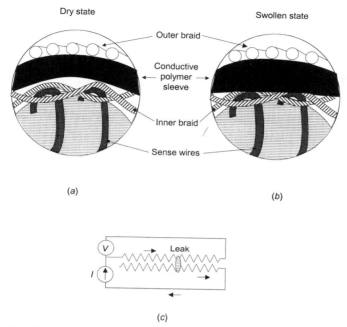

Figure 2.5 (*a*) Sensing cable to detect liquid hydrocarbons (TraceTek™, RayChem). (*b*) The conductive polymer swells when hydrocarbons diffuse into it, shorting two sensing wires. (*c*) The equivalent circuit shows that the leak behaves as virtual wiper.

each film, followed by a layer of pressure-sensitive ink. The unloaded sensor has a resistance about 200 kΩ. A full-scale load reduces resistance to about 20 kΩ. These sensors have been used for plantar-pressure measurement, for patient palpation, for force feedback in physical rehabilitation, to measure bite force in dentistry, for tactile sensors in robotics, to detect seat occupancy, as a virtual reality force sensor in gloves, to measure force on golf club grips, as a variable force control for computer joysticks, and so on.

Figure 2.5 shows a hydrocarbon sensor equivalent to a potentiometer, based on the swelling of polymers when exposed to solvents and fuels. There is a cable consisting of an outer porous containment braid, a conductive polymer sleeve, an inner separator braid, a sense wire, and an inner wire. The initial resistance is above 30 MΩ. The cable is buried in the monitored area. Liquid hydrocarbons penetrate the braid and diffuse into the conductive polymer sleeve that swells inwardly. The conductive polymer then shorts the two sense wires together, and the resistance decreases below 20 kΩ. This sensor has been proposed to detect and locate fuel leaks along buried pipes, tanks, and double contained pipes. There are also models sensitive to water, conductive liquids, and liquid organic solvents.

2.2 STRAIN GAGES

2.2.1 Fundamentals: Piezoresistive Effect

Strain gages are based on the variation of resistance of a conductor or semi-conductor when subjected to a mechanical stress. Lord Kelvin reported on this effect in conductors in 1856, and C. S. Smith studied the effect in silicon and germanium in 1954. The electric resistance of a wire having length l, cross section A, and resistivity ρ is

$$R = \rho \frac{l}{A} \tag{2.6}$$

When the wire is stressed longitudinally, each of the three quantities that affect R change and therefore R undergoes a change given by

$$\frac{dR}{R} = \frac{d\rho}{\rho} + \frac{dl}{l} - \frac{dA}{A} \tag{2.7}$$

The change in length that results when a force F is applied to a wire, within the elastic limit (Figure 2.6a), is given by Hooke's law,

$$\sigma = \frac{F}{A} = E\varepsilon = E\frac{dl}{l} \tag{2.8}$$

where E is the Young's modulus (T. Young, 1773–1829) (specific for each material and temperature-dependent), σ is the mechanical stress, and ε is the strain

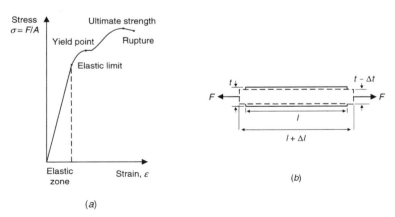

Figure 2.6 (*a*) Stress–strain diagram for mild steel. The elastic region has been greatly enlarged. (*b*) Action of Poisson's ratio: A longitudinal expansion implies a lateral contraction.

(unit deformation). ε is dimensionless, but to improve clarity it is usually given in "microstrains" (1 microstrain $= 1$ $\mu\varepsilon = 10^{-6}$ m/m). Strain in the elastic region is proportional to stress. The behavior in the plastic region is irreversible because there is strain after removing the applied force.

Consider a wire that in addition to a length l has a transverse dimension t (Figure 2.6b). A longitudinal stress changes both l and t. According to Poisson's law, we have

$$v = -\frac{dt/t}{dl/l} \tag{2.9}$$

where v is the Poisson ratio. The minus sign indicates that lengthening implies constriction. Usually, $0 < v < 0.5$; it is, for example, 0.17 for cast iron, 0.303 for steel, and 0.33 for aluminum and copper. Note that for the volume to remain constant, it should be $v = -0.5$, which is almost the case for rubber.

For a wire of circular cross section of diameter D, we have

$$A = \frac{\pi D^2}{4} \tag{2.10}$$

$$\frac{dA}{A} = \frac{2dD}{D} = -\frac{2vdl}{l} \tag{2.11}$$

The change in resistivity as a result of a mechanical stress is called the *piezoresistive effect*. This effect results from the amplitude change of vibrations in the metal lattice. A longitudinal extension causes that amplitude to increase, which reduces electron mobility, thus increasing resistivity. P. W. Bridgman showed that, in metals, the percent changes of resistivity and volume are proportional:

$$\frac{d\rho}{\rho} = C\frac{dV}{V} \tag{2.12}$$

where C is the Bridgman's constant. For the usual alloys from which strain gages are made, $1.13 < C < 1.15$. For platinum, $C = 4.4$. By applying (2.10), the change in volume can be expressed as

$$V = \frac{\pi l D^2}{4} \tag{2.13}$$

$$\frac{dV}{V} = \frac{dl}{l} + \frac{2dD}{D} = \frac{dl}{l}(1 - 2v) \tag{2.14}$$

Therefore if the material is isotropic, within the elastic limit, (2.7) leads to

$$\frac{dR}{R} = \frac{dl}{l}[1 + 2v + C(1 - 2v)] = G\frac{dl}{l} = G\varepsilon \tag{2.15}$$

where G is the gage factor, defined as the factor inside the square brackets. From the given values for v and C, $G \approx 2$. For isoelastic, $G \approx 3.2$; for platinum, $G \approx 6$.

Therefore for small variations, the resistance of the metallic wire is

$$R = R_0 + dR = R_0 \left(1 + \frac{dR}{R_0}\right) \cong R_0(1 + G\varepsilon) = R_0(1 + x) \tag{2.16}$$

where R_0 is the resistance when there is no applied stress, and $x = G\varepsilon$. Usually, $x < 0.02$.

Example 2.1   A 350 Ω strain gage having $G = 2.1$ is attached to an aluminum strut ($E = 73$ GPa). The outside diameter of the strut is 50 mm and the inside diameter is 47.5 mm. Calculate the change in resistance when the strut supports a 1000 kg load.

From (2.15),

$$\Delta R = RG\varepsilon = RG\frac{F/A}{E}$$

From geometry, the area supporting the force is

$$A = \frac{\pi(D^2 - d^2)}{4} = \frac{\pi \times (97.5 \text{ mm}) \times (2.5 \text{ mm})}{4} = 191 \text{ mm}^2$$

Therefore, with $R = 350$ Ω, $G = 2.1$, $F = 1000$ kg $= 9800$ N, and $E = 73$ GPa, we have

$$\Delta R = (350 \text{ }\Omega) \times 2.1 \times \frac{9800 \text{ N}}{(191 \times 10^{-6} \text{ m}^2) \times 73 \text{ GPa}} = 0.52 \text{ }\Omega$$

which is less than 0.15% of the initial resistance.

When a semiconductor is stressed, in addition to its dimensional change, both the number of carriers and their average mobility change. Unlike metals, the resistivity change under stress dominates over the dimensional change [5]. The magnitude and sign of the piezoresistive effect depend on the specific semiconductor, its carrier concentration, and the crystallographic orientation with respect to the applied stress. For simple tension or compression, if electrons flow along the stress axis, the relative change in resistivity is proportional to the applied stress,

$$\frac{\Delta\rho}{\rho_0} = \pi_L \sigma \tag{2.17}$$

where π_L is the longitudinal piezoresistive coefficient and ρ_0 is the resistivity for the unstressed material. The resulting gage factor

$$G = \frac{\Delta R / R_0}{\varepsilon} \tag{2.18}$$

is from about 40 to about 200. Semiconductors with a relatively low number of carriers yield large gage factors, but they are temperature-sensitive and depend on the stress; that is, they are nonlinear. Semiconductors with a relatively high number of carriers have smaller gage factors, but these are less temperature- and stress-dependent.

Thus the change in the electric resistance of a metal or a semiconductor is related to its strain. If the relationship between that strain and the force causing it is known [6], from the measurement of resistance changes it is possible to infer the applied forces and the quantities that produce those forces in a primary sensor. A resistor arranged to sense a strain constitutes a strain gage. This method has proved to be very useful for many years [7]. However, there are many limitations we must consider concerning this measurement principle to obtain valid information.

First, the applied stress should not exceed the elastic limit of the gage. Strain should not exceed 4% of gage length and ranges, approximately, from 3000 $\mu\varepsilon$ for semiconductor gages to 50,000 $\mu\varepsilon$ for metal gages.

Second, the measurement will be correct only if all the stress is transmitted to the gage. This is achieved by carefully bonding the gage with an elastic adhesive that must be also stable with time and temperature. At the same time, the gage must be electrically insulated from the object it adheres to and be protected from the environment.

We assume that all strains are in the same plane; that is, there is no stress in any direction perpendicular to the gage wires. To have a significant electric resistance for metal gages, they consist of a grid containing several longitudinal segments connected by shorter transverse segments having a larger cross section (Figure 2.7). Thus the transverse sensitivity is only from 1% to 2% of the longitudinal sensitivity. Figure 2.8 shows the conventional method for installing a strain gage.

Temperature interferes through several mechanisms. It affects the resistivity of the material, its dimensions, and the dimensions and Young's modulus of the support material. Thus once the gage is cemented, any change in temperature yields a change in resistance, hence an apparent strain, even before applying any mechanical force. In metal strain gages this change can be as large as 50 $\mu\varepsilon/^\circ$C.

Temperature interference may be compensated by dummy gages implementing the opposing-inputs method. Dummy gages are equal to the sensing gages and placed near them in order to experience the same temperature change but without experiencing any mechanical effort. Section 3.4.4 discusses their placement in the measuring circuit to compensate temperature-induced resistance changes. To avoid excessive differential strains, strain gages are available

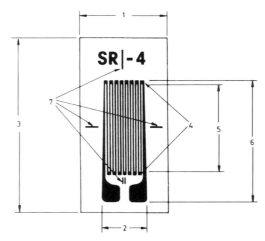

Figure 2.7 Parameters for a foil strain gage (from BLH Electronics): 1, matrix width; 2, grid width; 3, matrix length (carrier); 4, end loops; 5, active grid length; 6, overall gage length; 7, alignment marks. Typical thicknesses are 3.8 μm and 5 μm, depending on material type.

having a thermal expansion coefficient similar to that of different materials to be tested in the temperature range from $-45\,°C$ to $200\,°C$.

Temperature interference is stronger in semiconductor strain gages. In self-temperature-compensated gages, the increase in resistivity with increasing temperature is compensated by a decrease in resistance due to the expansion of the backing material. This method achieves thermal-induced apparent strain of only 5 με/°C over a temperature range of $20\,°C$.

Strain-gage resistance measurement implies passing an electric current through it, which causes heating. The maximal current is 25 mA for metal gages if the base material is a good heat conductor (steel, copper, aluminum, magnesium, titanium) and is 5 mA if it is a poor heat conductor (plastic, quartz, wood). The permissible power increases with the gage area and ranges from 770 mW/cm^2 to 150 mW/cm^2, depending on the backing. Maximal power dissipation in semiconductor strain gages is 250 mW.

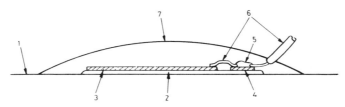

Figure 2.8 Installation of a foil strain gage (from BLH Electronics): 1, substrate material; 2, adhesive; 3, strain gage; 4, solder terminals; 5, solder; 6, lead wires; 7, environmental barrier.

Another interference is the thermoelectromotive force appearing at the junction between dissimilar metals (Section 6.1.1). If the gage is supplied by a dc voltage, metal junctions may produce a net voltage which adds to the voltage due to strain. Thermoelectric voltages can be detected by reversing the supply polarity: If they are present, the output voltage will change. Subtracting voltage readings for both polarities cancels added voltages because their polarity does not change, but that of the signal voltage changes with power supply polarity. Thermoelectric interference can be avoided by applying the intrinsic insensitivity method, selecting appropriate materials, filtering, or supplying the gages with ac voltage.

Strain gages should ideally be very small in order to measure strain at a given point. In practice, they have finite dimension and we assume that the measured "point" is at the gage geometric center. When measuring vibration, the wavelength must be longer than the gage. If, for example, the useful length of a gage is 5 mm and the measurement is done in steel where sound velocity is approximately 5900 m/s, then the frequency for one wavelength to equal one gage length is (5900 m/s)/0.005 m $\approx$ 1 MHz. To keep 10% of the wavelength equal to the gage length, the maximal measurable frequency is about 100 kHz (1 MHz/10).

In stress testing of a rough surface, such as concrete, we should measure an average strain in order to avoid any inaccuracies due to discontinuity in the surface. In this case a large gage should be used.

Silicon strain gages are also light-dependent, although optical effects are probably negligible under conventional illumination conditions [8].

In spite of all these possible limitations, strain gages are among the most popular sensors because of their small size, high linearity, and low impedance.

2.2.2 Types and Applications

Strain gages are made of different metals such as the alloys advance ($Cu_{55}Ni_{45}$), constantan ($Cu_{57}Ni_{43}$), karma ($Ni_{75}Cr_{20}Fe_xAl_y$), nichrome ($Ni_{80}Cr_{20}$), and isoelastic ($Ni_{36}Cr_8Fe_{55.5}Mo_{0.5}$), and they are also made of semiconductors such as silicon and germanium. The resistances of the selected metal alloys have low temperature coefficients because the reduced electron mobility is partially balanced by an increase in available conduction electrons. Constantan is the most common gage alloy. Karma is the preferred choice for static measurements over long periods of time (months or years) and has longer fatigue life and higher temperature range than constantan. Isoelastic has a relatively high temperature coefficient (145 $\mu\varepsilon/°C$) and good fatigue life, which makes it more suitable for dynamic than for static measurements. Platinum–tungsten gages have $G = 4.5$ and excellent fatigue life, and they operate from $-200 °C$ to $650 °C$. Screen-printable materials used in thick-film strain gages have $G > 10$.

Strain gages can be either bonded or unbonded (Figure 2.9). The backing or carrier of bonded strain gages is chosen according to the temperature of the material to test. Bonded metal strain gages can be made with paper backing

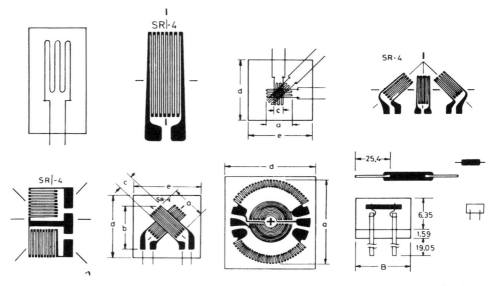

Figure 2.9 Bonded and unbonded metal and semiconductor strain gages can be simple or multiple (rosettes) and can be designed for specific elastic elements (from BLH Electronics).

from parallel wire or, currently, from photoetched metal foil on a plastic carrier. Figure 2.9 shows that there are strain gages for diaphragms, for torsion, and to determine minimal and maximal stress and its direction (rosettes). For tactile sensors in robots, conductive elastomers may form the strain gage. Liquid strain gages can measure large strains in biological tissue (muscle, tendons, and ligaments) [9]. They are composed of a silicone rubber tube filled with mercury or an electrolyte, such as saline. Thick-film strain gages become bonded to the deflecting substrate during thermal curing and withstand high temperature ($>250\,^\circ$C). Micromachined sensors use strain gages implanted in silicon.

Table 2.2 lists some typical characteristics of metal and semiconductor strain

TABLE 2.2 Typical Characteristics of Metal and Semiconductor Strain Gages

Parameter	Metal	Semiconductor
Measurement range	$0.1\ \mu\varepsilon$ to $50{,}000\ \mu\varepsilon$	$0.001\ \mu\varepsilon$ to $3000\ \mu\varepsilon$
Gage factor	1.8 to 4.5	40 to 200
Nominal resistance, Ω	$120, 250, 350, 600, \ldots, 5000$	1000 to 5000
Resistance tolerance	0.1% to 0.35%	1% to 2%
Active grid length, mm	0.4 to 150	1 to 5
	Standard: 3 to 10	

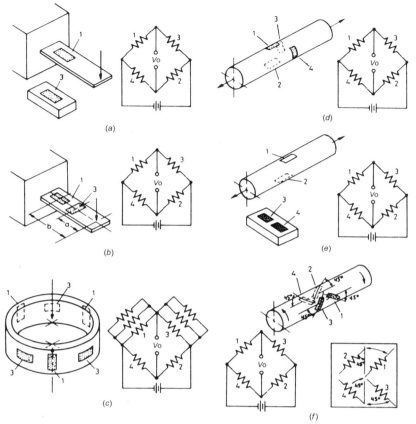

Figure 2.10 Different applications of strain gages to mechanical measurements (from BLH Electronics).

gages. The gage factor is determined by sampling because strain gages cannot be reused. The manufacturer specifies the probable value for G and its tolerance.

Strain gages can measure any quantity that by the use of an appropriate primary sensor we convert into a force capable of producing deformations of 10 μm and even smaller. Figure 2.10 shows several force and torque sensors based on elastic elements as primary sensors. Section 3.4 discusses the arrangement of strain gages in measurement bridges. Figure 2.10a shows a cantilever beam with an active gage; a separated dummy gage compensates temperature-induced resistance changes. Figure 2.10b includes an additional gage in the same cantilever but placed orthogonal to the first gage, which increases sensitivity and also compensates for temperature changes. Figure 2.10c shows a column-type load cell with three pairs of longitudinal and transversal strain gages. The strut in Figure 2.10d has two longitudinal and two transversal strain gages. In Figure 2.10e there are only two active gages and two dummy gages. In Figure 2.10f there are two sets of strain gages to sense shearing strain. Using similar primary sensors, it is possible to measure pressure, flow, acceleration, and so on (Section 1.7). Some pressure sensors use thin-film gages deposited on an electrical insulator film such as silicon monoxide deposited itself on the primary sensor (e.g., a diaphragm). Micromachined pressure sensors use ion-implanted gages in a silicon diaphragm. Strain gages can measure humidity by sensing the hygromechanical force—expansion or contraction according to relative humidity—on a suitable material such as a hair, nylon, and cellulose [23].

A unique application of the piezoresistive effect is the measurement of very high pressures (1.4 GPa to 40 GPa) through manganin gages. Manganin is an alloy ($Cu_{84}Mn_{12}Ni_4$) whose temperature coefficient is only 6×10^{-6}/K. Manganin wire subjected to a pressure from all directions exhibits a coefficient of resistance from 0.021 μΩ/Ω/kPa to 0.028 μΩ/Ω/kPa, with its change of resistance thus giving information about the applied pressure.

2.3   RESISTIVE TEMPERATURE DETECTORS (RTDs)

An RTD (resistance temperature detector) is a temperature detector based upon a variation in electric resistance. The commonest metal for this application is platinum, which is sometimes designated PRT (platinum resistance thermometer).

Figure 2.11 shows the symbol for RTDs. The straight line diagonally crossing the resistor indicates that it changes linearly. The label near that line indicates that the change is induced by the temperature and has a positive coefficient. Three- and four-wire resistors reduce measurement errors from connecting leads (Section 3.1).

RTDs rely on the positive temperature coefficient for a conductor's resistance. In a conductor the number of electrons available to conduct electricity does not significantly change with temperature. But when the temperature increases, the vibrations of the atoms around their equilibrium positions increase in amplitude. This results in a greater dispersion of electrons, which reduces their average speed. Hence, the resistance increases when the temperature rises. This relationship can be written as

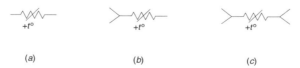

(a)   (b)   (c)

Figure 2.11   (a) Standard symbol for a resistor having a linear temperature dependence (IEC Publication 117-6). Resistive sensors with three (b) and four (c) terminals permit lead compensation.

$$R = R_0[1 + \alpha_1(T - T_0) + \alpha_2(T - T_0)^2 + \cdots + \alpha_n(T - T_0)^n] \quad (2.19)$$

where R_0 is the resistance at the reference temperature T_0. The coefficients can be determined from resistance measurements at fixed-point temperatures: 0.01 °C—(triple point of water), 100 °C (boiling water), 660.323 °C (freezing aluminum), and so on. The resistance changes because of both the change in resistivity and the change in dimensions caused by temperature. For a platinum wire, $\alpha_1 \approx 3.95 \times 10^{-3}$/K (it depends on metal purity) and $\alpha_2 = -5.83 \times 10^{-7}$/$K^2$. Therefore, for temperature increments up to about 650 °C, the linear term is more than 10 times larger than the quadratic term. For thin-film platinum, $\alpha_1 = 3.912 \times 10^{-3}$/K, $\alpha_2 = -6.179 \times 10^{-7}$/$K^2$, and $\alpha_3 = 1.92 \times 10^{-7}$/$K^3$.

Dynamically, an RTD behaves as a first-order low-pass system, because the resistor has a significant heat capacity (Section 1.5.2). A covered sensor—for example, for environmental protection—has a second-order low-pass over-damped response because of the additional heat capacity of the covering.

There are some restrictions on the use of (2.19) for temperature measurement. First, it is not possible to measure temperatures near the melting point of the conductor. Second, we must avoid any self-heating due to the measurement circuit. Otherwise, the sensor temperature would be higher than that of the surrounding medium. For a conductor in a given environment, the heat dissipation capability is given by the *heat dissipation constant* or *heat dissipation factor* δ (mW/K), which depends on the surrounding fluid and its velocity, because heat loss increases by convection.

Example 2.2 A given PRT has 100 Ω and $\delta = 6$ mW/K when immersed in air and $\delta = 100$ mW/K when immersed in still water. Calculate the maximal current through the sensor to keep the self-heating error below 0.1 °C.

The temperature increment above ambient temperature when dissipating a power P_D will be

$$\Delta T = \frac{P_D}{\delta} = \frac{I^2 R}{\delta}$$

Hence, the maximal current for a given temperature increment is

$$I = \sqrt{\frac{\Delta T \times \delta}{R}}$$

When the sensor is immersed in air, we obtain

$$I = \sqrt{\frac{(0.1\,°C)(0.006\ W/K)}{100\ \Omega}} = 2.4\ mA$$

When the sensor is immersed in water, we have

$$I = \frac{\sqrt{(0.1\ °C)(0.1\ W/K)}}{100\ \Omega} = 10\ mA$$

A dissipation factor more than 15 times higher in water than in air permits us to use a current just four times higher.

Note that $1\ °C = 1\ K$, but temperatures expressed in degrees Celsius and kelvins are different.

Mechanical strain similar to that encountered by strain gages also limits temperature sensing by RTDs because it also changes the electric resistance. This interference may inadvertently arise when measuring surface temperatures with a bonded sensor. In surface measurement, temperature gradients may also cause errors. To evaluate the possibility of temperature gradients, we use *Biot's modulus*, hl/k, where h is the heat transmission coefficient, l is the minor dimension of the measured object, and k is its thermal conductivity. If $hl/k > 0.2$, temperature gradients are probable; and therefore the dimensions, orientation, and placement of the sensor must be carefully chosen. On the contrary, if $hl/k < 0.2$, thermal gradients are improbable.

As for other sensors, RTDs must be stable. Time and thermal drifts, particularly at high temperature, limit temperature resolution. In addition, each of the metals used is linear over a limited temperature span.

The principal advantages of RTDs are their high sensitivity (ten times that of thermocouples), high repeatability, long-term stability and accuracy for platinum ($0.1\ °C/year$ in industrial probes, $0.0025\ °C/year$ in laboratory probes), and the low cost for copper and nickel. RTDs use low-cost copper connections, an advantage compared to thermocouples (Section 6.1). For metals used as RTD probes, in their respective linear range, (2.19) reduces to

$$R = R_0[1 + \alpha(T - T_0)] \tag{2.20}$$

where α is the *temperature coefficient of resistance* (TCR), calculated from the resistance measured at two reference temperatures (e.g., $0\ °C$ and $100\ °C$):

$$\alpha = \frac{R_{100} - R_0}{(100\ °C)R_0} \tag{2.21}$$

α is sometimes termed *relative sensitivity* and depends on the reference temperature (see Problem 2.3).

Example 2.3   A given PRT probe has $100\ \Omega$ and $\alpha = 0.00389\ (\Omega/\Omega)/K$ at $0\ °C$. Calculate its sensitivity and temperature coefficient at $25\ °C$ and $50\ °C$.

The sensitivity is the slope of the resistance–temperature curve, here a straight line, hence with constant slope. From (2.20), the sensitivity is

$$S = \alpha_0 R_0 = \alpha_{25} R_{25} = \alpha_{50} R_{50}$$

For the given sensor,

$$S = \frac{0.00389 \ \Omega/\Omega}{K} \times (100 \ \Omega) = 0.389 \ \Omega/K$$

At 25 °C,

$$\alpha_{25} = \frac{\alpha_0 R_0}{R_{25}} = \frac{\alpha_0 R_0}{R_0[1 + \alpha_0(25\,°C - 0\,°C)]} = \frac{\alpha_0}{1 + \alpha_0(25\,°C)}$$

$$= \frac{0.00389(\Omega/\Omega)/K}{1 + (0.00389/K) \times (25\,°C)} = 0.00355(\Omega/\Omega)/K$$

At 50 °C,

$$\alpha_{50} = \frac{\alpha_0}{1 + \alpha_0(50\,°C)} = \frac{0.00389(\Omega/\Omega)/K}{1 + (0.00389/K) \times (50\,°C)} = 0.00326(\Omega/\Omega)/K$$

Therefore, the temperature coefficient decreases for increasing temperature.

Table 2.3 gives some data for metals used in RTDs. Nickel offers a higher sensitivity but has smaller linear range than platinum. Copper has a broad linear range, but it oxidizes at moderate temperatures. Platinum offers the best performance; and the 100 Ω probe, designated as Pt100, is an industry standard. Tolerances in resistance range from about 0.1% to 1%. DIN-IEC-751 standard defines tolerance classes A and B for platinum, whose respective tolerances at 0 °C introduce ±0.15 °C and ±0.30 °C uncertainty. For comparison, carbon- and metal-film resistors used in electronic circuits have a temperature coefficient larger than, respectively, $-200 \times 10^{-6}/°C$ and $25 \times 10^{-6}/°C$. Resistivity should be high in order to have probes with high ohmic value (which allow the use of long connecting wires) and small mass (to have a fast response to temperature changes).

TABLE 2.3   Specifications for Some Different Resistance Temperature Detectors

Parameter	Platinum	Copper	Nickel	Molybdenum
Span, °C	−200 to +850	−200 to +260	−80 to +320	−200 to +200
α^a at 0 °C, $(\Omega/\Omega)/K$	0.00385	0.00427	0.00672	0.003786
R at 0 °C, Ω	25, 50, 100, 200, 500, 1000, 2000	10 (20 °C)	50, 100, 120	100, 200, 500, 1000, 2000
Resistivity at 20 °C, $\mu\Omega \cdot m$	10.6	1.673	6.844	5.7

^{a}Temperature coefficients depend on metal purity. For 99.999% platinum, $\alpha = 0.00395/°C$.

(a)   (b)

Figure 2.12   Platinum sensors for temperature probes use (*a*) wound or (*b*) film resistive elements.

For immersion in fluids, there are models that consist of a thin wire non-inductively wound around a ceramic form to allow some relative movement in order to make room for differential expansion (Figure 2.12*a*). An inert covering (stainless steel, glass), which may have one end of the resistance connected to it and grounded, protects the wire [10]. Nevertheless, thin wires may break when vibrating. There are hand-held probes for surface temperature measurement, for which it is essential to be flexible and a good electrical insulator. Surface temperature sensing RTDs can also be mounted similarly to strain gages, and they can also be formed by parallel wire, foil, or deposited metallic film (Figure 2.12*b*).

The most common application for RTDs is temperature measurement. William Siemens first proposed platinum thermometers in 1871. Platinum probes offer a stable and accurate output, and for that reason they are used as calibration standards to interpolate between fixed-point temperatures in the International Practical Temperature Scale (ITPS) from $-259.3467\,°C$ (triple point of hydrogen) to $961.78\,°C$ (freezing point of silver). Standard industrial probes are interchangeable with an accuracy from $\pm0.25\,°C$ to $\pm2.5\,°C$. Because platinum is a noble metal, it is not contamination-prone. In those applications where platinum would be too expensive, nickel and its alloys are preferable. At very high temperatures, tungsten is used. In order to reduce the nonlinear behavior of platinum thermometers at high temperature, a composite resistance thermometer has been proposed [11]. It consists of adding a second (noble) metal that compensates for α_2 in (2.19). Gold and rhodium are preferred. For cryogenic thermometry there are carbon–glass, germanium, and rhodium–iron thin-film resistive probes.

Thin-film platinum probes are 20 to 100 times smaller, cost less than wire-wound probes, and yield about the same performance, yet in a somewhat reduced temperature range. They are extensively used to control thermal processes in the chemical industry, in automobiles (exhaust emission control, engine management), in domestic appliances (ovens), and buildings (central heating systems). In cars, for example, if the temperature of the catalytic converter decreases below $250\,°C$, it can become contaminated. A PRT is immune to exhaust gases and can measure that temperature in order to control it. PRTs can also measure the temperature of intake air and that in the passenger area. A probe placed in the bumper can measure road temperature to warn of ice

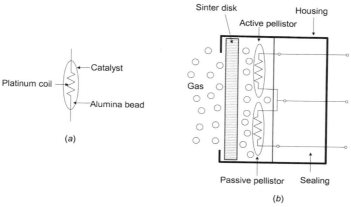

Figure 2.13   (*a*) Catalytic gas sensor based on a sintered bead with an embedded coil of platinum wire (*pellistor*) that (*b*) uses a passive pellistor connected to form a half-bridge for temperature and humidity compensation.

patches. Low-cost probes are used for temperature compensation of precision electronics such as trade weighing systems and brightness control of LCDs.

Platinum temperature probes are also used to measure fluid velocity in the so-called hot wire anemometer (Section 1.7.3). It is based on a very thin (25 μm in diameter) and short (0.2 mm to 25 mm) wire, held by its ends in a rigid support. An electric current passes through it in order to produce self-heating. When it is immersed in a fluid, the wire cools by convection, and consequently its electric resistance decreases. Obviously this probe would be useless if immersed in an electrically conductive fluid. In reference 12 there is a detailed mathematical analysis of probes using up to three wires.

Catalytic gas sensors use a fine coil of platinum wire and measure its temperature when heated at about 450 °C. The coil is embedded in a pellet (or "bead") of sintered alumina powder, called a *pellistor*, impregnated by a catalyst (platinum, palladium, etc.) (Figure 2.13*a*). If a flammable gas contacts the catalytic surface, it becomes oxidized (burns)—flameless—thus increasing the temperature of the wire by a few degrees, hence its resistance. In order to compensate for ambient temperature and humidity interference, a similar pellistor but without any catalyst is placed next to the active pellistor and series-connected to it (Figure 2.13*b*), thus forming a half-bridge (Section 3.4). The pellistor couple is housed in a can open to the environment through a sinter disk. Flammable gases can penetrate the disk and reach the pellistors; but in case of ignition inside the sensor, the disk would cool the flame so that it would not flash back into the atmosphere. The sensitivity is higher and the response time is shorter for small molecules (hydrogen, methane, ammonia) than for larger molecules (octane, toluene, xylene) because small molecules diffuse better into the sensor, but otherwise cannot discriminate among similar gases. Pellistors are stable, reliable, and last for several years. However, sulfur, phosphorous, silicon, and lead compounds and corrosive substances poison the catalyst.

Pellistors suit flammable gas concentrations below the lower explosive limit (LEL)—the minimal concentration able to continue burning even without a flame. Each substance has its specific LEL. The common range is from 0.05% to 5% in volume. High flammable gas concentrations reduce the concentration of oxygen, hence inhibiting the catalytic reaction. Thermal conductivity gas sensors overcome this limit. They measure the resistance change caused by convection cooling in a heated coil of platinum wire upon exposure to the target gas. The temperature decrease is a function of the thermal conductivity of the target gas [equation (1.55)], which is a unique physical property for a given gas. The sensing coil is embedded in a sintered alumina bead passivated with a silica glass coating. A similar coil is sealed in a cavity filled with a reference gas such as air or N_2. The difference in resistance between coils is measured by placing them in a Wheatstone bridge (Section 3.4), and each coil pair is calibrated for a target gas. Because the sensing principle is physical rather than chemical, there is no poisoning. However, gases whose thermal conductivity is similar to that of the reference gas used cannot be measured—for example, oxygen in air.

2.4 THERMISTORS

2.4.1 Models

Thermistor comes from "thermally sensitive resistor" and applies to temperature-dependent resistors that are based not on conductors as the RTD but on semiconductors. They are designated as NTC when having a negative temperature coefficient and as PTC when having a positive temperature coefficient. Figure 2.14 shows their respective symbols where the horizontal line at one end of the diagonal line indicates that the resistance variation is not linear. Michael Faraday described the first thermistor in 1833.

Thermistors are based on the temperature dependence of a semiconductor's resistance, which is due to the variation in the number of available charge carriers and their mobility. When the temperature increases, the number of charge carriers increases too and the resistance decreases, thus yielding a negative temperature coefficient. This dependence varies with the impurities; and when

(a)   (b)

Figure 2.14   Standard symbol for a resistor having a nonlinear temperature dependence, with positive (*a*) or negative (*b*) TCR (IEC Publication 117-6).

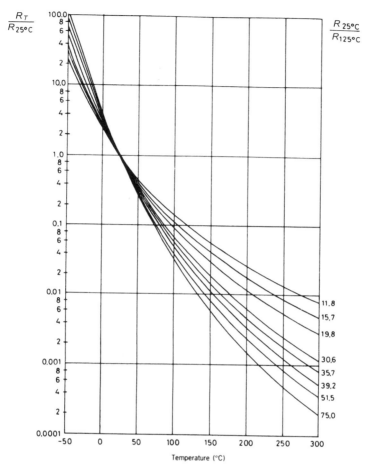

Figure 2.15 Resistance–temperature curve for several NTC thermistors (from Thermometrics).

the doping is very heavy, the semiconductor achieves metallic properties and shows a positive temperature coefficient over a limited temperature range. For NTC thermistors, over a 50 °C span the dependence is almost exponential:

$$R_T = R_0 e^{B(1/T - 1/T_0)} \qquad (2.22)$$

where R_0 is the resistance at 25 °C or other reference temperature, and T_0 is this temperature in kelvins. For $R_0 = 25$ °C, $T_0 = 273.15$ K $+ 25$ K $\simeq 298$ K. Figure 2.15 shows actual R_T versus T curves for several units from different materials.

B (or β) is called the *characteristic temperature* of the material; and its value, which is temperature-dependent, usually ranges from 2000 K to 4000 K. It increases with temperature. The Siemens THERMOWID®, for example, has $B(T_C) = B[1 + \gamma(T_C - 100)]$, where T_C is the temperature in degrees Celsius ($T_C = T - 273.15$ K) and $\gamma = 2.5 \times 10^{-4}$/K for $T_C > 100\,^\circ$C and $\gamma = 5 \times 10^{-4}$/K for $T_C < 100\,^\circ$C. B changes from unit to unit even for the same material, but there are interchangeable units available at extra cost.

From (2.22), the equivalent TCR, or relative sensitivity, is

$$\alpha = \frac{dR_T/dT}{R_T} = -\frac{B}{T^2} \tag{2.23}$$

which shows a nonlinear dependence on T. At $25\,^\circ$C and taking $B = 4000$ K, $\alpha = -4.5\%$/K, which is more than ten times higher than that of the Pt100 probe. In general, the higher resistance units have higher TCR.

Example 2.4   An alternative model to (2.22) is $R_T = Ae^{B/T}$. Determine A for a unit having $B = 4200$ K and 100 kΩ at $25\,^\circ$C. Calculate the value for α at $0\,^\circ$C and $100\,^\circ$C.

From (2.22) we deduce

$$A = R_0 e^{-B/T_0} = (100 \text{ k}\Omega)e^{(-4200 \text{ K})/(273.15 \text{ K}+25 \text{ K})} = 0.0762 \ \Omega$$

At $0\,^\circ$C ($= 273.15$ K),

$$\alpha_0 = \frac{-4200 \text{ K}}{(273.15 \text{ K})^2} = -0.0563/\text{K}$$

At $50\,^\circ$C ($= 323.15$ K),

$$\alpha_{50} = \frac{-4200 \text{ K}}{(323.15 \text{ K})^2} = -0.0402/\text{K}$$

Example 2.3 shows that the temperature dependence of α in Pt100 is linear and much smaller than that found here.

B can be calculated from the NTC thermistor resistance at two reference temperatures T_1 and T_2. If the measured resistances are, respectively, R_1 and R_2, successively replacing these values in (2.22) and solving for B yields

$$B = \frac{\ln(R_2/R_1)}{\dfrac{1}{T_1} - \dfrac{1}{T_2}} \tag{2.24}$$

B is then specified as B_{T_1/T_2}. For example, $B_{25/85}$ (see Problem 2.4).

Example 2.5   Calculate B for an NTC thermistor that has 5000 Ω at 25 °C and 1244 Ω at 60 °C.
From (2.25),

$$B = \frac{\ln \dfrac{1244 \ \Omega}{5000 \ \Omega}}{\dfrac{1}{(273.15 + 60) \ \text{K}} - \dfrac{1}{(273.15 + 25) \ \text{K}}} = 3948 \ \text{K}$$

Note that the result does not depend on the reference temperature.

For a typical thermistor, a two-parameter model yields a ± 0.3 °C accuracy for a 50 °C span. A three-parameter model reduces the error to ± 0.01 °C in a 100 °C span. The model is then described by the empirical equation of Steinhart and Hart,

$$R_T = e^{(A + B/T + C/T^3)} \tag{2.25}$$

or, alternatively, by

$$\frac{1}{T} = a + b \ln R_T + c(\ln R_T)^3 \tag{2.26}$$

Measuring R_T at three known temperatures and solving the resulting equation system yields a, b, and c. From these parameters, the value for R_T at a given temperature T is

$$R_T = \exp\left(\sqrt[3]{-\frac{m}{2} + \sqrt{\frac{m^2}{4} + \frac{n^2}{27}}} + \sqrt[3]{\frac{m}{2} - \sqrt{\frac{m^2}{4} + \frac{n^2}{27}}} \right) \tag{2.27}$$

where $m = (a - 1/T)/c$ and $n = b/c$.

A four-parameter model can further reduce the error to 0.00015 °C in the range from 0 °C to 100 °C by including a second-order term in (2.25) and measuring at a fourth known temperature in (2.26). Reference 13 compares models with up to five parameters.

Some applications rely on the relationship between thermistor current and the drop in voltage across it, rather than on the resistance–temperature characteristic. Figure 2.16 shows the $V = f(I)$ characteristic for a specific NTC thermistor. At low current levels, the drop in voltage is almost proportional to the current because thermistor self-heating is quite limited. When the current rises, the thermistor undergoes self-heating (point A in the curve), reaching a temperature higher than that of the ambient (for example, 50 °C at B, 100 °C at C, 200 °C at D); its resistance decreases and the drop in voltage across it decreases. The power available in the circuit determines when the steady state is

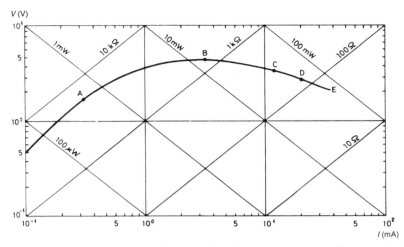

Figure 2.16 Voltage–current characteristic for a thermistor in still air at 25 °C.

attained. Point E limits the maximal nondangerous current. At higher ambient temperature the entire curve shifts downwards.

The applied electric power P equals the heat loss rate plus the heat accumulation rate,

$$P = V_T \times I_T = I_T^2 R_T = \delta(T - T_a) + C\frac{dT}{dt} \qquad (2.28)$$

where δ is the thermistor dissipation constant (mW/K), C is its thermal capacity (mJ/K)—mass times the specific heat, $C = M \times c$—and T_a is the ambient temperature. For a constant P, the thermistor temperature rises with time according to

$$T = T_a + \frac{P}{\delta}\left(1 - e^{-(\delta/C)t}\right) \qquad (2.29)$$

In steady-state condition, the time derivative becomes zero and we have

$$I_T^2 R_T = \delta(T - T_a) = \frac{V_T^2}{R_T} \qquad (2.30)$$

Substituting (2.22) for R_T into (2.30) yields the drop in voltage across the NTC thermistor:

$$V_T^2 = \delta(T - T_a) R_0 e^{B(1/T - 1/T_0)} \qquad (2.31)$$

To find the temperature corresponding to the maximal voltage we equate the derivative of (2.31) to zero:

$$0 = 1 - \frac{B}{T^2}(T - T_a) \tag{2.32}$$

This equation has two solutions. The solution corresponding to the maximal voltage (point B in Figure 2.16) is

$$T_{max} = \frac{B - \sqrt{B^2 - 4BT_a}}{2} \tag{2.33}$$

which depends on the material but not on the resistance.

Example 2.6 A P20 NTC thermistor (Thermometrics) has 10 kΩ, $\delta = 0.14$ mW/K in still air at 25 °C, and $R_{25}/R_{125} = 19.8$. Calculate the maximal drop in voltage across it when immersed in air at 35 °C.

We first calculate B from (2.24), then T_{max} from (2.33), and then R_T from (2.22). To calculate the drop in voltage we need to know in addition the current circulating through the thermistor.

$$B = \frac{\ln \dfrac{1}{19.8}}{\dfrac{1}{(273.15 + 125) \text{ K}} - \dfrac{1}{(273.15 + 25) \text{ K}}} = 3544 \text{ K}$$

$$T_{max} = \frac{3544 \text{ K} - \sqrt{(3544 \text{ K})^2 - 4 \times (3544 \text{ K}) \times (273.15 \text{ K} + 35 \text{ K})}}{2} = 341 \text{ K}$$

$$R_T = (10 \text{ k}\Omega)e^{(3544 \text{ K})(1/(341 \text{ K}) - 1/(273.15 \text{ K} + 35 \text{ K}))} = 3302 \text{ }\Omega$$

From (2.28), in steady state we have

$$I_T = \sqrt{\frac{\delta(T_{max} - T_a)}{R_T}} = \sqrt{\frac{(0.14 \text{ mW/K})(341 \text{ } K - 305.15 \text{ } K)}{3302 \text{ }\Omega}} = 1.18 \text{ mA}$$

This current yields a drop in voltage:

$$V_T = (1.18 \text{ mA}) \times (3302 \text{ }\Omega) = 3.9 \text{ V}$$

The actual load line for a circuit including an NTC thermistor can be obtained by considering the Thévenin equivalent circuit with respect to the NTC thermistor terminals. For an equivalent voltage V and resistance R,

$$V_T = V - I_T R \tag{2.34}$$

The intersection between this line and the actual voltage–current characteristic for the NTC thermistor determines the operating point.

If self-heating is negligible, (2.28) can be simplified to

$$\frac{dT}{dt} = -\frac{\delta}{C}(T - T_a) \tag{2.35}$$

whose solution is

$$T = T_a + (T_i - T_a)e^{-t/\tau} \tag{2.36}$$

where T_i is the initial temperature and $\tau = \delta/C$ is the *thermal time constant*, the time required to reach 63.2% of the temperature difference when subjected to a step change in temperature.

In the self-heating zone the thermistor is sensitive to any effect having an influence on heat dissipation rate. Thus we can use it to measure flow, level, and heat conductivity (vacuum, composition, etc.). If the dissipation constant δ is fixed, the thermistor is sensitive to the electric input power, thus being useful for voltage or power control.

Other applications rely on the current–time characteristic. Figure 2.17*a* shows the circuit used for this analysis. Figure 2.17*b* shows the typical curves

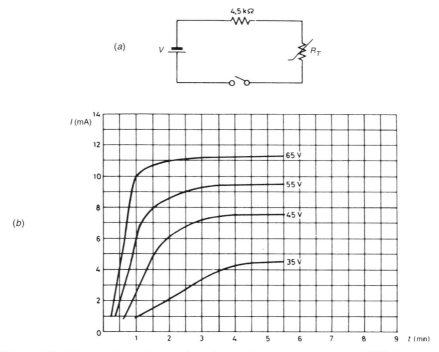

Figure 2.17 Current–time characteristic for a resistor in series with an NTC thermistor.

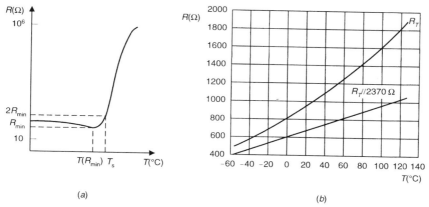

Figure 2.18 (a) Resistance–temperature characteristic for a posistor. (b) Resistance–temperature characteristic for a silistor (from Texas Instruments).

obtained at different applied voltages. For lower series resistance, the curves move upwards. Note that there is a time constant for self-heating that implies a delay between the applied voltage and the time when the steady current level is reached. This characteristic is useful in delay circuits and for transient suppression.

There are two different PTC thermistor characteristics depending on composition and doping level. Ceramic PTC thermistors—sometimes called *posistors* —show an abrupt change in resistance when they reach their Curie temperature (Figure 2.18a). Their TCR is positive—up to 100%/°C—in a narrow temperature span. Otherwise it is negative or negligible. The switching temperature is defined as that corresponding to a resistance twice its minimal value.

PTC thermistors based on doped silicon show a low slope with temperature (Figure 2.18b) and are called *tempsistors* or *silistors*. Over a temperature range of −60 °C to +150 °C, a silicon resistor obeys the law

$$R_T = R_{25} \left(\frac{273.15 \text{ K} + T}{298.15 \text{ K}} \right)^{2.3}$$ (2.37)

where T is in °C. Some units are available that include a linearizing resistor.

Example 2.7 Calculate the temperature coefficient of resistance for a silistor at 25 °C.

The TCR is defined as $(dR/dT)/R$. Therefore, taking the derivative of (2.37) yields

$$\frac{dR_T}{dT} = 2.3 R_{25} \left(\frac{273.15 \text{ K} + T}{298.15 \text{ K}} \right)^{1.3} \frac{1}{298.15 \text{ K}}$$

At 25°C,

$$\left.\frac{dR_T}{dT}\right|_{25°C} = 2.3R_{25}\left(\frac{273.15\text{ K} + 25\text{ K}}{298.15\text{ K}}\right)^{1.3}\frac{1}{298.15\text{ K}}$$

$$= \frac{2.3}{298.15\text{ K}} = (0.0077 \times R_{25})/\text{K}$$

Therefore,

$$\text{TCR}(25°C) = \left.\frac{dR_T/dT}{R_{25}}\right|_{25°C} = 0.77\%/\text{K}$$

As in RTDs, the dynamic behavior of a thermistor is (a) a low-pass first-order system if there is no protective coating and (b) an overdamped low-pass second-order system if there is protective covering.

The limitations to consider when using the above models in the application of thermistors to the measurement of temperature and other quantities are similar to those for RTDs. For thermistors, the limit set by melting is lower; and self-heating is a major problem, except for those applications that rely on it.

Thermistors are less stable than RTDs. Time stability is obtained by artificial aging. The YSI46000 series (Yellow Springs Instruments) drifts less than 0.01 °C in 100 months in the 0 °C to 70 °C span. Medium stability is obtained by covering the thermistor with glass. Interchangeability is another factor worth considering because it is guaranteed only for special models. Thus when a thermistor is replaced, it is usually necessary to readjust the circuit even if it is a unit of the same kind.

Because of their many advantages, thermistors are extensively used. Their high sensitivity yields a high resolution for temperature measurement. Their high resistivity permits small mass units with fast response and long connecting wires. Even if connecting wires undergo temperature (and resistance) changes, resistance changes in thermistors still dominate. They are readily molded into packages, which make them rugged and durable, and offer many different applications based on self-heating, usually at a very low cost.

2.4.2   Thermistor Types and Applications

NTC thermistors are manufactured by mixing and sintering doped oxides of metals such as nickel, cobalt, manganese, iron and copper, with an epoxy or glass package. Sintering consists of powder compression followed by firing at 1100 °C to 1400 °C without melting. The process is performed in a controlled environment and during the process thermistors are shaped in the desired form and size. The proportions of different oxides determine the resistance and temperature coefficients. The particular action of each of them can be found in reference 14. For temperatures higher than 1000 °C yttrium and zirconium are used.

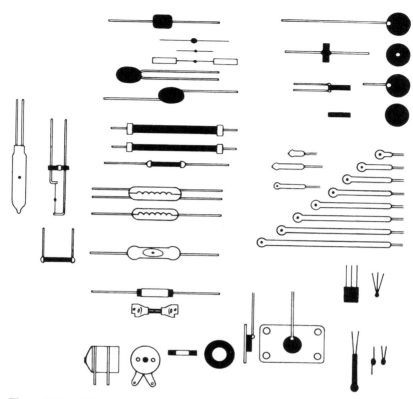

Figure 2.19 Different shapes for NTC thermistors (from Fenwal Electronics).

Switching PTC thermistors are ceramic disks based on barium titanate to which lead or zirconium titanates are added to trim the switching temperature. There are models available from $-80\,°C$ to $+350\,°C$. Silistors are based on doped silicon. They are often used for temperature compensation of semi-conductor devices and circuits requiring a 0.77%/°C temperature coefficient.

NTC thermistors are available in multiple forms, each suited to a given application. Figure 2.19 shows a variety of units available. Probe, foil, chip, bead, and some disk units are suitable for temperature measurement. Washer, rod, and other disk units are intended for temperature compensation and control, as well as for self-heating applications. There are also SMD (surface mount device) models and thermistor assemblies for biomedical applications (temperature measurement during induced hypothermia and general anesthesia, disposable fluid temperature sensors). Table 2.4 lists some parameters of ordinary NTC thermistors.

A first category of thermistor applications is those based on external heating of the thermistor, as used in temperature measurement, control and compensation. A second classification comprises applications based on a deliberate

TABLE 2.4   General Characteristics of Frequently Used NTC Thermistors

Parameter	
Temperature range	$-100\,°C$ to $450\,°C$ (not in a single unit)
Resistance at $25\,°C$	$0.5\ \Omega$ to $100\ M\Omega$ ($1\ k\Omega$ to $10\ M\Omega$ is common)
Characteristic temperature, B	2000 K to 5500 K
Maximal temperature	$>125\,°C$ ($300\,°C$ in steady state; $600\,°C$ intermittently)
Dissipation constant (δ)	1 mW/K in still air
	8 mW/K in oil
Thermal time constant	1 ms to 22 s
Maximal power dissipation	1 mW to 1 W

heating of the thermistor through the measurement circuit. This second group includes measurements of flow, liquid level, vacuum (Pirani Method), and gas composition analysis. In all these situations there is a change in the thermal conductivity of the environment surrounding the thermistor. This second group also includes automatic volume and power control, time delay applications, and transient suppression. Technical documentation from manufacturers usually includes some very useful ideas for different applications.

The circuit in Figure 2.20a is suitable for measuring a temperature over a limited range, for example that of cooling water in cars. It consists of a battery, a series adjustable resistor, a thermistor, and a microammeter. Current in the circuit is a nonlinear function of the temperature because of the thermistor, but the scale of the microammeter can be marked accordingly.

Figure 2.20b shows a thermal compensation application. Here the aim is to compensate for the undesired temperature sensitivity of a copper relay coil. Copper has a positive TCR. The series addition of a resistor with a negative

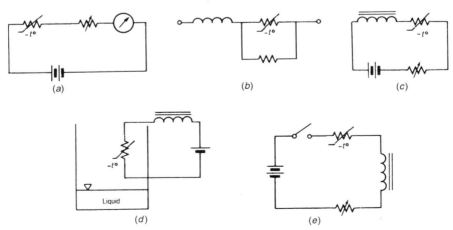

Figure 2.20 Some applications of NTC thermistors for the measurement and control of temperature and other quantities. (a) Temperature measurement. (b) Temperature compensation. (c) Temperature control. (d) Level control. (e) Time delay when connecting.

temperature coefficient results in the overall circuit exhibiting a negligible temperature coefficient. The same method can be used for deflecting coils in cathode ray tubes. Section 2.4.3 describes the function of the resistor shunting the thermistor. See also Problems 2.6 and 2.7.

Figure 2.20c shows a simple way to perform a temperature-dependent control action. When ambient temperature rises above a given threshold, the thermistor resistance decreases enough to allow the flow of a current capable of switching the relay. The adjustable resistor permits modification of the switching point. Similarly, a heat-transfer flowmeter (Section 1.7.3) uses a heater placed between two immersed thermistors. The flow produces a difference in temperature between the upstream and the downstream NTC thermistors.

The circuit in Figure 2.20d can control liquid level. The supply voltage must be high enough to heat the thermistor well above the ambient. When the liquid level reaches the thermistor that it cools, its resistance increases in value and the current is reduced, thus switching the relay. This method is applied to measure the level of lubricating oil in cars.

The circuit in Figure 2.20e is intended for time delay. The relay does not switch until the thermistor is hot enough to allow a higher current to flow. NTC thermistors can limit inrush currents through diodes, circuit breakers, and switches by placing them in series. The initial resistance of the NTC thermistor is high enough to limit the current to a safe level. As time passes on, the NTC thermistor heats by Joule effect and its value decreases, allowing the circuit rated current to flow with minimal burden.

Figure 2.21 shows several applications suggested for a switching PTC ther-

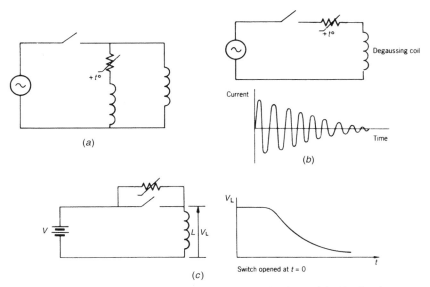

Figure 2.21 Some switching-PTC thermistors applications. (a) Single-phase motor starting. (b) Circuit for automatic degaussing. (c) Arc supression for switch contacts.

mistor. In Figure 2.21a it is used for starting a single-phase motor. When the switch is first closed the PTC thermistor has a low resistance and allows a high current to flow through the starting coil. When the PTC thermistor heats because of the current, its resistance increases to a very high level, thus reducing the current to a very low value.

The circuit in Figure 2.21b is commonly used for automatic degaussing—for example, in TV color sets. In these units a high degaussing current must flow when first turned on, and then it must reduce to a low value.

Transient suppression when a switch opens is useful to reduce both contact damage and transient propagation to any nearby susceptible circuits. In Figure 2.21c, when the switch is opened the PTC thermistor offers a low resistance because no current was flowing through it. But as time passes, its resistance increases and most of the power stored in the inductive load is dissipated in the PTC thermistor instead of being dissipated in an electric arc between switch contacts.

A series PTC thermistor protects from overtemperature—for example, in a stalled electric motor. As the current tries to rise, the PTC thermistor value increases, thus limiting the current increase. Unlike circuit breakers, PTC thermistors do not need any external action to restore the circuit after the conditions leading to overcurrent cease. PTC thermistors can also be used in self-heating mode—for example, for liquid level detection, stabilization, and self-regulating heating elements [15].

2.4.3 Linearization

To analyze an NTC thermistor in a circuit, we consider the equivalent Thévenin resistance R seen by the NTC thermistor between the terminals where it is connected. The parallel combination of both resistors is then

$$R_\mathrm{p} = \frac{RR_T}{R + R_T} \tag{2.38}$$

and its sensitivity with temperature is

$$\frac{dR_\mathrm{p}}{dT} = \frac{R^2}{(R_T + R)^2} \frac{dR_T}{dT} \tag{2.39}$$

R_p is not linear, yet its change with temperature is smaller than that of R_T because the factor multiplying dR_T/dT is smaller than 1. From (2.39) and (2.22), the equivalent TCR is

$$\frac{dR_\mathrm{p}/dT}{R_\mathrm{p}} = -\frac{B}{T^2} \frac{1}{1 + R_T/R} \tag{2.40}$$

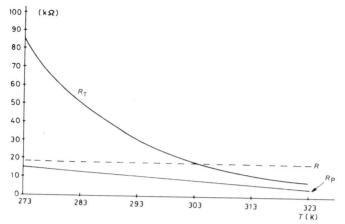

Figure 2.22 Resistance–temperature characteristic of an NTC thermistor shunted by a resistor R.

Thus the improved linearity has been gained at a cost of lower sensitivity. Figure 2.22 shows the result for the case $R_0 = 25$ kΩ, $B = 4000$ K, $R = 18500$ Ω.

Resistor R, or alternatively the NTC thermistor, can be chosen to improve linearity in the measurement range. An analytical method to calculate R is by forcing three equidistant points in the resulting resistance–temperature curve to coincide with a straight line. If $T_1 - T_2 = T_2 - T_3$, the condition is

$$R_{p1} - R_{p2} = R_{p2} - R_{p3} \tag{2.41}$$

From (2.38),

$$\frac{RR_{T1}}{R + R_{T1}} - \frac{RR_{T2}}{R + R_{T2}} = \frac{RR_{T2}}{R + R_{T2}} - \frac{RR_{T3}}{R + R_{T3}} \tag{2.42}$$

Solving for R, we obtain

$$R = \frac{R_{T2}(R_{T1} + R_{T3}) - 2R_{T1}R_{T3}}{R_{T1} + R_{T3} - 2R_{T2}} \tag{2.43}$$

This expression does not depend on any mathematical model for R_T. Thus this method can also be applied to PTC thermistors and other nonlinear resistive sensors (see Problems 2.8 and 2.10).

Another analytical method consists of forcing the resistance–temperature curve to have an inflection point just in the center of the measurement range (T_c). To obtain the necessary value for R we must take the derivative of (2.39) again with respect to the temperature and equate the result to zero. This gives for R a value of

$$R = R_{T_c} \frac{B - 2T_c}{B + 2T_c} \qquad (2.44)$$

The preferred method depends on the application. Equation (2.44) gives a better linearity around T_c and a worse linearity at the ends (see Problems 2.6 and 2.7). Equation (2.43) gives its best linearity at the zones near each one of the adjusting points. Section 3.2.2 describes an additional method based on the documentation provided by some manufacturers. By combining parallel and series resistors, it is possible to further linearize the resistance–temperature characteristic (see Problem 2.9). This is faster than software linearization.

Example 2.8   The circuit in Figure E2.8 linearizes an NTC thermistor over a limited measurement range. Calculate the value for R_1 and R_2 in order that at the temperature T_0 the equivalent resistance shows an inflection point and has a slope m.

We model the thermistor by (2.22). Let us call $R_1 = aR_0$, $R_2 = bR_0$. The equivalent resistor is then

$$R = (R_1 + R_T) \| R_2 = \frac{(aR_0 + R_T)bR_0}{aR_0 + R_T + bR_0}$$

To determine the inflection point, we must set the second derivative of R with respect to T to zero. We use the result in (2.23) to obtain

$$\frac{dR}{dT} = bR_0 \frac{(aR_0 + R_T + bR_0)(-B/T^2)R_T - (aR_0 + R_T)(-B/T^2)R_T}{(aR_0 + R_T + bR_0)^2}$$

$$= b^2 R_0^2 \frac{(-B/T^2)R_T}{(aR_0 + R_T + bR_0)^2}$$

$$\frac{d^2R}{dT^2} = b^2 R_0^2 \frac{R_T \dfrac{B^2}{T^4} \left[\left(\dfrac{2T}{B} + 1 \right)(aR_0 + R_T + bR_0) - 2R_T \right]}{(aR_0 + R_T + bR_0)^3}$$

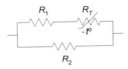

Figure E2.8 Thermistor linearization using two resistors. Two parameters of the (temperature-dependent) equivalent resistance can be chosen.

We wish

$$\frac{d^2R}{dT^2}\bigg|_{T=T_0} = 0$$

This condition is fulfilled when

$$\left(\frac{2T_0}{B} + 1\right)(aR_0 + R_0 + bR_0) = 2R_0$$

$$a + b = \frac{B - 2T_0}{B + 2T_0}$$

From the condition $dR/dT|_{T_0} = m$, we have

$$\frac{dR}{dT}\bigg|_{T=T_0} = b^2 R_0^2 \frac{(-B/T_0^2)R_0}{(aR_0 + R_0 + bR_0)^2} = m$$

$$b = \frac{2T_0}{B + 2T_0} \sqrt{\frac{-mB}{R_0}}$$

By substituting this value at the inflection point, we have

$$a = \frac{B - 2T_0}{B + 2T_0} - b$$

Some commercial linearized NTC units include one or more resistors in series and parallel combination with one or more thermistors using the preceding criteria. Obviously their "linearity" is limited to the measurement range specified by the manufacturer.

2.5 MAGNETORESISTORS

A magnetic field H applied to a current-carrying conductor exerts a Lorentz force on electrons:

$$F = ev \times H \tag{2.45}$$

where $e = -q = -0.16$ aC is the electron's charge and v is its velocity. This force deviates some electrons from their path. If the relaxation time due to lattice collisions is relatively short, electron drift to one side of the conductor yields a transverse electric field (Hall voltage, Section 4.3.2) that opposes further electron drift. If that relaxation time is relatively large, there is a noticeable increase in electric resistance, termed the magnetoresistive effect. Lord Kelvin first observed this effect in iron and nickel in 1856.

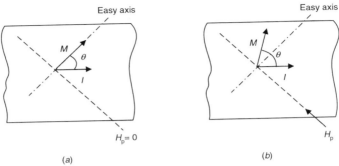

Figure 2.23   Anisotropic magnetoresistive effect. (*a*) The resistance of an anisotropic material depends on the direction of its magnetization, set during manufacturing along the so-called easy axis. (*b*) An external magnetic field rotates the magnetization, hence changing the resistance.

In most conductors the magnetoresistive effect is of a second order when compared to the Hall effect. But the resistance of anisotropic materials, such as ferromagnetics, depends on their magnetic moment (Figure 2.23*a*) according to

$$R = R_{min} + (R_{max} - R_{min}) \cos^2 \theta \tag{2.46}$$

The resistance R in the direction of the current is maximal for a magnetization parallel to the current and minimal for a magnetization transverse to the current. An external magnetic field causes the magnetization vector to rotate and changes angle θ, hence the resistance, depending on the field strength (Figure 2.23*b*). The relation between change in resistance and magnetic field strength is not linear. For a field normal to the current, that relation is quadratic [16, 17]:

$$R = R_{min} + (R_{max} - R_{min}) \left[1 - \left(\frac{H}{H_s} \right)^2 \right] \tag{2.47}$$

where H_s ($\geq H$) is the external field strength needed for a 90° rotation of the magnetization from the direction of current (saturation field). Nevertheless, biasing the element with a relatively large constant field linearizes the response. A bias field achieving a 45° rotation between the magnetization and the current direction yields a response

$$R = R_{min} + \frac{R_{max} - R_{min}}{2} + (R_{max} - R_{min}) \frac{H}{H_s} \sqrt{1 - \left(\frac{H}{H_s} \right)^2} \tag{2.48}$$

which is approximately linear for $H/H_s \ll 1$. The quotient $(R_{max} - R_{min})/R_{min}$ is termed the *magnetoresistive ratio*.

The magnetoresistive effect is also a first-order effect in semiconductors be-

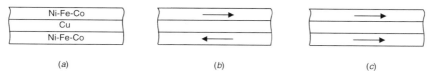

Figure 2.24 (*a*) The giant magnetoresistive effect appears in a multilayer of non-magnetic and magnetic thin layers. (*b*) The quiescent magnetization vector has opposite direction for the magnetic layers: They could also be transverse—that is, in and out of the paper plane. (*c*) An external magnetic field aligns the magnetic moment of both magnetic layers, thus reducing the electric resistance of the structure.

cause of the maxwellian distribution of electron velocities. Only electrons traveling at the average velocity have their Lorentz force balanced by the Hall field. Other electrons drift transversely, thus reducing the longitudinal current, hence increasing resistance. In weak magnetic fields, the change in resistance is proportional to the square of the magnetic field component perpendicular to the direction of the current [18].

The giant magnetoresistive effect was first observed in 1988 in multilayered structures made up from alternating layers of magnetic and nonmagnetic materials. In magnetic materials, conduction electrons with spin parallel to the magnetic moment of the material scatter much less than those whose spins are antiparallel to the magnetic moment. Figure 2.24 shows two ferromagnetic layers separated by a nonmagnetic conductor. The thickness of the layers is much less than the free path of conduction electrons in the bulk material. Therefore, the conductivity is determined by scattering at the boundaries rather than bulk scattering. If in the absence of an external field the two magnetic layers have opposite magnetic moments, electrons randomly moving from one layer to the other are scattered at one of the boundaries, because their spin is aligned with the field of either one or the other layer. The structure has high resistance. If an external field is strong enough to align the fields of the magnetic layers, and the atomic lattice of the nonmagnetic interlayer matches that of the magnetic layers, the resistance decreases because electrons with spin parallel to the field can freely move in both magnetic layers. The change in resistance is up to 70% in some experimental devices.

The colossal magnetoresistance effect is the semiconductor-to-metal transition undergone by some metal oxides when subjected to a magnetic field of a few teslas. The magnetoresistance ratio can be from 10 to 10^6.

If we ignore the need for linearization and the thermal dependence of the resistance, anisotropic magnetoresistors (AMRs) and giant magnetoresistors (GMRs) offer several advantages compared to other magnetic sensors. First, their mathematical model is a zero-order system. This differs from inductive sensors, whose response depends on the time derivative of magnetic flux density.

When compared with Hall effect sensors, which also have a zero-order model and measure without contact, magnetoresistors show increased sensitivity, temperature range ($-55\,°C$ to $200\,°C$), and frequency passband (from dc to 5 MHz

TABLE 2.5   General Characteristics of AMR, GMR, and Hall Effect Sensors

Parameter	AMR Sensor	GMR Sensor	Hall Effect Sensor
Input range	25 mT	2 mT	60 mT
Maximal output	2% to 5%[a]	4% to 20%[a]	0.5 V/T
Frequency range	Up to 50 MHz	Up to 100 MHz	25 kHz typical
			1 MHz feasible
Temperature coefficient	Fair	Good	Depends on model
Maximal temperature	200 °C	200 °C	150 °C
Cost (2000)	Medium–high	Low–medium	Low

[a] Maximal resistance change.

and even 100 MHz, compared with 25 kHz for common Hall effect sensors). Unlike Hall sensors, magnetoresistors are insensitive to mechanical stress, so they can be injection-molded to form subassemblies including electronics and magnet, if needed. Their greater sensitivity means that they can operate with larger air gaps than Hall sensors. On the other hand, they saturate at lower field strengths and are more expensive. Table 2.5 compares AMR, GMR, and Hall effect sensors.

AMR sensors are manufactured from permalloy ($Ni_{80}Fe_{20}$) thin film deposited on a silicon wafer in the presence of a magnetic field—which fixes the easy axis—and patterned as a resistive strip [16, 17]. The easy axis is parallel to the length of the resistor. The maximal change in resistance is about 2% to 3%. Also Ni–Fe–Co and Ni–Fe–Mo alloys have been tried. The element can be biased by a coil, thin-film permanent magnets deposited on top of the element, or the so-called barber-pole arrangement, formed by depositing strips of a conductor on the magnetoresistive film. The conductive strips rotate the current 45° with respect to the magnetization vector (easy axis), thus bringing the element into the linear zone. Table 2.6 lists some characteristics of two commercial AMR sensors that connect four elements in a Wheatstone bridge. A bridge

TABLE 2.6   Some Characteristics of Commercial Magneroresistive Sensors

Parameter	KMZ10A[a]	DM 208[b]	GMR B6[c]	NVS 5B50[d]
Field span, kA/m[e]	−0.5 to +0.5	—	−15 to +15	−4 to +4
Sensitivity, (mV/V)/(kA/m)	14.0	3.5	8	11 to 16
R_{bridge}, kΩ	1.2	0.65	0.7	5
Maximal operating voltage, V	10	13	7	24
Operating temperature, °C	−40 to +150	—	−40 to 150	−50 to 150

[a] AMR, Philips Semiconductors.

[b] AMR, Sony.

[c] GMR, Infineon (Siemens).

[d] GMR, Nonvolatile Electronics.

[e] In air, 1 kA/m corresponds to 1.26 mT.

configuration cancels out temperature effects (Section 3.4.4), which may be larger than those due to magnetic fields. Alternatively, we can use a bias coil and a controlled dc current source in a closed-loop circuit to obtain a wide range linear response by creating a field opposed to the field being measured. The current intensity needed is a measure of the external field [19].

Semiconductor magnetoresistors consist of narrow InSb or InAs stripes produced by photolithography, or NiSb needles precipitated in an InSb matrix, further etched to form meandering paths [18].

GMR sensors also consist of a deposited long, narrow stripe and differ in the number of layers and the method to set the direction of the quiescent magnetic moment [20]. The unpinned sandwich structure in Figure 2.24 uses a current of a few milliamperes per micrometer of stripe width. The typical magnetization ratio achieved is 4% to 9%, and the saturation field is from 2.4 kA/m to 5 kA/m. Structures with antiferromagnetic multilayers consist of the repetition of alternating conducting magnetic (Co) and nonmagnetic (Cu) layers of 1.5 nm to 2.0 nm—thinner than in sandwiches. For specific thickness the polarized conduction electrons cause antimagnetic coupling between the magnetic layers without needing any current. A large external field can overcome the coupling that causes that alignment and align the moments of all the layers. The magnetoresistive ratio is from 12% to 16%, and the saturation field is about 20 kA/m. An alternative device uses covering layers made from soft magnetic iron. Within the magnetic window where the soft magnetic layers rotate with an external field while the hard magnetic layers remain unchanged, the resistance depends only on the direction of the magnetic field [21]. Yet another structure uses spin valves or antiferromagnetic spin valves. They are similar to the sandwich in Figure 2.24 but with an additional layer of antiferromagnetic material (FeMn or NiO) at the top or bottom. This material couples to the adjacent magnetic layer and pins it in a fixed direction. The other magnetic layer is termed free layer. The magnetoresistive ratio is from 4% to 20%, and the saturation field is from 0.8 kA/m to 6 kA/m. GMR sensors are also manufactured, forming complete and half-bridge circuits. Table 2.6 lists some specifications of two commercial sensors.

The proposed applications for magnetoresistors can be divided into those related to the direct measurement of magnetic fields and those related to the measurement of other quantities through a magnetic field variation. The first group includes electric current measurement, compass navigation based on measuring two components of the Earth's magnetic field, magnetic audio recording (insensitive to tape speed fluctuations), computer disk drives, reading machines for credit cards, magnetically coded price tags, and airport and retail security systems.

The second group includes the measurement of linear and angular displacements, rotation, position, and angle, proximity switches, and ferromagnetic metal detection. In all these applications, the moving object must modify a magnetic field. To accomplish this it must either be a metallic object or an object with a metallic covering or identifier placed in a constant magnetic field,

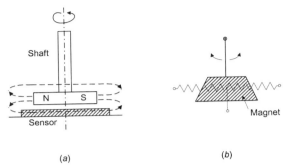

(a)   (b)

Figure 2.25   (*a*) Angle and (*b*) tilt measurement based on a magnetoresistor. The sensor in (*a*) is sensitive to the direction of the magnetic field.

or the moving element to be detected must incorporate a permanent magnet. Magnetoresistive sensors suit applications based on angular measurement. They are applied to automobile throttle position, accelerator pedal position, wheel, cam, and crankshaft speed measurement, antilock braking systems (ABS), antislip control (ASC), automatic headlight adjustment, tachometers, odometers, automobile detection in traffic control systems [based on the passing vehicle distorting the Earth's magnetic field (more than 1 μT)], control joysticks for tilting medical tables, and valve positioning. The detection of metal particles is applied to engine oil analysis and currency detection; some inks include ferromagnetic particles. Figure 2.25*a* shows an angle sensor. The tilt sensor in Figure 2.25*b* relies on a pendulum that supports a permanent magnet swinging in front of two series-connected magnetoresistors with a central tab. There are models for tilt angles up to ±30°.

2.6 LIGHT-DEPENDENT RESISTORS

Light-dependent resistors (LDRs)—photoresistors, photoconductors—rely on the variation in electric resistance in a semiconductor caused by the incidence of optical radiation (electromagnetic radiation with wavelength from 1 mm to 10 nm—300 GHz to 30 PHz). Figure 2.26 shows their symbol and a low-cost

(a)   (b)

Figure 2.26   Light-dependent photoresistor. (*a*) Standard symbol (IEC Publication 117-7). (*b*) Low-cost model encapsulated in transparent plastic (Philips). Resistor area is about 11 mm × 12 mm.

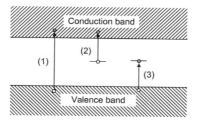

Figure 2.27 Three mechanisms that yield free carriers when illuminating doped semiconductors are (1) band-to-band transitions, (2) ionization of donor atoms, and (3) ionization of acceptor atoms.

LDR encapsulated in transparent plastic. Willoughby Smith first observed photoconductivity in selenium in 1873.

The electrical conductivity of a material depends on the number of charge carriers in the conduction band. Most of the electrons in a semiconductor at ambient temperature are in the valence band (Section 1.8). Thus it behaves like an electrical insulator. But when its temperature rises, electron vibrations increase; and because valence and conduction bands are very close in semiconductors (Figure 1.27*b* and Table 1.10), there is an increasing number of electrons raised from the valence band to the conduction band, thus increasing the conductivity. In a doped semiconductor, this raising of electrons is even easier because, in addition to band-to-band transitions, a donor atom can be ionized, thus contributing an electron to the conduction band, or an acceptor atom can be ionized, thus leaving behind a hole in the valence band (Figure 2.27). The sensitivity to incident radiation depends on how long these carriers remain free before recombining.

The energy needed to raise electrons from the valence to the conduction band can be provided by external energy sources other than heat—for example, by optical radiation or by an electric voltage. The energy *E* and frequency *f* of optical radiation are related by

$$E = h \times f \tag{2.49}$$

where $h = 6.62 \times 10^{-34}$ J · s is Planck's constant. If the incident radiation has enough energy to excite the electrons from one band to another, but without exceeding the threshold for them to leave the material, there is an *internal photoelectric effect*; and the greater the illumination (incident power per unit surface area), the higher the conductivity. If that threshold were exceeded, there would be an *external photoelectric effect*. In conductors, the conductivity by itself is so high that the change produced by the incident radiation is not noticeable.

Table 1.10 gives the band gap width (energy level difference between conduction and valence bands) for different semiconductors. The relation between photon energy and radiation wavelength λ is

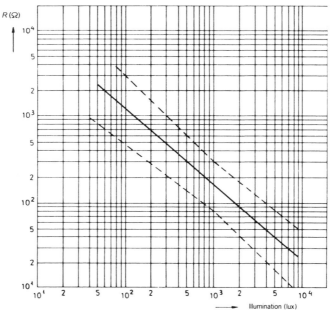

Figure 2.28 Resistance–illumination characteristic for a CdS photoconductor for a color temperature of 2850 K (from Philips). The color temperature refers to the industry standard light source that is a tungsten filament lamp operating at 2850 K, which determines the spectral output (amounts of blue, green, red, and infrared light).

$$\lambda = \frac{c \times h}{E} \tag{2.50}$$

where c is the velocity of light (≈ 300 Mm/s). If E is expressed in electron-volts (1 eV $= 0.1602$ aJ), (2.50) reduces to λ (μm) $= 1.24/E$ (eV).

The relationship between the resistance R for a photoconductor and the illumination E_v is strongly nonlinear. Reference 22 provides detailed models for the photoconductive effect. A simple model is

$$R = A \times E_v^{-\alpha} \tag{2.51}$$

where A and α depend on the material and on manufacturing parameters. For example, $0.7 < \alpha < 0.9$ for CdS. Figure 2.28 shows this relationship for a given CdS photoresistor and shows that, in addition to the nonlinearity, the ratio between the resistances when illuminated and when in the dark is larger than 10^4. The step of the resistance versus illumination curve is sometimes specified by the γ parameter, which is the ratio between R at two different E_v levels—for example, 10 lx and 100 lx. The actual resistance value depends not only on the

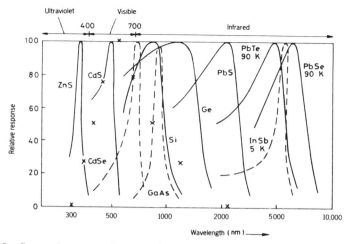

Figure 2.29 Spectral response for several photoconductors and the human eye (crosses).

current illumination but also on the illumination history. Hence, LDRs show hysteresis.

The response time of LDRs depends on the material, the illumination level, the illumination history, and the ambient temperature. The rise time is the time for the resistance to reach 63% of its final value when illuminated, and it is usually expressed in milliseconds. The fall time is the time required for the resistance to decay to 37% of its final value when darkened, and it is expressed in milliseconds or in kilohms per second. Storage in the dark slows the response. LDRs are also sensitive to temperature, which affects their sensitivity to incident radiation, especially for low-level illumination, because temperature causes thermal electron–hole generation. LDRs respond slower in colder temperatures. Temperature also causes the thermal noise that appears as current fluctuations when a voltage is applied to the photoresistor in order to measure it.

Because of the high sensitivity and spectral response, LDRs are the sensor of choice for applications involving visible light. Figure 2.29 shows that the spectral response of LDRs is narrow for various materials. Therefore the appropriate material depends on the wavelength of the radiation to be detected, taking also into account that the materials must be transparent to those wavelengths. In the visible range of the spectrum (400 nm to 700 nm) and in the near infrared (700 nm to 1400 nm), cadmium-based materials are used (CdS, CdSe, CdTe). CdS has the response closest to that of the human eye. In the infrared (1.4 μm to 3 μm), lead-based materials are used (PbS, PbSe, PbTe). In the medium (3 μm to 14 μm) and far (up to 1 mm) infrared, various indium-based materials (InSb, InAs), tellurium, tellurium–cadmium–mercury alloys (HgCdTe), and doped silicon and germanium are used. These long wavelengths are out of the range for photodiodes (Section 9.1.3). Table 2.7 compares some specifications for different photoresistors.

TABLE 2.7   Some Characteristics of Visible and Infrared Light-Dependent Resistors[a]

Parameter	2322 600 9500	P577-04	J15D5-M204-S01M
Sensor material	CdS	CdS	HgCdTe
Peak response λ	680 nm	570 nm	5 μm
Dark resistance	>10 MΩ	>3 MΩ	—
Light resistance	30 Ω to 300 Ω[b]	5 kΩ to 16 kΩ[c]	—[d]
Rise time	—	45 ms[e]	—
Fall time	>200 kΩ/s[f]	30 ms[g]	5 μs
Operating temperature	−20 °C to 60 °C	−30 °C to 70 °C	77 K
Dissipation	<0.2 W at 40 °C	0.3 W at 25 °C	20 mA bias current

[a] The 2322 600 9500 is from Philips, the P577-04 is from Hamamatsu, and the J15D5-M204-S01M is from Perkin Elmer Optoelectronics.
[b] At 1000 lx.
[c] At 10 lx.
[d] Typical sensitivity is 2×10^3 V/W.
[e] From darkness to 10 lx.
[f] From 1000 lx to darkness.
[g] From 10 lx to darkness.

Sensors used for long wavelengths (low energy) must be kept at low temperature by using the inverse Peltier effect (Section 6.1.1) or a cryostat for refrigeration in order to reduce thermal noise. Consequently, they are not available as simple two-lead resistors.

Ordinary photoconductors, useful at ambient temperatures, are CdS, PbS, and PbSe, with CdS being the most common because of its convenience for applications involving human light perception. They are manufactured in a very broad range of shapes by power mixing and sintering followed by interdigitated electrode deposition, lead mounting, and clear plastic coating. There are also symmetrical, differential, and multichannel cells for ease of application. Figure 2.26b shows a low-cost model with plastic encapsulation useful for moderate temperature and humidity. There are models with glass/metal package suited for high-humidity environments. Time constants range from 100 ms for some CdS models to 2 μs for some PbSe ones. The maximal accepted voltage when not illuminated is from 100 V to 600 V, and the maximal power dissipation at 25 °C is from 50 mW to 1 W. Maximal internal temperature is 75 °C; otherwise irreversible changes occur. Hence, soldering temperature should not exceed 250 °C for 5 s.

Applications for ordinary LDRs can be divided between those related to low-precision low-cost light measurement and those that use light as a radiation to be modified. Control applications need LDRs with a steep slope in their resistance-versus-illumination characteristic. Measurement applications need LDRs with shallow slopes. Some applications pertaining to the first group are automatic brightness and contrast control in TV receivers, diaphragm control in photographic cameras (exposure meters), dimmers for displays, automatic headlight dimmers in cars, flame detection, and street lamp switching. The second group includes presence and position detection, smoke detection, card readers, burglar alarms, object counters for conveyors, optocouplers, density of toner in photocopying machines, densitometers—(determining optical or photographic density), colorimetric test equipment, and tank level measurements that rely on a transparent tube. The first photocopiers used selenium. High-performance photoconductors such as HgCdTe are used for thermal imaging (Section 6.3.3), night vision, missile guidance (by tracking hot exhaust gases), CO_2 laser detection, and infrared spectroscopy.

2.7 RESISTIVE HYGROMETERS

Humidity is the amount of water vapor present in a gas. Moisture is the amount of water absorbed or adsorbed in a liquid or solid. The mass of water vapor contained in a given volume of gas is called *absolute humidity* (g/m^3). What is usually measured is the *relative humidity* (RH), defined as the partial pressure of the water vapor present as a percentage of that necessary to have the gas saturated at a given temperature.

Most electrical insulators show a marked decrease in resistivity (and an increase in electric permittivity) when their water content increases. If we add a hygroscopic medium, such as lithium chloride (LiCl), the decrease in resistivity is more pronounced. Measuring the variation of electric resistance yields a resistive hygrometer (or "humistor"). Measuring the change in electric capacitance yields a capacitive hygrometer (Section 4.1). F. W. Dunmore developed the first resistive hygrometer in 1938, which consisted of a bifilar winding of electrodes coated with a diluted LiCl paste. The salt absorbs and deabsorbs water to achieve equilibrium with the surrounding air. An increased presence of water increases electrolytic conductivity.

There are three basic types of resistive humidity sensors, based on the hygroscopic medium: salt (LiCl, BaF_2, P_2O_5), conductive polymer, or treated surface [24]. Conductive polymers have superseded salts. They become ionized when permeated by water, and the ions can move inside them. Sensors based on bulk polymer resistance changes are resistant to surface contamination because the contaminant cannot penetrate the polymer, accurate at high RH, and economical. They are less accurate at RH < 15% because the weak ionization present is difficult to measure. Their time response is slow because the water molecules must permeate the bulk of the material to fully affect resistance readings. Furthermore, they are susceptible to chemicals with properties similar to the polymer base. Sensors based on resistance changes on a treated surface are faster but prone to surface contamination. Compared to capacitive hygrometers, resistive hygrometers are more accurate at RH > 95%—they do not saturate—but they are a bit slower, less accurate at RH < 15%, and suit a narrower temperature range.

Because of the step change of resistance with RH, Dunmore elements vary the bifilar element spacing or the resistance properties of the film, or both, to provide linear resistance changes in specific humidity ranges. As a result, each Dunmore element covers an RH range from about 10% to 15%. Bulk resistive sensors accurately cover the range from 15% to 99% with a single resistor. They contain a grid of interdigitated electrodes deposited on an insulating ceramic substrate (alumina) and coated with a sensitive polymer resin. The resin is prepared by polymerizing a solution of quaternary ammonium bases. The sensor has a protective coating that is permeable to water vapor.

The Pope cell also consists of a conductive grid deposited on an insulating substrate. But instead of adding a hygroscopic film, the substrate—polystyrene —is sulfonated. Water adsorption contributes hydrogen ions and the sulfonate radical (SO_4^{2-}) detaches to combine with them. Mobile ions increase conductivity. The span is 15% RH to 95% RH. Surface contamination and hysteresis are the main shortcomings, and they have led to an increasing substitution of Pope cells by bulk polymer sensors.

Another type of humidity sensor uses hygroscopic polymers or gels deposited on a comb-like grid of electrodes and doped with conductive powder (i.e., carbon), or metal particles. As water is absorbed, the polymer expands and its resistance increases.

The relationship between the relative humidity and the resistance in polymer-based sensors is not linear, and it depends on the temperature: A temperature increase reduces resistance. The resistance of the model in Figure 2.30 changes by about four decades, almost exponentially. A logarithmic amplifier can yield an output voltage roughly proportional to RH. Alternatively, we can use a look-up table with a few coefficients to describe the approximate mathematical relationship between resistance, RH, and temperature. To prevent electrode polarization, the resistance must be measured with ac current having no dc component. The time constant (change to 63% of a step-change input) depends

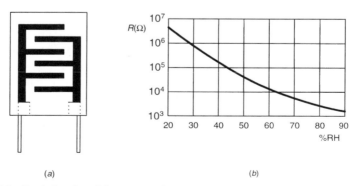

(a)   (b)

Figure 2.30 Resistive humidity sensor based on a bulk polymer and its resistance–humidity characteristic (from Ohmic Instruments).

TABLE 2.8 Specifications of Bulk Polymer Resistive Hygrometers

Parameter	EMD-2000	UPS-500
RH range	0% to 100%	15% to 95%
Accuracy	±1% RH	±2% RH
Hysteresis	±1% RH at 25 °C	<0.2% RH
Temperature coefficient	−0.3% RH/°C	−0.27% RH/°C
Long-term drift	—	<2% RH/5 years
Response time	10 s^a	5 s^b
Operating temperature	−40 °C to 100 °C	−30 °C to 70 °C
Excitation frequency	1 kHz to 10 kHz	33 Hz to 1 kHz
Excitation voltage	1 V (peak to peak)	1 V to 6 V (peak to peak)

^{a} Time to reach 90% or better of equilibrium rate for a step change from 11% RH to 93% RH.
^{b} 63% step change.

strongly on sensor size, but exceeds 10 s. Adsorption is quicker than desorption. Table 2.8 lists some specifications for RH resistive sensors.

Humidity is measured, for example, in HVAC equipment (heat, ventilation, and air conditioning), to control humidifiers and industrial dryers (wood kilns, textile, and paper dryers), for environmental monitoring, in medical equipment (anesthesia machines, ventilators, infant incubators), in food transportation and storage, in pharmaceutical production, and in electronic component manufacturing. Often, relative humidity and temperature are simultaneously measured.

2.8 RESISTIVE GAS SENSORS

Semiconductor resistive gas sensors rely on the change of surface or bulk conductivity of some metal-oxide semiconductors, depending on the concentration of oxygen in the ambient atmosphere. N. Taguchi started Figaro Engineering in 1962, which began to sell the first metal-oxide semiconductor sensor—the TGS (Taguchi gas sensor)—in 1968.

Crystal lattices of semiconducting oxides have defects, often oxygen-ion vacancies. At temperatures above about 700 °C, the adsorbed and absorbed O_2 molecules from the atmosphere dissociate to form O^- by extracting electrons from the metal oxide, thus reducing its conductivity. The relation between the electric conductivity of the oxide and the oxygen partial pressure takes the form [25]

$$\sigma = Ae^{-E_A/kT} p_{O_2}^{1/N} \tag{2.52}$$

where A is a constant, E_A is the activation energy for conduction, k is Boltzmann's constant, T is the absolute temperature, and N is a constant determined by the dominant type of bulk defect involved in the equilibrium between the oxide and oxygen. For TiO_2, for example, we have $-4 < N < -6$. An increase in oxygen concentration reduces the conductivity.

Metal oxides can also sense gases that react with oxygen—that is, combustible gases such as carbon monoxide (CO) and hydrogen (H_2). These reactions happen at temperatures in the range from 300 °C to 500 °C, and they reduce the amount of oxygen adsorbed in the surface. In an *n*-type oxide, adsorbed oxygen acts as a trap for electrons from the bulk, hence increasing the resistance of the material. Thus, oxygen reduction by a gas decreases that resistance. In a *p*-type oxide, adsorbed oxygen acts as a surface acceptor state that increases hole concentration, hence reducing the resistance of the material. Thus, oxygen reduction by a gas increases that resistance. Oxidizing gases such as chlorine (Cl_2) and nitrogen dioxide (NO_2) can also be detected because of their direct reaction with the oxide. The resistance of the material responds then in a way opposite to that for reducing gases. Therefore, if the atmosphere has a fixed O_2 concentration, as in air, minority gases increase or decrease the resistance of the material, depending on whether they are reducing or oxidizing gases and on the type of material (*n* or *p*). Water molecules (moisture) interfere with the process. Morrison [26] discusses how adsorbed gases affect the semiconductor resistance, combustible gas/oxygen reactions, and the effect of catalysts upon them.

Over a certain range of gas concentration, the relationship between sensor resistance and the concentration of deoxidizing gas can be expressed as

$$R = A[C]^{-\alpha} \qquad (2.53)$$

where A and α are constants and $[C]$ is the gas concentration. The sensitivity is often specified as the ratio between measured resistance at two reference gas concentrations. Figure 2.31 shows the resistance dependence on gas concentration and on temperature and humidity, for a CO sensor.

Conductive polymer gas sensors swell when absorbing gas vapor, which increases their resistance. The swelling depends on the chemical affinity of each vapor. The process is reversible, so that when the vapor is no longer present the

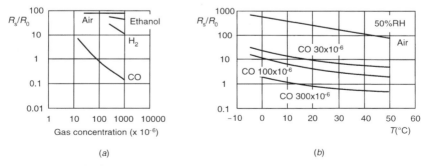

Figure 2.31 Sensitivity and temperature and humidity dependence of the TGS 2440 carbon monoxide sensor (from Figaro Engineering).

polymer shrinks and its resistance decreases to its original state. These sensors work at ambient temperature and can be more specific and faster than some metal-oxide sensors. However, they are very sensitive to humidity variations, have shorter lifetime (a few months), and are less reproducible. Some basic co-polymers used are pyrrole, aniline, and their respective derivatives. They are sensitive to vapors from species with polar molecules but not to alkanes and nonpolar molecules. This makes them complementary to other sensors rather than replacements.

Metal-oxide gas sensors are sensitive to temperature and environmental factors such as humidity. Microporous membranes can protect sensors exposed to moisture or dirty environments because they seal out liquids and heavy particulate substances while allowing unrestricted flow of air and gases. Resistive gas sensors lack selectivity (i.e., they are cross-sensitive), which improves with filters. Their recovery after exposure to high levels of the target gas (poisoning) takes several minutes. Their settling time is quite long, taking a few hours to settle to within 10% of their long-term zero resistance after switch-on, and they take a few days to really stabilize. This hinders accurate low-level measurements except for fixed systems, which can be calibrated after being left on for a few days. Metal-oxide gas sensors are very sensitive, stable, rugged, and economic. Their service life is up to 10 years. They are the sensors of choice for alarm control.

The commonest metal oxides for gas sensors are TiO_2 for bulk-type O_2 sensors and SnO_2 for surface-type sensors for O_2 and minority gases in air. Other sensors use Ga_2O_3 working above $950\,°C$ for oxygen detection, and between $600\,°C$ and $900\,°C$ for detecting different reducing gases. Some other metal oxides tested are ZnO, WO_3, and Fe_2O_3. Bulk-type sensors can be dense, but surface-type sensors should have a high surface-to-bulk ratio. Each sensor includes four basic elements: the sensing metal oxide and its support, metal contacts (electrodes) for resistance measurement, and a heater to achieve the operating temperature. The sensing material can be prepared in thick films using the screen printing method, in thin films, or as sintered elements. The operating temperature of the heating element is selected to tune the sensing material for a specific gas. In order to reduce humidity interference, the heater is sometimes operated by a pulsed voltage, so that first it reaches a high temperature able to remove water from the sensor and then the operating temperature. Table 2.9 lists some specifications of CO sensors.

The Taguchi gas sensor (TGS) is a sintered n-type semiconductor bulk device. Early models consisted of a porous SnO_2 paste over gold electrodes around the outside of a small insulating ceramic tube. For specific gas detection, platinum and palladium were added as catalysts. The heater coil ran inside the ceramic tube (Figure 2.32a). Recent models use an SnO_2 thick film screen-printed onto interdigitated platinum or gold electrodes on an alumina substrate (Figure 2.32b). The heater is a platinum or ruthenium oxide (RuO_2) track underneath the substrate and covered with an alumina layer.

The MGS1100 is based on a SnO_2 thin film patterned over a bulk-micromachined silicon diaphragm that includes a diffused heater (Figure 2.33).

TABLE 2.9   Specifications of Resistive Gas Sensors

Parameter	TGS2440	MGS1100
Target gas	CO	CO
Input range (volume fraction)	30×10^{-6} to 10^{-3}	30×10^{-6} to 100×10^{-6}
R, clean air	$>1.5 \ M\Omega$	$1 \ M\Omega$
$R, [CO] = 100 \times 10^{-6}$	$15 \ k\Omega$ to $150 \ k\Omega$	$30 \ k\Omega$ to $300 \ k\Omega^a$
Response time	—	2 min max.[b]
Heater resistance	$(17 \pm 2.5) \ \Omega$	83 Ω cold
		105 Ω operating
Heater supply	5 V max.	5 V max.
Heater power	14 mW	80 mW
Sensor driving voltage	5 V max.	5 V max.
Sensor power dissipation	—	1 mW max.

[a] $[CO] = 60 \times 10^{-6}$.
[b] 10% to 90% output change for a $[CO] = 100 \times 10^{-6}$ step change.

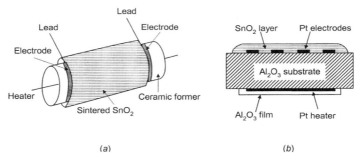

(a)   (b)

Figure 2.32   Tin dioxide gas sensor. (*a*) Tubular design. (*b*) Thick-film design.

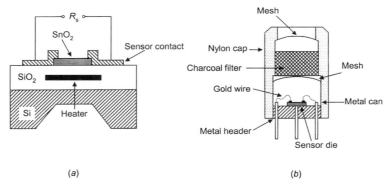

(a)   (b)

Figure 2.33   MGS1100 carbon monoxide sensor. (*a*) Cross-sectional schematic (from Motorola). (*b*) Package (from Motorola).

There are metal contacts for the heater and sensing resistor connected to a metal can package. An impregnated charcoal filter above the can package, secured by a nylon casing, enhances the specificity to CO. A wire mesh closes the top opening of the nylon case. To minimize interference from humidity, the manufacturer recommends driving the heater sequentially. The heater is driven by a 5 V, 5 s pulse to achieve a high temperature (400 °C) able to remove water and contaminants from the sensing film, and then 1 V for 10 s to detect the CO concentration in the air.

Conductive polymer gas sensors are fabricated by thin-film deposition between two electrodes on a silicon-type substrate, which includes a heater able to raise the film above ambient temperature in order to reduce the effects of humidity. They are mostly used in electronic noses.

The leading application fostering the development of low-cost gas sensors has been the need for detecting oxygen in exhaust gases from internal combustion engines, which depends on the lambda ratio—the ratio of air to fuel in the ignition mixture—hence the name *lambda sensors*. An appropriate lambda ratio reduces fuel consumption. Furthermore, emission levels allowed for carbon monoxide, hydrocarbons, and nitrogen oxides are declining, which requires cars to incorporate a catalyst that oxidizes carbon monoxide and unburned hydrocarbons and reduces nitrogen oxides. For the catalyst to work properly, the engine must operate near the lambda point, which requires monitoring oxygen in the exhaust gas. Oxygen concentration is also measured to monitor breathable atmospheres. Carbon dioxide (CO_2) is a suffocating gas measured to control greenhouses, infant incubators, and fermentation processes.

Flammable gases in air are measured to protect from unexpected fire or explosion. Such is the case of CO, H_2, alcohols, methane, propane, butane, and other hydrocarbons in petrochemical plants, utility plants, and waste-water treatment facilities. Toxic gases in air (CO, ammonia, hydrogen sulfide) are measured to monitor exposure limits. Alcohol testers are used to enforce driver regulations. Refrigerant gases, such as chlorofluorocarbons (CFCs), organic solvents, such as alcohol, toluene, and xylene, often are toxic or flammable, or both, and are measured for process and pollutant control. Gas sensors are also used to monitor the quality of air in enclosed spaces such as restaurants, toilets, bathrooms, offices, or motor vehicles.

Electronic noses (e-noses) use arrays of solid-state gas sensors, each sensitive to particular gaseous molecules. Metal-oxide sensors have been used in conjunction with quartz microbalance and SAW sensors (Section 8.2), as well as with conductive polymers. Each sensor type has different trade-offs. Metal-oxide sensors are less susceptible to poisoning than conductive polymers, but have reproducibility problems and need complex support electronics. Some developers combine different sensors and use statistical signal analysis or neural networks to identify the type, quality, and quantity of odors. Otherwise, using a specific sensor for each molecule would lead to an e-nose with an enormous number of sensors. Humans can discriminate among around 2000 odors. Trained personnel, such as wine makers and perfume evaluators, can discriminate among 10000 odors. Electronic noses can be applied to detect toxic substances in industry and home, in environmental monitoring (groundwater testing, oil and gas leaks), in the food industry for quality control and detecting spoilage, in the cosmetic and perfume industries for aroma quality control, and in indoor air quality control. Some proposed medical applications are (a) wound-infection detection (by recognizing the odor produced by a form of streptococcus) and (b) monitoring a patient's breath or excreted urine to diagnose infection or diseases such as pneumonia and diabetes.

2.9   LIQUID CONDUCTIVITY SENSORS

Sensors based on liquid conductivity use the dependence of electrolyte conductivity on the concentration of dissolved ionic species. Pure water, for example, has a theoretical conductivity (specific conductance) $\sigma = 0.038$ µS/cm at 25 °C because of the dissociation products of water itself (hydrogen ion H^+ and hydroxyl OH^-). A mere trace of electrolytic material increases σ sharply: Adding 1×10^{-6} by weight of table salt doubles σ; the same concentration of strong acid increases σ fivefold. The conductivity of concentrated solutions of strong acids is close to 1 S/cm. The development of electrolyte conductivity measurements is attributed to F. Kohlrausch (1840–1910).

The molecules of salt, acid, or base in an aqueous solution are dissociated into ions because of the high dielectric constant of water. Upon application of an electric field to an electrolyte, electrostatic force acts on the ions. The positive electrode (anode) attracts negative ions (anions), and the negative electrode (cathode) attracts positive ions (cations). The accelerated ions move inside the solution against the viscous drag of the dissolved ions of opposite charge, resulting in a constant migration speed. The voltage measured by a pair of electrodes immersed in the solution (Figure 2.34) is proportional to the current and

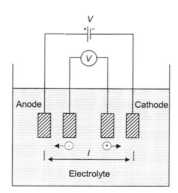

Figure 2.34   The current in an electrolyte when applying a voltage V across two immersed electrodes depends on the geometry of the container and on the conductivity of the electrolyte. The drop in voltage across two measuring electrodes is proportional to that current.

follows Ohms' law. The current through the cell, however, is not proportional to the applied voltage because of polarization phenomena at the anode and cathode that yields a voltage drop across the electrode-to-electrolyte interface (Section 6.5).

The total current will be the sum of charge transport by positive and negative ions,

$$I = \frac{(nzqv)_+ + (nzqv)_-}{l} \tag{2.54}$$

where n is the number of ions of one polarity, zq its charge, v its velocity, and l the distance between anode and cathode. v depends on the applied electric field (V/l) and the ion mobility μ, defined as the velocity in a unity applied field. μ is specific for each ion species, it is approximately independent of the mobility of other ionic species in the same solution, it is independent of the applied field below 100 MV/m, and it is independent of the frequency of the field below 10 MHz. μ increases at larger electric fields and higher frequencies. H^+, the smallest ion, has $\mu = 32 \times 10^{-8}$ (m/s)/(V/m). Larger ions are up to ten times slower. Some small ions are also relatively slow because they become hydrated, which increases their effective radius.

From (2.54), if the effective cross-sectional area of the conductivity cell is A, the conductivity is

$$\sigma = \frac{I}{V}\frac{l}{A} = \frac{(nzqv)_+ + (nzqv)_-}{l} = \frac{1}{V}\frac{l}{A} = (Nzq\mu)_+ + (Nzq\mu)_- \tag{2.55}$$

where $N = n/(lA)$ is the concentration (number per volume unit) of the ion species considered. This ion concentration differs from the solute concentration, depending on the degree of dissociation and the number of ions released per molecule dissociated. Hence, σ yields information about ion mobility and the ion concentration, not the solute concentration. Nevertheless, for dilute solutions they are proportional. For high solute concentrations, the relationship between σ and solute concentration is taken from empirical tables [27].

Conductimetry is a nonspecific method—all ions in the solution contribute to the current—and temperature-dependent (about 0.02/K), but quite simple. Because the measured resistance depends on both electrolyte conductivity and cell geometry, first a liquid of known conductivity is measured to determine the cell constant (geometry factor). Liquid potentiometers (Section 2.1) and liquid strain gages (Section 2.2) rely on the dependence of the geometry factor on the sensed quantity. Ac resistance measurement minimizes electrode polarization. Simultaneous temperature measurement permits the automatic correction to conductivity values at 25 °C. Electrodes are platinized in order to increase their effective surface, hence reducing their impedance, and to resist corrosion. Contactless conductivity cells avoid electrode polarization and corrosion problems by using coils or plates wound around a cylindrical cell as electrodes. Contactless cells can measure conductivity in solutions at high temperature.

Conductometry is useful to analyze binary water–electrolyte mixtures—for example, for chemical water monitoring. An increase in water conductivity indicates pollution by acids, bases, or other highly ionized substances. This method is used in pharmaceutical, food and beverage, semiconductor, power (boilers), and other high-purity water applications, to assess effluent and water treatment controls, to monitor the salinity of sea water, or to assess sea-water intrusion in fresh-water wells.

Conductometric gas analysis uses nebulizers, bubbler spirals, or impinger flasks to transfer the gaseous component into a solution. This method has been applied to monitor sulfur dioxide (SO_2) because when is absorbed in water it forms sulfuric acid, leading to a strong change in conductivity. Some ammonia and CO_2 analyzers use the same principle.

Conductometric titration is the measurement of the conductivity in an acid–base titration cell in order to determine the so-called *equivalence point* (EP). Titration is a method of determining the concentration of a dissolved substance in terms of the smallest amount of a reagent of known concentration required to bring about a given effect in reaction with a known volume of the test solution. In acid–base titration (neutralization titration), first the conductivity decreases when a strong base is added to the acidic solution, and then it increases when adding more base. The EP is the point of lowest conductivity. Because the conductivity increases linearly for both lower and higher titrant volumes around EP, this point can be determined by extrapolation from four conductivity measurements.

The Coulter counter for blood cells—introduced by W. H. Coulter in 1956 —relies on the lower electric conductivity of blood cells with respect to the solution in which they are suspended. A closed glass tube with a small orifice of 50 μm in diameter is suspended in a beaker (Figure 2.35) and connected to a suction pump. The negative pressure inside the tube causes the blood to flow through the aperture. Whenever a blood cell enters the aperture, the electric

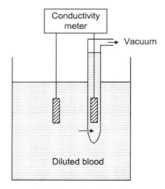

Figure 2.35 The Coulter counter for blood cells relies upon the decreased conductance between two electrodes on both sides of a small aperture whenever a blood cell enters the orifice, blocking the current because its insulating membrane makes it less conductive than the surrounding electrolyte.

resistance between two electrodes immersed in the highly conducting electrolyte on both sides of the aperture increases. If the device is supplied at constant current, cells flowing through the aperture yield voltage pulses that can be analyzed to infer blood cell count and volume [28].

2.10 PROBLEMS

2.1 A 2 W, 1 kΩ linear potentiometer is connected to a 20 kΩ circuit. What is the maximal supply voltage allowed in order not to exceed the power rating?

2.2 Consider a Wheatstone bridge where arm 1 is a 120 Ω advance strain gage $(G = 2.00)$, arm 4 (on the same side of the bridge) is a similar dummy gage intended for compensation, and arms 2 and 3 are fixed 120 Ω resistors. Maximal current through the gages is 30 mA.

  a. Calculate the maximal dc supply voltage.

  b. If the sensing gage is bonded on steel $(E = 210 \text{ GPa})$ and the bridge is supplied by 5 V, what is the bridge output voltage when the applied load is 70 kg/cm²?

  c. Calculate a calibrating resistor that placed in parallel to the unloaded active gage would produce the same bridge output voltage as 700 kg/cm² in a steel piece.

2.3 A given 500 Ω nickel RTD (Minco Products) has $\alpha = 0.00618$ (Ω/Ω)/K at 0 °C. It is used at temperatures around 100 °C, so we use the model $R_T = R_{100}[1 + \alpha_{100}(T - 100 \,°\text{C})]$. Calculate its sensitivity and temperature coefficient at 100 °C, and determine the resistance at 100 °C and 101 °C.

2.4 Using the model $R_T = Ae^{B/T}$, determine A and B for the MA200 NTC thermistor (Thermometrics), which has 5000.00 Ω at 25 °C and 1801.44 Ω at 50 °C. Calculate the sensitivity and α at 37.5 °C.

2.5 The 2322 640 90007 NTC thermistor (Philips) has $R_{25} = 12$ kΩ, $R_{90} = 1.3$ kΩ, and $\delta = 10$ mW/K in still water. We wish to use it in an application involving water from 0 °C to 100 °C. Calculate the maximal current allowable to keep the self-heating error below 0.5 °C.

2.6 A given ac tachometer whose copper winding has a dc resistance of 1500 Ω at 20 °C and a TCR = 0.0039/°C is used in environments with temperatures ranging from −20 °C to 60 °C. To compensate for resistance variations with temperature, an NTC thermistor shunted by a resistor is placed in series with the tachometer winding. For available thermistors with $B = 3367$ K (series B35 from Thermometrics), calculate the values for the thermistor resistance at 20 °C and for the shunting resistor.

2.7 A copper relay coil has 5000 Ω at 25 °C and pulls in at 1 mA. The relay should operate at a constant voltage from 0 °C to 60 °C. In order to compensate for the copper coil temperature coefficient, we add an NTC thermistor shunted by a fixed resistor R. If the copper coil has TCR = 0.0039/K at 0 °C, and the NTC thermistor has 5700 Ω at 0 °C and 810 Ω at 50 °C, determine R.

2.8 The dc amplifier in Figure P2.8 exhibits an increase in gain when there is an increase in temperature. The NTC thermistor has a resistance of 30 kΩ at 20 °C and $B = 4000$ K for the temperature range of interest. If at temperatures of 15 °C, 25 °C and 35 °C the gain should be, respectively, 0.9, 1, and 1.1, what values should the resistors R_s, R_p, and R_G have?

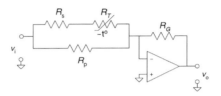

Figure P2.8   Dc amplifier that has specific gains at three different temperatures.

2.9 A given sensor with unipolar output has a temperature sensitivity α%/ °C. To obtain an output voltage constant with temperature, the circuit in Figure P2.9 is proposed. The NTC thermistor has $B = 3500$ K and $R_{25} = 10$ kΩ. Both the sensor and the NTC thermistor are at an ambient temperature of about 20 °C. Design the circuit in order to have a gain of 1000 at 20 °C and temperature compensation for the sensor; consider the op amps as ideal. If the NTC thermistor has $\delta = 1$ mW/K, what restriction does this parameter introduce?

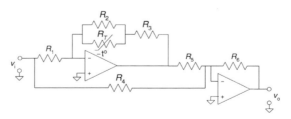

Figure P2.9   Dc amplifier with a gain–temperature characteristic tailored to the input signal temperature coefficient.

2.10 Calculate the parallel resistor able to linearize a PTC thermistor that in the range from 0 °C to 50 °C fulfills equation (2.37) and has $R_{25} = 1000$ Ω.

REFERENCES

[1] E. Gass and E. Holder. Conductive plastic potentiometers. *Measurements & Control*, Issue 151, 1992, 116–123.

[2] C. D. Todd (Bourns Inc.). *The Potentiometer Handbook*. New York: McGraw-Hill, 1975.

[3] S. Demurie and G. DeMey. Design of custom potentiometers. *IEEE Trans. Instrum. Meas.*, **38**, 1989, 745–747.

[4] K. Antonelli, J. Ko, and S. Ku. Resistive displacement sensors. Section 6.1 in: J. G. Webster (ed.), *The Measurement, Instrumentation, and Sensor Handbook*. Boca Raton, FL: CRC Press, 1999.

[5] Y. Kanda. Piezoresistance effect in silicon. *Sensors and Actuators*, **28A**, 1991, 83–91.

[6] Measurements Group. *Interactive Guide to Strain Measurements Technology*. Raleigh, NC: Vishay, 1999. Available on CD ROM and at www.measurementsgroup.com.

[7] P. Stein. Sixty years of bonded resistance strain gage. *Measurements & Control*, Issue 176, 1996, 131–140.

[8] D. A. Gorham and B. Pickthorne. Light sensitivity of silicon strain gages. *J. Phys. E: Sci. Instrum.*, **22**, 1989, 1023–1025.

[9] D. Meglan, N. Berme, and W. Zuelzer. On the construction, circuitry, and properties of liquid metal strain gages. *J. Biomech.*, **21**, 1988, 681–685.

[10] J. Burns. Resistive thermometers. Section 32.2 in: J. G. Webster (ed.), *The Measurement, Instrumentation, and Sensor Handbook*. Boca Raton, FL: CRC Press, 1999.

[11] J. M. Diamond. Linear composite resistance thermometers. *IEEE Trans. Instrum. Meas.*, **38**, 1989, 759–762.

[12] M. Acrivlellis. Determination of the magnitude and signs of flow parameters by hot-wire anemometry. Part I: Measurements using hot-wire X probes. Part II: Measurements using a triple hot-wire probe. *Rev. Sci. Instrum.*, **60**, 1989, 1275–1280 and 1281–1285.

[13] H. T. Hoge. Useful procedures in least squares and tests of some equations for thermistors. *Rev. Sci. Instrum.*, **59**, 1988, 975–979.

[14] E. D. Macklen. *Thermistors*. Ayr, UK: Electrochemical Publications, 1979.

[15] Thermometrics. *PTC Thermistors: Application Notes*. Edison, NJ: Thermometrics, 1998. Available at www.thermometrics.com.

[16] A. Petersen. The magnetoresistive sensor—a sensitive device for detecting magnetic field variations. *Electronic Components Appl.*, **8**, 1989, 222–239.

[17] U. Dibbern. Magnetoresistive sensors. Chapter 9 in: R. Boll and K. J. Overshott (eds.), *Magnetic Sensors*, Vol. 5 of *Sensors, A Comprehensive Survey*, W. Göpel, J. Hesse, J, N. Zemel (eds.). New York: VCH Publishers (John Wiley & Sons), 1989.

[18] R. Popovic and W. Heidenreich. Magnetoresistors. Section 3.4 in: R. Boll and K. J. Overshott (eds.), *Magnetic Sensors*, Vol. 5 of *Sensors, A Comprehensive Survey*, W. Göpel, J. Hesse, J, N. Zemel (eds.). New York: VCH Publishers (John Wiley & Sons), 1989.

[19] M. J. Carusso, T. Bratland, C. H. Smith, and R. Schneider. Anisotropic magneto-resistive sensors: theory and applications. *Sensors*, **16**, March 1999, 18–26.

[20] M. J. Carusso, T. Bratland, C. H. Smith, and R. Schneider. A new perspective on magnetic field sensing. *Sensors*, **15**, December 1998, 34–46.

[21] Siemens. *Magnetic Sensors, Giant Magneto Resistors.* Application Note 10.98. Munchen (Germany): Siemens, 1998. Available at www.infineon.com.

[22] C. I. Popescu, C. Popescu, and T. Stoica. Photoresistive sensors. Chapter 2 in: P. Ciureanu and S. Middelhoek (eds.), *Thin Film Resistive Sensors.* New York: IOP Publishing, 1992.

[23] R. L. Fenner. Cellulose crystallite-strain gage hygrometry. *Sensors*, **12**, May 1995, 51–59.

[24] P. R. Wiederwhold. Fundamentals of moisture and humidity, Part 3: Humidity measurement methods. *Measurements & Control*, Issue 190, September 1998, 131–143.

[25] P. T. Moseley. Solid state gas sensors. *Meas. Sci. Technol.*, **8**, 1997, 223–237.

[26] S. R. Morrison. Chemical sensors. Chapter 8 in: S. M. Sze (ed.), *Semiconductor Sensors.* New York: John Wiley & Sons, 1994.

[27] F. Oehme. Liquid electrolyte sensors. Chapter 7 in: W. Göpel, T. A. Jones, M. Kleitz, J. Lundström, and T. Seiyama (eds.), *Chemical and Biochemical Sensors*, Part 1, Vol. 2 of *Sensors, A Comprehensive Survey*, W. Göpel, J. Hesse, J, N. Zemel (eds.). New York: VCH Publishers (John Wiley & Sons), 1991.

[28] W. Groner. Cell counters, blood. In: J. G. Webster (ed.), *Encyclopedia of Medical Devices and Instrumentation.* New York: John Wiley & Sons, 1988, 624–639.

3

SIGNAL CONDITIONING FOR RESISTIVE SENSORS

There are many mechanisms that can modify the electric resistance of a material and also many signal conditioners for resistive sensors. Thus this group of sensors is the largest.

This chapter presents several methods to obtain from resistive sensors output voltages in a range suited to analog-to-digital converters (ADCs) or other electric measuring equipment. It also presents several methods of interference compensation and sensor linearization in the signal conditioner following the sensor.

Some of these sensors and signal conditioners enable us to introduce error analysis methods and circuit design aspects that are common to other more involved sensors.

The chapter first reviews resistance measurement methods, then analyzes conditioners for sensors with large magnitude resistive variations, then progresses to successively smaller magnitude resistive variations. Finally, interference types and their reduction by grounding and shielding are dealt with. Section 7.1 analyzes offset and drift, and Section 7.4 discusses noise. Chapter 8 describes oscillators that incorporate resistive sensors and direct sensor–microcontroller interfaces.

3.1 MEASUREMENT OF RESISTANCE

The general equation for a sensor whose resistance changes by a fraction x in response to a measurand is $R = R_0 f(x)$, assuming $f(0) = 1$. For linear sensors we have

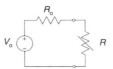

Figure 3.1 The Thévenin equivalent for the circuit seen from the terminals of a resistive sensor permits us to determine that the maximal dissipation is at $R = R_o$.

$$R = R_0(1 + x) \tag{3.1}$$

The range of values for x depends strongly on the type of sensor and on the measurand span. For linear sensors, x varies from 0 to -1 for linear potentiometers and can be up to 10 for conductive polymers and as small as 10^{-5} to 10^{-2} for strain gages. RTDs and measuring thermistors have intermediate values for x. The ratio between sensor resistance for extreme measurand values can be higher than 1000 in LDRs and humistors, and is less than 100 in magnetoresistors, gas sensors, and liquid conductivity sensors. Switching PTC thermistors increase their resistance by more than 10,000 for temperatures above the switching temperature.

There are two requirements for all conditioners for resistive sensors. First, they must drive the sensor with an electric voltage or current in order to obtain an output signal, because a change in resistance is not by itself a signal. Second, this supply, whose magnitude affects that of the output signal, is limited by sensor self-heating, which must be avoided unless the sensing principle uses sensor self-heating, as in some flowmeters and liquid level meters. Whatever the case, if the Thévenin equivalent circuit seen by the sensor has output voltage V_o and resistance R_o (Figure 3.1), the power dissipated by the sensor is

$$P = \left(\frac{V_o}{R_o + R}\right)^2 R \tag{3.2}$$

whose maximum with respect to R can be obtained by setting its first derivative to zero,

$$\frac{dP}{dR} = 2\frac{V_o}{R_o + R}\frac{-V_o}{(R_o + R)^2}R + \left(\frac{V_o}{R_o + R}\right)^2 = \left(\frac{V_o}{R_o + R}\right)^2\frac{R_o - R}{R_o + R} = 0 \tag{3.3}$$

This leads to $R = R_o$. The second derivative is negative at this point, hence it is a maximum. The corresponding power is

$$P_{\max} = \left(\frac{V_o}{R_o + R_o}\right)^2 R_o = \frac{V_o^2}{4R_o} \tag{3.4}$$

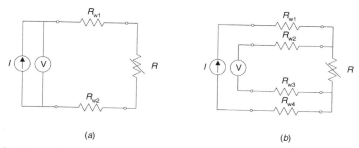

(a)   (b)

Figure 3.2 Measuring a resistive sensor by (a) a two-wire circuit yields an offset error that will drift with temperature. (b) The four-wire method is insensitive to lead resistance.

If the sensor resistance never meets the condition $R = R_o$, the maximal dissipation happens for the resistance value closest to R_o. If the sensor is driven at constant current, the dissipation is maximal for the highest sensor resistance.

Some sensors require particular circuits. Thermistors require linearization. Strain gages require interference cancellation. Sensors that yield small outputs require large gains in order for the dynamic range of the output signal to match the input range of the ADC (reference 1, Sections 1.3 and 1.4.2). Conditioners for remote sensors must be insensitive to connecting lead resistance or compensate for it. The voltage measured in Figure 3.2a is

$$V = I(R + R_{w1} + R_{w2}) \tag{3.5}$$

Hence, when $R = 0$ Ω, $V(0) \neq 0$ V and we have a zero error (offset). From Table 3.1, 10 m of No. 20 AWG wire would add 0.333 mΩ to R. This is about the resistance change of a Pt100 for a 1 °C increment. Because this is a zero error, it can be nulled out. However, ambient temperature changes would induce wire resistance variations that would not be cancelled. The voltage measured in the four-wire circuit in Figure 3.2b—also called a Kelvin circuit— is insensitive to wire resistance, provided that the output impedance of the current source and the input impedance of the voltage meter are large enough.

Methods for resistance measurement can be classified into deflection methods or null methods. Deflection methods sense the drop in voltage across the resistance to be measured or the current through it or both. Null methods are based on measurement bridges.

The simplest deflection method consists of supplying the resistance with a constant voltage source and then measuring the current through the circuit, or with a constant current source and then measuring the voltage. Figure 2.20a shows an NTC thermistor driven by a constant voltage; A current meter provides the output reading. Figure 3.3a shows a resistive sensor driven by a constant current. The output voltage for a linear sensor is

TABLE 3.1   Copper Conductor Data[a]

AWG[b]	Stranding	Diameter (mm)	Dc Resistance[c] (Ω/km)
10	Solid	2.600	3.28
	37/26	2.921	3.64
	49/27	2.946	3.58
20	Solid	0.813	33.20
	10/30	0.899	33.86
	26/34	0.914	32.97
30	Solid	0.254	340.00
	7/38	0.305	338.58

[a] Solid wire data are from Spectra-Strip, and stranded conductor data are from Alpha Wire. There is additional information at http://www.brimelectronics.com/AWGchart.htm.

[b] AWG stands for American Wire Gauge, also known as B&S gauge (from Brown and Sharp), which is a diameter specification. Metric gauge numbers correspond to 10 times the diameter in millimeters. For example, a metric wire No. 6.0 is 0.6 mm in diameter. AWG numbers do not correspond to diameter sizes proportionally; and the higher the gauge number, the smaller the diameter. Typical household wiring is AWG number 12 or 14. Telephone wire is usually 22, 24, or 26. Strand number n/m means that taking n strands of wire number m makes the desired wire number. For example, 10 strands of 30AWG or 26 strands of 34AWG make a 20AWG wire.

[c] At 20 °C.

$$v_o = I_r R = \frac{V_r}{R_r} R_0 (1 + x) \tag{3.6}$$

which is linear too. If $x \ll 1$, v_o will consist of small fluctuations (due to x) superimposed on a very large offset voltage corresponding to $x = 0$. Nevertheless, if we design $R_r = R_0$, subtracting the drop in voltage across R_r from v_o yields

$$v_s = v_o - I_r R_r = V_r (1 + x) - V_r = V_r x \tag{3.7}$$

which is zero for $x = 0$. If the relative accepted error is ε, the input offset voltage of the op amp must be $V_{io} < \varepsilon V_r$. Figure 3.3b shows a circuit to implement the same principle. Two identical current sources (e.g., ADT70, REF200) drive the sensor and a series-connected resistor, which furnishes a voltage to cancel the drop in voltage across the sensor when $x = 0$. If $R_z = R_0$, the output voltage is

$$v_o = I(R - R_z) = I R_0 x \tag{3.8}$$

and it is differential. Note that by using a four-wire sensor in Figure 3.3a, lead resistance does not contribute to the output voltage. Also, several sensors can be series connected and driven by the same current [2]. This signal conditioning method is often applied to RTDs (see Problems 3.1 and 3.2) The AD7711 and AD7113 are ADCs that integrate two 200 μA current sources for sensor excitation.

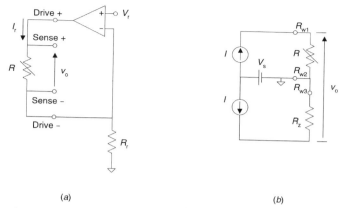

(a)   (b)

Figure 3.3 (a) Signal conditioner for a resistive sensor based on current excitation. (b) Using a dual current source and voltage subtraction cancels out the constant output voltage term.

Example 3.1   A temperature from 20 °C to 100 °C is to be measured with 0.1 °C resolution using the circuit in Figure 3.3a and a thin-film Pt100 (S245PD from Minco), which has 100 Ω and $\alpha = 0.00385$ Ω/Ω/K at 0 °C, and $\delta = 40$ mW/K in 0.4 m/s water. Calculate R_r if the reference voltage available is $V_r = 5$ V.

The temperature resolution is limited by self-heating because any change in heat dissipation would induce a change in resistance. We wish

$$\frac{I_r^2 R}{\delta} = \left(\frac{V_r}{R_r}\right)^2 \frac{R}{\delta} = \Delta T < 0.1\,°C$$

Because the sensor is driven at constant current, the maximal dissipation will be at 100 °C (maximal sensor resistance). Hence, the condition to fulfill is

$$R_r > V_r \sqrt{\frac{R_{100}}{\delta \times (0.1\,°C)}}$$

From (2.20),

$$R_{100} = (100\ \Omega)[1 + (0.00385/K)(100\,°C - 0\,°C)] = 138.5\ \Omega$$

Therefore

$$R_r > (5\ V)\sqrt{\frac{138.5\ \Omega}{(40\ mW/K) \times (0.1\,°C)}} = 930\ \Omega$$

where we have used the identity 1 K = 1 °C.

One additional criterion to select R_r could be to have a sensitivity of, say, 1 mV/°C. The output voltage would then be

$$v_o = \frac{V_r}{R_r} R_0 (1 + \alpha T)$$

and the sensitivity would be

$$S = \frac{dv_o}{dT} = \frac{V_r}{R_r} R_0 \alpha$$

Therefore,

$$R_r = \frac{V_r R_0 \alpha}{S} = \frac{(5 \text{ V})(100 \text{ } \Omega)(0.00385/\text{K})}{1 \text{ mV/K}} = 1925 \text{ } \Omega$$

If we select $R_r = 1.93$ kΩ, 0.1% tolerance, the actual sensitivity can be up to 0.36% smaller than 1 mV/°C, but this is a constant error that can be trimmed off. Also, at 20 °C, $R_{20} = 107.7$ Ω, leading to an offset voltage of about 279 mV. The current through the sensor would be about 2.5 mA.

Another deflection method for resistance measurement is the two-reading method shown in Figure 3.4. It consists of placing a known stable resistor in series with the unknown one. First a reading is taken across that resistor, which yields $V_r = IR_r$, and then another reading is taken across the unknown one, which yields $v_o = IR$. Afterwards the quotient between both readings is calculated to obtain

$$R = R_r \frac{v_o}{V_r} \qquad (3.9)$$

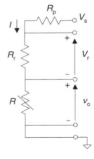

Figure 3.4 The two-reading method for resistance measurement: First the drop in voltage is measured across a known resistor, and then it is measured across the unknown resistor. R_p is a protection resistor that limits sensor self-heating and does not influence the result.

If $R_r \approx R$ the voltmeter error in both readings will be similar and will be cancelled when taking the quotient. We would typically select $R_r = R_{max}$ in the measurement span. Alternatively, a ratio ADC using V_r as reference directly provides the quotient in digital format (Sections 3.2.1 and 3.4.5) (see Problem 3.4). This method needs only a precision resistor. The current injection methods in Figure 3.3 further need a precision voltage.

Other methods to measure resistance are voltage dividers and Wheatstone bridges.

3.2 VOLTAGE DIVIDERS

Voltage dividers are commonly used to measure high-value resistances. In Figure 3.5a, if we assume the input resistance of the voltmeter is much higher than R, we have

$$v_o = \frac{V_r}{R_r + R} R_r \tag{3.10}$$

from which we can calculate the value for the unknown resistor

$$R = R_r \frac{V_r - v_o}{v_o} \tag{3.11}$$

Alternatively, if R_r and R switch their positions, we have

$$R = R_r \frac{v_o}{V_r - v_o} \tag{3.12}$$

Voltage dividers suit sensors with large variation in resistance and also nonlinear sensors such as NTC thermistors because the nonlinearity of the relationship between v_o and R permits thermistor linearization (Section 3.2.2) (see

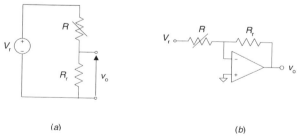

(a)   (b)

Figure 3.5 (a) Voltage divider method for resistance measurement. If one of the resistors is known, a single voltmeter reading allows us to calculate the unknown resistor. (b) Adding an op amp to a voltage divider yields a circuit whose output voltage is inversely proportional to the unknown resistance.

Problem 3.5). If we interpret R_r and R as the two parts of a potentiometer, voltage dividers can be applied to potentiometers (Section 3.2.1).

Example 3.2   The MGS1100 CO gas sensor (Motorola) has 1000 kΩ in air, from 30 kΩ to 300 kΩ (150 kΩ typical) for CO concentration of 60×10^{-6} (R_{60}), and a ratio $R_{400} = R_{60}/2.5 = 60$ kΩ typical, and 120 kΩ maximum. If the allowable voltage across the sensing resistor and power dissipation in it are 5 V and 1 mW, design a voltage divider according to Figure 3.5a for such a sensor if the expected CO concentration range is from 0 to 400×10^{-6}.

To guarantee the voltage limit, we can select $V_r = 5$ V. Power dissipation imposes the condition

$$\left(\frac{V_r}{R + R_r}\right)^2 R < 1 \text{ mW}$$

From (3.4), the maximal dissipation will happen when $R = R_r$, so that we must fulfill

$$R_r > \frac{V_r}{2\sqrt{P_{max}}} = \frac{5 \text{ V}}{2\sqrt{0.001 \text{ W}}} = 79 \ \Omega$$

The sensor resistance range is from 1000 kΩ to

$$R_{400} = 2.5 R_{60} = 375 \text{ k}\Omega$$

typical, and 750 kΩ maximum. Hence, the limit imposed by power dissipation will never be reached. We can thus select R_r equal, for example, to the sensor resting resistance for the atmosphere to monitor.

The voltage–resistance relation in voltage dividers is not linear because the current in the circuit depends on the unknown resistance. If that current were constant, the drop in voltage across a linear resistive sensor would be a linear function of the measurand as in Figure 3.3a. For sensors whose resistance decreases with the applied input, such as conductive polymer force sensors, the circuit in Figure 3.5b injects a constant current into the sensor and yields an output voltage

$$v_o = -V_r \frac{R_r}{R} \tag{3.13}$$

which increases with the measurand.

Example 3.3   Design the circuit in Figure 3.5b in order to obtain 0.5 V to 5 V output in response to a force of 0 kg to 45 kg when using a Flexiforce™ SSB-T sensor that has 200 kΩ when unloaded and 20 kΩ when applying 45 kg.

From (3.13),

$$v_o(0) = -V_r \frac{R_r}{200 \text{ k}\Omega} = 0.5 \text{ V}$$

$$v_o(45) = -V_r \frac{R_r}{20 \text{ k}\Omega} = 5 \text{ V}$$

Hence we need $V_r R_r = -100 \text{ V} \times \text{k}\Omega$. We can select $V_r = -5 \text{ V}$, which leads to $R_r = 20 \text{ k}\Omega$, which is standard. Its tolerance should be equal to the uncertainty accepted for the gain.

If the voltage divider circuit is applied to static measurements using a linear sensor whose percentage variations in resistance are small $(x \ll 1)$, the corresponding variations of the output voltage Δv_o are also small when compared with the voltage $v_o(0)$ obtained for zero input $(x = 0)$. This means that any error present when measuring $v_o = v_o(0) + \Delta v_o$ will result in a very high percentage error compared to Δv_o.

Because it is always easier to measure small voltages than to have a high resolution in the measurement of large voltages, the usual method to measure small changes in resistance consists of placing another voltage divider in parallel with the one incorporating the sensor. If both dividers are designed so that when there is no applied input both give the same voltage, the difference between their outputs is a signal that depends only on the measured variable. This arrangement is known as a *Wheatstone bridge*.

In addition to this fundamental advantage, in some instances the Wheatstone bridge increases the measurement sensitivity by using several sensors conveniently arranged in different arms. Also, some external interference can be canceled.

The Wheatstone bridge is a null measurement method, because the voltage in a voltage divider is compared with that in another divider incorporating the unknown resistance. But the output voltage or current can be measured by the null method or by the deflection method. For the null-type measurement, a known adjustable resistor is adjusted until both voltage dividers give the same output (Section 3.3). In the deflection method, we measure the voltage or current resulting from the imbalance between both voltage dividers when the sensor resistance changes (Section 3.4).

3.2.1 Potentiometers

Figure 3.6a shows the simplest signal conditioning for a potentiometer having a total resistance R_T. The linear or rotary movement from the device to be measured turns or slides the wiper. Figure 3.6b shows the equivalent electric circuit when the voltage meter has finite input resistance R_m. v_o is the open circuit voltage, and R_o is the output resistance. From (2.4) and (2.5), circuit analysis

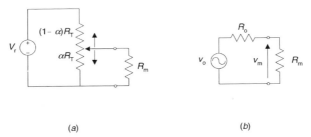

(a)   (b)

Figure 3.6 (a) Signal conditioning for a displacement measuring potentiometer and (b) equivalent circuit.

yields

$$v_m = \frac{v_o}{R_o + R_m} R_m = \frac{V_r \alpha}{1 + \frac{\alpha(1-\alpha)}{k}} \qquad (3.14)$$

where $k = R_m/R_T$. Therefore, the relationship between the measured voltage and wiper displacement is linear only when $k \gg 1$; that is, we need $R_m \gg R_T$.

If the theoretical (ideal) response is $v_o = V_r \alpha$, there is no error at the ends of the scale. At intermediate points, the relative error ε depends on k according to

$$\varepsilon = \frac{v_m - v_o}{v_o} = \frac{-\alpha(1-\alpha)}{k + \alpha(1-\alpha)} \qquad (3.15)$$

We can know the point where ε is maximal by determining when $d\varepsilon/d\alpha = 0$. From (3.15) we obtain

$$\frac{d\varepsilon}{d\alpha} = -\frac{k(1-2\alpha)}{[k + \alpha(1-\alpha)]^2} = 0 \qquad (3.16)$$

that is, $\alpha = 0.5$, namely the central point in the span. We evaluate $\varepsilon(0.5)$ by solving (3.15) at $\alpha = 0.5$, which yields

$$|\varepsilon_{max}| = \frac{0.25}{k + 0.25} \qquad (3.17)$$

Because in (3.15) α and $1 - \alpha$ can be interchanged with no effect, the relative error is symmetrical with respect to the central position. Also, because the minimal error is $\varepsilon = 0$, to know whether $\alpha = 0.5$ is a maximum or a minimum it is not necessary to take the second derivative. It is enough to note that according to (3.15), $\varepsilon \neq 0$ for $\alpha = 0.5$—and $v_m \neq 0$ V in (3.14)—and therefore this point will be a maximum.

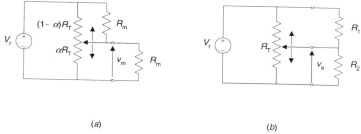

(a)   (b)

Figure 3.7 (a) Circuit to improve the loading-induced nonlinearity in a potentiometer. (b) Circuit to fine tune a voltage v_a using a potentiometer.

A simple way to reduce the loading error without increasing R_m is to place a resistor equal to R_m on the top side of the potentiometer, as shown in Figure 3.7a. The measured voltage is

$$v_m = V_r \frac{\alpha(k + 1 - \alpha)}{2\alpha(1 - \alpha) + k} \tag{3.18}$$

The added resistor forces $v_m = V_r/2$ at the central position ($\alpha = 0.5$), thus achieving zero error at that point. By using two different resistors (Figure 3.7b) we can obtain zero error at any desired point. This is useful for fine tuning a voltage around a given value [3].

Example 3.4 Calculate R_1 and R_2 in Figure 3.7b in order that a wiper displacement of $\pm 15\%$ of its stroke around the position corresponding to 1/4 of the full-scale value produces a change in voltage of only 10% with respect to the full-scale voltage.

By defining the parameters $a = R_T/R_1$ and $b = R_T/R_2$, the output voltage is

$$v_a = V_r \frac{\alpha(a - a\alpha + 1)}{1 + \alpha(1 - \alpha)(a + b)}$$

At $\alpha = 0.25 + 0.15$ the output voltage should be $v_a = (0.25 + 0.05) V_r$, and at $\alpha = 0.25 - 0.15$, $v_a = (0.25 - 0.05) V_r$. Then we have

$$0.3 = \frac{6a + 10}{25 + 6(a + b)}$$

$$0.2 = \frac{9a + 10}{25 + 9(a + b)}$$

Solving these two equations for a and b yields $a = 4.17$ and $b = 11.1$.

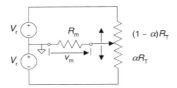

Figure 3.8 A symmetrical voltage supply reduces the loading error in a potentiometer.

Another method to reduce nonlinearity error due to loading effects is to use a symmetrical power supply connected as shown in Figure 3.8. Then the error at the central point and at the ends of the wiper travel is zero as in Figure 3.7a. The output voltage is

$$v_{\mathrm{m}} = \frac{V_{\mathrm{r}}(2\alpha - 1)}{1 + \dfrac{\alpha(1 - \alpha)}{k}} \tag{3.19}$$

For remote potentiometers, three-wire circuits yield zero and sensitivity (gain) errors. Figure 3.9a shows the connection. If in order to reduce loading effects we make $k \gg 1$, for $\alpha = 0$ we have

$$v_{\mathrm{m}}(0) = V_{\mathrm{r}} \frac{R_{\mathrm{w3}}}{R_{\mathrm{T}} + R_{\mathrm{w1}} + R_{\mathrm{w3}}} \tag{3.20}$$

This implies an offset error. For $\alpha = 1$ we have

$$v_{\mathrm{m}}(1) = V_{\mathrm{r}} \frac{R_{\mathrm{T}} + R_{\mathrm{w3}}}{R_{\mathrm{T}} + R_{\mathrm{w1}} + R_{\mathrm{w3}}} \tag{3.21}$$

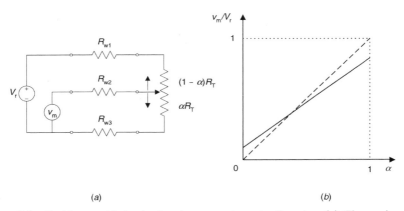

(a)   (b)

Figure 3.9 Problems with lead wires in a remote potentiometer. (a) Three-wire measurement circuit. (b) Effects: offset and sensitivity errors.

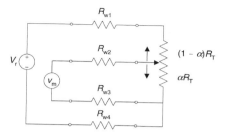

Figure 3.10 The four-wire measurement circuit for a potentiometer cancels offset error, but the sensitivity error remains unchanged with respect to the three-wire circuit.

Figure 3.9b shows that the actual voltage ratio has a smaller slope than the ideal ratio and does not go through zero.

The four-wire circuit in Figure 3.10 avoids the offset error because now $v_m(0) = 0$ V. However, the wiper at $\alpha = 1$ yields

$$v_m(1) = V_r \frac{R_T}{R_T + R_{w1} + R_{w3}} \tag{3.22}$$

Therefore, the sensitivity has not changed: When subtracting (3.21) from (3.20) we obtain (3.22). This is because the drop in voltage across leads makes the actual voltage across the potentiometer smaller than the supply voltage V_r.

The power supply for potentiometers should have a very low internal resistance so it is negligible with respect to other resistances in the circuit. It should have a very low temperature coefficient because (3.14) shows that its drifts would be directly reflected in v_m. But the ratio between the output voltage and the power supply is insensitive to power supply drifts. Analog-to-digital converters that accept an external reference voltage can implement ratio measurements. In Figure 3.11, if the input impedance of the ADC is much larger than R_T, the output digital word will be

$$D = (2^n - 1)\frac{v_m}{V_r} = (2^n - 1)\alpha \tag{3.23}$$

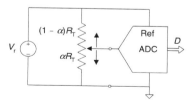

Figure 3.11 Measuring the ratio between the output voltage and the reference voltage of a potentiometer yields an output that is insensitive to supply voltage fluctuations.

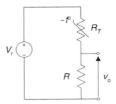

Figure 3.12 A linear temperature-dependent voltage divider based on an NTC thermistor.

Dc voltage supplies are best for potentiometers. Low-frequency ac voltage supplies are satisfactory. High-frequency ac supplies are not recommended because parasitic inductances and capacitances affect the output voltage, thus reducing the linearity the sensor has at dc.

3.2.2 Application to Thermistors

Section 2.4 noted that in a reduced temperature range an NTC thermistor can be modeled by two parameters and the equation

$$R_T = R_0 e^{B(1/T - 1/T_0)} = R_0 f(T) \tag{3.24}$$

where temperatures are in kelvins. This nonlinear behavior can be linearized to a certain extent by a voltage divider such as that in Figure 3.5a where according to (3.10), v_o does not change linearly with R.

The output for the voltage divider in Figure 3.12 is

$$v_o = V_r \frac{R}{R_T + R} = \frac{V_r}{1 + R_T/R} \tag{3.25}$$

From (3.24) we have

$$\frac{R_T}{R} = \frac{R_0}{R} f(T) = s f(T) \tag{3.26}$$

where we have defined $s = R_0/R$. v_o can be thus expressed as

$$v_o = \frac{V_r}{1 + s f(T)} = V_r F(T) \tag{3.27}$$

The shape of $F(T)$ depends on each particular material and on s. If a linear behavior of v_o with respect to T is desired, $F(T)$ should be a straight line. The appropriate value for s in order to have such a shape depends on the range of temperatures for the thermistor. For the material whose curves are shown in Figure 3.13, for example, in the range from 10 °C to 50 °C the best linearity is

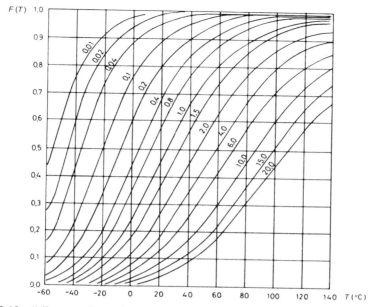

Figure 3.13 "s" curves for a given material used in thermistor manufacturing (courtesy of Thermometrics).

obtained for $s = 1.5$. But for the range from $90\,^\circ$C to $100\,^\circ$C a value of $s = 20.0$ is best.

These same curves can be applied to linearize an NTC thermistor by a shunting resistor R, thus offering an alternative method to the two analytical methods described in Section 2.4.3. The equivalent resistance for the parallel combination will be

$$R_\mathrm{p} = \frac{RR_T}{R + R_T} = R\left(1 - \frac{1}{1 + R_T/R}\right) = R[1 - F(T)] \qquad (3.28)$$

Therefore, if s is chosen for the desired temperature range so that $F(T)$ is approximately a straight line, then $1 - F(T)$ will also be a straight line and R_p will vary linearly with temperature.

3.2.3 Dynamic Measurements

The measurand, hence x in (3.1), may rapidly change with time periodically or nonperiodically. If we are only interested in its ac component, the voltage divider in Figure 3.5a is useful even if the values for x are small. We just need to couple the output signal to the measuring device through a capacitor that will form a high-pass filter with the input resistance of that device (Figure 3.14). The corner frequency should be tailored to the frequency band of interest. The

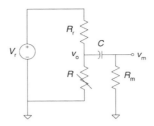

Figure 3.14   Ac-coupled voltage divider for dynamic measurements.

transfer function for the filter is

$$H(\omega) = \frac{j\omega R_m C}{1 + j\omega R_m C} = \frac{j\omega}{\omega_c + j\omega} \tag{3.29}$$

where $\omega_c = (R_m C)^{-1}$. The corresponding amplitude is

$$|H(f)| = \frac{1}{\sqrt{1 + \left(\dfrac{f_c}{f}\right)^2}} \tag{3.30}$$

where $2\pi f_c = \omega_c$. In order to have a relative amplitude error smaller than ε at frequency f_ε, the corner frequency should fulfill the condition

$$\frac{1}{\sqrt{1 + \left(\dfrac{f_c}{f}\right)^2}} > 1 - \varepsilon \tag{3.31}$$

which leads to

$$f_c < \frac{f_\varepsilon \sqrt{2\varepsilon - \varepsilon^2}}{1 - \varepsilon} \tag{3.32}$$

The sensitivity of the voltage divider is

$$S = \frac{dv_o}{dR} = V_r \frac{R_r}{(R + R_r)^2} \tag{3.33}$$

The value for R_r that optimizes this sensitivity is obtained by setting $dS/dR_r = 0$. This condition holds when $R_r = R$, and then $S = V_r/4R_r$. The second derivative, $d^2 S/dR_r^2$, is negative for that value for R_r, and therefore we are at a maximum. But R_r must have a constant value. An option is to choose it equal to that of R in the center of the measurement range. According to (3.4), this will also result in maximal power dissipation at that point. Sensors needing a given bias current may require a particular R_r, depending on the supply voltage.

Example 3.5   The resistance of a J15D mercury–cadmium–telluride photoresistor (EG&G) changes from 150 Ω to 10 Ω with incident light. Design the circuit in Figure 3.14 in order to supply 10 mA bias current when the available supply voltage is 12 V, the frequency of the input radiation is 1 Hz, and the maximal amplitude error allowed is 1%.

The required bias imposes the condition

$$R_r = \frac{V_r}{I_b} - R = \frac{12 \text{ V}}{1 \text{ mA}} - R = 1200 \ \Omega - R$$

Because R changes with light, a constant R_r implies that the bias will not be constant. Nevertheless, $R_r = 1$ kΩ, for example, will result in small bias change.

From (3.32), in order to have less than 1% error at 1 Hz we need

$$f_c < \frac{(1 \text{ Hz})\sqrt{2 \times 0.01 - 0.01^2}}{1 - 0.01} = 0.14 \text{ Hz}$$

If we select $R_m = 10$ MΩ, carbon film—assuming that the ensuing amplifier has much larger input resistance—we need $C = 114$ nF. We could select 120 nF, polyester, $\pm 10\%$ tolerance.

If an ac-supplied voltage divider is used, then the supply frequency must be at least 10 times higher than the maximal frequency of the measurand, as shown in Chapter 5.

3.2.4   Amplifiers for Voltage Dividers

Voltage dividers need high-impedance voltage meters. The noninverting amplifier in Figure 3.15a yields high input impedance and gain $1 + R_2/R_1$. C is added to limit bandwidth, hence noise (Section 7.4). If we consider the Thévenin equivalent circuit for the voltage divider and op amp error sources shown in Figure 3.15b—input offset voltage and currents—the output voltage is

$$v_o = v_s \left(1 + \frac{R_2}{R_1}\right) + V_{io}\left(1 + \frac{R_2}{R_1}\right) - I_p R_s \left(1 + \frac{R_2}{R_1}\right) + I_n R_2 \tag{3.34}$$

where

$$v_s = V_r \frac{R}{R_r + R} \tag{3.35}$$

$$R_s = \frac{R R_r}{R + R_r} \tag{3.36}$$

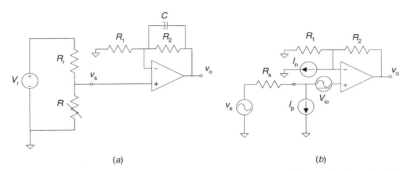

(a)   (b)

Figure 3.15 (*a*) Signal amplifier for a voltage divider and (*b*) equivalent circuit when considering op amp input offset voltage and current errors.

Hence, there is an *output zero error* (OZE),

$$\text{OZE} = \left(1 + \frac{R_2}{R_1}\right)\left(V_{io} - I_p R_s + I_n R_2 \frac{1}{1 + R_2/R_1}\right) \tag{3.37}$$

which, divided by the signal gain, yields the *input zero error* (IZE),

$$\text{IZE} = V_{io} - I_p R \| R_r + I_n R_1 \| R_2 \tag{3.38}$$

Op amp input currents may have signs opposite to those shown, but because the sign of V_{io} is unknown, their net contribution is always added to that from V_{io} in a worst-case condition. For common op amps, manufacturers do not separately specify I_p and I_n, but their average I_b. If the resistors in the amplifier are selected so that their parallel combination equals R_s, then

$$\text{IZE} = V_{io} + I_{io} R_1 \| R_2 \tag{3.39}$$

where $I_{io} = I_n - I_p$ is the op amp input offset current.

The values for V_{io}, I_p, I_n, and I_{io} in the above equations are those at the op amp working temperature, not at ambient temperature. Because of the internally generated heat, the op amp reaches a temperature above the actual ambient temperature T_a, depending on the thermal resistance between its junctions and the surrounding air. If the op amp dissipates P_d, then

$$T = T_a + P_d(\theta_{jc} + \theta_{cs} + \theta_{sa}) \tag{3.40}$$

where θ_{jc}, θ_{js}, and θ_{sa} are, respectively, the thermal resistances between the internal chip and amplifier case, which depends on the package type, between the case and the heat sink, if used, and between the heat sink and the ambient air. P_d includes the quiescent power (P_q), the power supplied to the feedback network, and the power dissipated by the current supplied to the ensuing stage (load), which usually has high input impedance. In signal amplifiers, often only

P_q matters. For split power supplies we have

$$P_q = |V_{s+}||I_{s+}| + |V_{s-}||I_{s-}| \qquad (3.41)$$

where I_{s+} and I_{s-} are the respective quiescent currents. This self-heating implies that air currents able to modify the thermal resistance from the IC to the environment will result in fluctuating offset voltages and currents (pseudonoise). Section 7.1 further delves in amplifier offset and drifts.

Example 3.6   The OPA177G is a precision op amp that has $V_{io} = 60$ μV with 1.2 μV/°C drift, $I_b = 2.8$ nA with 60 pA/°C drift, $P_q = 60$ mW when supplied at ±15 V (all maximal values), and $\theta_{ja} = 100\,°C/W$ for a plastic DIP package. Calculate its actual input offset voltage and currents when used in a non-inverting amplifier (Figure 3.15) supplied at ±15 V, with $R_2 = 100$ kΩ and $R_1 = 100$ Ω; the input voltage is 1 mV, the load resistance is 10 kΩ, and the ambient temperature inside the equipment is 35 °C.

We first estimate the power dissipated in the resistors and by the current supplied to the load in order to estimate the overall power dissipated by the op amp. For the given resistors, the gain will be 1001; hence the output voltage will be about 1 V. The drop in voltage across R_2 and across the load will be about 1 V, and that across R_1 will be about 1 mV. Therefore,

$$P_2 = \frac{1\ V^2}{100\ k\Omega} = 10^{-5}\ W$$

$$P_1 = \frac{(1\ mV)^2}{100\ \Omega} = 10^{-8}\ W$$

These powers are far below the quiescent power of the op amp.

To have 1 V across the 10 kΩ load, the op amp must supply 100 μA. This current goes from the 15 V power supply to the op amp output, hence dissipating some power inside the op amp:

$$P_L = (15\ V - 1\ V)(100\ \mu A) = 1.4\ mW$$

From (3.40), considering that there is only one thermal resistance involved, we obtain

$$T = T_a + (P_q + P_L)\theta_{ja} = 35\,°C + (60\ mW + 1.4\ mW) \times \frac{100\,°C}{1\ W} = 41\,°C$$

Therefore,

$$V_{io} = 60\ \mu V + (1.2\ \mu V/°C)(41\,°C - 25\,°C) = 79\ \mu V$$

$$I_b = 2.8\ nA + (60\ pA/°C)(41\,°C - 25\,°C) = 3.8\ nA$$

Voltage dividers whose output is ac-coupled must include a resistor such as R_m in Figure 3.14 in order to bias the op amp. Then R_m replaces R_s in (3.34) and (3.37).

Applications such as alarms do not need a continuous voltage corresponding to the measurand. A two-level voltage indicating whether the quantity is below or above a given threshold is enough. Then, in Figure 3.15 we use a *voltage comparator* instead of a voltage amplifier (reference 1, Section 6.1).

3.3   WHEATSTONE BRIDGE: BALANCE MEASUREMENTS

The Wheatstone bridge measurement method was first proposed by S. H. Christie in 1833 and reported by Sir Charles Wheatstone to the Royal Society (London) in 1858 as a method to measure small resistances. It is based on a feedback system, either electric or manual, in order to adjust the value of a standard resistor until the current through the galvanometer or other null indicator is zero (Figure 3.16). Once the balance condition has been achieved we have

$$R_3 = R_4 \frac{R_2}{R_1} \qquad (3.42)$$

That is, changes in R_3 are directly proportional to the corresponding changes we have to produce in R_4 in order to balance the bridge. This measurement method can be also used as a polarity detector because the output is positive or negative, depending on whether x is greater or less than a given threshold.

The condition (3.42) is reached independently of the power supply voltage or current and its possible variations. It does not depend on the type of detector or its impedance. Even more, it does not need to be linear because it must only indicate the balance condition. From (3.42) we can also deduce that the supply and the detector can interchange their positions without affecting the measurement. Figure 3.16*b* shows an arrangement for eliminating the influence that the

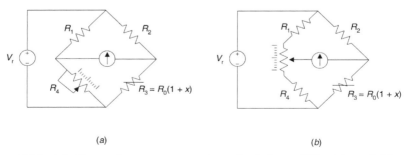

(a)   (b)

Figure 3.16   (*a*) Comparison measurement method for a Wheatstone bridge. (*b*) Arrangement to cancel the effect of contact resistance on the balance.

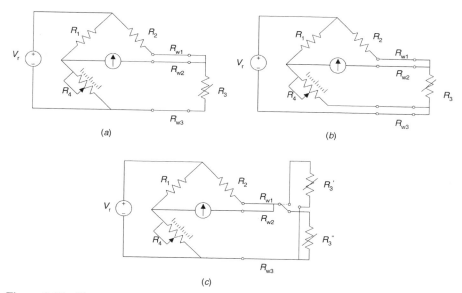

Figure 3.17 Siemens or three-wire method for measuring with a Wheatstone bridge when long leads are used.

contact resistance in the adjustable arm has on the measurement. It works by including that resistance in series with the central arm ("bridge"), through which there is no current when reaching the balance.

For remote sensors we must consider long leads whose resistance adds to the sensor resistance. Conductors with low TCR such as constantan and manganin have high resistivity ($\rho \approx 44 \ \mu\Omega \cdot cm$). Conversely, copper wires have lower resistivity ($\rho \approx 1.7 \ \mu\Omega \cdot cm$); but because its TCR is about $0.004 \ \Omega/\Omega/K$, temperature changes can result in important errors.

The Siemens or three-wire method (Figure 3.17a) solves this problem. Wires 1 and 3 must be equal and undergo the same temperature changes. The characteristics of wire 2 are irrelevant because in the balance condition there is no current through the bridge central arm. The relative error in the measurement of R_3 is

$$\varepsilon = \frac{R_4 R_2 / R_1 - R_3}{R_3} = \frac{R_w}{R_3} \left(1 - \frac{R_4}{R_1} \right) \tag{3.43}$$

Figure 3.17b shows an alternative circuit with the same objective. The error is similar. In both cases the error decreases when $R_3 \gg R_w$. Figure 3.17c shows how to apply this method to several sensors using a single set of three long wires.

The application of the null method to dynamic measurements depends on the availability of a fast enough automatic balancing system. Figure 3.18 shows

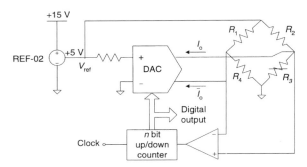

Figure 3.18 Wheatstone bridge using the comparison method with automatic balance through a digital-to-analog converter (DAC) and digital output.

such a method [4]. It is based on a digital-to-analog converter (DAC) that outputs two complementary current sources: a current corresponding to the digital input and another current corresponding to the complementary digital input. Any imbalance of the bridge output exceeding the comparator threshold modifies the converter outputs, via the up–down counter, so that one of them sinks the additional current necessary to keep the drop in voltage constant in both voltage dividers. At the same time, the other converter output reduces the amount of current it sinks, thus contributing to the voltage balance, which is reached independently of the sign of the change experienced by the sensor. The system output is then the digital word present at the input of the DAC needed to keep the bridge balanced.

3.4 WHEATSTONE BRIDGE: DEFLECTION MEASUREMENTS

3.4.1 Sensitivity and Linearity

Wheatstone bridges are often used in the deflection mode. Instead of measuring the action needed to restore balance on the bridge, this method measures the voltage difference between both voltage dividers or the current through a detector bridging them. Using the notation of Figure 3.19a, if the bridge is balanced when $x = 0$, which is the usual situation, we define a parameter k,

$$k = \frac{R_1}{R_4} = \frac{R_2}{R_0} \tag{3.44}$$

The voltage difference between both branches is

$$v_o = V_r\left(\frac{R_3}{R_2 + R_3} - \frac{R_4}{R_1 + R_4}\right) = V_r\frac{kx}{(k+1)(k+1+x)} \tag{3.45}$$

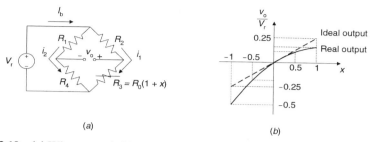

(a)

(b)

Figure 3.19 (a) Wheatstone bridge using the deflection method and (b) its ideal and real transfer characteristics when $k = 1$.

Thus the output voltage is proportional to the changes in R_3 only when $x \ll k + 1$; that is, the sensitivity depends on x (and k and V_r). For $x = 0$, the sensitivity is

$$S_0 = \left.\frac{dv_o}{d(xR_0)}\right|_{x=0} = \frac{V_r k}{R_0}\frac{1}{(k+1)^2} \qquad (3.46)$$

The maximal sensitivity as a function of k is obtained by setting $dS_0/dk = 0$, which yields $k = 1$. By calculating the second derivative we can verify that this point is a maximum. On the other hand, from (3.45) we infer that $k = 1$ yields a nonlinear output unless $x \ll 2$. Figure 3.19b shows how the actual output departs from a straight line through the origin for $k = 1$.

   If the bridge is supplied by a constant current I_r, the output voltage is

$$v_o = I_r R_0 \frac{kx}{2(k+1)+x} \qquad (3.47)$$

To have an approximately linear output we need $x \ll 2(k+1)$, and $x \ll 4$ when $k = 1$. Linearity is not necessary to achieve good accuracy. What matters is the repeatability of the results. But the output is easily interpreted when it is proportional to the measurand.

   For metal strain gages, x seldom exceeds 0.02. Therefore, usually $k = 1$ to improve sensitivity and, unless a very high linearity is desired, the presence of x in the denominator of (3.45) and (3.47) is ignored. Alternatively, the actual x can be calculated from the output voltage or current by solving for x in those equations [5]. For example, for a bridge supplied by constant voltage and $k = 1$, from (3.45) we obtain

$$x = \frac{4v_o}{V_r}\frac{1}{1 - \dfrac{2v_o}{V_r}} \qquad (3.48)$$

The second factor on the right-side member corrects the "uncorrected" output $4v_o/V_r$.

Example 3.7 A given uncorrected strain indicator based on a Wheatstone bridge with $k = 1$ uses a single gage as in Figure 3.19a, with $G = 2.0$. Determine the actual strain when the reading is 20,000 $\mu\varepsilon$ in compression ("negative").

According to (2.16), the fractional change in resistance is $x = G \times \varepsilon$. Therefore, when $\varepsilon = -20,000$ $\mu\varepsilon$, $x = -0.04$. However, the actual strain is not 20,000 $\mu\varepsilon$ because the indicator is uncorrected; that is, its reading corresponds to $4v_o/V_r$. Therefore, from (3.48),

$$x = -0.04\frac{1}{1 + 0.02} = -0.039216$$

The actual strain is

$$\varepsilon = \frac{x}{G} = \frac{-0.039216}{2} = -19608 \ \mu\varepsilon$$

The relative error is about 2%.

For resistance thermometers, x can be close to 1 and even higher, so that designing a bridge with $k = 1$ would result in a strongly nonlinear transfer characteristic. For a Pt100-based thermometer, for example, at 100 °C the resistance has changed from 100 Ω at 0 °C to 140 Ω. For these cases we can linearize the bridge by analog or digital techniques or work with a reduced sensitivity by making $k = 10$, or even higher, and compensating part of the sensitivity loss by increasing the supply voltage. This option is limited by sensor self-heating. Nevertheless, supplying short duty cycle rectangular voltages gives high peak output voltages with low rms value.

Example 3.8 A temperature ranging from -10 °C to $+50$ °C is to be measured, giving a corresponding output voltage from -1 V to $+5$ V, with an error less than 0.5% of the reading plus 0.2% of FSO. The sensor available is a Pt100 (100 Ω and $\alpha = 0.004$ $\Omega/\Omega/°$C at 0 °C), and $\delta = 5$ mW/K at the measurement conditions. The proposed solution is a voltage-supplied dc bridge whose output voltage is connected to an ideal amplifier. Calculate the value for bridge resistors and the supply voltage needed. Calculate the theoretical sensitivity for the bridge and the gain for the amplifier.

For the circuit in Figure 3.19a, if we take $R_3 = R_T = R_0(1 + \alpha T)$, where T is the incremental temperature above the temperature at which R_0 was measured, (3.45) takes the form

$$v_o = V_r\frac{k\alpha T}{(k + 1)(k + 1 + \alpha T)}$$

If we assume that the response is just

$$v_i = V_r \frac{k\alpha T}{(k+1)^2}$$

(tangent at the origin), the relative error due to nonlinearity will be

$$\varepsilon = \frac{v_o - v_i}{v_o} = \frac{-\alpha T}{k+1}$$

Therefore, the error increases with the measured temperature. Because we want a 0 V output at $0\,°C$, we choose $R_0 = 100\,\Omega$. Hence, $\alpha = 0.004/°C$ in the above equations. The maximal relative error will be at $T = 50\,°C$—that is, the maximum temperature. We want the relative error to be $\varepsilon < 0.005$; therefore

$$\frac{(0.004/°C)(50\,°C)}{k+1} < 0.005$$

This requires $k > 39$. Then the values for the other bridge resistors are $R_4 = 100\,\Omega$, $R_1 = R_2 = 3900\,\Omega$. Larger values for R_1 and R_2 would decrease the sensitivity.

The sensitivity also depends on the supply voltage for the bridge, which is limited by sensor self-heating. Recognizing that self-heating implies a nearly constant error yields the condition

$$P = \left(\frac{V_r}{R_2 + R_T}\right)^2 R_T < (0.002 \times 50\,°C) \times (5\,\text{mW}/°C) = 0.5\,\text{mW}$$

From (3.3) and Figure 3.1 we deduce that maximal sensor heating occurs when $R_T = R_2$. In the measurement range, R_T is always lower than R_2. Hence, maximal heating will occur at $50\,°C$ because then R_T reaches its maximal value, $120\,\Omega$. These conditions yield

$$V_r < \sqrt{\frac{0.0005\,\text{W}}{120\,\Omega}} \times (3900\,\Omega + 120\,\Omega) = 8.2\,\text{V}$$

If $V_r = 8$ V is chosen for convenience, then the sensitivity at $0\,°C$ will be $0.78\,\text{mV}/°C$. The gain needed to obtain a 5 V output when $T = 50\,°C$ will be

$$G = \frac{5\,\text{V}}{(0.78\,\text{mV}/°C) \times (50\,°C)} = 128.2$$

Note that according to Figure 3.19b, an end-point straight line may be closer to the actual transfer characteristic than the tangent to this characteristic at the origin, which is the assumed response.

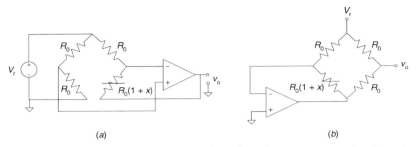

Figure 3.20 Forcing a constant current through a linear sensor and subtracting a constant voltage yields a linear voltage for (*a*) a resistance bridge with five terminals and (*b*) a common bridge with four terminals.

3.4.2 Analog Linearization of Resistive Sensor Bridges

A Wheatstone bridge that includes a single linear sensor is nonlinear because the current through the sensor depends on its resistance. To obtain a voltage proportional to any size change in one of the resistances in a Wheatstone bridge, we can modify the structure of the bridge to keep constant the current through it, similar to the circuit in Figure 3.5*b*. In Figure 3.20*a* we further subtract the resulting voltage drop across the sensor from that across a fixed resistor R_0. For an ideal op amp, the output is

$$v_{\mathrm{o}} = -V_{\mathrm{r}}\frac{x}{2} \tag{3.49}$$

This method, however, requires the bridge to have five terminals accessible; that is, the bridge must be opened at one of the junctions where the sensor is connected. The pseudobridge in Figure 3.20*b* overcomes that limitation. The op amp must have low offset voltage, input currents, and drifts (see Problem 3.19). If x can be negative, the op amp must operate on a dual (split) power supply. Problems 3.17 and 3.18 show additional linearization circuits.

3.4.3 Sensor Bridge Calibration and Balance

Equation (3.46) shows that the sensitivity of a sensor bridge depends on the supply voltage V_{r}, on the quiescent sensor's resistance R_0, and on the arms ratio k. To avoid the need to measure k, which may require opening the bridge junctions, we can determine S through the shunt calibration circuit in Figure 3.21. With the switch opened, for $x = 0$, the bridge is adjusted until $v_{\mathrm{o}} = 0$ V. After closing the switch and with the sensor at rest, the output deflection equals that obtained from a change x in R_3:

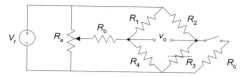

Figure 3.21 Shunt calibration of a resistive sensor bridge.

$$\frac{R_0 R_c}{R_0 + R_c} = R_0(1 + x)$$

$$x = -\frac{R_0}{R_0 + R_c}$$ (3.50)

The bridge sensitivity at $x = 0$ is then

$$S_0 = \frac{v_o}{xR_0} = -\frac{v_o}{R_0}\left(1 + \frac{R_c}{R_0}\right)$$ (3.51)

Hence, we need to measure only R_0 and the calibration resistor in order to calculate the bridge sensitivity from the measurement of v_o.

To calibrate the bridge for positive resistance variations, we shunt R_2 by a calibration resistance. If the bridge has more than one active arm, other calibration resistors can be connected one at a time by closing the respective switch—for example, with a relay in an automatic system.

If the calibration resistor or resistors cannot be placed close to the respective sensors, we must avoid placing lead wires for those resistors in series with the sensors. That is, we should use separate wires.

Because of resistor tolerance, often the balance condition in (3.44) is unfulfilled (see Problem 3.12). Figure 3.21 shows how to add a balance control through R_a and R_b. Measurement conditions may be different from calibration conditions, thus resulting in a nonzero bridge output at null conditions. This problem can be solved by a modified bridge [6], which consists of adding two known resistors R in series with each of R_3 and R_4. Either the junction between R and R_3 or between R and R_4 is driven by a current I so that the bridge is rebalanced at null at the measurement conditions. I is derived by a feedback system from the bridge output at null conditions, set either manually or under computer control.

3.4.4 Difference and Average Measurements and Compensation

Bridges have the additional advantage, as compared with voltage dividers, that permit the measurement of the difference between quantities or its average. Furthermore, by using several sensors they permit an increase in sensitivity and some interference compensation.

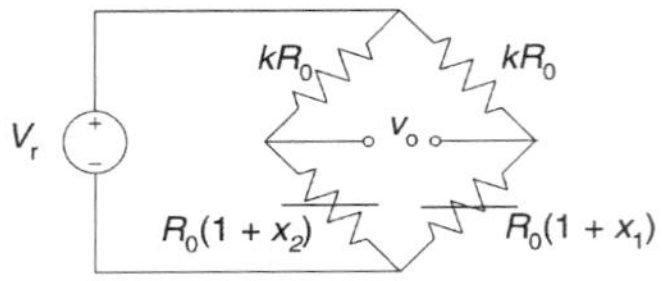

Figure 3.22 A resistance bridge with two sensors on opposite arms measures their difference.

The circuit in Figure 3.22 measures a difference by including two sensors in adjacent bridge arms in different branches. The output voltage is

$$v_o = V_r \frac{k(x_1 - x_2)}{(k + 1 + x_1)(k + 1 + x_2)} \qquad (3.52)$$

Whenever $x_1, x_2 \ll k + 1$, we can approximate

$$v_o = V_r \frac{k(x_1 - x_2)}{(k + 1)(k + 1)} \qquad (3.53)$$

For temperature sensors, this method yields temperature differences—useful, for example, to calculate thermal gradients or heat loss in pipes or to warn about freezing risks in agriculture: A fast, large temperature gradient indicates too large a heat loss from the soil to sky, ending in frozen soil. A comparison with (3.45) shows that the same compromise between sensitivity and linearity holds here, which influences the choice of k.

Example 3.9 A differential thermometer able to measure temperature differences from $0\,°C$ to $750\,°C$ in the range from $50\,°C$ to $800\,°C$ with an error smaller than $5\,°C$ is needed to measure the heat insulation capability of several materials. The sensors available are Pt100 having $100\ \Omega$ and $\alpha = 0.004/°C$ at $0\,°C$, and $\delta = 1\ \text{mW/K}$. Design a dc bridge supplied by a constant voltage, able to perform the desired measurements and whose output voltage is measured by an ideal voltmeter. Calculate the bridge sensitivity.

Using the notation in Figure 3.22 and taking $x_1 = \alpha T_1$ and $x_2 = \alpha T_2$, (3.52) leads to

$$v_0 = V_r \frac{k\alpha(T_1 - T_2)}{(k + 1 + \alpha T_1)(k + 1 + \alpha T_2)}$$

The ideal sensitivity is the quotient between output voltage v_o and temperature difference $T_2 - T_1$ if the output were linear. That is,

$$S = \frac{v_o}{T_2 - T_1} = \frac{V_r k\alpha}{(k + 1)^2}$$

The absolute nonlinearity error referred to the input (temperature difference) is

$$e = \left| \frac{v_o}{S} - (T_2 - T_1) \right| = (T_2 - T_1) \frac{\alpha(T_1 + T_2)(k + 1) + \alpha^2 T_1 T_2}{(k + 1 + \alpha T_1)(k + 1 + \alpha T_2)}$$

Therefore the error increases with the temperature difference and with the absolute temperature. The worst case is thus when $T_1 = 50\,°C$ and $T_2 = 800\,°C$. If we assume that the self-heating error is negligible, the condition to fulfill in order not to exceed the prescribed $5\,°C$ error is

$$e = (750\,°C) \frac{(0.004 \times 850) \times (k + 1) + (0.004)^2 \times 50 \times 850}{(k + 1 + 0.004 \times 50)(k + 1 + 0.004 \times 850)}$$

which leads to

$$k^2 - 504.6k - 601 = 0$$

The solutions are $k = 506$ and $k = -1.19$. Obviously the negative solution has no physical sense. The values for the bridge resistors are then $R_1 = R_2 = 50.6\,k\Omega$.

The supply voltage must be low enough to avoid significant self-heating. Because the measurement is differential, there will be a self-heating error whenever each probe dissipates a different amount of power. The absolute error, in temperature, is

$$e_s = \left(\frac{V_r}{R_1 + R_4} \right)^2 \frac{R_4}{\delta} - \left(\frac{V_r}{R_2 + R_3} \right)^2 \frac{R_3}{\delta}$$

$$= \frac{V_r^2}{R_0 \delta} \left[\frac{1 + \alpha T_2}{(k + 1 + \alpha T_2)^2} - \frac{1 + \alpha T_1}{(k + 1 + \alpha T_1)^2} \right]$$

Because k is large, we can approximate

$$e_s \approx \frac{V_r^2}{R_0 \delta} \frac{\alpha(T_2 - T_1)}{(k + 1)^2}$$

If we want this error to be for example only $0.05\,°C$, from this equation the supply voltage must be less than $20.7\,V$. Then the sensitivity would be $163\,\mu V/°C$. A smaller V_r would decrease this error, and a slightly larger k would reduce the total absolute error below the $5\,°C$ target.

Several strain gages arranged in the same bridge also offer many advantages. Two strain gages bonded to an element as shown in Figure 3.23a and connected in a bridge as in Figure 3.23b yield an output voltage:

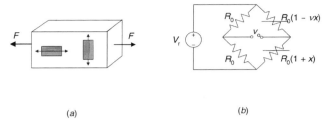

(a)   (b)

Figure 3.23   Using two active strain gages, one longitudinal and the other one transverse, in a measurement bridge increases the sensitivity by the Poisson ratio v.

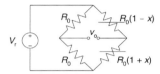

Figure 3.24   Using two active strain gages undergoing opposite variations doubles the sensitivity and yields a linear output.

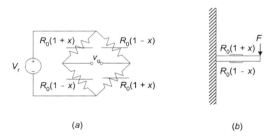

(a)   (b)

Figure 3.25   Two dual rosettes connected as shown yield a linear voltage.

$$v_\mathrm{o} = V_\mathrm{r} \frac{x(1+v)}{2[2+x(1-v)]} \approx V_\mathrm{r} \frac{x(1+v)}{4} \tag{3.54}$$

Hence, the additional gage has increased the sensitivity by the Poisson ratio (v).

Two gages undergoing strains of the same magnitude but opposite in sign and connected as shown in Figure 3.24 yield an output voltage:

$$v_\mathrm{o} = V_\mathrm{r} \frac{x}{2} \tag{3.55}$$

Thus not just the sensitivity increases, but the output is linear.

The bridge in Figure 3.25*a* combines two equal dual rosettes bonded on each side of the cantilever beam in Figure 3.25*b*. The output voltage is

TABLE 3.2 Output Voltage for Different Strain Gage Connections of Quarter Bridge, Half Bridge, and Full Bridge (Figure 3.19a) Supplied by a Constant Voltage or Current

R_1	R_2	R_3	R_4	Constant V_r	Constant I_r
R_0	R_0	$R_0(1+x)$	R_0	$V_r \dfrac{x}{2(2+x)}$	$I_r R_0 \dfrac{x}{4+x}$
$R_0(1+x)$	R_0	$R_0(1+x)$	R_0	$V_r \dfrac{x}{2+x}$	$I_r R_0 \dfrac{x}{2}$
R_0	R_0	$R_0(1+x)$	$R_0(1-x)$	$V_r \dfrac{2x}{4-x^2}$	$I_r R_0 \dfrac{x}{2}$
R_0	$R_0(1-x)$	$R_0(1+x)$	R_0	$V_r \dfrac{x}{2}$	$I_r R_0 \dfrac{x}{2}$
$R_0(1-x)$	R_0	$R_0(1+x)$	R_0	$V_r \dfrac{-x^2}{4-x^2}$	$I_r R_0 \dfrac{-x^2}{4}$
$R_0(1+x)$	$R_0(1-x)$	$R_0(1+x)$	$R_0(1-x)$	$V_r x$	$I_r R_0 x$

$$v_o = V_r x \qquad (3.56)$$

which is also linear and yields a sensitivity four times that of a single strain gage. Figure 2.10 shows additional connections for multiple strain gages.

These different measurement connections are respectively called quarter bridge (one sensor), half bridge (two sensors), and full bridge (four sensors). Table 3.2 lists the corresponding output voltages for constant voltage or constant current supplies. Bridges with constant current excitation are more linear than bridges with constant voltage excitation. Full bridge connections are common in load cells and piezoresistive pressure sensors.

Example 3.10 A given strain-gage-based load cell uses a full bridge apparently damaged because it does not yield a zero output when unloaded. To find a possible explanation, the voltage supply leads (1 and 2) and amplifier leads (3 and 4) are disconnected. Then several measurements between leads yield the following results: between 1 and 2, 127 Ω; between 1 and 3, 92 Ω; between 1 and 4, 92 Ω; between 2 and 3, 92 Ω; between 2 and 4, 106 Ω; between 3 and 4, 127 Ω. All the measurements are performed with open connections for the nonmeasured lead pair. Determine the damaged gage and give a possible explanation for its damage.

Using the terminology in Figure 3.19a for resistors, the measurements performed correspond to the following resistance combinations:

$$R_{12} = \frac{(R_1 + R_4)(R_2 + R_3)}{R_1 + R_2 + R_3 + R_4} = 127 \ \Omega$$

$$R_{13} = \frac{R_1(R_4 + R_2 + R_3)}{R_1 + R_2 + R_3 + R_4} = 92 \ \Omega$$

$$R_{14} = \frac{R_2(R_4 + R_1 + R_3)}{R_1 + R_2 + R_3 + R_4} = 92\ \Omega$$

$$R_{23} = \frac{R_4(R_1 + R_2 + R_3)}{R_1 + R_2 + R_3 + R_4} = 92\ \Omega$$

$$R_{24} = \frac{R_3(R_1 + R_2 + R_4)}{R_1 + R_2 + R_3 + R_4} = 106\ \Omega$$

$$R_{34} = \frac{(R_1 + R_2)(R_3 + R_4)}{R_1 + R_2 + R_3 + R_4} = 127\ \Omega$$

From the first and sixth results we deduce $R_2 = R_4$. From the second and third, $R_1 = R_2$. The fourth confirms that $R_1 = R_4$. The fifth indicates that R_3 is different. Thus the problem is to find two resistances, R and R_3, for which two equations are enough:

$$\frac{R(2R + R_3)}{3R + R_3} = 92\ \Omega$$

$$\frac{3RR_3}{3R + R_3} = 106\ \Omega$$

By solving these equations we obtain $R = 120\ \Omega$, $R_3 = 150\ \Omega$. This large value for R_3 and the inability of nulling the zero may be due to an overload of R_3, leading to an irreversible (permanent) deformation.

Wheatstone bridges can measure average values by connecting several sensors as shown in Figure 3.26. The three (or more) sensors are assumed equal but measure different values of the same quantity—for example, temperature. The output voltage is

$$v_o = V_r \frac{k(x_1 + x_2 + x_3)/3}{[k + 1 + (x_1 + x_2 + x_3)/3](k + 1)} \tag{3.57}$$

hence proportional to x average if $k + 1$ is large enough.

Strain gages are temperature sensitive and a bridge minimizes this problem. Figure 3.27 shows that a dummy gage in the same bridge branch as the active gage compensates the relative resistance change y suffered because of a tem-

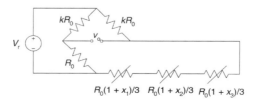

Figure 3.26 Several equal sensors placed in the same bridge arm yield an output voltage that is proportional to the average of the measurand.

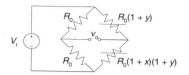

Figure 3.27   A dummy gage in the same branch as the active gage compensates for temperature interference.

perature change. The dummy gage is not bonded to the stressed material, so that it undergoes the same temperature-induced change but not the strain-related change x (Fig. 2.10a). Half- and full-bridge connections have inherent thermal compensation. AMR sensors (Section 2.5) are also placed in bridges to cancel temperature interference. Voltage dividers also compensate temperature interference.

Lead wire resistance adds to sensor resistance and cannot be completely compensated by a bridge connection. Nevertheless, three-wire sensors connected as shown in Figure 3.17 have reduced errors as compared to two-wire sensors [7]. Half-bridge circuits using two-wire sensors also compensate wire resistance only partially.

3.4.5   Power Supply of Wheatstone Bridges

To obtain an output signal from a Wheatstone bridge that includes one or more sensors we must supply the bridge with a voltage or current, dc or ac. This supply must be stable with time and temperature because otherwise its drift would propagate to the output. In a resistive bridge supplied by a dc voltage, for example, the output voltage is given by (3.45). If x remains constant but the supply voltage drifts, we have

$$\frac{dv_o}{v_o} = \frac{dV_r}{V_r} \tag{3.58}$$

which means that the output undergoes the same percent change. This may preclude, for example, the use of an ordinary power voltage supply with a 0.1 %/°C drift or that of some monolithic voltage regulators with thermal drifts in excess of 1 %/°C.

Nevertheless, as in potentiometers (Figure 3.11) and voltage dividers, the ratio between the output voltage and the reference voltage is insensitive to supply drift. Figure 3.28 shows a circuit for ratiometric measurement applied to a sensor bridge. If the voltage supply for the bridge were ac, the same method could be used but the ac voltage would have to be rectified in order to yield the reference voltage, and rectification would contribute additional errors. The voltage supply for the amplifier does not require high stability because the amplifier is able to reject its variations, as given by the specified power supply rejection ratio (PSRR).

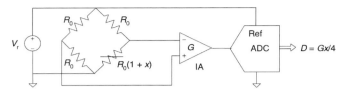

Figure 3.28 Ratiometric measurements based on an analog-to-digital converter eliminate the need for high stability of the bridge supply.

High-precision applications that do not use ratiometric ADCs need a stable
voltage or current supply. This can rely on precision voltage references, like
those used in multiplying DACs or ADCs with external reference. Table 3.3
gives the stability characteristics for several precision voltage reference ICs.
However, their maximal output current is limited to less than 20 mA. For a
10 V supply, this means that the bridge should have more than 500 Ω. Some
strain gage bridges have only 120 Ω. Figure 3.29 shows a method to supply a
large, stable current. Most of the current comes from the power supply through
R, but the drop in voltage across the bridge is controlled by the voltage reference IC. If the average current sunk by the bridge is I_b (maximum I_{max}, minimum
I_{min}) and the IC regulator sources I_{o+} and sinks I_{o-}, the circuit equations are

$$R = \frac{V_s - V_r}{I_b} \tag{3.59}$$

$$I_{max} = I_b + I_{o+} \tag{3.60}$$

$$I_{min} = I_b + I_{o-} \tag{3.61}$$

Some IC voltage references have $I_{o-} = 0$ mA (or, alternatively, $I_{o+} = 0$ mA). If
the reference voltage available lacks the accuracy needed, we can use a ratiometric measurement.

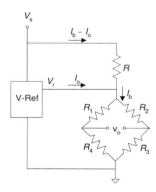

Figure 3.29 An IC voltage reference keeps the voltage driving the bridge constant even
though the current comes from a voltage supply having poor stability.

TABLE 3.3 **Stability of Several Components that Provide a Constant Output Voltage**

Parameter	Designation					
	AD581L	LM399A	LT1021A	MAX671C	REF10A	REF102C
Output (V/mA)	10/10	6.95/10	10/10	10/10	10/20	10/10
Time drift (10^{-6}/1000 h)	25	20	15	50	50	5
Thermal drift (10^{-6}/K)	5	0.6	2	1	8.5	2.5
Supply (10^{-6}/V)	50	10	4	50	100	100
Load (10^{-6}/mA)	50	3	25	1	800	10
Noise (0.1–10) Hz (μV, p–p)	40	6	6	50	30	5

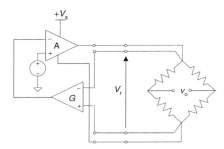

Figure 3.30   The four-wire measurement method applies the desired voltage across a remote bridge but does not compensate for the drop in voltage along supply cables.

In remote, low-resistance bridges, lead wire resistances cannot be neglected and the drop in voltage across the bridge differs from the power supply output voltage. The four-wire circuit in Figure 3.30 permits us to accurately determine the voltage applied to the bridge. Two wires apply the voltage and a different wire pair senses the actual drop in voltage across the bridge. The detected voltage is used to adjust the voltage output from the source through an amplifier (A). Note that this method does not avoid the drop in voltage along the supply lead wires, but only yields the desired supply voltage across the bridge. Measuring the bridge output requires two additional wires, thus making a total of six wires. Remote bridges supplied at constant current do not have this problem and need only four wires.

An additional problem in strain gage bridges is that their stress sensitivity depends on temperature because of the temperature dependence of Young's modulus. Adding a split series resistor to the voltage supply leads can compensate for that change. Fraden [8] analyzes this and other techniques of temperature compensation of resistive bridges. Because computation power is available at low cost, a common solution is to measure temperature and to compensate for thermal sensitivity digitally.

Bridge excitation can be dc or ac. If a dc signal is used, the thermoelectromotive effects (Section 6.1.1) appearing in junctions of dissimilar metals and amplifier drifts cause errors that restrict the physical layout of the circuit. Thermoelectromotive interference adds an offset voltage to the detected signal. Because the bridge output depends on the polarity of the supply voltage, reversing this polarity reverses that of the bridge output but not the interference. Subtracting the readings for each polarity cancels the interference. Nevertheless, thermoelectromotive interference depends on environmental conditions (such as air movement) that can easily change during the supply reversal procedure. The MIC4427 is a switch pair that easily reverses excitation polarity. An ac supply avoids thermoelectromotive effects, but stray capacitances may imbalance the bridge. The capacitance between a strain gage and the structure it is bonded on is about 100 pF. The impedance of stray capacitances has a more pronounced effect at high frequencies. But the supply frequency must be at least ten times higher than the maximal frequency of the measurand, as we will find in Chapter 5. Furthermore, if the measurement range includes both positive and negative values, we will require a phase-sensitive detector in order to know the sign of the bridge output signal. As a result, ac supplies are used only in those applications where the available sensor favors that kind of supply, or when we desire the low noise of ac amplifiers and interference-rejection capability of phase-sensitive demodulation (Section 5.3).

3.4.6 Detection Methods for Wheatstone Bridges

The type of detecting device for the output signal of a sensor bridge depends on the intended application. Most situations call for an analog-to-digital conversion. Therefore, the detector must amplify the bridge output voltage or current to match its dynamic range and level to those of the ADC and must have an adequate input impedance: high to sense voltage, low to sense current. Furthermore, its input configuration must be compatible with the signal type (differential or single ended, grounded or floating) provided by the bridge output (reference 1, Section 1.2.2).

Figure 3.31 shows alternative circuits depending on power supply grounding.

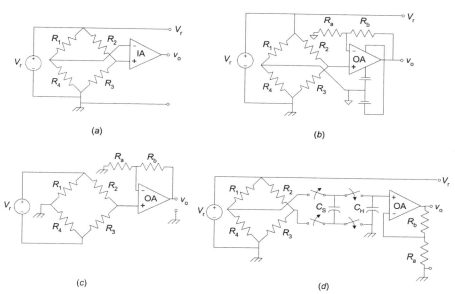

(a)   (b)

(c)   (d)

Figure 3.31 Alternative methods to detect the output voltage of a sensor bridge depending on power supply grounding. A grounded excitation needs (*a*) a differential amplifier or (*b*) a single-ended amplifier with floating power supply. (*c*) A floating excitation permits us to use a single-ended, grounded amplifier. (*d*) A floating capacitor permits the connection of a bridge with grounded excitation to a single-ended, grounded amplifier.

If the power supply driving the bridge is grounded, the amplifier must be differential (Figure 3.31*a*) or single-ended with floating power supply (Figure 3.31*b*). Conversely, if the bridge power supply is floating as in Figure 3.31*c*, the amplifier can be single-ended, but the ratio meter for ratiometric measurement needs a differential or floating input for V_r. It is also possible to connect a grounded bridge to a single-ended, grounded amplifier by converting the bridge differential output voltage into a single-ended voltage through a floating capacitor (Figure 3.31*d*). Here input switches first connect the sampling capacitor C_S to the bridge, and they charge it to the output voltage. Then input switches open and output switches close so that the holding capacitor C_H shares the charge acquired by C_S. Switches are clocked at a frequency much higher than that of the measurand. The linearized bridge in Figure 3.20*b* yields a single-ended output in spite of having a grounded excitation because of the linearizing op amp.

Common instruments and PC add-on cards can be used as bridge detectors, provided that they fulfill the gain, level, impedance, and input terminal compatibility discussed above. Digital panel meters and multimeters often have floating input and can implement the method in Figure 3.31*b*. Oscilloscopes have single-ended input and therefore can only be applied to bridges with a floating power supply. PC add-on cards have grounded power supplies (unless plugged in battery-supplied portable computers) and can often be configured for differential or single-ended inputs, so that they can implement the circuits in Figures 3.31*a* and 3.31*c*.

There is an increasing availability of ICs that include several functions for sensor bridge (and voltage-divider) conditioning, including excitation, amplification, A/D conversion, calibration, and compensation, achieving 1% accuracy and better—for example, AD280, AD693 (4 to 20 mA output), AD7711/3, MAX1450/7/8 piezoresistive signal conditioners, MLX90308 (Melexis), and UTI (Smartech) [9]. There is also a variety of modules implementing signal conditioning functions. Each October issue of *Measurements & Control* lists manufacturers of signal conditioning equipment.

3.5 DIFFERENTIAL AND INSTRUMENTATION AMPLIFIERS

3.5.1 Differential amplifiers

Voltage amplifiers yield an output voltage proportional to that at their input. A differential amplifier processes the voltage difference between two input terminals, neither of which is connected to the reference voltage of its power supply. In addition, we will show later that it is best for amplifier input terminals to have high and similar impedance to ground. Common resistive sensor bridges supplied by a grounded voltage or current source cannot have any output terminal grounded. Therefore, differential amplifiers suit them. Differential amplifiers were first proposed by B. H. C. Matthews in 1934—using vacuum tubes—for electrophysiological studies.

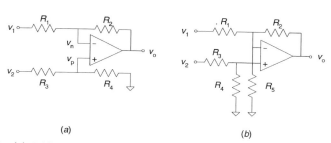

(a)   (b)

Figure 3.32 (a) Differential amplifier based on a single op amp and four matched resistors. (b) Adding a resistor increases the input common mode voltage range.

Figure 3.32a shows a simple differential amplifier. If we first assume that the op amp has a frequency-dependent differential gain A_d and negligible common mode gain ($A_c = 0$), then the (Laplace transform of the) output voltage is (reference 1, Section 2.4.2)

$$V_o = \frac{1}{1 + \dfrac{1}{A_d\beta}} \left[V_2 \frac{R_4}{R_3 + R_4} \left(1 + \frac{R_2}{R_1}\right) - V_1 \frac{R_2}{R_1} \right] \qquad (3.62)$$

where $\beta = R_1/(R_1 + R_2)$ is the *feedback factor* for the op amp. To illustrate the differential properties of the circuit, it is convenient to write the output as a function of the differential input voltage $v_d = v_2 - v_1$. In order to this, we substitute in (3.62)

$$v_d = v_2 - v_1 \qquad (3.63a)$$

$$v_c = \frac{v_1 + v_2}{2} \qquad (3.63b)$$

where v_c is termed common mode voltage. The result is

$$V_o = \frac{1}{1 + \dfrac{1}{A_d\beta}} \left\{ v_d \frac{1}{2}\left[\frac{R_4}{R_3 + R_4}\left(1 + \frac{R_2}{R_1}\right) + \frac{R_2}{R_1}\right] + v_c \frac{R_4 R_1 - R_2 R_3}{R_1(R_3 + R_4)} \right\} \qquad (3.64)$$

The factor multiplying v_d is the differential gain, and that multiplying v_c is the common mode gain; that is,

$$G_d = \left.\frac{V_o}{V_d}\right|_{v_c=0} = \frac{1}{1 + \dfrac{1}{A_d\beta}} \frac{1}{2}\left[\frac{R_4}{R_3 + R_4}\left(1 + \frac{R_2}{R_1}\right) + \frac{R_2}{R_1}\right] \qquad (3.65)$$

$$G_c = \left.\frac{V_o}{V_c}\right|_{v_d=0} = \frac{1}{1 + \dfrac{1}{A_d\beta}} \frac{R_4 R_1 - R_2 R_3}{R_1(R_3 + R_4)} \qquad (3.66)$$

To amplify v_d but not v_c we must have $G_c = 0$, which is obtained when

$$\frac{R_4}{R_3} = \frac{R_2}{R_1} = k \tag{3.67}$$

Then

$$G_d = \frac{k}{1 + \dfrac{1}{A_d \beta}} = \frac{k}{1 + \dfrac{k+1}{A_d}} \tag{3.68}$$

and $G_d \approx k$ when $A_d \gg k + 1$. For gains below 1000, most op amps meet this condition at low frequencies. Otherwise, a too low A_d introduces a gain uncertainty because G_d would depend on A_d, which is not well known.

Because the matching expressed by (3.67) is difficult to fulfill exactly, the circuit's ability to reject common mode signals will be limited rather than infinite. It is quantified by the *common mode rejection ratio* (CMRR), defined as the differential gain divided by the common mode gain. For Figure 3.32a it is

$$CMRR_R = \frac{G_d}{G_c} = \frac{1}{2} \frac{R_1 R_4 + R_2 R_3 + 2 R_2 R_4}{R_1 R_4 - R_2 R_3} \tag{3.69}$$

where the subscript R indicates that the finite CMRR results only from resistor mismatching. The CMRR is usually expressed in decibels by taking the decimal logarithm of (3.69) and multiplying the result by 20. For resistors with tolerance t_R, in a worst-case condition we have

$$CMRR_R = \frac{k(1 - t_R^2) + 1 + t_R^2}{4 t_R} \approx \frac{k+1}{4 t_R} \tag{3.70}$$

If the op amp in Figure 3.32a has finite common mode gain A_c, $CMRR_{oa} = A_d/A_c$, and the overall CMRR is [10]

$$CMRR = \frac{1}{CMRR_R^{-1} + CMRR_{oa}^{-1}} + \frac{1}{4(CMRR_R + CMRR_{oa})} \tag{3.71}$$

which can normally be approximated by

$$CMRR \approx \frac{1}{CMRR_R^{-1} + CMRR_{oa}^{-1}} \tag{3.72}$$

That is, the CMRR for resistors and for the op amp combine in "parallel"— their reciprocals add. Each CMRR must be expressed as a fraction, not in decibels. If $CMRR_R = -CMRR_{oa}$, the resulting CMRR is infinite. This con-

dition can be achieved at low frequencies—at which CMRRs are real numbers
—by trimming R_4 or R_4/R_3 (by adding a potentiometer between R_3 and R_4,
connected to the noninverting input of the op amp). Nevertheless, the TCR for
trimming potentiometers differs from that for common resistors (Section 7.5),
and therefore the optimizing condition will not be fulfilled when the temperature changes. The CMRR decreases for increasing frequency. The output voltage is

$$v_o = G_d v_d + G_c v_c = G_d \left(v_d + \frac{v_c}{\text{CMRR}} \right) \tag{3.73}$$

Therefore, if v_c is constant, the finite CMRR adds a zero error; but if v_c depends on the measurand, the CMRR introduces a gain or a nonlinearity error.

The circuit in Figure 3.32b extends the common mode voltage range beyond
that of the op amp because two voltage dividers formed by R_3 and R_4 and
by R_1 and R_5 reduce the input common mode voltage. If $R_1 = R_3$ and
$R_3/R_4 = R_1/(R_2 \| R_5) = k$, then

$$v_o = (v_2 - v_1)\frac{R_2 + R_5}{kR_5} \tag{3.74}$$

The differential amplifiers in Figure 3.32 are very useful in signal conditioning and are both available in ICs that include the op amp and matched resistors
(Table 3.4). Their major shortcoming is that, even assuming an ideal op amp,
the input resistance from each input is very limited. For input 1,

$$Z_{i1} = \frac{V_1}{\dfrac{V_1 - V_n}{R_1}} = \frac{V_1}{\dfrac{V_1 - V_p}{R_1}} = \frac{R_1}{1 - \dfrac{V_2}{V_1}\dfrac{R_4}{R_3 + R_4}} \tag{3.75a}$$

For input 2,

$$Z_{i2} = \frac{V_2}{\dfrac{V_2 - V_p}{R_3}} = R_3 + R_4 \tag{3.75b}$$

In differential mode ($v_c = 0$ V, $v_1 = -v_2$), $Z_{i1} = (k + 1)R_1/(2k + 1)$, in common
mode ($v_d = 0$ V, $v_1 = v_2$), $Z_{i1} = (k + 1)R_1$, and $Z_{i2} = R_3 + R_4$ in both cases.
Therefore, R_2 and R_4 will have to be very large resistors if high input impedance and high gain are required. Furthermore, changing the differential gain
requires the modification of two resistors without degrading their matching.

Example 3.11 Calculate the minimal input differential and common mode
resistances and the CMRR of a differential amplifier to be connected to the

TABLE 3.4 Basic Parameters of Some Differential Amplifiers

Amplifier	G_d	ε_d^a max (%)	R_1, R_3 (kΩ)	R_2, R_4 (kΩ)	CMRR min (dB)	f_G (kHz)	V_{io} max (μV)	$\Delta V_{io}/\Delta T$ typ (μV/°C)	Comments
AD626B[b]	10, 100	0.01	n.a.	n.a.	80	100	2500	6	Single supply
INA117BM	1	0.02	380	380, 20	86	200	1000	8.5	±200 V common mode
INA132U	1	0.075	40	40	76	300	250	1	Single supply
INA133U	1	0.05	25	25	80	1500	450	2	Low power
INA143U	0.1, 10	0.05	10	100	86	150[c]	250	1	Low power
INA146U[b]	0.1–100	0.1	100	10	70	50[d]	5000	10	Programmable gain
INA154U	1	0.05	25	25	80	3100	750	2	
INA157U	0.5, 2	0.05	12	6	86	4000	500	2	
MAX4198ESA	1	0.1	25	25	74	175	500	0.5	Low power, single supply
MAX4199ESA	10	0.3	25	250	84	45	300	3	Low power, single supply

[a] Gain error.

[b] Internal structure different from that in Figure 3.32a.

[c] $G = 10$.

[d] $G = 1$.

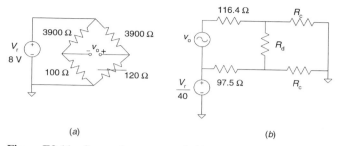

Figure E3.11 Sensor in a quarter bridge and equivalent circuit.

Wheatstone bridge in Example 3.8 so that their finite values yield an error negligible with respect to the specified errors for the design of the bridge (0.5% of reading plus 0.2% of FSO—5 V).

Figure E3.11 shows the sensor bridge at its largest imbalance (at 50 °C the Pt100 has 120 Ω) and the equivalent circuit when connected to a differential amplifier with input resistance R_d (differential mode) and R_c (common mode). The approximated common mode voltage is $V_r/40 = (8\ \mathrm{V})/40 = 200$ mV, and it will result in an almost constant error because of the common mode gain G_c. If we wish this error to be 10% of the constant error allowed, which is 0.2% of 5 V, the maximal common mode gain allowed is

$$G_c = \frac{(0.002 \times 5\ \mathrm{V})/10}{200\ \mathrm{mV}} = 0.005$$

Because the differential gain required is 128.2, we need

$$\mathrm{CMRR} > \frac{128.2}{0.005} = 25640 = 88\ \mathrm{dB}$$

Furthermore, v_c will produce a differential voltage at the input of the amplifier because of the unbalanced attenuators formed by the bridge output resistances and R_c. If we wish this contribution to be limited as that of G_c, the condition to fulfill is

$$(200\ \mathrm{mV})\left(\frac{116.4\ \Omega}{R_c + 116.4\ \Omega} - \frac{97.5\ \Omega}{R_c + 97.5\ \Omega}\right) \times 128.2 < \frac{0.002 \times (5\ \mathrm{V})}{10}$$

which leads to $R_c > 485$ kΩ.

The finite input differential resistance attenuates v_o,

$$v_d = v_o \frac{R_d}{R_d + 116.4\ \Omega + 97.5\ \Omega} = v_o \frac{R_d}{R_d + 213.9\ \Omega}$$

therefore resulting in a relative error

$$\varepsilon = \frac{213.9 \ \Omega}{R_{\rm d} + 213.9 \ \Omega}$$

If we wish this error to be ten times smaller than the relative error allowed (0.5 %), then we need $R_{\rm d} > 428$ kΩ. Note that $R_{\rm d}$ and $R_{\rm c}$ are not those seen by a differential or common mode voltage.

This example shows that the differential impedance must be high to prevent signal loading and that the common mode impedance must be high to prevent the common mode signal from producing a differential input voltage, hence reducing the effective CMRR (see Problem 3.13). The common-mode to differential-mode conversion increases for large source impedance unbalance. Sensors in quarter-bridges and half-bridges yield signals with unbalanced output resistances. Sensors in full bridges yield balanced signals. If the differential signal has output resistance $R_{\rm o}$ and $R_{\rm o} + \Delta R_{\rm o}$, and the amplifier has common mode input resistance $R_{\rm c}$, the common mode signal yields an output voltage

$$v_{\rm o}|_{v_{\rm c}} = v_{\rm c} \left(\frac{R_{\rm o} + \Delta R_{\rm o}}{R_{\rm c} + R_{\rm o} + \Delta R_{\rm o}} - \frac{R_{\rm o}}{R_{\rm c} + R_{\rm o}} \right) + v_{\rm c} G_{\rm c}$$

$$\approx v_{\rm c} G_{\rm d} \left(\frac{\Delta R_{\rm o}}{R_{\rm c}} + \frac{1}{\text{CMRR}} \right) = \frac{v_{\rm c} G_{\rm d}}{\text{CMRR}_{\rm e}} \tag{3.76}$$

where the approximation is valid when $R_{\rm c} \gg R_{\rm o}$ and CMRR$_{\rm e}$ is the effective CMRR.

If the op amp in Figure 3.32a has a first-order frequency response,

$$A_{\rm d} = A_{\rm d0} \frac{f_{\rm a}}{f_{\rm a} + jf} = \frac{f_{\rm T}}{f_{\rm a} + jf} \tag{3.77}$$

then from (3.68) the differential gain is

$$G_{\rm d} = \frac{k}{1 + (k+1)\dfrac{f_{\rm a} + jf}{f_{\rm T}}} \approx \frac{k}{1 + (k+1)\dfrac{jf}{f_{\rm T}}} = \frac{k}{1 + \dfrac{jf}{f_{G}}} \tag{3.78}$$

where the approximation is valid when $f_{\rm T} \gg (k+1)f_{\rm a}$, and $f_{G} = f_{\rm T}/(k+1)$. For a dynamic measurand, if we wish a gain with amplitude error smaller than ε, we need

$$\frac{\left| \dfrac{k}{\sqrt{1 + \left(\dfrac{f}{f_{G}}\right)^{2}}} - k \right|}{k} < \varepsilon \tag{3.79}$$

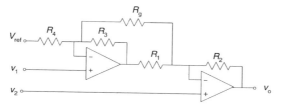

Figure 3.33   Instrumentation amplifier based on two op amps.

which limits the maximal frequency to

$$f_\varepsilon \le \frac{f_G \sqrt{2\varepsilon - \varepsilon^2}}{1 - \varepsilon} \tag{3.80}$$

For a given op amp, the larger the desired differential gain k, the lower f_G, hence f_ε.

3.5.2   Instrumentation Amplifier Based on Two Op Amps

An instrumentation amplifier (IA) is a differential amplifier that simultaneously yields high input impedance and high CMRR. In addition, it usually offers high stable gain that can be adjusted by a single resistor, low value and low drift offset voltage and currents, and low output impedance.

Figure 3.33 shows an IA built from two op amps. We consider the op amps ideal, then repeat the same steps leading to (3.62) to (3.66). The necessary condition to obtain an infinite CMRR is also that expressed by (3.67). The output voltage is then

$$v_o = v_d \left(1 + k + \frac{R_2 + R_4}{R_g} \right) + V_{ref} \tag{3.81}$$

Therefore, although it is also necessary to match four resistors, now R_g allows gain adjustment without affecting the matching of those four critical resistors. But it is not possible to obtain unity gain. If the CMRR is finite, (3.73) applies—by adding V_{ref} to the right-side member. A shortcoming of this circuit is the possible saturation of the first op amp when the common mode input signal is large. The condition to fulfill to avoid that saturation in the usual case with $v_c \gg v_d$ is $v_c(1 + R_3/R_4) < V_{sat}$, where V_{sat} is the op amp saturation voltage. Furthermore, because of the asymmetrical gain path for v_1 and v_2, the common mode gain will never be zero, even if op amps and resistors are perfectly matched. Nevertheless, the CMRR can be very high below 10 Hz.

Example 3.12   The piezoresistive pressure sensor in Figure E3.12 consists of a bridge of four silicon strain gages of 4000 Ω that has 1 mV/psi sensitivity and

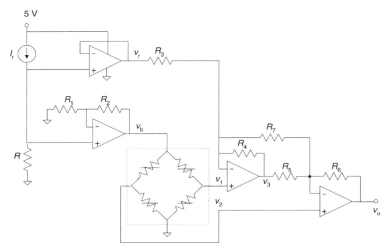

Figure E3.12   Signal conditioner for a piezoresistive pressure sensor.

1 mV maximal offset when supplied at 12 V. $I_r = 100\ \mu\text{A}$ is highly stable. If the op amps are supplied at 0 V and 5 V, calculate the resistors in order for a pressure range from 0 psi to 100 psi to yield an output voltage from 0.5 V to 4 V.

The right-most op amps make up a two-op-amp instrumentation amplifier. By comparing with Figure 3.33, from (3.68) the matching condition for optimal CMRR is

$$\frac{R_3}{R_4} = \frac{R_6}{R_5} = k$$

Then the output is

$$v_\text{o} = (v_2 - v_1)\left(1 + \frac{R_6}{R_5} + \frac{R_3 + R_6}{R_7}\right) + V_\text{r} = (v_2 - v_1)G + V_\text{r}$$

In order to have 0.5 V for zero pressure ($v_1 = v_2$) we need $V_\text{r} = 0.5$ V. Hence, $R = (0.5\ \text{V})/(100\ \mu\text{A}) = 5\ \text{k}\Omega$. In order to have 4 V for 100 psi, we need

$$4\ \text{V} = (v_2 - v_1)G + 0.5\ \text{V} = (100\ \text{psi}) \times (1\ \text{mV/psi}) \times \frac{V_\text{b}}{12\ \text{V}} \times G + 0.5\ \text{V}$$

where V_b is the bridge supply voltage,

$$V_\text{b} = (0.5\ \text{V})\left(1 + \frac{R_2}{R_1}\right)$$

These two equations yield

$$G\left(1 + \frac{R_2}{R_1}\right) = \frac{3.5 \times 12}{0.1 \times 0.5} = 840$$

We can separately select G and the gain for the noninverting amplifier supplying the bridge, but V_b is limited by the op amp saturation voltage and the maximal supply voltage for the bridge. If we select $V_b = 3.5$ V, then we need $G = 120$ and $R_2/R_1 = 6$. If we select $R_3 = R_4 = R_5 = R_6 = 10$ kΩ to simplify the design, then we need

$$1 + 1 + \frac{20 \text{ k}\Omega}{R_7} = 120$$

which requires $R_7 = 169.5$ Ω. We can select $R_7 = 169$ Ω ($\pm 1\%$ tolerance). If we select $R_1 = 10$ kΩ, then R_2 can be the series combination of 59 kΩ and 1 kΩ ($\pm 1\%$).

3.5.3 Instrumentation Amplifiers Based on Three Op Amps

The circuit in Figure 3.34 is the classic implementation for an instrumentation amplifier. It is built from a fully differential amplifier and a differential amplifier acting as differential to single-ended converter. When the three op amps are assumed ideal, the outputs of the first stage are

$$v_a = v_1\left(1 + \frac{R_2}{R_1}\right) - v_2\frac{R_2}{R_1} \tag{3.82}$$

$$v_b = v_2\left(1 + \frac{R_2}{R_1}\right) - v_1\frac{R_2}{R_1} \tag{3.83}$$

By setting $v_1 = v_2$ we find that the common mode gain is 1; hence there is less saturation risk than in the two-op-amp IA. The second stage yields

$$v_o - V_{\text{ref}} = (v_b - v_a)\frac{R_4}{R_3} = \left(1 + \frac{2R_2}{R_1}\right)\frac{R_4}{R_3}(v_2 - v_1)$$

$$= (1 + G)k(v_2 - v_1) \tag{3.84}$$

where $G = 2R_2/R_1$. Because R_1 does not have to fulfill any matching condition, we can control the differential mode gain through R_1 without affecting the CMRR.

In practice we have neither perfect resistor matching nor ideal op amps. This does not have any serious repercussions on input impedances that always reach very high values both in common mode and differential mode. The CMRR_{IA}, however, depends on (a) resistor matching and CMRR_{oa} in the second stage and (b) matching of input op amps (but not resistors R_2). The result is [10]

$$\frac{1}{\text{CMRR}_{\text{IA}}} = \frac{1}{\text{CMRR}_{\text{i}}} + \frac{1}{G+1}\left(\frac{1}{\text{CMRR}_{\text{oa}}} + \frac{1}{\text{CMRR}_{\text{R}}}\right) \quad (3.85)$$

CMRR_{i} is infinite when input op amps have matched differential and common mode gains (A_{d} and A_{c}). Resistor imbalance is quantified by (3.70). It follows that a cost-effective way of implementing an IA is to use a dual op amp at the input stage (to enhance the chances of having them matched) and an IC differential amplifier at the second stage. A large G increases CMRR_{IA} but reduces bandwidth. Equation (3.73) yields the actual output—by adding V_{ref} to the right-side member.

Example 3.13   If we desire a differential mode gain of 1000 and use 5% tolerance resistors, determine how the values for G and k influence the maximal CMRR that can be achieved in a three-op-amp IA.

From (3.85), $1000 = (1 + G)k$. If the three op amps are considered to be ideal, from (3.86) and (3.71) it follows

$$\text{CMRR}_{\text{IA}} = \frac{(1 + G)(k + 1)}{4t_{\text{R}}}$$

Therefore $\text{CMRR}_{\text{IA}} = 1000(k + 1)/0.2k = 5000(k + 1)/k$. When $k = 1$, we achieve 80 dB; when $k = 10$, we achieve about 74 dB. Because op amps can be assumed ideal only at low frequencies, these CMRR values are valid only at low frequencies.

When the circuit in Figure 3.34 is built from discrete parts, we should consider that bipolar input op amps are usually more linear and have lower offset voltage and drifts than FET input op amps. However, FET input op amps have lower bias currents and higher input impedances, which are instrumental in achieving a high CMRR when considering the output impedance of the signal source (see Example 3.11). FET inputs must be protected with a series current-limiting resistor.

The finite offset voltage of op amps yields a zero error. Rewriting (3.84) when $v_1 = V_{\text{io1}}$ and $v_2 = V_{\text{io2}}$, and considering V_{io3} for the op amp at the output stage yields

$$v_{\text{o}}(0) - V_{\text{ref}} = (V_{\text{io2}} - V_{\text{io1}})(1 + G)k + V_{\text{io3}}(k + 1) \quad (3.86)$$

which reinforces the need for matching the op amps of the input stage. Because of the dependence described by (3.86), the specified offset voltage for integrated IAs includes two terms, one related to the input stage and the other related to the output stage (Section 7.1.5).

Table 3.5 lists some IC instrumentation amplifiers that implement the circuits in Figures 3.33, 3.34, and others. Many of them permit gain control

TABLE 3.5 Basic Parameters of Some Instrumentation Amplifiers[a]

Amplifier	V_{io} (µV)	$\Delta V_{io}/\Delta T$ (µV/°C)	I_b (nA)	$\Delta I_{io}/\Delta T$ (pA/°C)	e_G (%)	e_{nIG} (%FSO)	CMRR (dB)	$f_G{}^b$ (kHz)	$t_{st}{}^c$ (µs)
AD621A	35	0.3	0.5	1.5	0.15[d]	0.0002[d]	130	200	20
AD623A	27	0.1	17	25	0.1	0.0005	110	10	20
AD624A	250[d]	2.5[d]	50[d]	50	0.25[d]	0.005[b]	100	150	15
AD627A	50	0.1	3	20	0.15	0.0002	125	3	290
AMP02E	20	0.5	2	9	0.3[d]	0.006	120	200	10
AMP04E	80	3[d]	17	28	0.2	0.025	105	700[g]	—
INA103KP	42	1.2	8000	4×10^4	0.07	0.006	129	800	3.5
INA110KP	110	2.2	0.02	—[e]	0.02	0.004	110	470	4
INA114AP	25.3	1.1	0.5	8	0.05	0.0005	110	10	120
INA116P	500	—	3×10^{-6}	—	0.35	0.001	94	70	145
INA125P	50	0.25	10	60	0.05	0.001	114	4.5	375
INA128P	11	0.2	2	30	0.05	0.001	125	200	9
INA131AP	25	0.25	0.5	8	0.05	—	110	70	100
INA141P	20	0.2	2	30	0.03	0.005	100	110[f]	15[f]
INA155E	200	0.2	2	30	0.03	0.005	125	200	9
LTC1101AM	50	0.4	0.8	0.5	0.008	0.0007	112	3.5	—
LTC1167AC	15	0.05	0.05	0.3	0.025	0.0002	125	120	14
MAX4194	50	0.5[h]	6	15	0.05	0.001	115	1.5	5000[i]

[a] Values are typical for $G = 100$, unless otherwise noted, at 25 °C ambient temperature, but measured at different conditions.

[b] Corner frequency at −3 dB.

[c] Settling time to 0.01 % of the final value.

[d] Maximal value.

[e] Bias current doubles every 10 °C increase.

[f] $G = 50$.

[g] $G = 1$.

[h] $G = 10$.

[i] Settling time to 0.1 %.

181

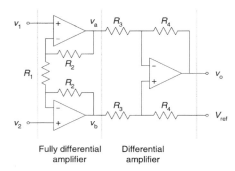

Fully differential   Differential
amplifier   amplifier

Figure 3.34   An instrumentation amplifier based on three op amps has two stages: an input stage with differential input and output, and an output stage that is a differential to single-ended converter.

through R_1. The larger the gain, the narrower the bandwidth. To achieve a large gain without bandwidth reduction, a fully differential amplifier stage such as that in Figure 3.34 can precede the IA. This yields a larger CMRR than adding a single-ended gain stage after the IA. The ADS1250 (Burr–Brown) is a 20 b ADC with a programmable gain IA. The GS 9001 (Goal Semiconductor) includes an instrumentation amplifier, a programmable-gain amplifier, and a variety of support components all on one chip.

Example 3.14   The sensor bridge in Figure E3.14 includes four 350 Ω advance strain gages (gage factor 2.0). The REF102 voltage reference sets a constant 10.0 V between its terminals. The op amp is assumed to be ideal. Calculate the maximal current through each gage if the allowable power dissipation is 250 mW. Calculate R and its power rating in order to limit the bridge current to 25 mA. If the gages are bonded onto steel with $E = 210$ GPa and change in

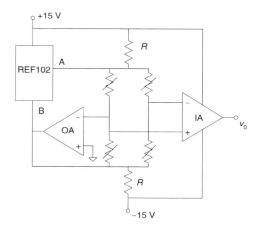

Figure E3.14   Sensor bridge signal conditioning based on an instrumentation amplifier.

opposite directions, calculate the gain G for the IA in order to have 0.5 V when the load is 50 kg/cm². If the IA has offset voltage $(250 + 900/G)$ μV at 25 °C, offset drift $(2 + 20/G)$ μV/°C, thermal resistance $\theta_{ja} = 102$ °C/W, and supply current 8.5 mA (each supply terminal), calculate the error (kg/cm²) when $G = 1000$ and the ambient temperature is 30 °C.

The circuit supplies 10 V to the bridge by using a single voltage reference in spite of the symmetrical power supply because the op amp sets an output terminal to 0 V. To limit power dissipation to 250 mW at each strain gage we need

$$I^2 R < 250 \text{ mW}$$

$$I < \sqrt{\frac{250 \text{ mW}}{350 \ \Omega}} = 26.7 \text{ mA}$$

Because REF102 keeps 10 V between its terminals, the current supplied to the bridge is

$$I_b = \frac{30 \text{ V} - 20 \text{ V}}{2R} = 25 \text{ mA}$$

Therefore, we need $R = 400 \ \Omega$. The power dissipated in each resistor will be

$$P = \frac{V^2}{R} = \frac{(15 \text{ V} - 5 \text{ V})^2}{400 \ \Omega} = 250 \text{ mW}$$

We need $\frac{1}{4}$ W resistors.

If strain gages are arranged as in Figure 3.25a, the bridge output will be

$$v_s = (10 \text{ V}) \left(\frac{1+x}{2} - \frac{1-x}{2} \right) = 10x \text{ V}$$

The fractional change in resistance at each gage will be

$$x = k\varepsilon = k\frac{\sigma}{E} = 2\frac{50 \times 9.8 \times 10^4}{210 \times 10^9} = 46.7 \times 10^{-6}$$

Hence, for the IA we need a gain

$$G = \frac{0.5 \text{ V}}{467 \times 10^{-6} \text{ V}} = 1071$$

The offset voltage of the IA depends on the temperature and the temperature depends on self-heating. The internal temperature will be

$$T_j = T_a + \theta_{ja} \times P_{IA} = 30 \,°\text{C} + (102 \,°\text{C/W})(2 \times 15 \text{ V} \times 8.5 \text{ mA}) = 56 \,°\text{C}$$

and the actual offset voltage

$$v_{io}(56\,^{\circ}C) = \left(250 + \frac{900}{1000}\right)\mu V + \left(2 + \frac{20}{1000}\right)\frac{\mu V}{^{\circ}C} \times (56 - 25)\,^{\circ}C = 313\ \mu V$$

which causes an error

$$\sigma_e = \frac{313\ \mu V}{(467\ \mu V)/(50\ \text{kg/cm}^2)} = 33.5\ \text{kg/cm}^2$$

3.6   INTERFERENCE

3.6.1   Interference Types and Reduction

Interference is defined in Section 1.3.1 as those signals that affect the measurement system as a consequence of the measurement principle used. Here we are concerned with electronic signal conditioning, and therefore interference is any electric signal present at the output of the system or circuit being considered and coming from a source external to it. Interference problems are not exclusive for electronic measurement systems but are also present in any electronic system. In reference 11 there is an excellent analysis of interference problems in general, and reference 12 discusses interference in measurement circuits.

The appropriate technique to reduce interference depends on the coupling method for the undesired signals. Depending on whether the coupling method is through a common impedance, an electric field or a magnetic field, we will respectively speak of resistive, capacitive, and inductive interference.

Figure 3.35 illustrates *resistive interference*. A signal v_s is measured that is ground-referred at a point far from the reference ground for the amplifier. These reference points may be connected to Earth at the respective locations. Therefore, because the ground is used as a return path for leakage currents from electronic equipment, there is always a voltage difference v_i between different grounds. In industrial environments, at least 1 V to 2 V is to be expected. In printed circuits, ground paths can be common to signal and supply currents, hence resulting in a drop in voltage along them.

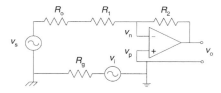

Figure 3.35   Resistive interference due to the drop in voltage produced by stray currents between two distant reference (ground) points or by return currents along a shared impedance.

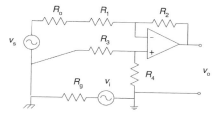

Figure 3.36 Reduction of resistive interference by applying a differential amplifier.

A differential amplifier connected as shown in Figure 3.36 solves resistive interference if the effective CMRR reduces the interference to an output level below that desired. We assume that the common mode voltage at the op amp inputs due to v_i does not exceed the maximal allowed value. An IA would provide increased CMRR.

It may happen, however, that either the available CMRR is not high enough or the common mode voltage is too high or just that in addition to the input amplifier there are other circuits connected to the same reference. All these situations call for other solutions that will be described in the following sections.

Figure 3.37 shows the general problem of *capacitively coupled interference* [11]. Between any pair of conductors there is a finite capacitance. Whenever one conductor is at a certain voltage with respect to a third conductor (the ground plane in Figure 3.37), the second conductor will also increase its voltage with respect to the third conductor. The drop in voltage across the equivalent input resistor R presented by the circuit encountering the interference is

$$V_R = \frac{j\omega RC_{12}}{1 + j\omega R(C_{12} + C_2)} V_1 \tag{3.87}$$

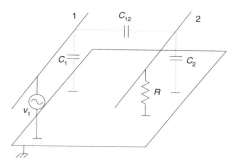

Figure 3.37 Model to describe capacitive coupling between conductors 1 and 2. (From H. W. Ott, *Noise Reduction Techniques in Electronic Systems*, copyright 1988. Reprinted by permission of John Wiley & Sons, New York.)

In low-impedance circuits, $R \ll 1/[\omega(C_{12} + C_2)]$, and

$$V_R \approx j\omega R C_{12} V_1 \tag{3.88}$$

On the other hand, in high-impedance circuits, $R \gg 1/[\omega(C_{12} + C_2)]$, and

$$V_R \approx \frac{C_{12}}{C_{12} + C_2} V_1 \tag{3.89}$$

That is, for low R the interference increases at increasing frequencies, whereas for large R the interference is frequency-independent and larger than when R is low. In both cases the interference increases with C_{12}, which is proportional to conductor length. In measurement systems the usual interference sources are the 60 (or 50) Hz power lines that couple into sensor cables, and therefore the situation is better described by (3.88), particularly when the interfered circuit measures voltage from a source with low output impedance or measures current. To reduce interference coupled to power supply lines, connect a high-value (electrolytic) capacitor (1 μF to 10 μF) shunted by a ceramic capacitor (10 nF to 100 nF) between the output of voltage regulators and ground. These capacitors keep the circuit impedance low in spite of the increasing output impedance of voltage regulators with frequency. Because of their geometry, electrolytic capacitors have increasing impedance from frequencies of about 100 kHz and higher. Ceramic capacitors behave as actual capacitors up to higher frequencies but are not available in large values.

Separating conductors 1 and 2 reduces C_{12}, but not very effectively [13]. Shielding either conductor 1 or 2 is more effective. *Shielding* a conductor or circuit consists of wholly enclosing it by an electrically conductive material connected to a constant voltage. Figure 3.38a shows conductor 2 shielded with a grounded shield. Actually, conductor 2 is not totally enclosed, which is the real situation when there is at least one input and one output with galvanic (ohmic) connection. If $R \gg 1/(\omega C_2)$ at the frequencies considered, the equiva-

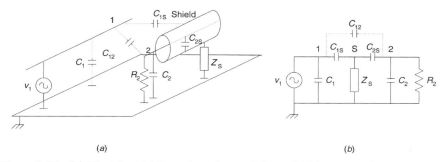

(a)   (b)

Figure 3.38 (a) Electric shielding of conductor 2 by a shield connected to a constant voltage (ground in this case) and (b) equivalent circuit for its analysis when $R \gg 1/(\omega C_2)$. (From H. W. Ott, *Noise Reduction Techniques in Electronic Systems*, copyright 1988. Reprinted by permission of John Wiley & Sons, New York.)

lent circuit is that in Figure 3.38b. If the impedance of the shield-to-ground connection Z_S is low enough, we have

$$V_R \approx \frac{C_{12}}{C_{12} + C_{2S} + C_2} V_1 \tag{3.90}$$

where C_{12} is now much smaller than when no shield is used because it concerns only those segments outside of the shield (which is considered as perfect). Then the final interference will be greatly reduced. In practice, conductors are enclosed in a wire mesh whose effective shielding or coverage factor depends on how closely it is woven. In view of the simplifications leading to (3.90) and by considering Figure 3.38b, we conclude that shields are effective when $Z_S \ll 1/(\omega C_{1S})$.

If $R \leq 1/(\omega C_2)$ but Z_S is small enough, we obtain

$$V_R \approx \frac{j\omega R C_{12}}{1 + j\omega R(C_{12} + C_{2S} + C_2)} V_1 \tag{3.91}$$

For $R \ll 1/[\omega(C_{12} + C_{2S} + C_2)]$ we have

$$V_R \approx j\omega R C_{12} V_1 \tag{3.92}$$

That is, the interference is directly proportional to C_{12}, which is now very small.

Shields are effective only if connected to a constant voltage. Otherwise, even if C_{12} were zero, interference would result. For the case analyzed, if we take $Z_S = \infty$ and we suppose, for example, the situation where R is large, we have

$$V_R \approx V_S \approx V_1 \frac{C_{1S}}{C_{1S} + C_S} \tag{3.93}$$

That is, if C_{1S} is large, the resulting interference may exceed that without shielding. The shield must thus be connected to a constant voltage. We must decide which end of the shield to connect to which voltage. The following sections answer these questions.

There is an *inductive coupling* or a *magnetic interference* when the magnetic field produced by the current in a circuit induces a voltage in the signal circuit being considered. The relationship between the current in a circuit and the magnetic flux it produces in another is given by the mutual inductance M,

$$M = M_{12} = M_{21} = \frac{\Phi_{12}}{I_1} = \frac{\Phi_{21}}{I_2} \tag{3.94}$$

In case of a variable magnetic flux $\boldsymbol{B}$, the voltage v_2 induced in a loop with area $\boldsymbol{S}$ is

$$v_2 = -\frac{d}{dt}\int_S \boldsymbol{B} \cdot d\boldsymbol{S} \tag{3.95}$$

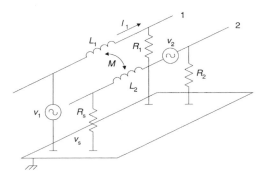

Figure 3.39   Model to describe inductive coupling between circuit 1 and circuit 2. (From H. W. Ott, *Noise Reduction Techniques in Electronic Systems*, copyright 1988. Reprinted by permission of John Wiley & Sons, New York.)

If the loop is static and **B** changes sinusoidally at frequency ω, we have

$$V_2 = j\omega BS \cos \theta \tag{3.96}$$

where θ is the angle between **B** and **S**.

Therefore, in a way similar to the case of capacitive interference, a current I_1 circulating along a conductor induces an interfering voltage V_2 in a circuit such as that in Figure 3.39, as given by (3.96). But now the interference is always proportional to the frequency (for capacitive coupled interference there is proportionality only at low frequencies) and is independent of the impedance presented by the receiving circuit (capacitive interference increases with increasing circuit impedance).

If a reduction in **B** is not possible, the usual solution to reduce magnetic interference is by reducing the area **S**. This is done by twisting leads or by placing the conductor close to the return path, if the return path is not a wire conductor. Sometimes it is also possible to reduce the $\cos \theta$ term by reorienting the circuit. Note that a conductive shield around conductor 2 does not solve the problem: The shield will be raised to a voltage level $V_S = j\omega M_{1S}I_1$, or we will have $V_S = 0$ if one end is tied to ground, but V_2 will not decrease.

3.6.2   Signal Circuit Grounding

A *ground* is a point or equipotential plane that serves as a reference for the voltages in a circuit or system. When grounding a circuit or system, we must minimize the noise voltages generated by currents flowing between circuits through a shared impedance. We must avoid ground loops, because they are susceptible to magnetic interference and to voltage differences between different grounding points. Figure 3.40 shows three different grounding methods and the respective circuits to analyze them.

In the series ground connection method, supply currents for each circuit produce drops in voltage that result in a different voltage reference for each

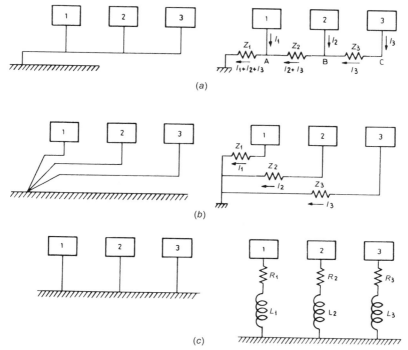

Figure 3.40 Different grounding methods and equivalent circuits to analyze them: (*a*) single-point series grounding; (*b*) single-point parallel grounding; (*c*) multipoint parallel grounding. (From H. W. Ott, *Noise Reduction Techniques in Electronic Systems*, copyright 1988. Reprinted by permission of John Wiley & Sons, New York.)

circuit, namely,

$$V_A = (I_1 + I_2 + I_3)Z_1 \tag{3.97a}$$

$$V_B = (I_1 + I_2 + I_3)Z_1 + (I_2 + I_3)Z_2 \tag{3.97b}$$

$$V_C = (I_1 + I_2 + I_3)Z_1 + (I_2 + I_3)Z_2 + I_3Z_3 \tag{3.97c}$$

Because the output signals for each circuit are voltage-referenced to different points, this interference source may be important. Therefore this grounding method should not be used whenever there are circuits with dissimilar supply currents. In any case, the more susceptible stages should be placed close to the common reference point.

Parallel grounding at a single point (Figure 3.40*b*) requires a more involved physical layout but overcomes the problem pointed out for series grounding. Therefore it is the preferred method for low-frequency grounding.

For high-frequency circuits (>10 MHz), multiple grounding points (Figure 3.40*c*) are preferred to single-point grounding because a lower ground impedance is obtained. Ground plane impedance can be further reduced by plating its surface.

3.6.3   Shield Grounding

Section 3.6.1 points out that the shield of a conductor is effective only when it is connected to a constant voltage. When shielding amplifiers, the shield must be connected to the reference voltage for the enclosed circuit, whether it is grounded or not. Figure 3.41a shows the correct connection.

If the shield were not connected or connected to a different voltage, there would be a parasitic feedback from the amplifier output to its input that could even lead to oscillations. Figure 3.41b shows the case where the shield is left unconnected. Figure 3.41c shows the equivalent circuit for its analysis. The circuit may oscillate because of the stray feedback path through C_{1S}.

Figure 3.41d shows that grounding the reference point for an amplifier when its shield is not connected does not reduce external interference v_i. The equivalent circuit in Figure 3.41e shows that coupling from v_i to the shield is minimal when C_{2S} is very small; that is, it must be short-circuited.

When grounding amplifier shields, the internal circuit must be connected to

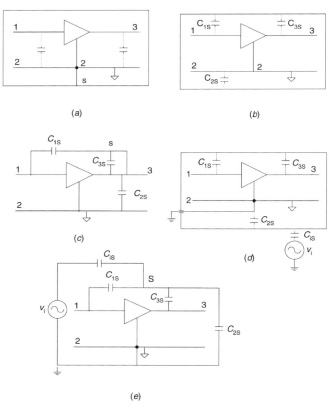

Figure 3.41   Amplifier shielding: (a) correct shield connection; (b) incorrect situation (shield unconnected); (c) circuit to analyze the previous case; (d) grounding does not solve the problem; (e) circuit to analyze the previous case.

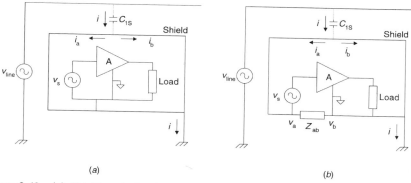

Figure 3.42 (a) Shield and circuit must be connected at a single point; (b) otherwise resistive interference $v_a - v_b$ may appear.

the shield at a single point—for example, as shown in Figure 3.42a for a grounded shield intended to reduce power line interference. Otherwise, interference currents along the shield may produce resistive interference (Figure 3.42b).

We must choose the single connection point carefully in order to avoid currents coupled to the shield from circulating along the same path as signal currents. For example, if the signal were grounded and the shield–amplifier connection were as shown in Figure 3.43a, then the interfering voltage v_i would couple a current to ground through C_{iS} via S–2–b; that is, it would share the path 2–b with the signal. Thus we should choose a grounding scheme such as the one in Figure 3.43b, where the reference point for the amplifier ("2") is connected to the shield not directly in the amplifier but at the signal source. Then external interference does not share any path with the signal. Note that the solution in Figure 3.43b needs an amplifier with a "floating" input; that is, point 2 is not grounded within the amplifier.

When grounding a cable shield with a single ground connection, we must decide which end to connect: the one at the signal end or the one at the ampli-

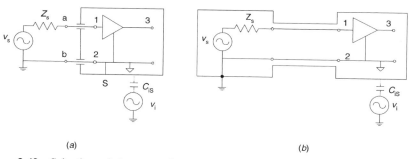

Figure 3.43 Selection of the grounding point for a shield. In case (a), the interference induces a current that shares a path 2–b common to the signal; in case (b), external interference follows a path different from that of the signal.

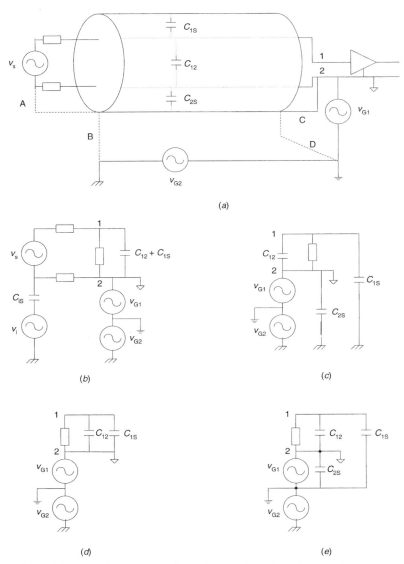

Figure 3.44   (*a*) Ground connection for a shield cable when the signal is not grounded and the amplifier is. The appropriate connection is indicated by a solid line. (*b–e*) Equivalent circuits.

fier end. If the signal is not grounded and the amplifier is, the best solution is to connect the shield to the input reference terminal for the amplifier (Figures 3.44*a* and 3.44*d*). If the shield were connected to the reference terminal at the signal side (connection A, dashed), all interference currents coupled to the shield would flow to ground along the signal lead wire connected to terminal 2

(Figure 3.44b)—the amplifier is assumed to have a high input impedance. If connection B were used, from Figure 3.44c the interfering voltage at input of the amplifier would be

$$V_{12} = (V_{G1} + V_{G2}) \frac{Z_{12}}{Z_{12} + Z_{1S}} = (V_{G1} + V_{G2}) \frac{C_{1S}}{C_{12} + C_{1S}} \qquad (3.98a)$$

If the shield were grounded at the amplifier side, connection D, from Figure 3.44e the interfering voltage would be

$$V_{12} = V_{G1} \frac{Z_{12}}{Z_{12} + Z_{1S}} = V_{G1} \frac{C_{1S}}{C_{12} + C_{1S}} \qquad (3.98b)$$

Therefore, if the signal source is not grounded but the amplifier is, the shield must be connected to the reference terminal for the amplifier, even if it is not grounded.

If the signal is grounded but the amplifier input is not, it is better to ground the shield on the signal source end as shown in Figures 3.45a and 3.45b. If instead of that it were connected to ground at the signal end (connection B, Figure 3.45c), we would have the same interference given by (3.98b). We should not connect the shield to the reference terminal at the amplifier input (connection C, Figure 3.45d), because then all currents coupled to the shield would flow to ground along one of the signal lead wires. If connection D were used, from Figure 3.45e the input interfering voltage would be that given by (3.98a).

Note that the situation for connection A in Figure 3.45 is similar to that in Figure 3.43, but including a nonperfect grounding connection and an interfering voltage between the signal reference point and ground, which are connected by a low-value impedance.

If both the signal source and the amplifier are grounded, perhaps the compromise solution is to connect the shield to ground at both ends. But depending on the difference in voltage between grounding points and on the magnetic coupling to the newly created ground loop, the resulting interference may be increased. If this were the case, the loop must be opened by using differential input amplifiers or isolation amplifiers.

3.6.4 Isolation Amplifiers

An isolation amplifier is an amplifier that offers an ohmic isolation between its input and output terminals. This isolation must have low leakage and a high dielectric breakdown voltage—that is, high resistance and low capacity. Typical values for these are, respectively, 1 TΩ and 10 pF.

Isolation amplifiers are interesting because instrumentation amplifiers withstand a limited common mode voltage, usually about 10 V. Measurement situations encountering high common mode voltages arise in obvious cases such as

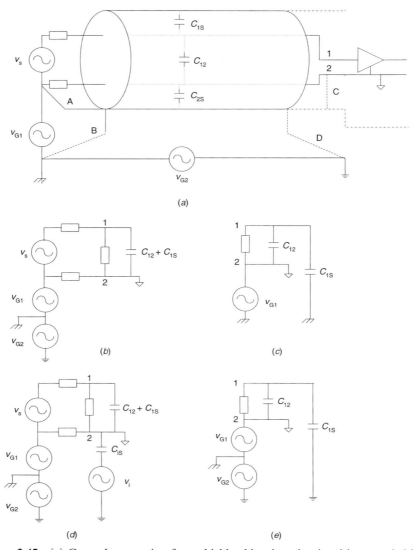

Figure 3.45   (*a*) Ground connection for a shield cable when the signal is grounded but the amplifier is not. The appropriate connection is indicated by a solid line. (*b–e*) Equivalent circuits.

in a high-voltage device. They may also arise in unsuspected cases as a sensor bridge supplied by more than 20 V or when two grounding points are involved whose voltage difference amounts to several tens of volts. In medical electronics, safety standards forbid making any connection to the patient's body leading to a dangerous current through him or her in the event of a contact with a live conductor.

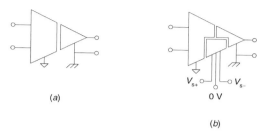

Figure 3.46   The different symbols used for isolation amplifiers indicate that there is no ohmic continuity from the input to the output.

In isolation amplifiers, there is no ohmic continuity from the input reference terminal (input common, input ground) to the output reference terminal (output common, output ground). The input common is also independent of the reference terminal of the power supply (supply common, supply ground). In some cases, the power supply is also independent from the output common. Figure 3.46 gives some of the symbols used for these.

Signals and supply power are magnetically coupled from one part to another of isolation amplifiers by transformers with low interwinding capacitance. Signals can be also coupled by optocouplers or series capacitors. A modulated carrier is used through the isolation barrier in order to improve linearity. The ability for rejecting those voltages appearing between the input common and the other common terminals is quantified by the *isolation-mode rejection ratio* (IMRR) defined in a way similar to the CMRR as the voltage across the isolation barrier times the gain divided by the output voltage it produces. The IMRR is usually expressed in decibels and decreases at a rate of 20 dB/decade starting from values as high as 160 dB at 1 Hz.

Note that an isolation amplifier is not an op amp, a differential amplifier, or an instrumentation amplifier. In fact, there are IC models whose input stage is an uncommitted op amp that can be connected as needed; other models have an input stage that is an instrumentation amplifier. IC isolation amplifiers are not usually precision devices. Nevertheless, they suit precision sensor signal conditioning when provided an isolated supply voltage able to supply a high-quality preamplifier and to excite a sensor bridge or voltage divider (see Problem 3.22). The AD102/4, AD202/4, AD210, AD215 (Analog Devices) and the ISO series (Burr–Brown) are isolation amplifiers.

Table 3.6 shows that isolation amplifiers are compatible with any signal type if they have a high enough isolation impedance. Differential amplifiers are compatible with single-ended and differential signals but must withstand the common mode voltage and reject it to a level compatible with the desired resolution. Single-ended amplifiers are only compatible with floating signals. They may also be compatible with single-ended grounded signals, provided that there is no conducted interference.

TABLE 3.6 Compatibility Between Signal Sources and Conditioners

Conditioner Input \ Signal Source	(single-ended grounded input)	(differential input, guarded)	(differential input)	(differential input, grounded)
(grounded source)	Incompatible unless grounds are very close	Compatible if CMRR is large	Compatible	Compatible
(floating/driven source)	Compatible	Compatible	Compatible	Compatible
(isolated source)	Incompatible unless grounds are very close	Compatible if CMRR is large	Compatible for large Z_i	Compatible

	Incompatible	Compatible	Compatible for large Z_i	Compatible
	Compatible	Compatible	Compatible	Compatible
	Incompatible	Compatible if CMRR is large	Compatible for large Z_i	Compatible

Note: Ground points for grounded signal sources and amplifiers are assumed to be different but connected to each other. Isolation impedance is assumed to be very high for (floating) signal sources but finite (Z_i) for conditioners.

Source: J. G. Webster (ed.), *The Measurement, Instrumentation, and Sensors Handbook*, copyright 1999. Reprinted by permission of CRC Press, Boca Raton, FL.

3.7   PROBLEMS

3.1 The circuit in Figure P3.1 is part of a thermometer that uses a Pt100
as sensor (100 Ω and $\alpha = 0.004\ \Omega/\Omega/K$ at $0\,°C$). The reference current
(200 µA) is highly stable, and the op amp and FET amplify it without
any significant error. The ADC is a 24 b sigma–delta model with differential input and 800 mV input range. If the temperature span is from
$-50\,°C$ to $+150\,°C$, calculate R_1/R_2 in order to obtain the desired
800 mV range and determine the theoretical temperature resolution.

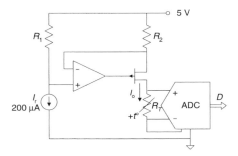

Figure P3.1   Thermometer circuit based on an RTD excited by a constant current and a
high-resolution ADC that does not need signal amplification.

3.2 Figure P3.2 shows a signal conditioner for a remote Pt100 (100 Ω and
$\alpha = 0.003912\ \Omega/\Omega/K$ at $0\,°C$) supplied at constant current and intended
to measure from $15\,°C$ to $250\,°C$. Calculate R in order to supply 1 mA to
the probe. If lead wires have finite and equal resistance, determine the
condition to fulfill by R_1, R_2, and R_3 in order to have FSO = 5 V, independent of wire resistance.

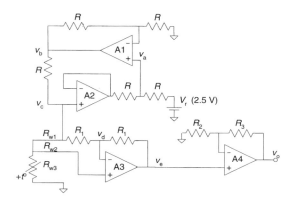

Figure P3.2   Thermometer circuit based on a three-wire Pt100 probe excited by constant current.

3.3 The PTC thermistor in Figure P3.3 has 813.5 Ω at 0 °C, 1000 Ω at 25 °C, and 1211 Ω at 50 °C. R linearizes its response and the circuit around the first op amp supplies a constant current from a single 3.3 V power supply. If the range of the temperature to measure is from 0 °C to 50 °C, calculate R to have an approximate linear response. Determine R_b to limit the sensor current to 400 µA. If the desired output is 0.1 V at 0 °C and 3.1 V at 50 °C, calculate R_1, R_2, and R_3. If the ambient temperature at the circuit location can rise up to 40 °C and the op amp has a type D package, determine the error (°C) resulting from op amp offset voltage.

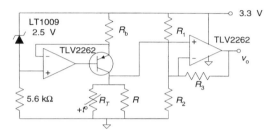

Figure P3.3 Thermometer circuit based on a linearized PTC thermistor supplied by a constant current.

3.4 The data acquisition system IC in Figure P3.4 accepts a differential input and also a differential reference voltage, which permits the direct implementation of the two-reading resistance measurement method. The sensor is a thin-film Pt1000 that has 1000 Ω and $\alpha = 0.00385$ $\Omega/\Omega/K$ at 0 °C, and $\delta = 7$ mW/K. If the ADC has 12 b resolution and the span of the temperature to measure is from 0 °C to 600 °C, determine R_r, the end-of-scale digital outputs, and the maximal self-heating error.

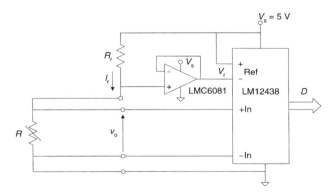

Figure P3.4 Two-reading resistance measurement method implemented by an IC data-acquisition system with differential input and (external) reference voltage.

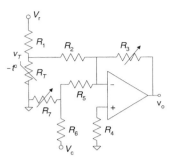

Figure P3.5 Thermometer circuit based on an NTC thermistor and two voltage dividers whose outputs are added and amplified.

3.5 The NTC thermistor in Figure P3.5 has 10,000 Ω at 25 °C, 29,490 Ω at 0 °C, and 3893 Ω at 50 °C. Design the circuit in order to have 0 V at 0 °C, 0.5 V at 50 °C, and less than 0.5 mA in the NTC thermistor when $V_c = -15$ V and $V_r = 5$ V.

3.6 We wish to measure a temperature from 0 °C to 50 °C with 0.25 °C resolution using the circuit in Figure P3.6. The platinum sensor has 1000 Ω and $\alpha = 0.00375$ $\Omega/\Omega/K$ at 25 °C, and the ADC following the amplifier has a 0 to 2 V input range. Determine R_1, R_2, R_p, and V_{ref} in order to limit the sensor current to 50 µA and to achieve the desired output voltage.

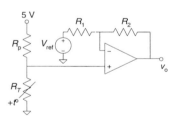

Figure P3.6 Temperature measurement by an RTD placed in a voltage divider.

3.7 The output signal of a potentiometer is connected to a recorder whose input resistance is 10 kΩ. The nonlinearity error due to loading effect must be lower than 1% FSO. A series of 5 W potentiometers with resistance from 100 Ω to 10,000 Ω in 100 Ω increments is available. What unit would give the maximal sensitivity without exceeding any of the imposed restrictions? What would its sensitivity be if they were single-turn models (360°)?

3.8 A method of reducing the nonlinearity error due to meter loading effect in a potentiometer is by placing a resistor in series with the power supply and the potentiometer. Determine the wiper position where the nonlinearity error is maximum and give the expression for the error as a function of resistance ratios.

3.9 Show that the potentiometer circuit with split power supplies in Figure 3.8 has a reduced absolute error as compared to the circuit in Figure 3.7a.

3.10 A given quantity x ranging from $x = 0$ to $x = 10$ is to be measured by means of a linear resistance sensor such that for $x = 0$ its resistance is 1000 Ω and for $x = 10$ it is 1100 Ω. In order to obtain an electric output signal corresponding to x, the sensor is placed in a resistance bridge supplied by a dc voltage, whose value is limited by the maximal power dissipated by the sensor specified at 25 mW.

a. Assume that for $x = 0$ the bridge is balanced and that bridge resistors are chosen for the maximal bridge sensitivity for a given supply voltage. Calculate the maximal relative error that would be produced when the bridge output is considered to be linearly dependent on x with sensitivity equal to that for $x = 0$.

b. Assume a balance condition for $x = 0$ and that the relative error must be kept below 1%. What values should the bridge resistors have?

c. Assume that x is a force, that the bridge is supplied by the maximal acceptable voltage, and that its output is linear. What would the sensitivity be for the previous case?

d. Assume the bridge output is linear. What would the sensitivity be if the four bridge resistors were equal? Explain why it is different from the sensitivity in the previous case.

3.11 Assume that in the previous problem the three fixed resistors are valued 1000 Ω, that the sensitivity is 25 mV/N, and that the output voltage is measured with an instrumentation amplifier where a 0 V to 5 V output should correspond to the range $x = 0$ to $x = 10$.

a. Calculate the gain G for an ideal amplifier.

b. Assume an amplifier with CMRR = 70 dB + 20 lg$(G+1)/2$ and equal input differential and common mode resistances. Assume that other error sources (offset, drifts, noise) are negligible. Then the value for G calculated in the previous point will not give 5 V when $x = 10$ but error voltages will be present. If the bridge supply and the amplifier have a common reference terminal, calculate the relative error when $x = 10$ as a function of the input differential mode resistance. Would this error be zero if that resistance were infinite? Why?

3.12 A given load cell has four 250 Ω strain gages bonded on steel ($E = 210$ GPa) and connected in full bridge as in Figure 3.25a. The gage factor is 2.0 and resistance tolerance 0.3%. If the bridge is supplied at 10 V, determine the output voltage when the applied load is 100 kg/cm². If there is no provision for bridge balance at rest, what would the maximal error (kg/cm²) be because of resistor tolerance?

3.13 Calculate the effective CMRR for a differential amplifier connected to a full bridge built from 350 Ω strain gages whose maximal change is 1% if the common-mode input resistance is 110 kΩ and CMRR = 86 dB. If the bridge is supplied at 12 V, and the amplifier has $G = 10$, calculate the output voltage at null condition and when the strain gages undergo their maximal change.

3.14 A given platinum RTD probe has a resistance of 1000 Ω and $\alpha = 0.004 \ \Omega/\Omega/K$ at 25 °C, and $\delta = 5$ mW/K. Use it to design a thermometer for the range from 0 °C to 100 °C having the maximal possible sensitivity but without exceeding a 1% relative error of the output voltage. Use a bridge circuit and assume that its output is measured with an ideal voltmeter.

3.15 The circuit in Figure P3.15 is proposed for a thermometer based on the TSP 102 sensor (a linearized PTC thermistor) whose resistance at 25 °C is 1000 Ω and its approximate temperature coefficient is 0.007/°C. Design the values for R and R_2 in order to measure temperatures from -10 °C to $+50$ °C.

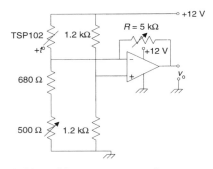

Figure P3.15   Pseudobridge with current output for temperature measurement.

3.16 Use the circuit in Figure P3.16 to measure temperatures in the range from 0 °C to 40 °C, with a corresponding output voltage from 0 V to

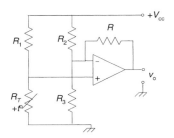

Figure P3.16   Pseudobridge with current output for temperature measurement.

10 V. The sensor is a linearized PTC thermistor having $\alpha = 0.0075/\text{K}$, a resistance of 2000 Ω at 25 °C, and a maximal acceptable current of 1 mA. The op amp is assumed to be ideal. Design the circuit components in order to obtain the output signal desired. Determine the temperature where the nonlinearity error is maximal, and calculate this error.

3.17 The circuit in Figure P3.17 is a pseudobridge based on two equal linear resistance sensors. Assume the op amp is ideal and show that the output voltage is directly proportional to the measured quantity.

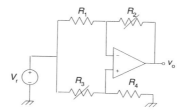

Figure P3.17   Wheatstone bridge linearization by an op amp.

3.18 The pseudobridge in Figure P3.18 includes a 350 Ω strain gage, gage factor 2.0, bonded on steel ($E = 210$ GPa), able to dissipate up to 20 mW. If we wish a zero output in null condition, calculate the maximal V_r in order not to exceed the self-heating limit. If we select $V_r = 2.5$ V, and op amps are assumed to be ideal, determine the gain in order to have a 10 mV output when the applied load is 100 kg/cm^2. If each op amp has a maximal offset voltage of 100 μV, calculate the maximal error (kg/cm^2) in a worst-case condition.

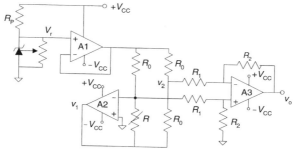

Figure P3.18   Wheatstone bridge linearization with an op amp and differential output.

3.19 The pseudobridge in Figure P3.19 includes a 350 Ω, isoelastic strain gage (gage factor 3.5) bonded on aluminum ($E = 70$ GPa), which accepts up to 15 mA. The op amps have low offset voltage and drift and their

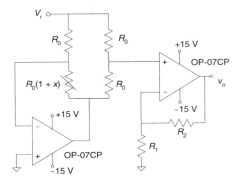

Figure P3.19 Wheatstone bridge linearization with an op amp and single-ended output.

maximal output current is 10 mA. If $V_r = 1.5$ V, calculate the gain G in order to obtain 1 V when the load is -100 kg/cm² (compression). If we select $G = 1000$, calculate the error resulting from offset voltages when the ambient temperature is 30 °C.

3.20 The pressure sensor in Figure P3.20 has a sensitivity of 0.04 mV/V/kPa. R_8 is used for zero balance and will be ignored here. Determine the gain for the amplifier in order to obtain an output from 0.5 V to 4.5 V when the span of the input pressure is from 0 kPa to 100 kPa. Determine the condition to be fulfilled so that the output is independent of the common mode voltage. Design resistors R_1 to R_7.

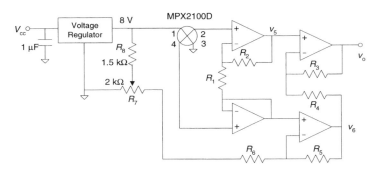

Figure P3.20 Signal conditioner for a piezoresistive pressure sensor.

3.21 The five-terminal bridge in Figure P3.21 includes four 350 Ω strain gages able to dissipate up to 250 mW, bonded on a diaphragm and connected so that to increase the pressure sensitivity. Their maximal fractional change is 0.02. Calculate V_r and the resistors so that a 0 to 0.02 fractional change yields a 0 to 10 V output. Determine the nonlinearity error (in

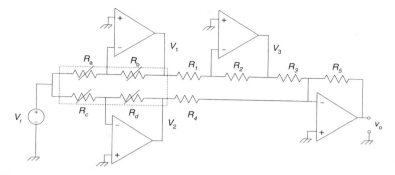

Figure P3.21   Signal conditioner for a piezoresistive pressure sensor.

voltage) and the error resulting from resistor tolerance (R_1 to R_5) when $x = 0.02$. Use 1 % resistors.

3.22 Measuring the supply current in a dc motor involves high voltages. Figure P3.22 shows how to apply an isolation amplifier to prevent any damage or accident when measuring the drop in voltage in series with the motor winding. To achieve high accuracy, the ISO102 ($G = 1$) is preceded by a precision op amp (OPA27), supplied from a ±15 V isolated power supply (not shown), which has 5 % ripple. If the maximal drop in voltage across R is 50 mV and we wish a -10 V output, determine R_1, R_2, and the absolute error resulting from offset voltage, bias current, PSRR, and gain errors for the op amp and the ISO102. (Consult their specifications at Burr–Brown's web site www.burr-brown.com).

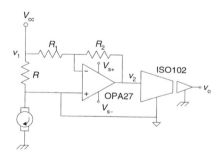

Figure P3.22   Supply current measurement by a series resistor and an isolation amplifier preceded by a precision op amp.

REFERENCES

[1] R. Pallàs-Areny and J. G. Webster. *Analog Signal Processing*. New York: John Wiley & Sons, 1999.

[2] K. F. Anderson. The new current loop: an instrumentation and measurement circuit topology. *IEEE Trans. Instrum. Meas.*, **46**, 1997, 1061–1067.

[3] C. D. Todd (Bourns Inc.). *The Potentiometer Handbook*. New York: McGraw-Hill, 1975.

[4] Anonymous. Differential and multiplying digital-to-analog converter applications. Application Note AN–19. Norwood, MA: Analog Devices. Available at www.analog.com.

[5] Anonymous. Wheatstone bridge nonlinearity. Measurements Group Technical Note 507. Raleigh, NC: Vishay, 1999. Available at www.measurementsgroup.com.

[6] C. D. Johnson and C. Chen. Bridge-to-computer data acquisition system with feedback nulling. *IEEE Trans. Instrum. Meas.*, **39**, 1990, 531–534.

[7] M. C. Headley. Effects of lead wires in 2- and 3-wire quarter-bridge circuits. *Measurements & Control*, Issue 190, September 1998, 148–154.

[8] J. Fraden. *Handbook of Modern Sensors, Physics, Design, and Applications*, 2nd ed. Woodbury, NY: American Institute of Physics, 1997.

[9] F. M. L. Van der Goes and G. C. M. Meijer. A universal transducer interface for capacitive and resistive sensor elements. *Analog Integrated Circuits and Signal Processing*, **14**, 1997, 249–260.

[10] R. Pallàs-Areny and J. G. Webster. Common mode rejection ratio for differential amplifier stages. *IEEE Trans. Instrum. Meas.*, **40**, 1991, 669–676.

[11] H. W. Ott. *Noise Reduction Techniques in Electronic Systems*, 2nd ed. New York: John Wiley & Sons, 1988.

[12] R. Morrison. *Instrumentation Fundamentals and Applications*. New York: John Wiley & Sons, 1984.

[13] C. S. Walker. *Capacitance, Inductance, and Crosstalk Analysis*. New York: Artech House, 1990.

4

REACTANCE VARIATION AND ELECTROMAGNETIC SENSORS

Reactance variations in a component or circuit offer alternative measurement methods from those available with resistive sensors. Many reactance variation measurement methods do not require any physical contact with the system to be measured, or when they do, exert a minimal mechanical loading effect. In particular, they offer alternative solutions to those described in Chapter 2 for the measurement of linear or rotary displacements of ferromagnetic materials and for humidity measurement.

The inherent nonlinearity of some of the measurement methods used in this kind of sensors is overcome through differential sensors. On the other hand, these methods limit the maximal frequency for the measurand because it must be at least ten times lower than the excitation frequency, which must be an alternating voltage or current.

Some electromagnetic sensors are in fact self-generating sensors, but they are discussed in this chapter because of the similarity between their output signal and that of some variable reactance sensors.

4.1 CAPACITIVE SENSORS

4.1.1 Variable Capacitor

A capacitor consists of two electric conductors separated by a dielectric (solid, liquid, or gas) or a vacuum. The relationship between the charge Q and the difference in voltage V between them is described by means of its capacitance, $C = Q/V$. This capacitance depends on the geometrical arrangement of the conductors and on the dielectric material between them, $C = C(\epsilon, \mathrm{G})$.

For example, for a capacitor formed by n equal parallel plane plates having an area A, with a distance d between each pair, and an interposed material with a relative dielectric constant ϵ_r, the capacitance is

$$C \approx \epsilon_0 \epsilon_r \frac{A}{d}(n-1) \tag{4.1}$$

where $\epsilon_0 = 8.85$ pF/m is the dielectric constant for vacuum. Therefore, any measurand producing a variation in ϵ_r, A, or d will result in a change in the capacitance C and can be in principle sensed by that device. Baxter [1] gives the expressions for the capacitance of several electrode arrangements useful for sensor design. The condenser microphone described by E. C. Wente in 1917 is perhaps the earliest capacitive sensor.

The relative permittivity ϵ_r for air is nearly 1, and for water it changes from 88 at 0 °C to 55.33 at 100 °C. Therefore the substitution of water for air as dielectric results in a noticeable change. This can be applied, for example, to level measurement for water in a tank or to humidity measurement by using a dielectric that absorbs and exudes water without hysteresis. The capacitance of the HC1000 (E+E Electronik) humidity sensor, for example, is

$$C = C_{76}[1 + \alpha_{76}(\text{RH} - 76)] \tag{4.2}$$

where $C_{76} = 500$ pF $\pm$ 50 pF and $\alpha_{76} = (2900 \pm 150) \times 10^{-6}/(\% \text{ RH})$. The linearity error is less than 2% RH.

The dielectric constant for ferroelectric materials above the Curie temperature (T_C) is proportional to the reciprocal of the temperature according to

$$\epsilon = \frac{k}{T - T_C} \tag{4.3}$$

where k is a constant. Thus we can measure the variation in temperature by measuring the change in the capacitance of a device that includes such a material.

The use of a variable capacitor as a sensor has several limitations. In first place, fringe effects are usually neglected in the expression for the capacitance, and this may not always be acceptable. For a parallel plate capacitor, fringe effects are negligible if the distance between plates is far smaller than their linear dimensions. Otherwise, (4.1) is no longer valid. Correction factors depend on electrode geometry [2, 3].

Figure 4.1 shows how to reduce fringe effects without changing geometrical relations. It consists of using guards that are connected to a constant voltage so that electric field flux lines remain confined into the volume defined by the sensing electrode. Capacitance correction because of the finite gap width w

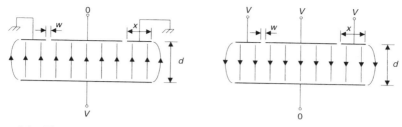

Figure 4.1 The outer guard ring in a capacitive sensor is kept at the same voltage as one of the two electrode plates to reduce fringing fields.

depends on w/d (d is the electrode distance) and on electrode thickness [4]. The gap width x needed to achieve a relative error lower than a is $w = -(d \ln a)/\pi$ [5].

The insulation between plates must be high and constant. For example, varying humidity may introduce leakage resistance in parallel with C as a result of insulation resistance changes in the dielectric. Then there would be impedance changes in the capacitor not attributable to a change in capacitance. Hence, measurement methods sensitive only to impedance modulus but not to its phase could cause important errors. Polar dielectrics, such as water, acetone, and some alcohols, have a relatively high conductivity. Power dissipation in the equivalent resistance may lead to thermal interference. Nonpolar dielectrics such as oils have a very low conductivity.

Because only one of the two conductive surfaces can be grounded, there is a risk of capacitive interference (as in Figure 3.37). Shielding of sensor plates and wires connected to them reduces that interference (Section 5.2).

Connecting wires are another possible error source. Shielding them to prevent capacitive interference adds a capacity in parallel with the sensor. This results in a loss of sensitivity because the measured quantity will change only the capacitance of the sensor that now is only a part of the total capacity. A relative movement between cable wires and the interposed dielectric can become a source of error if there is a noticeable change in geometry or if the dielectric in the cable has piezoelectric properties (Section 6.2.1).

Capacitive sensors are linear or nonlinear, depending on the parameter that changes and whether we measure the capacitive impedance or admittance. In a parallel plate capacitor, for example, the output voltage is linear when we measure the admittance (proportional to C) if ϵ_r or A change, but it is nonlinear if the measurand changes the separation between plates, be it of the form $C = \epsilon A/z$ or $C = \epsilon A/(d + z)$. In this second case we have

$$C = \epsilon \frac{A}{d(1 + x)} \qquad (4.4)$$

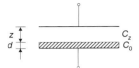

Figure 4.2   Interposition of an additional dielectric in a parallel plate capacitive sensor reduces nonlinearity.

where $x = z/d$. By taking the derivative of this equation to find the sensitivity, we have

$$\frac{dC}{dz} = \frac{-\epsilon A}{d^2(1+x)^2} = \frac{-\epsilon C_0}{d(1+x)^2} \approx -\frac{C_0}{d}(1 - 2x + 3x^2 - 4x^3 + \cdots) \qquad (4.5)$$

Hence, the sensor is nonlinear because the sensitivity instead of being constant depends on z and increases when d and z are small. This might suggest capacitors with a very small d, but there is a minimal separation determined by dielectric breakdown, which is 30 kV/cm for air.

For a sensor of the kind $C = \epsilon A/z$, the sensitivity is $-\epsilon A/z^2$, also nonlinear. Adding a dielectric as shown in Figure 4.2, yields for each part of the capacitor, $C_z = \epsilon_0 A/z$ and $C_0 = \epsilon_r \epsilon_0 A/d$. The total capacitance will be the series combination of both parts:

$$C = \frac{C_0 C_z}{C_0 + C_z} = \epsilon_r \epsilon_0 \frac{A}{d + \epsilon_r z} \qquad (4.6)$$

The sensitivity is now

$$\frac{dC}{dz} = -\frac{\epsilon_r \epsilon_0 A \epsilon_r}{(d + \epsilon_r z)^2} = -\frac{\epsilon_r^2 \epsilon_0 A}{d^2} \frac{1}{\left(1 + \frac{\epsilon_r z}{d}\right)^2}$$

$$\approx -\frac{C_0}{d}\epsilon_r[1 - 2\epsilon_r x + 3(\epsilon_r x)^2 - \cdots] \qquad (4.7)$$

which is more linear than $-\epsilon A/z^2$. Equation (4.6) also shows the effect of a dielectric covering an electrode plate—for example, for electrical insulation.

Another method to obtain a linear voltage from a sensor when the distance between plates varies is to measure, instead of its admittance, its impedance (Section 5.1). Differential capacitors also yield an output linearly dependent on the measurand (Section 4.1.2).

Capacitive sensors have high output impedance. This certainly decreases when the supply frequency increases, but stray capacitances also cause impedance decrease at higher frequencies. Possible solutions are to place signal conditioning circuits close to the sensor and to use an impedance transformer. Also we can measure the current through the sensor instead of the drop in voltage across it, thus eliminating the need for an amplifier with high input impedance as shown in Section 5.1.

In spite of the preceding limitations, capacitive sensors have several advantages that render them attractive for many applications. As mechanical displacement sensors, for example, their loading error is minimal. Unlike potentiometers, capacitors have no direct mechanical contact, friction, or hysteresis errors. Furthermore, no significant force is needed in order to displace the moveable element. By considering that the energy E stored in a capacitance C is $E = (CV^2)/2$, then for a parallel plate capacitor the force needed to move a plate is about

$$F \approx \frac{E}{d} = \frac{1}{2} \frac{\epsilon A}{d^2} V^2 \tag{4.8}$$

If, for example, $A = 10 \text{ cm}^2$, $d = 1 \text{ cm}$, and $V = 10 \text{ V}$, we need

$$F \approx \frac{8.85 \text{ pF/m}}{2} \frac{10 \text{ cm}^2}{1 \text{ cm}} (10 \text{ V})^2 = 4.45 \text{ nN}$$

which is a negligible force. Furthermore, the plates can be lightweight, thus reducing their inertia.

Capacitive sensors are highly stable and reproducible because the capacitance C is independent of the electrical conductivity of the plates. Hence, temperature changes interfere only through dimensional changes, and aging or time drift effects are minimal. If the dielectric material is air, ϵ_r changes only slightly with temperature according to

$$\epsilon_r(\text{air}) = 1 + \frac{p}{T} \left[28 + \frac{\text{RH} \times p_w}{p} \left(\frac{135}{T} - 0.0039 \right) \right] \tag{4.9}$$

where T is the absolute temperature, p is the pressure, RH is the relative humidity, and p_w is the partial pressure of water at temperature T:

$$\lg p_w = 7.45 \frac{T - 273}{T - 38.3} + 2.78 \tag{4.10}$$

Materials other than air undergo larger changes in permittivity with temperature, but their resistivity usually changes more and therefore resistive sensors are more sensitive to temperature than capacitive sensors.

The high resolution available for capacitance measurements makes a high resolution also available for capacitive sensors, particularly for displacement

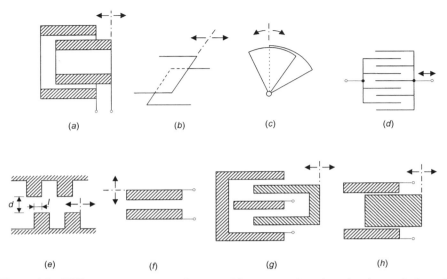

Figure 4.3   Different arrangements for capacitive sensors based on $(a\!-\!e)$ a variation of area, (f) plate separation, and (g, h) dielectric.

measurement where 10 pm resolution has been achieved. Photolithography techniques common in the semiconductor industry have opened many application areas for capacitive sensors.

Finally, although capacitive sensors must be shielded against external electric fields, they themselves do not produce large magnetic or electric fields. This is an advantage when compared with inductive sensors that can produce intense stray magnetic fields.

Figure 4.3 shows some sensor configurations for capacitive displacement sensors based on a change of area $(a\!-\!e)$, electrode separation (f), or the dielectric (g, h). Change in dielectric is not common because of the mechanical problems posed by its manufacturing and operation. The configuration based on a variation of the distance between electrodes is common for measuring large and very small displacements. The configuration based on a variation of area is more common for medium-range displacement measurement, 1 cm to 10 cm.

For many of these configurations, there are models consisting of multiple plates, whose capacitance is given by (4.1) if they are parallel plates. It is important to note that for multiple plate sensors, if for example the variable parameter is A, the sensitivity increases because we have

$$\frac{dC}{dA} = \frac{\epsilon}{d}(n-1) \tag{4.11}$$

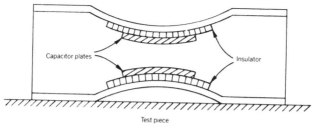

Figure 4.4 The capacitance strain gage consists of two arched flexible strips bonded on the test piece. The horizontal stress in the test piece changes the bowing of the strips and hence the vertical gap between the capacitor plates.

but the relative sensitivity remains the same, $dC/C = dA/A$. Thus they provide a larger capacitance but the same percent change.

C usually ranges from 1 pF to 500 pF and the supply frequency is normally selected higher than 10 kHz in order to reduce the sensor output impedance.

From Figure 4.3 we deduce that common applications for capacitive sensors are the measurement of linear and rotary displacements. Each February issue of *Measurements & Control* lists manufacturers of capacitive displacement and proximity sensors. Capacitive proximity detectors have a range twice that of inductive sensors and detect not only metal objects but also dielectrics such as paper, glass, wood, and plastics. They can even detect through a wall or cardboard box. Because the human body behaves as an electric conductor at low frequencies, capacitive sensors have been used for human tremor measurement and in intrusion alarms. Capacitive sensors suit silicon implementation and integration. They can sense any quantity that a primary sensor converts to a displacement—for example, pressure (using a diaphragm that displaces an electrode or works as an electrode itself as in pressure sensors and condenser microphones), force and torque (using elastic elements as one electrode), and acceleration (using an inertial mass) (Figure 1.10b). Their high resolution even allows strain measurement as in Figure 4.4. Two bowed metal strips are bonded to the test piece. The horizontal elongation bends the strips and separates the electrodes. Capacitive strain gages withstand high temperature and have lower temperature coefficient than resistive strain gages, but have larger dimensions (from 1 cm to 2 cm). Reference 6 describes several capacitive micrometers. In reference 7 there is a rotary potentiometer whose sliding contact is an electrode capacitively coupled to the resistive element. A single-chip fingerprint imaging sensor by ST Microelectronics uses capacitance sensing.

Variations in dielectric constant are used for example for humidity measurement using polymer film as the dielectric material sandwiched between two electrodes. They operate correctly even below 40% RH—thus surpassing resistive hygrometers—and are more accurate up to 70% RH but less accurate above 95% RH and fail under condensation. They suit applications such as

TABLE 4.1 Some Specifications of Two Capacitance Humidity Sensors

Parameter	HS1100 (Humirel)	H1 (Philips)
Humidity range (RH)	1% to 99%	10% to 90%
Ambient temperature	$-40\,°C$ to $100\,°C$	$0\,°C$ to $85\,°C$
Nominal capacitance (25 °C)	180 pF at 55% RH	122 pF at 43% RH
Average sensitivity	0.34 pF/% RH[a]	(0.4 ± 0.05) pF/% RH[b]
Temperature coefficient	0.04 pF/°C	0.01% RH/K
Response time (33% RH to 76% RH)	5 s[c]	<5 min
Humidity hysteresis	$\pm1.5\%$	3%
Long-term stability	0.5% RH/year	—
Supply voltage	5 V, 7 V max.	15 V max.
Leakage	1 nA	$\tan\delta < 0.035$ at 100 kHz

[a] From 33% RH to 75% RH.
[b] From 33% RH to 43% RH.
[c] From 33% RH to 76% RH.
[d] From 43% RH to 90% RH.

office products, humidifiers, dryers, fan controllers, and brakes. Table 4.1 summarizes some specifications of two commercial humidity sensors.

Chemical analysis of binary mixtures of nonconducting fluids having rather different dielectric constants (such as water and oil) also relies on sensing changes in capacitance, as does flow imaging using capacitance tomography [8]. Other capacitive sensors based on dielectric variation are (a) temperature sensors that use a mixture of ferroelectric materials in order to yield the desired Curie temperature [e.g., $SrTiO_3$ $(T_C = -240\,°C)$ and $BaTiO_3$ $(T_C = +120\,°C)$] and (b) thickness gages for thin dielectric materials whose permittivity undergoes only small changes with humidity.

Some level gages for conductive and nonconductive liquids (oil, gasoline) also rely on capacitance change. Figure 4.5a shows a level sensor for conductive liquids (water, mercury) based upon the variation of area. The capacitance of the system consisting of two cylindrical concentric electrodes is

$$C = \frac{2\pi\epsilon h}{\ln\dfrac{d_2}{d_1}} \qquad (4.12)$$

The metal container is grounded to avoid electric discharges and stray capacitance. The level sensor in Figure 4.5b is based on the variation in distance, and it works when the conductivity for the liquid is very high (mercury, water, etc.) so its surface acts as an "electrode plate." The resulting capacitive voltage divider yields an output voltage

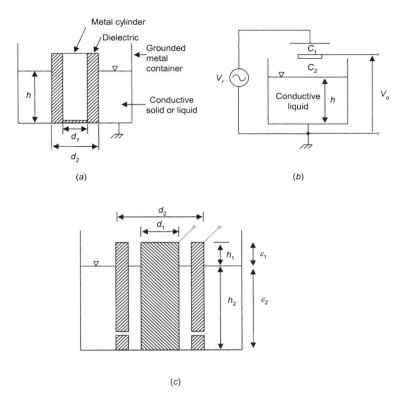

Figure 4.5 Capacitive level sensors for (a, b) conductive liquids and (c) nonconductive liquids.

$$v_o = v_r \frac{C_1}{C_1 + C_2} \tag{4.13}$$

where C_1 is constant and C_2 changes inversely with the liquid height h. The output is therefore nonlinear, but it can be linearized by means of a feedback system that displaces the measuring and reference electrodes, so that their distance to the liquid remains constant. Then the output is the displacement of the measuring electrode. The level sensor shown in Figure 4.5c is based on a variation in the dielectric material. If the conductive cylinders are coaxial, the total capacitance will be

$$C \approx \frac{2\pi(\epsilon_1 h_1 + \epsilon_2 h_2)}{\ln \dfrac{d_2}{d_1}} \tag{4.14}$$

Therefore, in the absence of stray capacitance, C increases linearly with h_1.

Example 4.1   A given capacitive level sensor consists of two concentric cylinders with diameter 40 mm and 8 mm. The storage tank is also cylindrical, 50 cm in diameter and 1.2 m in height. The stored liquid has $\epsilon_r = 2.1$. Calculate the minimal and maximal capacitance for the sensor and its sensitivity (pF/L) when used in the storage tank.

If in Figure 4.5c we call $h_2 = h$ and $h_1 = H - h$, then (4.14) leads to

$$C \approx \frac{2\pi}{\ln \dfrac{d_2}{d_1}} [\epsilon_0 (H - h) + \epsilon_0 \epsilon_r h] = \frac{2\pi}{\ln \dfrac{d_2}{d_1}} [\epsilon_0 H + \epsilon_0 (1 - \epsilon_r) h]$$

If $d_2 = 40$ mm, $d_1 = 8$ mm, $H = 1.2$ m, and $\epsilon_r = 2.1$, we have

$$C_{min} = \frac{2\pi\epsilon_0 H}{\ln 5} = \frac{2\pi \times (8.85 \text{ pF/m}) \times (1.2 \text{ m})}{\ln 5} = 41.46 \text{ pF}$$

$$C_{max} = \frac{2\pi\epsilon_0 \epsilon_r H}{\ln 5} = (41.46 \text{ pF}) \times 2.1 = 87.07 \text{ pF}$$

The volume of the storage tank is

$$V = \frac{\pi d^2}{4} H = \frac{\pi (0.5 \text{ m})^2}{4} (1.2 \text{ m}) = 235.6 \text{ L}$$

Hence, the sensitivity is

$$S = \frac{C_{max} - C_{min}}{V} = \frac{87.07 \text{ pF} - 41.46 \text{ pF}}{235.6 \text{ L}} = 0.19 \text{ pF/L}$$

Note that because the minimal capacitance is not zero, the sensitivity cannot be calculated by dividing the maximal capacitance by the volume. The minimal capacitance must first be subtracted from the maximal capacitance.

4.1.2   Differential Capacitor

A differential capacitor consists of two variable capacitors so arranged that they undergo the same change but in opposite directions. For example, the arrangement in Figure 4.6 yields

$$C_1 = \frac{\epsilon A}{d + z} \tag{4.15a}$$

$$C_2 = \frac{\epsilon A}{d - z} \tag{4.15b}$$

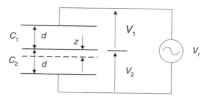

Figure 4.6 Differential capacitor based on the variation of the distance between plates.

The respective drop in voltage across each capacitor is

$$V_1 = \frac{V_r}{\frac{1}{j\omega C_1} + \frac{1}{j\omega C_2}} \frac{1}{j\omega C_1} = V_r \frac{C_2}{C_1 + C_2} \qquad (4.16a)$$

$$V_2 = \frac{V_r}{\frac{1}{j\omega C_1} + \frac{1}{j\omega C_2}} \frac{1}{j\omega C_2} = V_r \frac{C_1}{C_1 + C_2} \qquad (4.16b)$$

Substituting for the capacitances their values as given by (4.15a) and (4.15b) yields

$$V_1 = V_r \frac{1/(d-z)}{1/(d+z) + 1/(d-z)} = V_r \frac{d+z}{2d} \qquad (4.17a)$$

$$V_2 = V_r \frac{1/(d+z)}{1/(d+z) + 1/(d-z)} = V_r \frac{d-z}{2d} \qquad (4.17b)$$

When we subtract both voltages we obtain

$$V_1 - V_2 = V_r \left(\frac{d+z}{2d} - \frac{d-z}{2d} \right) = V_r \frac{z}{d} \qquad (4.18)$$

Therefore, an appropriate output signal conditioning yields a linear output that has an increased sensitivity compared to a single capacitor.

If the measurand changes the area of C_1 and C_2 instead—for example, as in Figure 4.7a—we have

$$C_1 = \epsilon \frac{w(z_0 - z)}{d} = \epsilon \frac{w}{d} z_0 \frac{z_0 - z}{z_0} = C_0 \frac{z_0 - z}{z_0} \qquad (4.19a)$$

$$C_2 = \epsilon \frac{w(z_0 + z)}{d} = \epsilon \frac{w}{d} z_0 \frac{z_0 + z}{z_0} = C_0 \frac{z_0 + z}{z_0} \qquad (4.19b)$$

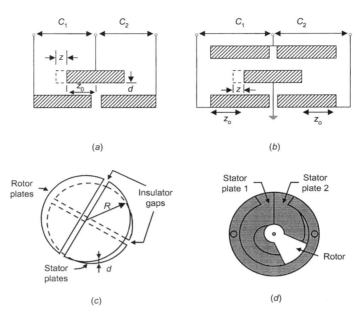

Figure 4.7   Differential capacitors based on the variation of effective plate area.

Here too, measuring the difference between capacitances yields a result proportional to z. The sensor in Figure 4.7b has the same equations.

Differential capacitors are used for displacement measurement in the range from 0.1 pm to 10 mm, with capacitance values from about 1 pF to 100 pF. The seismic accelerometer in Figure 1.10b is differential. The proof mass is the central electrode and the outer electrodes are fixed to the housing. Acceleration in the direction perpendicular to the plates moves the central electrode closer to an outer electrode and separates it from the other electrode.

Figure 4.7c shows a linear rotary differential capacitance sensor (LRDC) [9]. It consists of two equal-size parallel circular plates, each divided by an insulating gap along a diameter. One pair of the resulting two-pair set of semicircular plates is the rotor, and the other pair is the stator. The area between the plates is proportional to the angular displacement θ measured with respect to the zero displacement position, which occurs when the insulating gaps of the stator and rotor are perpendicular. If stray capacitances are ignored, we have four capacitors whose respective values are

$$C_1 = C_3 = \frac{\epsilon_0 \pi R^2}{4d}\left(1 + \frac{2\theta}{\pi}\right) \tag{4.20a}$$

$$C_2 = C_4 = \frac{\epsilon_0 \pi R^2}{4d}\left(1 - \frac{2\theta}{\pi}\right) \tag{4.20b}$$

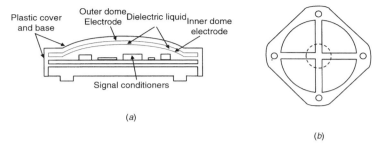

Plastic cover and base | Outer dome Electrode | Dielectric liquid | Inner dome electrode

Signal conditioners

(a)

(b)

Figure 4.8   Capacitive inclinometer based on the variation of capacitance resulting from dielectric displacement. The dashed line shows the air bubble, which is centered when the sensor is horizontal (courtesy of Lucas Sensing Systems).

Arranging these four capacitors in a bridge, with C_1 and C_3 (and C_2 and C_4) in opposite arms, yields an output voltage proportional to θ.

The angular displacement sensor in Figure 4.7d has two fixed plates and one movable plate (rotor). This defines two capacitors whose sum is constant but whose values increase and decrease by the same amount when the rotor turns.

The inclinometer in Figure 4.8 includes two pairs of electrodes whose cross-capacitance changes when the dielectric moves inside the vial because of a tilt. The external surface is aluminum and shields the four sensing electrodes from external electric fields. The resolution is 0.01° and the span is ±20°.

Differential capacitor sensors have the same limitations described for the variable capacitor, except nonlinearity, because we can obtain a proportional output even when the variable is the distance between plates. A particularly important source of error here is the capacitance of output cables because they shunt C_1 and C_2, thus resulting in a nonlinearity and in a loss in sensitivity.

Example 4.2   The capacitive sensor in Figure E4.2 is intended to measure displacements up to ±50 mm from the central position. There is a sliding metal electrode (A, dotted lines) centered between two flat electrodes. One of these is rectangular (C, drawn apart for clarity). The other electrode consists of two trapezoids (B and B′). Calculate the approximate capacitance between electrode A and each of the other electrodes, neglecting edge effects, when $L = 110$ mm, $w = 8$ mm, $h = 10$ mm, $d = 0.5$ mm, and $q = 1$ mm.

Because the distance between electrodes and the dielectric does not change, capacitances are determined by effective electrode areas. The area between A and C does not change, hence

$$C_{AC} = \epsilon_0 \frac{wh}{d}$$

The area between B and B′ is the same when $z = 0$ mm. When A moves to the left, the area between A and B increases and that between A and B′ decreases

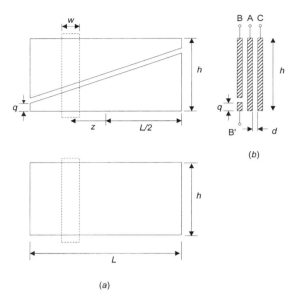

(b)

(a)

Figure E4.2   (*a*) Displacement sensor based on a differential capacitor formed by a sliding electrode (A) and two electrodes, one of them split in two equal trapezoids (B and B'). (*b*) End view.

by the same amount. If we divide each of those areas in a rectangle corresponding to $z = 0$ plus a trapezoid, we have

$$C_{AB} = \epsilon_0 \frac{w}{d} \left[\frac{h}{2} + \frac{z}{L} (h - 2q) \right]$$

$$C_{AB'} = \epsilon_0 \frac{w}{d} \left[\frac{h}{2} - \frac{z}{L} (h - 2q) \right]$$

Note that $C_{AB} + C_{AB'} = C_{AC}$.

4.2   INDUCTIVE SENSORS

4.2.1   Variable Reluctance Sensors

The reluctance of a circuit indicates the amount of magnetic flux it links, due to an electric current. If it is a current flowing along the circuit itself, we call it *self-inductance L*. Otherwise we call it *mutual inductance M*. Sensors based upon a variation of mutual inductance are described in Sections 4.2.3 and 4.2.4.

The inductance can be expressed as

$$L = N\frac{\Phi}{i} \qquad (4.21)$$

where N is the number of turns in the circuit, Φ is the magnetic flux, and i is the current in the circuit. Magnetic flux is related to the magnetomotive force F_m and to the magnetic reluctance $\Re$ by

$$\Phi = \frac{F_\mathrm{m}}{\Re} \qquad (4.22)$$

Because $F_\mathrm{m} = Ni$, finally, we have

$$L = \frac{N^2}{\Re} \qquad (4.23)$$

For a coil having a cross section A and a length l much longer than its transverse dimensions, $\Re$ is given by

$$\Re = \frac{1}{\mu_0 \mu_\mathrm{r}}\frac{l}{A} + \frac{1}{\mu_0}\frac{l_0}{A_0} \approx \frac{1}{\mu_0 \mu_\mathrm{r}}\frac{l}{A} \qquad (4.24)$$

where μ_r is the magnetic permeability for the magnetic core inside the coil, l_0 is the path of field lines through the air (outside the coil), and A_0 is the path cross section. The approximation made is valid when A_0 is very large, which is the usual case.

If the magnetic circuit includes paths through the air and paths through a ferromagnetic material placed in series, then the general equation for the reluctance is

$$\Re = \sum \frac{1}{\mu_0}\frac{l_0}{A_0} + \sum \frac{1}{\mu_0 \mu_\mathrm{r}}\frac{l}{A} \qquad (4.25)$$

For the magnetic circuit shown in Figure 4.9, for example, if leakage flux in the armatures and air gaps are considered negligible, the total reluctance is

$$\Re = \frac{\Re_1}{2} + \frac{\Re_2}{2} + \frac{\Re_3}{2} + \Re_4 \qquad (4.26)$$

Therefore, any variation in N, μ (magnetic permeability of the material inside and around the coil) or in the geometry (l or A) can in principle be applied to sensing. Nevertheless, most inductive sensors are based on a variation of reluctance and it is a displacement that modifies it, usually modifying l_0 or μ. Those that modify l_0 are called *variable gap sensors*, and those that modify μ are called

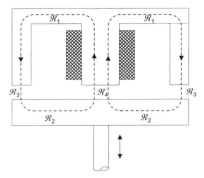

Figure 4.9 Variable reluctance sensor with magnetic flux paths in air and in a ferromagnetic material.

moving core sensors. $\Re$ can also change because of eddy currents, as described in next section.

The application of a variable inductance to sensing has several limitations. First of all, stray magnetic fields also affect the value of L. Therefore it may be necessary to place a magnetic shield around the sensor to ensure that any measured change is due only to the measurand.

The relationship between L and $\Re$ is not constant but changes near the ends of the device because the field is no longer uniform there. Fringing magnetic fields are usually larger than fringing electric fields in capacitors. This limits the measurement range for a given sensor length and interferes with nearby devices or circuits.

According to (4.23), L and $\Re$ are inversely proportional. If the variable parameter is l, the sensor impedance will be inversely proportional to l. When it is μ that changes, the changes in impedance are directly proportional to those in μ.

As in other sensors driven by an alternating current or voltage supply, when there is a central position with null output, the output is bidirectional and we need a carrier amplifier in order to detect the phase of the output voltage. It is not enough just to measure its amplitude (Section 5.3.1).

All devices based on magnetic properties of materials work only at temperatures below their respective Curie temperatures. This restricts their temperature range.

One main advantage of inductive sensors is that they are not affected by the ambient humidity or by other contaminants that can have a noticeable influence on capacitive sensors. Also their mechanical loading effect is very small, although higher than that of a variable capacitor. In addition, they have a very high sensitivity.

Figure 4.10 shows several configurations used in measurement sensors. In Figures 4.10*a*, *b*, *c*, and *d*, a wiper changes the number of turns of the coil defined between a fixed contact and the sliding or rotary contact. In Figures

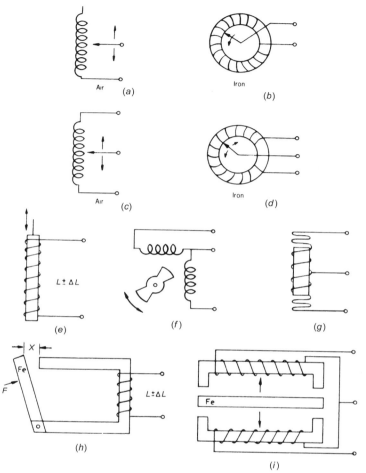

Figure 4.10   Different configurations for variable reluctance sensors. (a) to (d) rely on a variation in the number of coil turns; (e) to (g) rely on a magnetic core movement; (h) and (i) rely on a gap variation. The sensors in (c), (d), (f), (g), and (i) are differential.

4.10e, f, and g, L changes because of the displacement of the magnetic core. In Figures 4.10h and i, L changes because of a gap variation. Differential models (Figures 4.10c, d, f, g, and i) are less sensitive to external magnetic fields, temperature variations, and drifts in the supply voltage and frequency.

The properties of variable reluctance sensors strongly depend on the kind of core. Sensors with an *air core*—meaning no magnetic core material—work at frequencies higher than those available when using an *iron core*, but inductance variations are smaller. Sensors with a core of iron or other ferromagnetic material should work below about 20 kHz in order to avoid increased losses in the core. Furthermore, μ changes with current intensity, thus limiting the rms

Table 4.2   Some Specifications for the WT5TK Differential Displacement Sensor

Parameter	Value	Unit
Nominal displacement	± 5	mm
Precision class	0.4 (or 0.2)	
Full-scale nominal output voltage (FSO)	$\pm 80 \pm 0.01$	mV/V
Interchangeability error	$< \pm 1$	%
Linearity error (as percent of FSO)	$< \pm 0.4$ (or 0.2)	%
Thermal drift of the nominal output voltage, each 10 K (as percent of FSO)	$< \pm 0.5$	%
Nominal temperature range	-20 to $+80$	°C
Supply voltage	2.5 ± 0.125	V
Supply frequency	5	kHz
Total inductance	10	mH
Total resistance	90	Ω

Source: Courtesy of Hottinger Baldwin Measurements.

supply voltage to about 15 V. Ferromagnetic cores better define the magnetic circuit, thus resulting in reduced interfering fields and also in a reduced susceptibility to external fields. Furthermore, the variations in inductance are larger than those for air core sensors. Nominal inductance values for these sensors range from 1 mH to 100 mH. Coil windings are costly and bulky, thereby preventing miniaturization. Integrated microcoils overcome these limitations.

As in the case for capacitive sensors, Figure 4.10 suggests that common applications for variable reluctance sensors are displacement and position measurement and proximity detectors for metal objects, particularly in wet and dirty industrial environments and also under vibration. Table 4.2 gives some specifications of a commercial differential inductive sensor. The detectors placed under road pavements to count the number of passing vehicles are also inductive.

Variable reluctance sensors can also sense other quantities if an appropriate primary sensor converts them into a displacement. This is the case for the pressure sensor shown in Figure 4.11, where the diaphragm is assumed to be ferromagnetic. An inductive measurement of the displacement of the central point of the diaphragm could still be used to sense pressure if the diaphragm were not ferromagnetic. Inductive sensors can be applied to detect the movement of the free end of a Bourdon tube as well. Ferromagnetic diaphragms allow the measurement of high-frequency pressures because they can be placed in direct contact with the fluid. Other diaphragms cannot contact some fluids and need a stainless steel diaphragm and interposed oil, or other fluid, to transmit the sensed pressure. This added compliance reduces the frequency response. As an improvement to diaphragms with strain gages, here there are no wires connecting the diaphragm to external electronics, hence yielding sturdy sensors.

Variable reluctance force sensors can sense the deflection of an elastic ele-

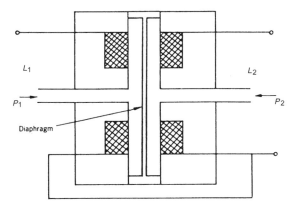

Figure 4.11 Variable reluctance pressure sensor based on a ferromagnetic diaphragm and two fixed coils.

ment. They are used, for example, in the landing gear of airplanes for weight control and to calculate the center of gravity before takeoff.

Thickness measurement using variable reluctance sensors is based on the variation of the magnetic flux depending on the thickness of the part. The reluctance changes depend on the dimensions of the flux path and these can be made thickness dependent. For a magnetic part, configurations such as those shown in Figure 4.12 are common. The core must be ferromagnetic and have a reluctance well below that of the part to be measured. Further it must be laminated in order to reduce eddy currents. Nonmagnetic parts can be placed on a thick ferromagnetic base as shown in Figure 4.13. Measurement ranges for steel, for example, are from 0.025 mm to 2.5 mm.

4.2.2 Eddy Current Sensors

The inductive reactance of a coil supplied by alternating current decreases if a nonmagnetic conductive target is placed inside its magnetic field. This is because the varying magnetic field induces eddy currents on the target surface that produce a secondary magnetic field, which induces an opposing voltage in

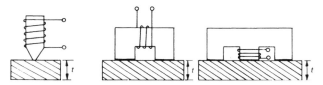

Figure 4.12 Different methods for thickness measurement of a ferromagnetic part based on a variation in reluctance.

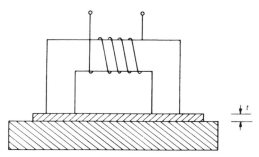

Figure 4.13   Thickness measurement of a nonferromagnetic part based on a variation in reluctance.

the coil. The closer the target is to the coil, the larger the change in impedance. Ferromagnetic target materials first increase the coil reactance because of the magnetic field increase resulting from their higher permeability. However, if eddy currents are strong enough to overcome that effect, the coil inductive reactance decreases.

To use this principle as a measurement method, the target in which eddy currents are induced must be thick enough compared with the penetration or skin depth of those currents, as given by

$$\delta = \frac{1}{\sqrt{\pi f \mu \sigma}} \tag{4.27}$$

where σ is the conductivity for the target, μ its permeability and f the frequency of the supply voltage. A thick target avoids variations due to nonhomogeneity. Nonconductive targets can be sensed by fixing a thin aluminum tape ($\rho = 28$ nΩ·m, $\mu_r = 1$) to their surface and working at high frequency. At 1 MHz, for example, $\delta = 84$ µm for aluminum.

The relationship between coil impedance and distance to the target approximates an exponential function [10]. The change in impedance is higher for closer targets and also depends on the conductivity and permeability of the target. Therefore it will also be sensitive to their changes, for example due to temperature variations, thus leading to thermal interference. Target size relative to coil diameter also affects the output, but surface roughness has no effect— unless it is very prominent.

Eddy current sensors do not need any magnetic material in order to work. Therefore they can be applied at high temperatures, exceeding Curie temperatures. Some commercial models work up to 600 °C. They are unaffected by nonconductive intervening materials such as oil, grease, dirt, water, and steam because these do not modify magnetic fields. Furthermore, some models do not need any mechanical link, thus resulting in a reduced loading effect as compared with variable reluctance sensors and improved reliability because they are

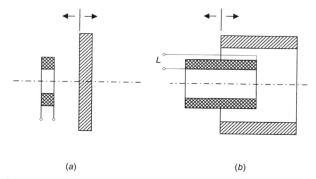

Figure 4.14 (*a*) Eddy current proximity sensor. (*b*) Eddy current displacement sensor.

wear-free. They also permit the design of sealed packages able to operate in corrosive environments, or suited for explosive areas.

Figure 4.14 shows two usual configurations. One of them uses a coil perpendicular to a metallic target. The other is based on a conductive sleeve that slides to cover the coil. There are also differential models based on two active coils and models with one active coil and a passive coil used for compensation in the measurement bridge. Some detectors are mounted in a harmonic oscillator whose frequency depends on the closeness of the target (ECKO, eddy current killed oscillator), and often contain the electronics circuitry needed to drive electromagnetic or electronic switches. Fixture materials are aluminum, stainless steel, or ceramics. Supply frequencies are high, usually 1 MHz or higher, so that according to (4.27), only thin metal targets are needed. Typical measurement ranges go from 0.5 mm to 60 mm with a maximal resolution of 0.1 μm. There are precision models with a 0.05 mm measurement range and long-stroke models with 630 mm range. Table 4.3 summarizes some specifications of two different models. Each December issue of *Measurements & Control* lists the

Table 4.3 Some Specifications of Two Families of Eddy Current Sensors with Added Electronics[a]

Parameter	NCDT100	Series EDS
Measuring range	0.5/1/2/3/6/15 mm	100/160/250/300/400/630 mm
Linearity	±0.5% FSO	±0.2% FSO
Resolution	0.01% FSO	0.02% FSO
Frequency response (−3 dB)	10 kHz	150 kHz
Temperature range	−50 °C to +150 °C	−40 °C to +85 °C
Temperature stability (10 °C to 65 °C)	0.036% FSO	—

[a] Micro-epsilon Messtechnik.

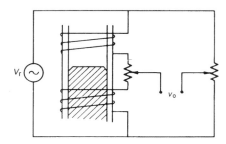

Figure 4.15   Liquid metal level measurement based on eddy currents.

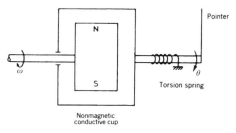

Figure 4.16   Drag-cup-type eddy current tachometer.

manufacturers of proximity switches, under the heading Inductive/Magnetic Sensors.

Some applications of eddy current sensors are similar to those of variable reluctance sensors: proximity and alignment detectors, position, displacement and thickness gaging, acceleration measurement based on spring–mass systems where the mass displacement is measured, dimensioning, and parts sorting. Proximity sensors are used in machine tools, transfer lines, elevator car positioning, railroad yard position sensing, and closed barrier indicators; they are also used to count metal parts in conveyor belts, to detect control valve position, to sort ferrous and nonferrous can tops, and so on. The displacement and position of piston and valves in hydraulic and pneumatic cylinders relies on long-stroke models that use an aluminum tube as the measuring target. This tube is attached to the moving object (e.g. piston rod) and moves concentrically over the metal-encapsulated sensor coil.

Other applications are unique for this measurement principle. Figure 4.15 shows a system for liquid metal level measurement. The tube walls are from nonmagnetic steel. The inductance of each coil depends on eddy currents induced in the liquid and therefore changes when the level does.

Figure 4.16 shows a drag cup tachometer, where we measure the velocity of a shaft that spins a magnet. The magnet induces eddy currents in the nonferromagnetic conductive cup, which produces its own magnetic field that interacts with that of the magnet. The cup is held by a torsion spring that twists

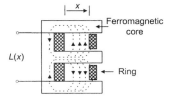

Figure 4.17 Sliding ring displacement sensor based on eddy currents. The conductive ring defines the path for magnetic flux lines because of the currents induced in it.

to an angle at which its torque balances the dragging torque, thus converting a speed into a torque. The twist angle is given on a scale. This sensor has a second order dynamic response.

Figure 4.17 shows a sliding ring sensor based upon the same principle. There is a copper ring that slides along an axis so that its position is at the edge of the magnetic field produced by another coil having a magnetic core (with an E-shape in this case). Eddy currents induced in the ring create an opposite field. This way the ring acts as a magnetic insulator and its position determines coil inductance. One application is the measurement of linear positions and angles in cars in laboratories. In reference 11 there is an eddy current displacement sensor for microweighing based on a magnetic suspension balance.

4.2.3 Linear Variable Differential Transformers (LVDTs)

The linear variable differential transformer shown in Figure 4.18a is usually designated by the acronym LVDT and was patented by G. B. Hoadley in 1940 [12]. It is based on the variation in mutual inductance between a primary winding and each of two secondary windings when a ferromagnetic core moves along its inside, dragged by a nonferromagnetic rod linked to the moving part to sense. When the primary winding is supplied by an ac voltage, in the center position the voltages induced in each secondary winding are equal. When the core moves from that position, one of the two secondary voltages increases and the other decreases by the same amount. Usually both secondary windings are connected in series opposition as shown in Figure 4.18a to yield a linear output as shown in Figure 4.18b.

Figure 4.19 shows the equivalent circuit. If the total resistance in the secondary windings is designated R_2,

$$R_2 = R_{c2} + R'_{c2} + R_L \tag{4.28}$$

then in the primary winding we have

$$E_1 = I_1(R_1 + sL_1) + I_2(-sM_1 + sM_2) \tag{4.29}$$

and in the secondary

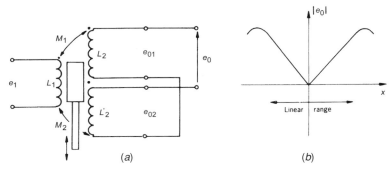

Figure 4.18   (*a*) Circuit diagram for an LVDT. The secondary windings are normally connected in series opposition, but some models have the four secondary terminals separated. (*b*) Output voltage for a series-opposition connected LVDT.

$$0 = I_1(-sM_1 + sM_2) + I_2(R_2 + sL_2 + sL_2' - sM_3) \tag{4.30}$$

From (4.29) and (4.30) we have

$$I_2 = \frac{s(M_2 - M_1)E_1}{s^2[L_1(L_2+L_2'-2M_3)-(M_2-M_1)^2] + s[R_2L_1+R_1(L_2+L_2'-2M_3)]+R_1R_2} \tag{4.31}$$

The output voltage is then

$$E_o = I_2 R_L \tag{4.32}$$

In the center position, $M_2 = M_1$, and according to (4.31) and (4.32), $e_o = 0$ V, as predicted. For other core positions, L_1, L_2, L_2', M_3, and $M_2 - M_1$

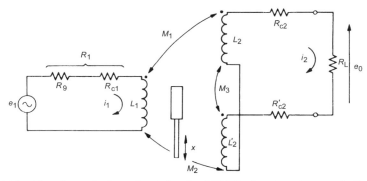

Figure 4.19   Electric equivalent circuit for the LVDT when its primary winding is supplied by a constant voltage.

undergo the following approximate variations: L_1 and M_3 change slowly as the core deviates from x_0; $M_2 - M_1$ displays a fast and linear change on both sides of x_0 [$M_2 - M_1 = k_x(x - x_0)$]; and $L_2 + L_2'$ is nearly constant.

The relation between output voltage and core position depends on load resistance R_L. If there is no load connected to the secondary, the output voltage reduces to

$$E_o = \frac{s(M_1 - M_2)E_1}{sL_1 + R_1} = s(M_1 - M_2)I_1 \tag{4.33}$$

where

$$I_1 \approx \frac{E_1}{sL_1 + R_1} \tag{4.34}$$

is the primary current, which is nearly constant regardless of core position. Therefore, e_o is proportional to $M_2 - M_1$, hence to core position, and it is $90°$ out of phase with respect to the primary current. From (4.33) we also deduce that E_o/E_1 has a high-pass response with respect to the excitation frequency. That is, the sensitivity increases with the excitation frequency. When the excitation frequency is R_1/L_1, the sensitivity is 70% (-3 dB) of that obtained at frequencies 10 times higher.

If there is a load connected to the secondary but we accept that $L_2 + L_2' - 2M_3$ remains nearly constant independent of the core position, and we designate it $2L_2$, and if we also assume that $2L_2L_1 \gg (M_2 - M_1)^2$, then the output voltage is

$$E_o = \frac{s(M_1 - M_2)E_1R_L}{s^2 2L_1L_2 + s(R_2L_1 + 2R_1L_2) + R_1R_2} \tag{4.35}$$

Thus the sensitivity increases with the load resistance. It also increases initially when the excitation frequency does, but then, from a given frequency, it starts to decrease. Figure 4.20 shows this behavior for a given model.

From (4.35) we can also deduce that there is a phase shift between primary and secondary voltages, which depends on the excitation frequency. This phase shift is zero at a frequency

$$f_n = \frac{1}{2\pi}\sqrt{\frac{R_1R_2}{2L_1L_2}} \tag{4.36}$$

which is the same frequency where the sensitivity starts to decrease. If the primary winding is excited at a frequency f_n, then the output voltage does not depend on the excitation frequency and is

$$E_o = \frac{(M_1 - M_2)R_L}{R_2L_1 + 2R_1L_2}E_1 \tag{4.37}$$

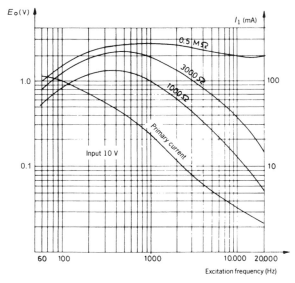

Figure 4.20   Output voltage for full-scale displacement as a function of the frequency for a typical LVDT excited by a 10 V primary voltage. Different curves correspond to different load resistances (courtesy of Schaevitz Engineering).

Thus at a given frequency the output voltage is proportional to the difference in mutual coupling between the primary and each of the secondary windings. If the difference in mutual coupling is proportional to the core position, then the output voltage will be also. Note that even though the device produces a change in mutual impedance as a result of a displacement, the output signal is actually an amplitude-modulated voltage, not an impedance change as was the case with the differential sensors described in the preceding sections.

Example 4.3   The Model 226-0040 LVDT (RDP) has a primary winding with a dc resistance of 67 Ω and two series-opposition connected secondary windings that have 2800 Ω in total. At 1 kHz, the primary has impedance of 290 Ω, the series-connected secondary windings have 4800 Ω, and the sensitivity— normalized to the exciting voltage—is 270 $(\mu V/V)/\mu m$ when the load resistance is 500 kΩ. Calculate the respective inductance for the primary and each secondary winding (assume that the output resistance of the exciting oscillator is negligible). Calculate the excitation frequency that yields zero phase shift between the primary and secondary voltages when the load resistance is 500 kΩ. Calculate the normalized sensitivity when the LVDT is excited at 60 Hz, first when the load resistance is 500 kΩ and then when the load resistance is 10 kΩ.

The inductance can be calculated from impedance data, assuming that coil resistance does not change from dc to 1 kHz. For the primary winding we have

$$|Z_1| = \sqrt{R_1^2 + (2\pi f L_1)^2}$$

$$L_1 = \frac{\sqrt{|Z_1|^2 - R_1^2}}{2\pi f} = \frac{\sqrt{(290\ \Omega)^2 - (67\ \Omega)^2}}{2\pi(1000\ \text{Hz})} = 45\ \text{mH}$$

Similarly, for the secondary windings we have

$$L_2 = \frac{1}{2}\frac{\sqrt{|Z_2|^2 - (R_{c2} + R_{c2}')^2}}{2\pi f} = \frac{\sqrt{(4800\ \Omega)^2 - (2800\ \Omega)^2}}{2\pi(1000\ \text{Hz})} = 310\ \text{mH}$$

From (4.36), the phase shift will be zero at

$$f_n = \frac{1}{2\pi}\sqrt{\frac{(67\ \Omega)(2800\ \Omega + 500\ \text{k}\Omega)}{2 \times (45\ \text{mH})(310\ \text{mH})}} = 5514\ \text{Hz}$$

When the load resistance is 500 kΩ, we can assume that the secondary is in open circuit. If we consider $M_1 - M_2 = k_x x$, we can apply (4.33) to obtain

$$S = \frac{|E_0/E_1|}{x} = \frac{2\pi f k_x}{\sqrt{R_1^2 + (2\pi f L_1)^2}}$$

Since at 1 kHz, $S = 270\ \mu\text{V/V/}\mu\text{m}$, we infer

$$k_x = \frac{270 \times 10^{-6}}{10^{-6}\ \text{m}}\frac{\sqrt{R_1^2 + (2\pi f L_1)^2}}{2\pi f}$$

$$= \frac{270}{1\ \text{m}}\frac{\sqrt{(67\ \Omega)^2 + (2\pi \times 1\ \text{kHz} \times 45\ \text{mH})^2}}{2\pi \times 1000\ \text{rad}}$$

$$= 12.5(\Omega/\text{m})/(\text{rad/s})$$

Therefore, the sensitivity at 60 Hz will be

$$S = \frac{(2\pi \times 60\ \text{rad}) \times (12.5\ \Omega \cdot \text{s/m} \times \text{rad})}{\sqrt{(67\ \Omega)^2 + (2\pi \times 60\ \text{rad} \times 45\ \text{mH})^2}} = 68.2(\mu\text{V/V})/\mu\text{m}$$

If the load resistance is reduced to 10 kΩ, then we need (4.35) to calculate the sensitivity. We obtain

$$S = \frac{E_o}{x - x_0} = \frac{k_x 2\pi f R_L}{\sqrt{[R_1 R_2 - (2\pi f)^2 L_1 L_2]^2 + (2\pi f)^2 (R_2 L_1 + 2R_1 L_2)^2}}$$

$$= \frac{4.71 \times 10^7}{\sqrt{7.2 \times 10^{11} + 5.4 \times 10^{10}}}\frac{\mu\text{V/V}}{\mu\text{m}} = 53.5(\mu\text{V/V})/\mu\text{m}$$

The decreased load resistance has reduced the sensitivity, as expected from Figure 4.20.

Note several limitations for the ideal behavior described above. The first one is that in the center position of real devices the output voltage is not zero but reaches a minimal value. This is due to stray capacitance between primary and secondary windings, which is independent of core position, and also to a lack of symmetry in windings and magnetic circuits. This error is usually lower than 1 % FSO.

Another limitation is the presence of harmonic components in the output voltage, particularly at the null position. The most relevant harmonic is the third one, which is caused by magnetic material saturation. Low-pass filtering the output voltage reduces this interference.

Temperature is another interference source because of its influence on the electric resistance of the primary winding. An increase in temperature increases the resistance, thus reducing the primary current and the output voltage if the excitation is a constant ac voltage. Hence, it is better to excite with a constant current rather than a constant voltage. If the excitation frequency is high enough, the impedance of L_1 predominates over that of R_1 and temperature effects are smaller. The thermal drift of the output voltage can be expressed as

$$E_T = E_{25}[1 + \alpha(T - 25) + \beta(T - 25)^2] \tag{4.38}$$

where T is the temperature in degrees Celsius, and α and β are frequency-dependent constants.

A self-compensating LVDT has been proposed that uses dual secondary coils instead of a single pair [13]. The voltages of one pair are subtracted as usual $(e_1 - e_2)$ and those of the other pair, which are respectively equal to those of the first pair, are added $(e_1 + e_2)$. The ratio $(e_1 - e_2)/(e_1 + e_2)$ is then proportional to the core displacement, but, otherwise, it is highly insensitive to variations in excitation current and frequency and changes in ambient and coil temperatures.

The LVDT has many advantages that explain its extended use. First, its resolution is infinite in theory and better than 0.1 % in practice. Friction between the core and windings is zero. The magnetic force exerted on the core is proportional to the square of the primary current. It is zero in the center position and increases linearly with the displacement. This force is larger than that in a capacitive sensor, but the output voltage is larger here. Because of the lack of friction they have a nearly unlimited life and a high reliability. Some models have a MTBF exceeding 2×10^6 h (228 years!).

Another advantage of LVDTs is that they offer electrical isolation between the primary and secondary windings, which permits different reference voltages or grounding points. This helps in avoiding ground loops (Section 3.6). They also offer electrical isolation between the sensor (core) and the electric circuit,

Table 4.4   General Characteristics for the Model 210A-0050 LVDT When Its Primary Winding Is Excited by a 5 V, 2000 Hz Sinusoidal Voltage

Parameter	Minimum	Nominal	Maximum	Unit
Linear range	−1.3		+1.3	mm
Linearity			±0.25	% FSO
Optimal frequency		2000		Hz
FSO (each winding)	225	250	275	mV
Primary winding impedance	440	490	540	Ω
	+62	+67	+72	°
Secondary winding impedance	159	177	195	Ω
	+57	+62	+67	°
Primary resistance	113.8	133.9	154.0	Ω
Secondary resistance	63.1	74.2	85.3	Ω
Phase shift (primary to secondary)	+4	+9	+14	°
Output at center position			0.5	% FSO
Temperature coefficient		$\alpha = -0.5 \times 10^{-4}$		$°C^{-1}$
		$\beta = -2 \times 10^{-7}$		$°C^{-2}$

(Documentation Transicoil/RDP)

because the coupling between these is magnetic. This is of particular interest for applications in explosive environments because the energy that can be delivered inside the measured system remains limited.

Other advantages are high repeatability (especially of the zero position) due to its symmetry; directional sensitivity; high linearity (up to 0.05%); high sensitivity, though dependent on the excitation frequency; and broad dynamic response.

LVDTs are manufactured by winding the primary along the center of the core and then the secondary windings at symmetrical positions on each side of the center. The three windings are covered by a waterproof coating, thus allowing them to work in high humidity environments. To solve the problem of a restricted linear range of only 30% of the total length, several special arrangements are in use that yield a range/length ratio of 0.8 [12]. Reference 14 describes an LVDT intended for position detection in linear dc motors that is flat instead of cylindrical and uses coreless coils.

The core of common LVDTs is an iron–nickel alloy, longitudinally laminated in order to reduce eddy currents. The rod that drags the core must be nonmagnetic. The whole device can be enclosed by a magnetic shield that makes it insensitive to external fields.

Measurement ranges go from ±100 μm to ±25 cm. Typical rms supply voltages range from 1 V to 24 V, with frequencies from 50 Hz to 20 kHz. Sensitivities range from about 0.1 V/cm to 40 mV/μm for each volt of excitation voltage. Resolutions of even 0.1 μm are feasible. Table 4.4 gives some specifications of a commercial unit.

Some models integrate the electronic circuitry enabling them to be supplied by a dc voltage. They include the oscillator, amplifier, and demodulator (Section 5.3) and give a dc output voltage. These devices are named DCLVDTs.

There are also models for rotary displacements (RVDT). Their linear range is about $\pm 20°$ and their sensitivity is of the order of 10 mV/°, but in general they have lower performance when compared with the linear models.

The equivalent circuit for the LVDT is an alternating current generator whose frequency is that exciting the primary and whose amplitude is modulated by the core displacement. The output impedance is constant and normally lower than 5 kΩ.

From (4.33) the phase shift between the exciting voltage on the primary and the secondary voltage outputs is, under no-load conditions,

$$\phi = 90° - \arctan \frac{\omega L_1}{R_1} \qquad (4.39)$$

When the secondary is loaded, from (4.35) we have

$$\phi = 90° - \arctan \frac{\omega(R_2 L_1 + 2R_1 L_2)}{R_1 R_2 - 2L_1 L_2 \omega^2} \qquad (4.40)$$

When it is not possible to work at the zero phase shift frequency (f_n), we can adjust the phase shift by adding one of the circuits in Figure 4.21.

LVDTs are commonly used for measuring displacement and position. They are frequently used as null detectors in feedback positioning systems in airplanes and submarines. By placing a spring between the frame of the sensor and the far end of the dragging rod, LVDTs can be used in machine tools as an input system that follows the shape or outline we are interested in duplicating.

We can use LVDTs to measure other quantities that can be converted to a core movement. Figure 4.22a shows how to apply the LVDT to acceleration and inclination measurements through a mass–spring system. Figure 4.22b

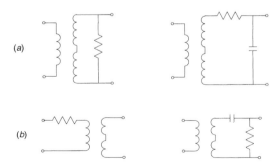

Figure 4.21 Different circuits for (a) phase lag or (b) phase lead for an LVDT, when it is not excited at its nominal frequency.

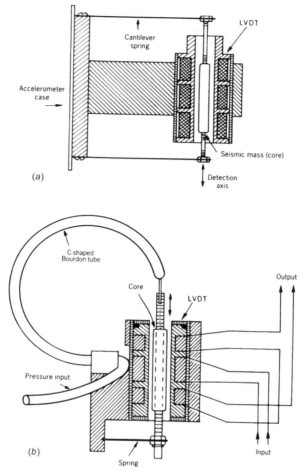

Figure 4.22 Application of a LVDT to (*a*) acceleration measurement and (*b*) pressure measurement (courtesy of Schaevitz Engineering).

shows pressure measurement using a Bourdon tube. It can also use a diaphragm, bellows, or capsule [12]. Each June issue of *Measurements & Control* lists LVDT manufacturers, and each September issue lists the manufacturers of reluctance and LVDT pressure sensors.

LVDTs with sealed windings can be applied to float-based instruments. The float drags the rod or even the float itself is the core, and its movement is detected in the form of a difference in voltage between both secondary windings. Rotameters (Section 1.7.3) and level detectors (Section 1.7.4) suit this application. Load cells and torque meters, where a very small displacement is produced, can also rely on an LVDT [12].

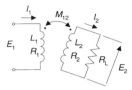

Figure 4.23   Variable transformer sensor where there is a change in the relative position between primary and secondary windings.

4.2.4 Variable Transformers: Synchros, Resolvers, and Inductosyn

One or several transformer windings can be designed to slide or rotate with respect to the other windings, thus changing the primary-to-secondary coupling. Then its mutual inductance, as well as the voltage induced in the windings, will also change if the primary is excited by an ac voltage. Figure 4.23 shows this situation when there is only a primary and a secondary winding. The mutual inductance between them is

$$M_{12} = N_2 \frac{\Phi_2}{i_1} \tag{4.41}$$

where N_2 is the number of turns of the secondary winding and i_1 is the primary current. The flux linked by the secondary winding, Φ_2, is

$$\Phi_2 = \boldsymbol{B} \cdot \boldsymbol{S} = BS \cos \alpha = \mu HS \cos \alpha = \mu \frac{N_1 i_1}{l} S \cos \alpha \tag{4.42}$$

where S is the secondary cross section, N_1 is the number of turns of the primary winding, l its length, μ the magnetic permeability of the core, and α is the geometrical angle between primary and secondary windings. Therefore,

$$M_{12} = N_2 N_1 \frac{\mu}{l} S \cos \alpha = M \cos \alpha \tag{4.43}$$

When the secondary is open circuited and the primary is excited by a sinusoidal voltage of angular frequency ω, resulting in a current $i_1 = I_\mathrm{p} \cos \omega t$, in the secondary we have

$$e_2 = M_{12} \frac{di_1}{dt} \tag{4.44}$$

$$E_2 = j\omega I_1 M_{12} = j\omega I_\mathrm{p} M (\cos \alpha)(\cos \omega t) = k(\cos \alpha)(\cos \omega t) \tag{4.45}$$

Therefore, the output voltage has the same frequency as the excitation voltage, but its amplitude depends on the angle between windings, although not proportionally.

This measurement principle suits those applications aiming to determine an

Table 4.5   Approximate Maximal Accuracy Attainable by Different Angular Position Sensors

Sensor Type	Accuracy
Rotary Inductosyn™	$1''$ to $4''$
High-resolution synchros/resolvers	$5''$
Absolute optical encoders	$20''$
Incremental optical encoders	$30'$
Standard synchros/resolvers	$5'$ to $0.5'$
Potentiometers (with 14 b ADC)	$7'$
Contact encoders	$26'$

Source: Adapted from reference 15.

angular position or displacement. The relative advantages and shortcomings must be compared with those for other rotary sensors, particularly potentiometers (Section 2.1) and digital encoders (Section 8.1).

Because of their small moment of inertia, variable transformer sensors introduce a mechanical load to the turning shaft lower than that of digital encoders, which require large disks in order to achieve a high resolution. Because of their rugged construction, they also stand temperature, humidity, shock, and vibration better than digital encoders and some potentiometers. Thus they suit aerospace and military applications. Units with magnetic coupling instead of slip rings and brushes to contact rotating parts have improved reliability, particularly under extreme environments.

We will later show that variable transformers can send analog information farther than 2 km by using appropriate cables, to feed them to an analog-to-digital converter. Digital encoders, on the other hand, undergo serious interference problems if their output signal is directly transmitted, in particular when there are large environmental magnetic fields such as in radar positioning. Another advantage is the inherent electrical isolation between input (excitation) and output voltages, which helps to prevent conducted interference. Table 4.5 compares the maximal accuracy that can be attained by different angular position measurement systems [15].

Because of the many advantages of variable transformers, several physical configurations have found a broad range of applications to the point that some registered trademark names have become the typical designation for similar devices from different manufacturers.

One of the simplest configurations is the induction potentiometer, shown in Figure 4.24. It consists of two coaxial plane windings with a respective ferromagnetic core. The moveable rotor can turn within the fixed stator. When one of the windings is excited by a sinusoidal voltage, the voltage induced in the other, assuming no load impedance, is given by (4.45).

4.2.4.1   *Three-Phase Synchronous Transformers (Synchros).* The three-phase synchronous transformer or synchro is also a variable transformer. It consists

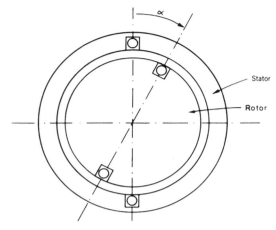

Figure 4.24   An induction potentiometer has concentric stator and rotor with ferromagnetic core and plane windings.

of a cylindrical stator of ferromagnetic material with three windings placed at 120° and star-connected. It has an H-shaped rotor also of ferromagnetic material with one or three windings, turned by the shaft in which we measure rotation. Electric contacts with the rotor can be made with slip rings and brushes. Usually a 50, 60, 400, or 2600 Hz voltage is applied to the rotor and the stator acts as a secondary. Using the terminology in Figure 4.25, the voltages induced in each of the stator windings are

$$e_{S10} = k_1 \cos(\omega t + \phi_1) \cos(\alpha + 120°) \tag{4.46}$$

$$e_{S20} = k_2 \cos(\omega t + \phi_2) \cos\alpha \tag{4.47}$$

$$e_{S30} = k_3 \cos(\omega t + \phi_3) \cos(\alpha - 120°) \tag{4.48}$$

If we assume that the transformer couplings are the same for all windings, $k_1 = k_2 = k_3$, that the phase shifts are all the same, $\phi_1 = \phi_2 = \phi_3$, that the sta-

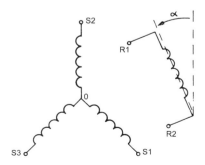

Figure 4.25   Electric circuit for a synchro showing the typical notation.

tors are open circuited, and that the rotor is excited by a voltage generator, then the voltages between each pair of stator terminals are

$$e_{S13} = e_{S30} - e_{S10} = K \cos \omega t[\cos(\alpha - 120°) - \cos(\alpha + 120°)]$$

$$= \sqrt{3}K \cos \omega t \sin \alpha \qquad (4.49)$$

$$e_{S32} = e_{S20} - e_{S30} = K \cos \omega t[\cos \alpha - \cos(\alpha - 120°)]$$

$$= \sqrt{3}K \cos \omega t \sin (\alpha + 120°) \qquad (4.50)$$

$$e_{S21} = e_{S10} - e_{S20} = K \cos \omega t[\cos(\alpha + 120°) - \cos \alpha]$$

$$= \sqrt{3}K \cos \omega t \sin(\alpha + 240°) \qquad (4.51)$$

We thus obtain a three-phase geometrical, not temporal, system. That is, the three voltages are in phase and only their envelope changes, with the amplitudes being proportional to the sine of $\pm 120°$. The set of equations (4.49), (4.50), and (4.51) represents the angle α in synchro format.

The frequency of the induced voltages is the same as that of the excitation voltage supplying the rotor. Rotor terminal 2 is usually the reference, so we have $e_{R21} = e_{R1} - e_{R2}$. Standard values for this voltage are 11.8 V, 26 V, and 115 V. Table 4.6 lists some characteristics of two commercial units.

TABLE 4.6 Some Characteristics for Two Commercial Synchros of Different Sizes

Parameter	26V08CX4c[a] Control Transmitter	CGH11B2[b] Torque Transmitter
Frequency	400 Hz	400 Hz
Input voltage (rotor)	26 V	26 V
Maximal input current	153 mA	170 mA
Nominal input power	0.7 W	0.58 W
Input impedance with open circuit output	192 Ω, 79°	$(20 + j150)$ Ω
Output impedance with open circuit input	39.3 Ω, 70.5°	$(4.3 + j24.6)$ Ω
Dc resistance, rotor	—	10.5 Ω
Dc resistance, stator	—	3.6 Ω
Output voltage	11.8 V	11.8 V
Transformer ratio	0.454 ± 0.009	—
Sensitivity	206 mV/°	206 mV/°
Phase shift (lead)	8.5°	4°
Maximal output at zero position	30 mV	30 mV
Maximal error	7′	12′
Friction at 25°C	30 μN·m	40 μN·m
Rotor moment of inertia	82 μg·m²	330 μg·m²
Mass (max.)	48 g	48 g

[a] Singer Kearfott.
[b] Clifton Precision.

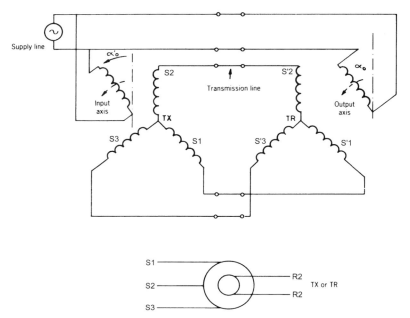

Figure 4.26 Torque transmitter (TX) and torque receiver (TR) for angle information transmission, along with symbol for these devices.

For applications there are two kinds of synchros: torque synchros and control synchros. Even though their name may suggest just the opposite, torque synchros offer a very low torque at their output. They are also called selsyns and are connected as shown in Figure 4.26. One of the units consists of a torque transformer (TX) and the other a torque receiver (TR). They are used to transmit angle information from one shaft to another and provide power high enough to position this second shaft (usually that of an analog indicator) without requiring a feedback system.

They work as follows: Both rotors are supplied by the same excitation. Then in the stator of TX, voltages are induced that produce a current along the connecting lines. Currents in the stator of TR are due to the stator of TX and also to the position of its own rotor. As a result, a torque acts on this rotor; and because it is free to turn, as opposed to the rotor of TX that has a fixed position, it turns until currents in the stator of TR are zero. A stable equilibrium position is then reached. Nevertheless, a mechanical damping system is added to minimize oscillations when changing from one position to another.

Figure 4.27 shows the differential torque transmitter (TDX) and its symbol. This device has an electric input, usually coming from a torque transmitter, and also a mechanical input. It provides an electric output signal in synchro format, usually to a following torque receiver. The output power comes from the electric input, because the TDX has no input reference voltage (power lines). It consists of three windings on the rotor and three on the stator, so a reference

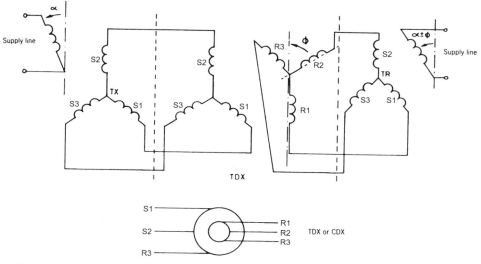

Figure 4.27 Differential torque transmitter (TDX) connected in order to perform the algebraic addition of an input angle α in synchro format to another manually entered angle ϕ (by turning a shaft).

angle (manually adjustable) is added or subtracted to the input angle represented in electric synchro format.

Control synchros, or just synchros, are smaller than torque synchros and can act either as control transmitters (CX) or as control transformers (CT). Figure 4.28 shows their connection. They are only angular position sensors, and therefore they do not need to supply power to the mechanical load to be moved. A feedback system assists the positioning, thus providing a large output torque if necessary.

The control transmitter is a high-impedance version of the TX. Control transformers are high-impedance versions of the TR, usually with a stator made of thinner wire and with more turns in order not to load the stator of the CX to which it is connected. Its rotor is cylindrical and its windings are at 90° with respect to those of the TX. Figure 4.28 shows its connection. A null voltage results when the rotor of the control transformer is in the same direction as that of the control transmitter; the voltage increases when the rotor is deflected from that position. Thus we can use this device as a null detector. The voltage changes according to the sine of the relative angle, but for angles smaller than 20° the output can be considered proportional to angle. Similarly to the TDX, there is also a differential control transformer, CDX. The same symbol is used for both.

Synchros are used in feedback systems for angular positioning in radar, cockpit indicators, navigation inertial reference units, robots, solar cell panels, machine tools, automatic plotters, and so forth.

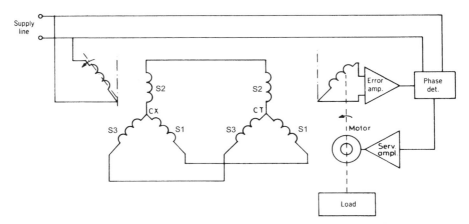

Figure 4.28 Control transmitter and control transformer connected to rotate a mechanical load to a given angle by means of a feedback system.

4.2.4.2 Resolvers. Resolvers are another type of variable transformer, similar to synchros but with windings at 90° both in the stator and in the rotor. The format for angle representation is different, and it uses two voltages instead of three. Depending on the intended application, they are connected in different forms, and the device receives different names. Because windings at 90° are easier to manufacture than windings at 120°, resolvers are less expensive than synchros. Brushless resolvers use a transformer to couple the signals from the stator to the rotor and are more rugged and reliable than common resolvers with brushes or slip rings.

In *vector resolvers*, Figure 4.29, there is a winding on the rotor that acts as a primary and two windings on the stator that act as secondaries. The voltages induced are

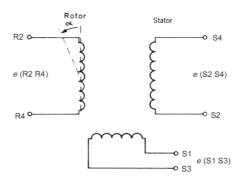

Figure 4.29 Sine and cosine generator (vector resolver). The rotor acts as a primary winding, and both windings in the stator act as secondaries.

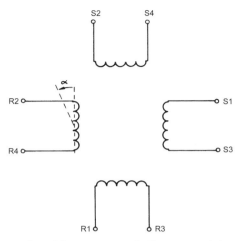

Figure 4.30 Electric resolver. There are two windings around the rotor and two around the stator, but one of them remains idle in some applications (it is either short-circuited or left open-circuited depending on the application).

$$e_{S13} = K \cos \omega t \sin \alpha \tag{4.52}$$

$$e_{S42} = K \cos \omega t \cos \alpha \tag{4.53}$$

In the *electric resolver* there are also two windings on the rotor at 90° and two windings around the stator, also at 90°. We can use only one of the primary windings, on the rotor or on the stator. Usually a winding on the stator is short-circuited and the two rotor windings yield the angle in resolver format. With the terminology in Figure 4.30, output voltages are

$$e_{R24} = K(e_{S13} \cos \alpha + e_{S24} \sin \alpha) \tag{4.54}$$

$$e_{R13} = K(e_{S24} \cos \alpha - e_{S13} \sin \alpha) \tag{4.55}$$

In addition to its straight application for angle measurement, resolvers are also used to perform calculations, particularly those related to axis rotation and coordinate transformation. For example, in order to convert from polar to rectangular coordinates—that is, from (M, α) to (x, y)—we need only to short-circuit a stator winding (connect S2 to S4 in Figure 4.30). Rotor winding voltages will be

$$e_{R13} = -E_1 \sin \alpha \tag{4.56}$$

$$e_{R24} = E_1 \cos \alpha \tag{4.57}$$

If we make $E_1 = E_M \cos \omega t$, where E_M is proportional to the modulus M, then the maximal amplitude of voltages in the rotor yield the values for x and y. The

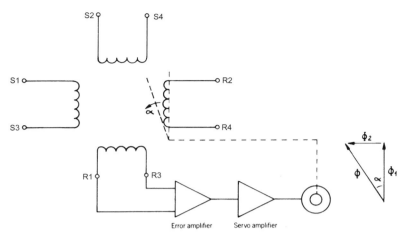

Figure 4.31   Transformation from rectangular (x, y) to polar (M, α) coordinates by means of a resolver and feedback. Input voltages are applied to the stator windings. The output angle must be sensed by an additional sensor.

net electromotive force in the short-circuited winding is zero because

$$e_{S24} = e_{R13} \cos \alpha + e_{R24} \sin \alpha = -E_1 \sin \alpha \cos \alpha + E_1 \cos \alpha \sin \alpha = 0$$

Therefore, it is not necessary in theory to short-circuit it. In practice, however, it is short-circuited in order to avoid any residual voltages induced in it by the transformer effect.

In order to convert from rectangular to polar coordinates, from (x, y) to (M, α), we need a feedback system as shown in Figure 4.31. The input voltages are applied to the stator and can be expressed as

$$e_{S13} = E \cos \alpha \tag{4.58}$$

$$e_{S24} = E \sin \alpha \tag{4.59}$$

In winding R1–R3 a voltage will be induced until its position is perpendicular to the net flux resulting from voltages in the stator, Φ. Then the other rotor winding links all the flux Φ, and therefore its output reaches its maximal value E:

$$e_{R24} = e_{S13} \cos \alpha + e_{S24} \sin \alpha = E \tag{4.60}$$

In order to obtain the angle α in electric form, we need to sense the axis rotation.

Another application is the rotation of rectangular axes. With the terminology in Figure 4.32, after rotating the axis by an angle α, the new coordinates for

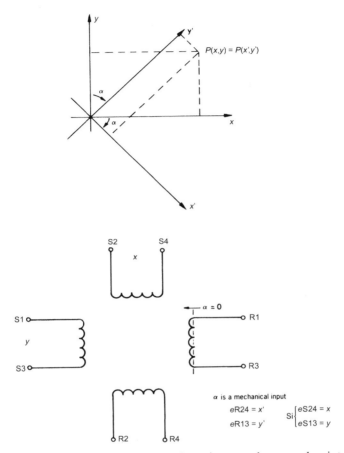

Figure 4.32 A resolver can rotate a rectangular axis system by an angle α introduced by turning the rotor.

a given point are

$$x' = x \cos \alpha - y \sin \alpha \tag{4.61}$$

$$y' = x \sin \alpha + y \cos \alpha \tag{4.62}$$

By applying a voltage proportional to x and to y to each of the stator windings and by mechanically turning the rotor by an angle α, we obtain a voltage proportional to x' in one of the rotor windings and a voltage proportional to y' in the other.

A similar application is the introduction of a time phase shift in a sinusoidal voltage. By applying a voltage $e_{S13} = E \sin \omega t$ to one of the stator windings and a voltage $e_{S24} = E \cos \omega t$ to the other, when the rotor is turned by an angle α,

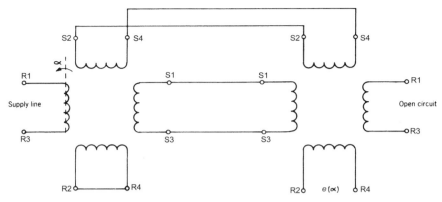

Figure 4.33 A resolver can transmit an angle to another resolver by connecting their stators.

the voltages induced in its windings are

$$e_{R24} = E(\sin \omega t \cos \alpha + \cos \omega t \sin \alpha) = E \sin(\omega t + \alpha) \qquad (4.63)$$

$$e_{R13} = E(\cos \omega t \cos \alpha - \sin \omega t \sin \alpha) = E \cos(\omega t - \alpha) \qquad (4.64)$$

For angle measurement, data transmission resolvers are used. Figure 4.33 shows the configuration that transmits an angle α. The system works similarly to that of synchros. There are also differential transmitters. Table 4.7 lists some characteristics of two commercial models.

4.2.4.3 Inductosyn. Inductosyn™ is the trademark (Farrand Industries) of a kind of variable transformer that differs from the previous ones. It can be realized not only in rotary forms but also in linear forms. It consists of precision printed circuit patterns with parallel hairpin turns repeated along a flat bar (Figure 4.34a) or radial hairpin turns on the flat surface of a disk. A similar pattern is attached to a ruler sliding on the scale (shown in Figure 4.34a for clarity), with a small air gap. An ac current in one element induces a voltage in the other element that depends on their relative position because of the added contributions of winding conductors. The voltage is maximum when both windings are aligned (Figure 4.34b), it goes through zero when the winding conductors of one element are midway between the conductors of the other winding (Figure 4.34c), and it is minimum when both windings are in their next aligned position (Figure 4.34d). Intermediate positions yield a sinusoidal voltage according to

$$V_o = k V_e \cos 2\pi \frac{x}{P} \qquad (4.65)$$

where P is the pitch—that is, the length of one cycle of the hairpin pattern.

TABLE 4.7   Some Characteristics for Two Commercial Resolvers of Different Sizes

Parameter	11R2N4r 100[a] Data Transmission	HZC-8-A-1/A008[b] Computing Resolver
Frequency	10 kHz	400 Hz
Input voltage (rotor)	40 V	15 V
Maximal input current	14 mA	22 mA
Nominal input power	0.173 W	0.11 W
Input impedance with open circuit output	3900 Ω, 65°	$(230 + j640)$ Ω
Output impedance with open circuit input	892 Ω, 65°	$(176 + j806)$ Ω
Dc resistance, rotor	—	220 Ω
Dc resistance, stator	—	158 Ω
Output voltage	18.5 V	15.0 V
Transformer ratio	0.462 ± 0.014	—
Sensitivity	323 mV/°	—
Phase shift (lead)	0.4°	18.8°
Maximal output at zero position	78 mV	1 mV/V
Maximal error	5′	0.01 %
Friction at 25 °C	40 μN·m	—
Rotor moment of inertia	200 μg·m²	—
Mass (max.)	133 g	—

[a] Singer Kearfott.
[b] Clifton Precision.

Usually there is an additional reading winding displaced $P/4$ from the first one. The voltage induced in this second winding will be

$$V_{o2} = kV_e \sin 2\pi \frac{x}{P} \qquad (4.66)$$

In order for the coupling between the fixed and sliding windings to be inductive but not capacitive, an electrostatic shield is placed between them. Each reading winding consists of several "cycles" so that the output signal is the average of several of the "fixed cycles," thus averaging any random errors in their dimensions. If the base materials for scale and slider have the same temperature coefficient, thermal interference is minimal.

For the ordinary linear models, $P = 2$ mm, the total length ranges from 250 mm to 36 m, and the gap between scale and slider is 0.178 mm. Circular models are available with 18 to 1024 cycles. The usual nonlinearity error is 2.5 μm for linear models and $\pm 1''$ to $\pm 4''$ for circular models.

The output voltage is smaller than 100 mV, and its frequency is that of the carrier (excitation), which can be from 200 Hz to 200 kHz. They can also work as null detectors if the two sliders are supplied by 90° out-of-phase voltages.

The inductosyn is used for position control in computer disk memories (with a repeatability of ± 0.5 μm), in machine tools with numeric control, in laser fire

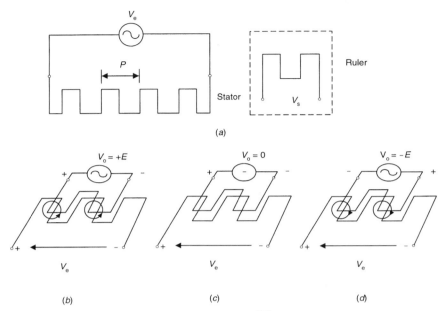

Figure 4.34   Working principle for the Inductosyn™ (Farrand Industries). (*a*) The ruler slides on the scale, but it is shown moved aside for clarity. Because of the addition of the flux from each conductor, the output voltage goes (*b*) from a maximum for aligned conductors, (*c*) through zero when the ruler displaces by $P/4$, and (*d*) it reaches a minimum when conductors are aligned again.

control directors, radar, scanners, antennas, radiotelescopes, riveting and drilling control of aircraft wing spars, and in the remote manipulator of the space shuttle.

4.2.5   Magnetoelastic and Magnetostrictive Sensors

Magnetoelastic and magnetostrictive sensors are inductive sensors that differ from sensors described in the previous sections because they are not based upon a change of geometry or on the position of conductive or magnetic materials but on the effect of the measurand on the magnetic permeability.

Magnetoelastic sensors rely on the Villari effect (discoverd by E. Villari in 1865), consisting of reversible changes in the magnetization curves of a ferromagnetic material when it is subjected to a mechanical stress [16], as shown in Figure 4.35. Some alloys display a linear relationship between the mechanical stress σ and the magnetization curve when undergoing traction or compression (but not for both kinds of stress). That is, $\sigma = k/\mu_r$, where k depends on the material. This dependence results from the internal mechanical stress that prevents the growing of magnetic domains and their orientation along the applied external field.

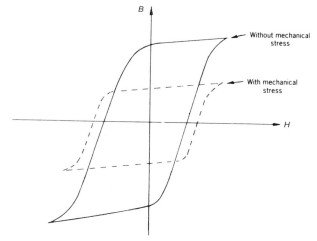

Figure 4.35 Villari effect: The magnetization curve changes depending on the mechanical load applied to the material.

The Villari effect is applied to the measurement of physical quantities through three different configurations. The configuration in Figure 4.36 has a predefined distribution of magnetic flux, and the permeability changes because of the mechanical strain in one direction. The change in coil inductance is proportional to the load. The configuration in Figure 4.37 has a magnetic flux distribution that changes in two directions in the same plane when exerting stress. This is due to the different modification that an isotropic magnetic material experiences in the direction of the applied effort and in the transverse direction, thus becoming magnetically anisotropic. One coil creates an ac magnetic flux, and a different coil placed at 90° detects any asymmetry and yields an output voltage. Alternatively, we can use a ring of ferromagnetic material as a

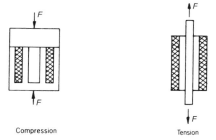

Figure 4.36 Magnetoelastic sensors based on a constant magnetic flux distribution. The mechanical load changes the permeability.

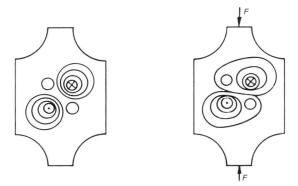

Figure 4.37 Magnetoelastic sensors based on the variation of the magnetic flux distribution when a mechanical load is applied, which makes the material magnetically anisotropic.

coil core and measure the change in inductance of the coil. A third configuration, used for torque sensing in shafts, relies on the change in permeability of the shaft surface or of a surface layer. As torque is applied to the shaft, each shaft element undergoes tensile stress as well as an equal and perpendicular compressive stress, whose magnitude depends on the distance of the element to the shaft axis. The resulting difference in permeability in different directions can be sensed by different methods. This noncontact method is more reliable than using strain gages bonded to the shaft and slip rings. It is also inexpensive because many shafts are themselves ferromagnetic.

Materials suitable for magnetoelastic sensors need good mechanical and magnetic properties. Crystalline material with soft magnetic properties (small hysteresis loop, which is what interests us when applying alternating currents) are also mechanically soft. This prohibits, for example, high permeability and high Young's modulus in the same device. Most magnetoelastic sensors use amorphous metals (metallic glasses), which consist of alloys of iron, nickel, chromium, cobalt, silicon, boron, and others.

Common applications for these sensors are the measurement of force, torque, and pressure in cars and mechanical industries. The arrangement in Figure 4.36 is used in some commercial load cells. The arrangement in Figure 4.37 is used in load cells and in pressure and torque sensors respectively called *press-ductors* and *torductors* [16].

Magnetostriction is the change in dimensions of a ferromagnetic material when subjected to a magnetic field. James P. Joule discovered this effect in 1842, hence the name Joule effect. In 1858 Gustave Wiedemann further observed that a ferromagnetic rod immersed in a longitudinal magnetic field twists when the rod carries a current (Figure 4.38). This twist results from the combined longitudinal and circular magnetic fields, with the latter being produced by the current. Because of the skin effect, current in a cylindrical conductor is minimal at

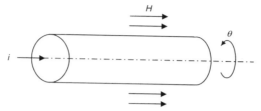

Figure 4.38 The Wiedemann effect is the torsion experienced by a rod immersed in a longitudinal magnetic field **H** when a current i flows through it.

the center and maximal at the surface; hence the magnetic field is also nonuniform, and its interaction with the external field produces a local distortion.

The Wiedemann effect is the basis of magnetostrictive position sensors, Figure 4.39 [17]. There is a magnetostrictive tube acting as an acoustical waveguide that senses the position of a permanent magnet placed around it and mounted to the moving member being sensed. An interrogation pulse applied to a conductor inside the tube creates a circular magnetic field. This field combines with that of the permanent magnet; and the Wiedemann effect twists the tube, producing a torsional strain pulse that travels along the tube in both directions at the speed of sound—it is a mechanical wave. This twist is transformed into a lateral stress in thin magnetoelastic strips that generate a voltage in a sensing coil, or the twist is sensed by a coil or piezoelectric strip (Section 6.2) around the tube. The system measures the time interval between the interrogation pulse and the received signal, which permits us to calculate the distance to the magnet and also velocity. Rubber dampers prevent reflections from the ends of the tube.

This sensor is limited by the Curie temperature of the waveguide material,

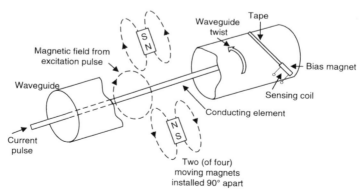

Figure 4.39 Magnetostrictive position sensor based on detecting the position of a moveable magnet by measuring the time interval between conductor excitation and twist pulse detection by a magnetoelastic sensor.

and it is affected by the temperature coefficient of sonic velocity in that material. Alloys with high Curie temperature usually have a high temperature coefficient. Resolution in time interval measurement limits frequency response for a given position resolution. In order to achieve a high time resolution without increasing clock frequency to a high value, we can add several readings and divide by the number of readings, but the position must remain constant during the counting time.

The permanent magnet in Figure 4.39 does not contact the tube. Hence, there is no mechanical wear and the sensor is rugged. The MTBF is 4×10^6 h. Moreover, power interruption does not reset the sensor. There are commercial sensors with displacements up to 7620 mm, ± 0.05 mm nonlinearity, and 0.02 mm hysteresis. Frequency response is narrower for longer strokes—for example, 200 Hz for 305 mm and 50 Hz for 2.5 m. There are models with a hermetic package able to withstand high pressure. Each September issue of *Measurements & Control* lists manufacturers of magnetostrictive position sensors.

Magnetostrictive position sensors are used in industrial applications involving high cycle rates, such as injection molding machines and hydraulic cylinders. They also suit long-stroke applications such as those found in material handling, grinding machines, lumber mills and machine tools. Liquid level is sensed by embedding a position magnet within a float. Using two floats with different buoyancy, it is possible to measure the respective level for immiscible liquids such as water and oil in a storage tank. The sensor housing is selected according to the application.

4.2.6 Wiegand and Pulse-Wire Sensors

Wiegand and pulse-wire sensors are based on the reversal switching of the magnetization of the core of a ferromagnetic wire when applying a strong magnetic field. Placing a one-thousand-turn coil around the wire yields voltage pulses of 3 V, 20 µs, regardless of the rate of field change dH/dt [18, 19], whereas the voltage induced in passive magnetic pulse sensors goes to zero with dH/dt.

Wiegand sensors rely on a manufacture process pattented by J. R. Wiegand in 1981. The basis is the effect discovered by K. J. Sixtus and L. Tonks in 1931 that hard-drawn nickel–iron wires magnetized and under tension form a single magnetic domain with the saturation magnetization along the wire axis. Reversing the magnetization field causes no effect until a critical value H_c is exceeded. Wiegand twisted Vicalloy wire ($Co_{52}Fe_{38}V_{10}$) with a diameter of 0.25 mm or 0.30 mm while drawing it in several steps. Then age-harden the wire to hold in the tension built up during the cold-working process. This procedure yields (a) an inner core magnetically soft (coercivity from 10 A/cm to 20 A/cm) because of grain orientation under stress and (b) an outer shell magnetically hard (coercivity 20 A/cm to 30 A/cm). Furthermore, the outer shell undergoes plastic deformation so that it keeps the core under tensile stress. Therefore, if the wire is immersed in a cycling longitudinal magnetic field, the

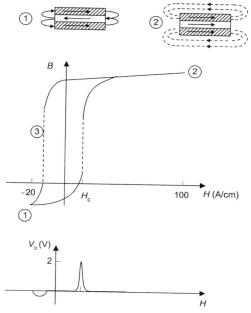

Figure 4.40 Asymmetrical switching of a Wiegand wire and voltage pulses induced in a surrounding coil.

core and the shell reverse their magnetization at different field intensities. The resulting discontinuities in the hysteresis loop are called Barkhausen jumps. Changes in magnetic flux induce a voltage in a surrounding or close coil.

Figure 4.40 describes the switching process for an asymmetric field from -20 A/cm to 100 A/cm. At the starting point the shell and core of the wire are saturated with opposite magnetization (point 1 in Figure 4.40). Flux lines from the shell keep the (soft) core magnetized, and the external magnetic field is negligible. When the intensity of an external longitudinal magnetic field parallel to the magnetization of the shell reaches H_c, the core switches its magnetization because its domains reorient. Magnetic flow lines now extend around the wire and can induce a voltage in a surrounding coil. If the field is strong enough, shell and core are magnetized in the same direction (point 2 in Figure 4.40). To enable the next magnetic switching we need to return the wire to its initial condition by applying a (relatively small) magnetic field able to reverse the magnetization polarity of the core but not that of the shell. This magnetic switching yields a small-voltage pulse with reverse polarity (point 3 in Figure 4.40). A large, symmetrical external field would yield a symmetrical hysteresis loop and two large voltage pulses, as well as two small voltage pulses. Output voltage for symmetrical fields is lower than that for asymmetrical fields.

The sensitivity decreases for increasing temperature. A given model has constant sensitivity from $-44\,^\circ$C to $+55\,^\circ$C, but the pulse amplitude decreases by 18% at 200 °C. The Curie temperature of external magnets also limits the working temperature. Magnets and polar pieces must be designed so as to avoid stray magnetic fields that may produce erratic switching.

Wiegand sensors yield a voltage pulse without any external supply. Hence, they are self-generating sensors and need only two connecting wires. They are contactless. Their sensitivity is constant for variable magnetic fields from dc to 20 kHz. The working temperature of Wiegand wires is from $-200\,^\circ$C to 260 °C (in different models).

Wiegand wire is 7.5 mm to 32 mm long. The larger the wire, the higher the output. The output voltage also increases with the number of turns of the pickup, but its electric resistance increases too. That voltage can directly switch a CMOS or TTL gate. Switching fields are about 20 A/m.

Pulse wires use the same working principle in Wiegand sensors but are produced by methods other than torsion. Some use composite wires made of two different alloys. By careful alloy selection it is possible to achieve either (a) properties similar to those of Wiegand wires or (b) different switching field intensities.

Wiegand and pulse-wire sensors are applied to noncontact measurement of magnetic fields and measurands able to modify them such as position and motion. Many applications rely on a moving Wiegand wire and a stationary pickup that includes the coil and a permanet magnet. They are used in cars for crankshaft position detection, speed sensing, and position indication, as well as in antilock braking systems. They are also used as rotational or linear counting pulsers, in bounceless computer keyboards, and for flow measurement using a turbine as primary sensor with Wiegand wire attached to its vanes and a stationary read head external to the pipe. They permit sensing low rotation speeds in tachometers. In some identifying cards each bit is an embedded Wiegand wire. Some antitheft security systems also use Wiegand sensors.

4.2.7 Saturation-Core (Flux-Gate) Sensors

The basic flux-gate sensor consists of a ferromagnetic core and a pickup coil. An external magnetic field H_e produces a magnetic flux density $B_e = \mu_0 H_e$ and a magnetic flux BA in the core of (average) area A. For low-intensity fields, B is proportional to B_e, $B = \mu_a B_e$, where μ_a is the apparent or effective permeability. Changes in B induce a voltage in the N-turn pickup coil according to

$$v_o = NA\frac{dB}{dt} \tag{4.67}$$

From (1.65) we obtain

$$B = \mu_0(H + M) = \mu_0\mu_r H = \mu_a B_e \tag{4.68}$$

which leads to

$$M = H(\mu_r - 1) \tag{4.69}$$

The internal field can be expressed as

$$H = H_e - D \times M \tag{4.70}$$

where D is the *demagnetization factor* [20, 21]. Solving for μ_a in the above equations yields

$$\mu_a = \frac{\mu_r}{1 + D(\mu_r - 1)} \tag{4.71}$$

If μ_r changes with time, by taking the derivative in (4.67) we finally obtain

$$v_o = NAB_e \frac{1 - D}{[1 + D(\mu_r - 1)]^2} \frac{d\mu_r}{dt} \tag{4.72}$$

One method to make μ_r variable is by periodically saturating the core by an ac excitation: In Figure 1.29, the slope of the hysteresis loop, and hence μ_r, decreases when the material is close to saturation. Because a material with high μ_r concentrates magnetic flux lines, and a material with low μ_r does not, the ac excitation turns a magnetic flux concentrator or flux gate on and off. If the driving current in Figure 4.41 is sinusoidal with frequency f, because the core saturates each half-cycle, the picked voltage consists of even harmonics of f and, according to (4.72), its amplitude is proportional to the (low-frequency) flux density along the sensing axis B_e. The ensuing electronics acquires the second harmonic and demodulates its amplitude, hence the name second-harmonic sensors. H. Aschenbrenner and G. Goubau used a ring-core flux-gate magnetometer in 1928, but the first patent application was from H. P. Thomas in 1931, to whom it was granted in 1935.

Flux-gate magnetometers cannot measure ac magnetic fields whose frequency is close to that of the driving current, which is limited by the core magnetic properties. The upper frequency limit for the external field is about 10 kHz. The geometry of the core and drive coil determine the power consumption, which is higher than in search-coil magnetometers.

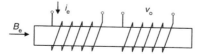

Figure 4.41 A flux-gate sensor includes at least one driving and one pickup coil and a soft ferromagnetic core.

In contrast to other sensitive magnetometers, such as proton precession and optically pumped magnetometers, flux-gate magnetometers directly measure the components of the magnetic field. They are also more sensitive and stable than Hall effect magnetometers. They offer a wide measurement range (100 pT to 1 mT) and low noise level (10 pT resolution), and they can measure up to ± 500 nT superposed on a background field of up to about 60 μT. Because they have no moving parts, they are rugged and reliable.

Materials with a square hysteresis loop yield the highest sensitivity. Low power consumption, however, requires low coercivity and saturation fields. Common sensors use ferrite ceramics and permalloy. In parallel sensors (Figure 4.41), the measured and excitation fields have the same direction. In orthogonal sensors, they are perpendicular. Two-core sensors have symmetrical halves excited by the same signal but with driving coils wound in opposite directions, hence achieving equal but opposite magnetization. By so doing, the mutual inductance between the exciting coil and the (single) pick-up coil is negligible, hence reducing the transformer effect between the excitation and sensing windings. Some sensors use square- or triangular-wave excitation. Others measure pulse height instead of second harmonic amplitude.

Flux-gate sensors were initially developed to detect submarines in the 1930s and 1940s because of their ability to sense feeble changes in the Earth's magnetic field. They are extensively used in magnetometers in space craft, such as low-altitude satellites and in deep-space missions—with magnetic fields as low as 1 nT—and in electronic compasses for aircraft and vehicle navigation. They are also used in geological prospecting, in orientation sensors for virtual reality systems, and to detect ferrous objects in security systems.

4.2.8   Superconducting Quantum Interference Devices (SQUIDs)

SQUIDs, also called Josephson interferometers, are magnetic-flux-to-voltage converters that combine two phenomena, namely, flux quantization and Josephson tunneling [22, 23]. Flux quantization is the fact that the magnetic flux Φ in a superconducting loop is an integer multiple of the magnetic flux quantum $\Phi_0 = h/2q = 2.068$ fWb. Superconductivity means zero resistance and was discovered in 1911 by H. K. Onnes in mercury at 4.2 K (liquid helium). Josephson tunneling, named for B. D. Josephson, who predicted it in 1962, is the electron flow by tunnel effect between two superconductors that are separated by a thin (1 nm to 10 nm) insulating layer (Josephson junction) without any voltage drop. Above a critical current I_c, however, the junction has some resistance. The magnitudes of the superconducting current and I_c depend on the magnetic flux at the junction and are periodic with it. The maximal current happens for $n\Phi_0$ (n is any integer), the minimal current happens for $(n + \frac{1}{2})\Phi_0$, and the period is Φ_0. A large flux reduces I_c. SQUIDs were first demonstrated by R. C. Jaklevick, J. Lambe, A. H. Silver, and J. E. Mercerau in 1964.

If a superconducting material forms a loop, any line of variable magnetic field linked by the loop induces a superconducting current. If the loop is inter-

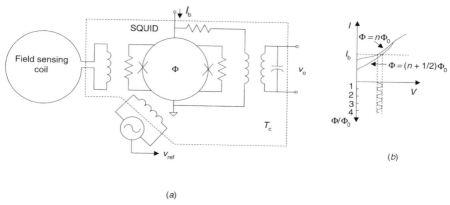

(a)

(b)

Figure 4.42 SQUIDs are magnetic-flux-to-voltage converters that include a supercon-ductor loop with one or two Josephson junctions ($\times$) at cryogenic temperature T_C. The I–V characteristic of the junction depends on the magnetic flux Φ linked by the loop. A large-area coil gathers the flux to measure, and the output signal is ac-coupled to an external amplifier.

rupted by a Josephson junction and biased above I_c (typically at $2I_c$) this cur-rent can be detected from the periodic voltage across the junction because of the Josephson effect (Figure 4.42). Depending on that bias current there are dc SQUIDs and RF SQUIDs. RF SQUIDs use a single Josephson junction and were developed first. Dc SQUIDs use two matched Josephson junctions con-nected in parallel on a superconducting loop. In both cases a resistance shunt-ing each junction eliminates the hysteresis of its I–V characteristic, and there is an input coil inductively coupled to the SQUID (Figure 4.42). This coil can be connected to a field-sensing coil to gather flux over a large area as shown, or it can be connected to the voltage or current to measure.

Most dc SQUIDs work in a negative feedback loop that applies a modulat-ing flux of frequency f_m (100 kHz to 500 kHz) to the SQUID. If this flux is $\Phi_0/2$, when the quasi-static flux in the SQUID is $n\Phi_0$, the output voltage is a rectified version of the input signal, i.e. its frequency is $2f_m$, so that an amplifier sensing only signals of frequency f_m yields a zero output. If the quasi-static flux is $(n + \frac{1}{4})\Phi_0$, the output voltage has frequency f_m and the output is maximum. The output voltage is coupled via a cooled transformer, as shown in Figure 4.42, or an LC circuit, to an ac amplifier followed by a coherent demodulator (Section 5.3).

SQUIDs do not detect the absolute magnetic field but instead detect its change from some arbitrary level. They need a cryogenic chamber for their operation, but otherwise they are extremely sensitive and linear, have dc re-sponse, and have a bandwidth up to some megahertz. Their low-frequency re-sponse is limited by $1/f$ noise (Section 7.4). SQUIDs with negative feedback achieve a dynamic range of up to 180 dB. For a typical area of 10 mm^2, the flux period Φ_0 corresponds to a flux density period of 200 pT.

Josephson junctions are made from thin films. The loop is a few millimeters in diameter. Low-temperature SQUIDs operate at 4.2 K and high temperature SQUIDs operate at 77 K (liquid nitrogen). The first dc SQUIDs used niobium-based thin films and RF SQUIDs used bulk-mechanized niobium. However, in 1986 it was shown that some ceramic oxides become superconductive at temperatures above 130 K, and they are currently applied to SQUIDs.

SQUIDs can measure magnetic field (magnetometers), magnetic field gradient (gradiometers), magnetic susceptibility (susceptometers), and any quantity that can be converted to a magnetic flux, such as current, voltage, and displacement [24]. SQUIDs have received much attention because of their use for submarine communication. They are commonly used in paleomagnetics (measuring magnetic field orientation of rock samples) and magnetotellurics (measuring the resistance of the Earth's crust from its magnetic and electric fields for oil searching), as well as to analyze the properties of high-temperature superconductors. Because of their dc response and higher sensitivity, SQUIDs surpass eddy current sensors for nondestructive testing.

SQUIDs are used in medical research to provide functional rather than structural information by sensing the faint magnetic fields generated by the internal ion currents in the body in magnetocardiography and magnetoencephalography (mapping brain functions to localize disordered neuronal activity). They can also trace magnetic materials ingested by the body and measure the magnetic susceptibility of iron in the liver.

4.3 ELECTROMAGNETIC SENSORS

Sensors discussed so far in this chapter, other than Wiegand sensors and SQUIDs, can be described by one or two variable capacitors or by one or two variable inductances or mutual inductances. There are other devices where a physical quantity can result in a change in a magnetic or electric field without implying a change in inductance or capacitance. This section describes some of the more frequently used sensors.

4.3.1 Sensors Based on Faraday's Law

In 1831 Michael Faraday reported that in any circuit or coil consisting of N turns linking a magnetic flux Φ, whenever Φ changes with time, a voltage or electromotive force e is induced that is proportional to the rate of change of Φ according to

$$e = -N\frac{d\Phi}{dt} \qquad (4.73)$$

The flux Φ can be variable in itself (e.g., when it is produced by an alternating current), or the position of the circuit can be made to change with respect to a constant magnetic flux. Ac tachometers are devices of the first kind, whereas dc tachometers, linear velocity meters, search coil magnetometers, and electromagnetic flowmeters are of the second kind.

4.3.1.1   Generating Tachometers. A generating tachometer (ac tachometer) is similar, in its working principle, to an electric power generator. The voltage induced in a circuit with N turns moving with an angular speed n (r/s) with respect to a constant magnetic field with a flux density B is

$$e = -N\frac{d\Phi}{dt} = -N\frac{d(BA\cos\theta)}{dt} = NBA\sin\theta\frac{d\theta}{dt} \qquad (4.74)$$

Because $\omega = 2\pi n = d\theta/dt$, we have

$$e = NBA\omega\sin\int\omega\,dt \qquad (4.75)$$

When n is constant,

$$e = NBA2\pi n\sin 2\pi nt \qquad (4.76)$$

The output is thus a voltage whose amplitude and frequency are both variable. This arrangement is impractical because at low rotating speeds, the amplitude would be very low.

The arrangement in Figure 4.43 yields a variable amplitude but a constant frequency. It has two windings placed at 90° as in a two-phase induction motor, but it works as a single-phase motor. One winding is for excitation and the other for detection. The rotor has the form of a squirrel cage, with all wire turns around a drum short-circuited. The rotor of some models is just an aluminum drum. Applying an ac voltage having a constant amplitude and a frequency

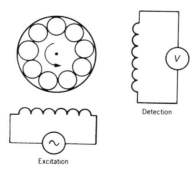

Figure 4.43   Scheme for an ac tachometer having an output voltage with constant frequency and an amplitude proportional to the rotating speed.

$f(=\omega/2\pi)$ to the excitation coil yields a current of frequency f that creates a magnetic flux density $\boldsymbol{B}$. According to Faraday's law, $\boldsymbol{B}$ induces a voltage e_r in the rotor. Because the rotor is short-circuited, that voltage creates a current i_r, hence a magnetic flux density $\boldsymbol{B}_r$. The detection winding links part of this flux— and none of the excitation flux $\boldsymbol{B}$ because of the windings' relative position. When the rotor turns at speed n, the voltage in the output winding is

$$e = k\omega n \sin(\omega t + \phi) \tag{4.77}$$

That is, now the output has the same frequency as the excitation voltage and its amplitude is proportional to the angular speed. We have assumed that the input impedance for the measuring instrument is very high. Otherwise, the current flowing in the output winding would produce a magnetic field, resulting in a nonlinear behavior.

Because the information of interest is the voltage amplitude, if the angular speed is not constant, its frequency of variation must be lower than that of the exciting signal, which acts as a carrier. In order to avoid the need for very complex filters when demodulating, the carrier frequency should be at least 10 times higher than that of the speed to be measured. In many instances, power line frequency (50, 60, or 400 Hz) is used because it is easily available.

Typical sensitivities for ac tachometers range from 3 V/(1000 r/min) to 10 V/(1000 r/min). They are temperature-sensitive because winding resistance changes with temperature, thus resulting in a change in the excitation current. To reduce this interference, some models include a compensating linearized NTC thermistor in series with the primary winding so that the total temperature coefficient is very small (see Problem 2.6).

Dc tachometers or tachometer dynamos are similar to ac units, but the output voltage is rectified as in dc voltage generators. They consist of a permanent magnet based on proprietary alloys and formed by sintering, which creates a constant magnetic flux, and on a multiturn coil turning inside the field at the speed to measure. The variable magnetic flux seen by the coil induces a voltage in it. The polarity of the output connection is periodically switched to obtain an output voltage whose polarity depends on the turning direction and whose amplitude is proportional to the turning speed.

In practice, however, the output voltage is not strictly constant but displays a ripple due to mechanical asymmetries (eccentricity, noncylindrical rotor, etc.), magnetic anisotropy, or electric problems (such as in slip ring contacts). The ratio of the difference between the maximal and minimal output voltages to the mean output voltage is called *undulation* and is one of the figures of merit for these sensors. Sometimes, the undulation is also given in rms values. This ripple can in fact be eliminated by low-pass filtering, but when measuring low speeds in a feedback system the delay introduced by the filters may be unacceptable because of the risk of oscillation. Then we must use models with small ripple, which are more expensive.

Temperature affects the magnetization of permanent magnets and can there-

TABLE 4.8 Characteristics for Two Tachometers

Parameter	SA-470A-7[a]	22C14-204.36[b]
Sensitivity	2.6 V/(1000 r/min)	5.5 V/(1000 r/min)
Linearity	9.36 mV	±0.01% output
Ripple	3% output (rms)	10% output (peak–peak)
Resistance	38 Ω	450 Ω
Inductance	24 mH	—
Temperature coefficient (20–70 °C)	0.01% output/°C	−0.02% output/°C
Reversibility error	0.25% output	±0.01% output
Maximal ac voltage between each terminal and the shaft for 1 s	1250 V	—
Maximal speed	12,000 r/min	—
Slip ring life with 1 mA at 3600 r/min	100,000 h	—
Mass	85 g	54 g
Moment of inertia	850 μg·m²	150 μg·m²
Friction torque	1.8 mN·m	0.05 mN·m

[a] Servo-Tek.
[b] Portescap.

fore be another error source. To compensate for it, use is made of an arrangement of several materials having a Curie temperature falling inside the temperature range where compensation is to be obtained. By so doing, the reluctance for the magnetic circuit changes when temperature does so that the magnetic flux varies less than 0.005%/K.

The ordinary sensitivity for dc tachometers ranges from 5 V/(1000 r/min) to 10 V/(1000 r/min), and the measurement range reaches 8000 r/min. Table 4.8 gives some characteristics of two commercial models. Dc tachometers are obviously used for speed measurement and also in velocity or position feedback systems, in the latter case to provide a velocity feedback.

4.3.1.2 Linear Velocity Sensors. To measure a linear velocity, it not always possible to convert it into an angular speed and then use a tachometer. This happens, for example, when measuring vibration movements and in spacecraft. Linear velocity sensors directly measure linear velocity. In industry they are termed LVTs, from linear velocity transducers.

A conductor having a length l and moving with a linear velocity v defines a time-varying area. If the conductor is perpendicular to a magnetic field having a flux density B and moves in a direction perpendicular to l and B, the voltage induced on that conductor as calculated from (4.73) is

$$e = Blv \tag{4.78}$$

which implies a direct proportionality between voltage and velocity.

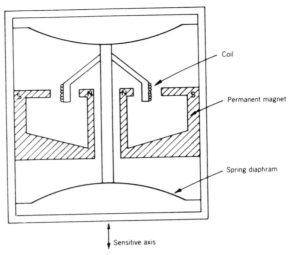

Coil

Permanent magnet

Spring diaphram

Sensitive axis

Figure 4.44   Moving coil linear velocity sensor.

This measurement principle is applied in two different arrangements. Moving coil devices are similar to electrodynamic loudspeakers. The velocity to be measured moves the coil inside a fixed permanent magnet (Figure 4.44). In order to increase the length of the conductor, and therefore the sensitivity, a very thin wire is used. This implies an increased output resistance and therefore requires the input impedance of the meter to be high. The ordinary sensitivity is about 10 mV/(mm/s), and the bandwidth is from 10 Hz to 1000 Hz.

Moving core sensors are based on an arrangement similar to moving core differential inductive devices (Figure 4.10g). But now the core is a permanent magnet rather than a simple ferromagnetic material. Figure 4.45a shows one of these sensors [12]. Commercial units include a stainless steel cover and a

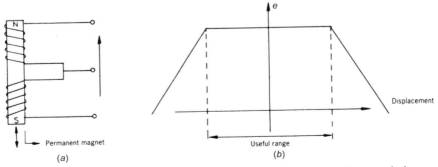

Figure 4.45   (a) Moving core linear velocity sensor. (b) Output voltage variation as a function of core displacement when it moves at a constant velocity.

magnetic shield. Note that this sensor includes two series-opposition connected coaxial coils. If only one coil were used, once the entire core had entered inside the coil, the voltage induced would be zero as long as the velocity remained constant because opposite magnet poles would induce opposed voltages in the winding that would cancel each other. By using two windings, it is possible to add the voltages induced in each one through the series-opposition connection.

This arrangement allows for an increased measurement range (up to 25 cm) as compared with the moving coil sensor. The travel allowed influences the output impedance. This can, for example, be (a) 8 kΩ in series with 0.9 H for short travel and (b) 17 kΩ in series with 2.8 H for long travel. Figure 4.45b shows the relationship between core displacement and output voltage for a given velocity. Sensitivity is about 20 mV/(mm/s).

The application of LVTs to velocity measurement relies on a mass–spring system as primary sensor (Figure 1.10a). We measure the velocity of the mass $\dot{x}_o$ using the LVT of Figure 4.45a. According to (1.57), the system has a high-pass response. Note that with this method we can measure $\dot{x}_i$ from the measurement of $\dot{x}_o$; but while $\dot{x}_i$ is an absolute velocity and may be associated with a very large displacement, $\dot{x}_o$ is a relative velocity, and it is associated with a very small displacement.

4.3.1.3 Search Coil Magnetometers.
Search-coil or induction-coil magnetometers are simple magnetic field sensors based on Faradays' law. They consist of a coil around a ferromagnetic core (a rod) that gathers magnetic flux lines to increase flux density as in Figure 4.41. According to (4.73) the output voltage depends on the number of turns, the relative permeability of the core, the area of the coil, and the rate of change of the magnetic flux.

Search coils have no dc response but can detect dc fields by rotating the sensor. Some units use an air (no) core (loop antenna) that offers higher linearity than sensors with rod core and work at higher frequency but yield lower sensitivity and are bulky and heavy, particularly at low frequencies, which limits their spatial resolution. Rod cores are from high-permeability soft magnetic materials such as ferrites, permalloy, or amorphous glass alloys. To prevent electric fields from influencing the output voltage, the coil is shielded by a nonmagnetic conductor.

Figure 4.46 shows the equivalent circuit for search-coil magnetometers [25]. The inductance, resistance, and interwinding capacitance increase with the number of turns, and they limit the frequency response. Electrostatic shields increase C. Measuring the short-circuit current removes the effect of capacitance, and for frequencies beyond R/L the output is independent of the frequency of the magnetic field. At frequencies well below resonance, the open circuit voltage equals the induced voltage; hence it depends on the frequency of the magnetic field. Therefore, voltage detection suits low-frequency and tuned-frequency magnetic field measurements, and current detection suits broad-band measurements from about 10 Hz to 1 MHz.

Search coils are extensively used in geophysics to observe micropulsations of

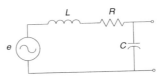

Figure 4.46   Equivalent circuit for a search coil magnetometer.

the Earth's magnetic field, in magnetotelluric measurements for oil explorations, and to measure magnetic fields in electromagnetic compatibility testing.

4.3.1.4   Electromagnetic Flowmeters. Electromagnetic flowmeters rely on a conductive, nonmagnetic liquid that moves inside an exciting magnetic field created by two external coils. Then a small induced voltage (1 mV at 1 m/s) resulting from (4.73) is detected by two electrodes placed at 90° with respect to the flow and the field, as shown in Figure 4.47.

   The output voltage is proportional to liquid flow only when the velocity profile is symmetric with respect to the flow axis and the magnetic field is uniform in the cross section that includes the electrodes. The dependence of the output voltage on the velocity profile depends on the size of the electrodes. In principle, the larger the electrodes, the better the system works. To prevent the electrodes from getting dirty and from deteriorating, they may be covered by an insulating material, thus resulting in a capacitive coupling to the liquid and an increased output impedance. Alternatively, they can be externally cleaned by ultrasound. The pipe must be completely filled for the measure to be valid.

   The pipe must be nonmetallic and nonmagnetic so that the exciting magnetic field is not distorted, and it must have an inner lining with a wear-resistant material. The lining must also be electrically insulating in order to avoid short-circuiting the induced signal detected by the electrodes. Some of the lining materials used are Teflon^{TM}, polyurethane, neoprene^{TM}, rubber, and ceramics. Electrodes are flush-mounted and made from stainless steel, platinum-iridium, zirconium, titanium, or tantalum.

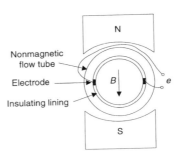

Figure 4.47   In an electromagnetic flowmeter, the magnetic field is vertical, the output electric field is horizontal, and the flow is into the page.

The exciting magnetic field is several teslas, and it can be dc or ac. Ac fields avoid the electrochemical problems in electrodes and thermoelectric interference at the connections of electric lead wires. But if the supply current for the exciting magnetic field is sinusoidal, such as from 60 Hz, then the varying magnetic field itself induces parasitic voltages in any conductive loop, including that formed by electrode leads and the liquid. A solution for this problem is to supply the magnet with a square or trapezoidal wave and then to measure the induced voltage on the electrodes only during the time when the magnetic field is constant. Some units use a pulsating magnetic field at a fraction of the power line frequency and calibrate the zero output between pulses.

This measurement principle only works when the flowing liquid is conductive and nonmagnetic. Conductivities of 100 μS/m are high enough, and some models work for alcohols with $\sigma = 0.05$ μS/m. It does not work for hydrocarbons or for gases. The output does not depend on liquid, density, viscosity, or temperature. The method is noninvasive and is well-suited, for example, for wastewater, corrosive liquids, or liquids with suspended solid matter such as slurries. It is also used in the pharmaceutical and food industries, as well as for blood flow measurement during open thorax surgery and in artificial kidneys. Each April issue of *Measurements & Control* lists manufacturers of electromagnetic flowmeters.

4.3.2 Hall Effect Sensors

The Hall effect consists of the generation of a difference in electric potential across a current-carrying conductor or semiconductor while in a magnetic field perpendicular to the current. Edwin H. Hall discovered this effect in gold in 1879.

Figure 4.48 shows for a semiconductor the sense of the generated voltage,

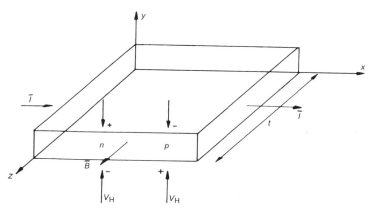

Figure 4.48   The direction of the Hall voltage in a semiconductor depends on the type of majority carriers.

which depends on the type of majority current carriers. A Lorentz force acts on these carriers, $F = qv \times B$, where $v = \mu E_L$, with μ being the carrier mobility and E_L the longitudinal electric field. The force on the charge carriers leads to a charge accumulation on the surface that results in a transversal electric field such that its force on the charge carriers balances that exerted by the magnetic field. Because the force direction depends on the charge of majority carriers, the Hall voltage will have opposite signs for p and n materials.

The Hall voltage generated V_H depends on the thickness t for the material, on the primary current I, on the applied magnetic field B, and on the electrical properties of the material (charge density and carrier mobility). These dependences are described by the Hall coefficient A_H,

$$A_H = \frac{V_H t}{IB} \tag{4.79}$$

which shows that the thinner the element, the larger the voltage for a given material, but also the higher the element resistance.

The application of this principle to the measurement of physical quantities is therefore very simple as long as those magnitudes produce a change in the magnetic flux B. But (4.79) describes an ideal behavior. The Hall voltage depends in practice on other factors such as the mechanical pressure p and the temperature T. The dependence on the mechanical pressure (piezoresistive effect) is a factor to be considered mainly for the manufacturer when encapsulating the device. It is not of great concern for the user.

The temperature has a double influence. On the one hand, it affects the electric resistance for the element, so that if we supply a constant voltage, then the "bias" I changes with temperature and this will change the output voltage V_H. It is therefore much better to supply a constant current than a constant voltage. On the other hand, the temperature affects the mobility of majority carriers, thus also the sensitivity. Given that these two effects have opposite signs, it is possible to compensate for them (see Problem 4.6). Nevertheless, it is always advisable to limit the supply current to reduce self-heating.

Another limitation in precision applications is the presence of an offset voltage—that is, an output voltage even in the absence of any magnetic field in spite of having well-centered electrodes. This offset results from physical inaccuracies and material nonuniformities, and they can be as high as 100 mV for a 12 V supply voltage. Some sensors solve this problem by an additional control electrode that injects the necessary current to obtain a null output when $B = 0$ T. Other sensors include two Hall elements in parallel with opposite bias currents. Still other sensors reduce offset by chopper techniques (Section 7.1.2). Reference 26 discusses problems and trade-offs in Hall sensor design.

Compared with other magnetic field sensors, Hall elements have the advantage of producing an output voltage that is independent of the rate of variation of the detected field. On the contrary, inductive sensors yield a very small output when the rate of flux variation is low. However, Hall sensors do not reach beyond 1 MHz.

Compared with sensors based on an emitter–detector optical pair, Hall effect sensors have the advantages of being insensitive to some ambient conditions (dust, humidity, vibration) and of having characteristics constant with time. In photoelectric sensors, emitter light decreases with age. Some Hall sensors withstand up to 200 °C and near 0 K.

Because they are contactless when applied to movement detection, Hall effect sensors are more robust than those sensors whose contacts wear and become an interference source because of arcing.

Hall effect sensors are based on semiconductors rather than metals because their conductivity is smaller and the drift velocity of the charge carriers in semiconductors is larger than in metals, hence yielding a larger Hall voltage. Also, carrier mobility in semiconductors can be controlled by adding impurities, making it possible to obtain a repeatable Hall coefficient. Because the Hall effect depends only on carrier mobility, there are no perturbations due to surface effects (as happens in p–n junctions and bipolar elements); thus they are easily reproducible and highly reliable.

Some of the materials used for Hall elements are InSb, InAs, Ge, GaAs, and Si. Silicon has the advantage that signal-conditioning circuits can be placed on the same chip to provide either an analog or digital output. Added electronics, however, limits the temperature range. III–V semiconductors yield higher sensitivity because of their larger carrier mobility. InSb has a sensitivity of 1.6 V/ (T·mA) [27]. One type of Hall sensor IC yields a differential output voltage superimposed on a common mode voltage, whereas a second type yields a single-ended output superimposed on a quiescent output voltage (corresponding to $B = 0$ T). Some IC sensors include a voltage regulator; others add overvoltage protection. Digital models include a Schmitt trigger and, often, an open collector output (Figure 4.49). Some linear models offer a ratiometric output and even programmable sensitivity and other parameters. Hall elements are manufactured in different shapes: rectangular, butterfly (which concentrates the flux in the central zone), and also as a symmetrical cross, which permits the interchange of electrodes. Single-plate models sense the absolute magnetic field,

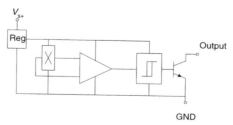

Figure 4.49 Simplified structure of a Hall sensor with digital output.

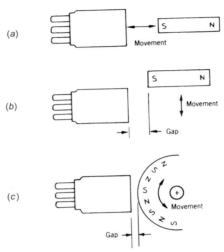

Figure 4.50 Different arrangements for movement sensing using Hall effect sensors (courtesy of Micro Switch–Honeywell).

whereas dual-plate models sense the difference between the magnetic field through each plate. There is also a wide range of packages available.

From (4.79) we deduce that common applications for Hall effect sensors are the measurement of magnetic flux density (gaussmeters)—for example, in compasses—and also the multiplication of any two quantities that we can convert into a current and into a magnetic field—for example, for electric power measurement (wattmeters). It is also possible to measure an electric current intensity by placing the Hall element in the gap of an open toroidal core where a current on a winding around it produces a proportional magnetic field.

Nevertheless, other arrangements are necessary in order to sense other physical quantities. Applications can be either switching or linear, and they are quite common in cars, aircraft, appliances, tools, and computer keyboards. Figure 4.50 shows, for example, several methods to apply Hall elements to movement measurements and proximity detectors. In case (a)—head-on mode—the movement results in a variation in the distance between a permanent magnet and the detector. If the Hall element interrupts the electric circuit to act as a switch, then we have a proximity detector. The arrangement in case (b)—slide-by mode—is also used in proximity detectors. Proximity detectors are used, for example, in seatbelt, airbag ejection, power-window, door-ajar, and refrigerator-door sensors. The arrangement in case (c) suits rotating speed measurement if a switching sensor is used. In automotive ignition systems, a ferromagnetic vane changes the magnetic reluctance of a circuit where both the permanent magnet and the Hall element are stationary. In magnetic potentiometers intended to measure angular position, there is a permanent magnet that turns around the center of a fixed Hall element carrying a current out of the page; the output voltage is proportional to the magnetic flux density perpendicular to the current, hence to sin ϕ. Applications that include a permanent magnet must consider the inconvenience of attracting ferromagnetic dust.

Magnetic flux densities detected by switching Hall sensors range from 100 mT to 500 mT. Linear elements yield a sensitivity of about 10 V/T. Tables 4.9 and 4.10 list the respective characteristics for some models of both kinds.

TABLE 4.9   Characteristics of Some Digital Hall Sensors

Parameter	A3421LKA[a]	HAL114SO-A[b]	HS-220-40
Supply voltage	4.5 V to 18 V	4.5 V to 24 V	4.5 V to 24 V
Supply current	5.0 mA to 18 mA	6 mA to 11 mA	14 mA[e]
Operate point, B_{OP}[d]	16 mT	2.13 mT	15 mT
Release point, B_{RP}[d]	−17.5 mT	1.76 mT	10 mT
Hysteresis, B_{hys}[d]	33.5 mT	0.37 mT	2 mT
Output sink current	30 mA	20 mA	—
Operating temperature	−40 °C to +150 °C	−40 °C to +170 °C	0 °C to +70 °C

[a] Allegro Microsystems.
[b] Micronas Intermetall.
[c] Honeywell.
[d] At 25 °C.
[e] At 24 V.

TABLE 4.10   Characteristics of Some Linear Hall Sensors

Parameter	A3515LUA[a,b]	HAL400SO-A[c]	SS495B[b,d]
Supply voltage, V_s	4.5 V to 5.5 V	−12 V to +12 V	4.5 V to 10.5 V
Supply current	7.2 mA[e]	14.5 mA	7 mA[e]
Magnetic range	±80 mT[f]	±75 mT	±67 mT
Output voltage span	0.2 V to 4.7 V	−0.3 V to 12 V	0.2 V to V_s −0.2 V
Sensitivity	50 mV/mT	42.5 mV/mT	31.25 mV/mT
Nonlinearity error	—	0.5%	1% of span
Temperature null drift	—	25 µV/K max.	±0.08%/°C
Sensitivity drift	2.5% at T_{max} −1.3% at T_{min}	—	+0.05%/°C
Bandwidth	30 kHz	10 kHz	—
Operating temperature	−40 °C to +150 °C	−40 °C to +150 °C	−40 °C to +150 °C

[a] Allegro Microsystems.
[b] Micronas Intermetall.
[c] Honeywell.
[d] Ratiometric.
[e] At 5 V, 25 °C.
[f] Higher fields yield nonlinear output but are safe.

4.4   PROBLEMS

4.1   In order to measure the inclination of a crane arm, an LVDT is placed on it with a 10 kg mass linked to its rod. The LVDT is clamped on the arm and a spring is placed connecting the sensor frame to the mass, so that the mass can slide in the longitudinal direction and drag the rod as indicated in Figure P4.1.

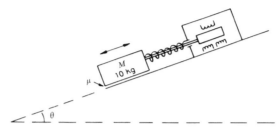

Figure P4.1   Application of an LVDT to inclination measurement.

a. By assuming that the coefficient of friction for the mass M is μ, the LVDT sensitivity is 100 mV/(mm/V), and the spring constant is $K = 200$ N/cm, derive the equation for the output voltage for the LVDT when its primary is supplied by a 5 V rms voltage. What can we conclude about the value for μ?

b. Assuming that the changes in the measured angle θ are rather slow, the power line frequency of 60 Hz can be used to supply the primary winding. The LVDT has a specified zero phase shift at 2.5 kHz. The transfer function relating the primary and secondary winding voltages when the load resistance is 100 kΩ is critically damped. What is the phase shift when it is excited at 60 Hz? How could this phase shift be corrected?

4.2   An LVDT having the characteristics given below is excited at 400 Hz and used to measure displacements at frequencies up to 20 Hz. If the output voltage of the two series-opposed connected secondary windings is measured with a device having an input impedance of 100 kΩ||100 pF, design a correcting network that yields a zero phase shift between the primary and output voltage.

Model	Range (mm)	Supply Frequency (Hz)	Sensitivity (mV/mm/V)	Input Z (Ω)	Output Z (Ω)	Phase Shift (°)
S40	2	60	72	72	1000	+75
		1000	274	325	4250	+6

4.3 A given LVDT able to measure ±50 mm has 250 mV (rms) FSO when excited by 5 V, 2 kHz. At 2 kHz its primary winding has 3500 Ω, +71°. Calculate the FSO when the primary is excited by 12 V (peak), 20 kHz. Assume that the parameters modeling the impedance of the primary winding remain constant with frequency.

4.4 The circuit in Figure P4.4 is used to introduce a phase shift in a constant frequency voltage through a resolver. By assuming that the output voltage is applied to a device with a very high input impedance, what is the condition to be fulfilled by R and C so that the output amplitude is constant independent of rotor position? Then what is the relation between the relative input–output phase shift?

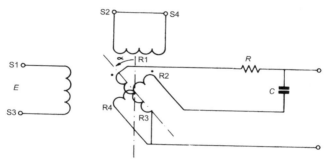

Figure P4.4 Phase shifter (for a constant frequency) based on a resolver.

4.5 Derive (4.77) for ac tachometers. (*Hint*: Obtain the voltage induced in a single wire loop in the rotor and the output voltage from the resulting rotor current. Consider then that the rotor consists of N loops and add their contribution).

4.6 A Hall effect sensor has a positive temperature coefficient of resistance $\alpha = 0.6\%/°C$, and a negative temperature coefficient of sensitivity $\beta = -0.08\%/°C$. The circuit in Figure P4.6 is designed to reduce these

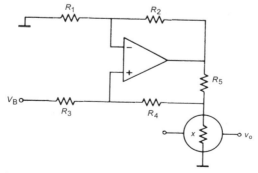

Figure P4.6 Circuit to compensate temperature interference in Hall effect sensors.

temperature effects. If the op amp is ideal, what condition must the resistors fulfill to compensate for temperature variation? Give values for the components when the sensor has an internal resistance of about 700 Ω and requires a supply voltage between 5 V and 10 V.

REFERENCES

[1] L. K. Baxter. *Capacitive Sensors Design and Applications*. New York: IEEE Press, 1997.

[2] W. C. Heerens. Application of capacitance techniques in sensor design. *J. Phys. E: Sci. Instrum.*, **19**, 1986, 897–906.

[3] H. J. Wintle and S. Kurylowicz. Edge corrections for strip and disc capacitors. *IEEE Trans. Instrum. Meas.*, **34**, 1985, 41–47.

[4] D. G. W. Goad and H. J. Wintle. Capacitance corrections for guard gaps. *Meas. Sci. Technol.*, **1**, 1990, 965–969.

[5] F. N. Toth, D. Bertels and G. C. M. Meijer. A low-cost, stable reference capacitor for capacitive sensor systems. *IEEE Trans. Instrum. Meas.*, **45**, 1996, 526–530.

[6] R. V. Jones and J. C. S. Richards. The design and some applications of sensitive capacitance micrometers. *J. Phys. E: Sci. Instrum.*, **6**, 1973, 589–600.

[7] X. Li and G. C. M. Meijer. A novel smart resistive-capacitive position sensor. *IEEE Trans. Instrum. Meas.*, **44**, 1995, 768–770.

[8] R. A. Williams and M. S. Beck (eds.). *Process Tomography, Principles, Techniques, and Applications*. Boston MA: Butterworth-Heinemann, 1995.

[9] R. D. Peters. Linear rotary differential capacitance transducer. *Rev. Sci. Instrum.*, **60**, 1989, 2789–2793.

[10] S. D. Welsby and T. Hitz. True position measurement with eddy current technology. *Sensors*, **14**, November 1997, 30–40.

[11] Chen Huai-ning. An investigation of microweighing with an eddy current transducer. *Rev. Sci. Instrum.*, **59**, 1988, 2297–2299.

[12] E. E. Herceg. *Handbook of Measurement and Control*. Pennsauken, NJ: Schaevitz Engineering, 1976. Fourth printing, 1986.

[13] S. C. Saxena and S. B. Lal Seksena. A self-compensated smart LVDT transducer. *IEEE Trans. Instrum. Meas.*, **38**, 1989, 748–753.

[14] Y. Kano, S. Hasebe, C. Huang, and T. Yamada. New type linear variable differential transformer position transducer. *IEEE Trans. Instrum. Meas.*, **38**, 1989, 407–409.

[15] G. S. Boyes (ed.). *Synchro and Resolver Conversion*. Surrey, UK: Memory Devices Ltd., 1980.

[16] G. Hinz and H. Voigt. Magnetoelastic sensors. Chapter 4 in: R. Boll and K. J. Overshott (eds.), *Magnetic Sensors*, Vol. 5 of *Sensors, A Comprehensive Survey*, W. Göpel, J. Hesse, J. N. Zemel (eds.). New York: VCH Publishers (John Wiley & Sons), 1989.

[17] D. Nyce. Magnetostriction-based linear-position sensors. *Sensors*, **11**, April 1994, 22–26.

[18] G. Rauscher and C. Radeloff. Wiegand and pulse-wire sensors. Chapter 8 in: R. Boll and K. J. Overshott (eds.), *Magnetic Sensors*, Vol. 5 of *Sensors, A Comprehensive Survey*, W. Göpel, and J. Hesse, J. N. Zemel (eds.). New York: VCH Publishers (John Wiley & Sons), 1989.

[19] D. Dlugos. Wiegand effect sensors theory and applications. *Sensors*, **15**, May 1998, 32–34.

[20] P. Ripka. Review of fluxgate sensors. *Sensors and Actuators A*, **33**, 1992, 129–141.

[21] W. Bornhöfft and G. Trenkler. Magnetic field sensors: flux gate sensors. Chapter 5 in: R. Boll and K. J. Overshott (eds.), *Magnetic Sensors*, Vol. 5 of *Sensors, A Comprehensive Survey*, W. Göpel, J. Hesse, J. N. Zemel (eds.). New York: VCH Publishers (John Wiley & Sons), 1989.

[22] J. Clarke. Principles and applications of SQUIDs. *Proc. IEEE*, **77**, 1989, 1208–1223.

[23] H. Koch. SQUID Sensors. Chapter 10 in: R. Boll and K. J. Overshott (eds.), *Magnetic Sensors*, Vol. 5 of *Sensors, A Comprehensive Survey*, W. Göpel, J. Hesse, and J. N. Zemel (eds.). New York: VCH Publishers (John Wiley & Sons), 1989.

[24] R. L. Fagaly. Superconducting sensors: instruments and applications. *Sensors*, **13**, October 1996, 18–27.

[25] G. Dehmel. Magnetic field sensors: induction coil (search coil) sensors. Chapter 6 in: R. Boll and K. J. Overshott (eds.), *Magnetic Sensors*, Vol. 5 of *Sensors, A Comprehensive Survey*, W. Göpel, J. Hesse, and J. N. Zemel (eds.). New York: VCH Publishers (John Wiley & Sons), 1989.

[26] R. S. Popovic. Hall-effect devices. *Sensors and Actuators A*, **17**, 1989, 39–53.

[27] B. Drafts. Understanding Hall effect devices. *Sensors*, **14**, September 1997, 72–74, 77.

5

SIGNAL CONDITIONING FOR REACTANCE VARIATION SENSORS

In order to obtain a useful signal from the variation of a capacitance or an inductance, we need at least an ac excitation voltage or current for the sensor and some method for detecting the variations due to the measurand. If the intended application includes an ADC, its input signal must be dc and within a standard amplitude range.

This chapter proposes several circuits to interface variable reactance sensors. Its contents parallel those of Chapter 3, but here we emphasize the new concepts that arise from working with alternating voltages, such as coherent detection, without repeating concepts similar to those for dc measurements such as those relating to interference. Variable oscillators, which can also be applied to resistive sensors, are analyzed in Section 8.3.

Signals from variable transformer sensors intended for angle measurement are digitized by specific converters that are not usually described in ADC books. For this reason, we will discuss them here.

5.1 PROBLEMS AND ALTERNATIVES

The end uses of measurement signals are immediate analog presentation, conversion to digital, conversion to a variable frequency signal, voltage telemetry, and current telemetry.

Variable reactance sensors can consist of the following: a single varying capacitance or inductance ($C_0 \pm \Delta C$ or $L_0 \pm \Delta L$); a varying inductance plus a reference inductance ($L_0 \pm \Delta L, L_0$) (e.g., in eddy current proximity detectors); a differential capacitance or inductance ($C_0 + \Delta C, C_0 - \Delta C; L_0 + \Delta L, L_0 - \Delta L$);

or a variable transformer that yields an amplitude-modulated signal (e.g., LVDTs, synchros, and resolvers).

Signal conditioning for all of these sensors must include a supply of exciting alternating current. Capacitive sensors usually have capacitance smaller than 100 pF. The supply frequency must then be from 10 kHz to 100 MHz in order to yield reasonable values for their impedances. In order to avoid capacitive interference because of their high output impedance, we frequently connect capacitive sensors with shielded cables. But this adds a capacitance in parallel with that of the sensor, which reduces sensitivity and decreases linearity. Furthermore, any relative movement between cable conductors and the insulating dielectric can increase errors. The usual solution is to place the electronic circuits as close as possible to the sensor, thus using short cables and even rigid cables, and to apply driven shield techniques or impedance transformers. The trade-offs of different methods for measuring small capacitances have been reviewed in reference 1.

When the measurement system requires all the signals to be converted to dc voltages, some available options for the sensors working at alternating frequencies are peak detection, rms measurement, and, most commonly, mean value calculation after rectification (Section 5.2).

A common solution to obtain an electric signal from a variable reactance sensor is just to apply Ohm's law. A change in impedance can be detected by measuring the change in current when a constant ac voltage is supplied to it, or by the change in the drop in voltage across it when driven by a constant alternating current.

For any impedance, we define the quality factor Q as its reactance divided by its resistance at a given frequency. The direct application of Ohm's law to sensors whose Q is not very high implies the measurement of two components of the output signal: the one in phase and the one 90° out of phase—in quadrature—with respect to the supply signal. However, only the in-phase signal carries information about the measurand. Moreover, the actual impedance variations are often very small, and stray capacitances usually interfere with the changes to be measured. Thus reactance measurement techniques must consider those problems.

The circuit in Figure 5.1a applies the constant current supply method to a capacitive displacement sensor based upon the variation of the separation of plates in a parallel plate capacitor. When the capacitance changes according to (4.4), we obtain

$$C_x = \epsilon \frac{A}{d(1+x)} = \frac{C_0}{1+x} \tag{5.1}$$

If the op amp is assumed ideal and R is disregarded, the output voltage is

$$v_o = -v_e \frac{Z_x}{Z} = -v_e \frac{C}{C_0}(1+x) \tag{5.2}$$

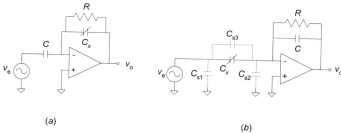

(a)   (b)

Figure 5.1   Signal conditioning for single capacitive sensors to obtain a linear output. (a) For sensors with linear admittance changes. (b) For sensors with linear impedance changes. The resistor provides a bias current path.

Hence, it is proportional to the measured distance, in spite of the nonlinear relationship between the capacity and the distance. R is added to bias the op amp and should be much higher than the sensor impedance at the excitation frequency. Any stray capacitance shunting C_x contributes an output error. Hence, it must be reduced, for example, by shielding the leads connected to the capacitor plates.

The circuit in Figure 5.1b, termed a charge amplifier (Section 7.3), applies a constant voltage to the sensor and measures the resulting current by converting it to a voltage through C. Disregarding R (op amp bias) and stray capacitance C_{s3}, the output voltage is

$$v_o = -v_e \frac{C_x}{C} \tag{5.3}$$

and thus it is proportional to the sensor capacitance. Note that stray capacitances C_{s1} and C_{s2} do not contribute to the output. C_{s1} is in parallel with a voltage source and C_{s2} has both ends at the same voltage because of the op amp. Nevertheless, a large C_{s2} may cause oscillation. Shielding sensor leads reduces C_{s3}.

Example 5.1   Figure E5.1 shows a signal conditioner for the differential capacitive sensor in Figure E4.2. C_1 and C_2 are the variable capacitors and C_3 is constant. The excitation voltage is a 10 V (peak), 10 kHz square waveform.

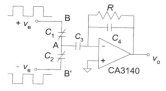

Figure E5.1   Signal conditioner for the differential capacitive sensor in Figure E4.2.

Design C_4 and R to obtain an output range from $+1$ V to -1 V for an input displacement from $+50$ mm to -50 mm. Use the results from Example 4.2 for sensor capacitances.

Since R is just to bias the op amp, we will disregard it. The output voltage is

$$v_o = -v_A \frac{C_3}{C_4}$$

where v_A is the voltage at the movable (sliding) electrode. Kirchhoff's current law at this electrode yields

$$j\omega C_1(V_e - V_A) = j\omega C_3 V_A + j\omega C_2(V_e + V_A)$$

$$V_A = V_e \frac{C_1 - C_2}{C_1 + C_2 + C_3}$$

From Example 4.2 we have

$$C_1 = C_{AB} = \epsilon_0 \frac{w}{d}\left[\frac{h}{2} + \frac{z}{L}(h - 2q)\right]$$

$$C_2 = C_{AB'} = \epsilon_0 \frac{w}{d}\left[\frac{h}{2} - \frac{z}{L}(h - 2q)\right]$$

$$C_3 = \epsilon_0 \frac{wh}{d}$$

Replacing these expressions in those for v_A and v_o yields

$$v_o = -v_e \frac{C_1 - C_2}{2C_4} = -v_e \frac{\epsilon_0 w}{d}(h - 2q)\frac{x}{L}\frac{1}{C_4}$$

In order to have a FSO $= 1$ V we need

$$(10 \text{ V})\epsilon_0 \frac{w}{d}\frac{h - 2q}{L}x_{max}\frac{1}{C_4} = 1 \text{ V}$$

which for $w = 8$ mm, $d = 0.5$ mm, $h = 10$ mm, and $q = 1$ mm leads to $C_4 = 5$ pF.

In order for R not to affect the output waveform, we select its impedance to be, say, 10 times that of C_4 at the excitation frequency, thus implying an amplitude error of about 0.5% and a $6°$ phase shift. Hence, we need $R = 33$ MΩ. The op amp should have an FET input stage; otherwise bias currents would yield a large offset voltage that would reduce the dynamic range for the output signal.

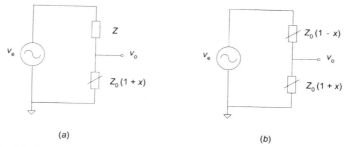

Figure 5.2 (*a*) A voltage divider yields a nonlinear output for a single sensor but (*b*) yields a linear output for a differential sensor.

Voltage dividers are an alternative solution to interface variable reactance sensors. But a voltage divider that includes a sensor with linear impedance change $Z_0(1 + x)$ and a fixed impedance $Z = Z_0$ (Figure 5.2*a*) yields an output voltage

$$v_o = v_e \frac{Z_0(1 + x)}{Z + Z_0(1 + x)} = v_e \frac{1 + x}{2 + x} \tag{5.4}$$

which is nonlinear with respect to x. Stray capacitance shunting the sensor will produce an output error. Nevertheless, for differential sensors (Figure 5.2*b*) we have

$$v_o = v_e \cdot \frac{Z_0(1 + x)}{Z_0(1 - x) + Z_0(1 + x)} = v_e \frac{1 + x}{2} \tag{5.5}$$

Now the output changes linearly with x, though there is a constant term that has a relatively high amplitude whenever $x \ll 1$. That is, the output includes the excitation voltage v_e. This output component could be filtered out if the frequency components of x were much higher than dc but well below the excitation frequency, but this is seldom the case. Note that voltage dividers cancel interference that causes a multiplicative error in both sensor elements, such as the temperature coefficient of coil resistance or the change in dielectric constant because of humidity.

5.2   ac BRIDGES

5.2.1   Sensitivity and Linearity

The classical solution to cancel the constant term appearing at the output of a voltage divider (5.4) and (5.5) is a bridge circuit. Because the bridge involves reactive impedances, it must be supplied by an alternating current or voltage.

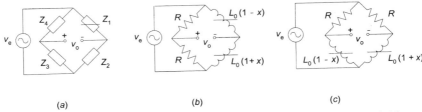

(a)   (b)   (c)

Figure 5.3 (a) General ac bridge with one reactive sensor; (b) linear ac bridge with resistive arms and a differential inductive sensor; (c) ac bridge with resistive arms and a differential sensor to double sensitivity at the cost of reduced linearity.

When the bridge includes only one sensor whose impedance changes linearly with the measurand, $Z_1 = Z_0(1 + x)$ (Figure 5.3a), and $Z_0 = Z_2 = Z_3 = Z_4$, the output voltage is

$$v_o = v_e \frac{x}{2(2 + x)} \tag{5.6}$$

which shows a nonlinear relationship with x. But for a differential sensor placed in adjacent arms, $Z_2 = Z_0(1 - x)$ in Figure 5.3a, the output is

$$v_o = v_e \frac{x}{2} \tag{5.7}$$

and therefore v_o is proportional to x. Furthermore, the same as in voltage dividers (and dc bridges, Section 3.3.4), this circuit cancels changes that are simultaneous for both sensor elements (such as the changes due to temperature). This makes ac bridges the most attractive solution for differential sensors.

The impedances for the two remaining bridge arms not occupied by the sensor are chosen, depending on the type of sensor. For a differential inductive sensor, resistors can be used because they have low to medium impedance. Whenever resistive losses in sensor coils are small (high Q), changes in sensor resistance can be neglected and the circuit in Figure 5.3b yields a linear output. The circuit in Figure 5.3c has double sensitivity, but it is nonlinear.

Because single and differential capacitive sensors have large impedance, using resistors for the remaining bridge arms may produce large errors due to parasitic impedances to earth (which may have similar or lower impedance than the resistors). Using a bridge with two tightly coupled inductive arms having an accurate winding ratio and a central terminal reduces errors from stray capacitance. These bridges are known as Blumlein or transformer bridges [2].

They consist of a transformer (Figure 5.4a) or autotransformer (Figure 5.4b) with a central terminal. This yields three terminals able to form the two fixed arms of a bridge. When the device 1 is an oscillator (for bridge excitation) we have a voltage transformer and the detector (device 2) is connected between the central terminals of the (differential) sensor and the transformer. Conversely, if

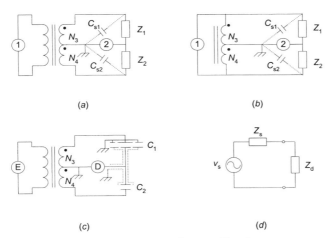

(a)   (b)

(c)   (d)

Figure 5.4 Blumlein bridges: (*a*) using a transformer; (*b*) using an autotransformer. The oscillator and detector can be either device 1 or 2. (*c*) Shield connection for a guarded sensor. (*d*) Equivalent circuit for the bridge connected to the detector.

we place the excitation oscillator as device 2 and the detector as device 1, we have a current comparator, usually designed as current transformer.

For capacitive sensor conditioning, the transformer's central terminal is usually grounded. By so doing, stray capacitances to ground C_{s1} and C_{s2} have a negligible influence on bridge balance. This is because in a voltage transformer the number of coil turns N_3 and N_4 determine the voltage ratio across Z_1 and Z_2: The output impedance of the equivalent voltage generator is very small compared to that of stray capacitances. In a balanced current transformer, the flux density in the core is zero and therefore there is no drop in voltage across capacitances C_{s1} and C_{s2}. Hence, these have no effect. This immunity to stray capacitance allows the detection of very small changes in capacity even in the presence of much larger stray capacitances. Reference 3 reports the detection of changes of 0.1 fF in 50 pF capacitors, in the presence of 1 nF stray capacitances. Figure 5.4*c* shows how to connect in a transformer bridge a sensor with two guards and shielded leads. Guards minimize edge capacitance. The stray capacitance of the shielded cable reduces the equivalent input impedance of the detector but does not affect the sensor capacitance C_1 or that of the reference capacitor C_2.

Transformer bridges have the additional advantage that the voltage or current ratio is very constant both with time and temperature, because it only depends on N_3/N_4. Furthermore, this ratio can be changed very accurately along a broad range of values by placing intermediate terminals in the transformer. Nevertheless, transformer characteristics degrade above 100 kHz. Nulling the bridge by mechanically tuned capacitors or a switched capacitor array is quite inconvenient. Manual nulling is acceptable for laboratory measurements involving static quantities.

Three-winding transformers provide galvanic isolation between the oscillator and detector. This permits them to be grounded at different points without requiring differential measurements.

Figure 5.4d shows the equivalent circuit for transformer bridges, where Z_d is the detector input impedance and v_s and Z_s are, respectively, the Thévenin equivalent source voltage and impedance. For a voltage transformer having $N_3 = N_4$, the coil impedances are so small that we have $Z_s = Z_1 \| Z_2$ and

$$v_s = v_e \frac{Z_2}{Z_2 + Z_1} - \frac{v_e}{2} = \frac{v_e}{2} \frac{Z_2 - Z_1}{Z_2 + Z_1} \tag{5.8}$$

Because Z_s depends on the sensor, in order to have a linear result it is not enough to have a linear v_s. Rather, the input impedance of the detector must be also considered.

Suppose that sensor impedance variations are linear, as happens, for example, for a differential capacitive sensor based on the change in plate separation where, from (4.4), $Z_1 = Z_0(1 - x)$, $Z_2 = Z_0(1 + x)$. Therefore, $Z_s = Z_0(1 - x^2)/2$. If the input impedance of the detector is high, the detected voltage is

$$v_d = v_s = v_e \frac{x}{2} \tag{5.9}$$

which is linear. However, stray capacitance from the bridge output to ground will reduce the input impedance of the detector.

In contrast, for a differential capacitive sensor based on the variation of effective plate area, from (4.19a) and (4.19b) we have $Z_1 = Z_0/(1 - x)$, $Z_2 = Z_0/(1 + x)$. From (5.8) the bridge output voltage is $v_s = -v_e x/2$, and $Z_s = Z_0/2$. Therefore, it is better to use a low input impedance detector because this yields

$$i_d = -\frac{v_e}{Z_0} x \tag{5.10}$$

which is linear with x. Stray capacitance from the bridge output to ground does not influence the system, provided that the detector has low input impedance.

The corresponding equations for current transformers are much more involved and show that capacitive sensors may induce resonance [4]. This makes current transformers less attractive so that they are mainly used for differential inductive sensors whose impedance is so high that stray capacitances should be considered and their effects canceled by means of a transformer ratio bridge.

Ac bridges can also be applied to resistive sensors—for example, to avoid drift and low-frequency noise in dc amplifiers (Section 7.1.1) and thermoelectromotive interference from parasitic thermocouples (Section 6.1). Resistive sensors based on electrolytes must be ac-supplied because a dc supply would produce electrolysis. Sensors attached to rotating parts—for example, strain gage

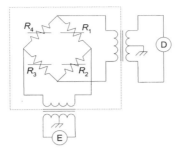

Figure 5.5 Resistive sensor bridge in a shaft (dashed line) with excitation (E) and detection (D) coupled through rotating transformers.

bridges in shafts—can be supplied by an ac voltage through rotating transformers (Figure 5.5), thus avoiding slip rings and brushes that create interference and reduce reliability.

5.2.2 Capacitive Bridge Analog Linearization

As for resistance bridges, ac bridges including a single sensor yield a nonlinear output even if the sensor is linear. Pseudobridges preserve the advantages of bridges for interference canceling and ratio measurements, but in addition yield a linear output. They are simpler to build than transformer bridges and are very common for capacitive sensors.

When the sensor is a single capacitor, the circuit in Figure 5.6a yields a linear output for a choice of parameters that may change the capacitance. The output is

$$v_o = v_e \frac{Z_3/Z_4 - Z_2/Z_1}{1 + Z_3/Z_4} \qquad (5.11)$$

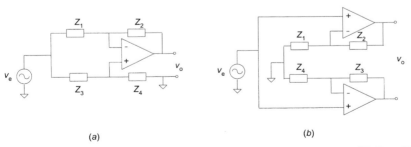

(a)   (b)

Figure 5.6 Capacitive pseudobridges. (a) For single capacitive sensors. (b) For differential capacitive sensors. In this second case the output voltage is differential.

When the measurand changes the plate separation, the sensor must be placed at Z_2. When permittivity or area change, the sensor must be placed at Z_1. In both cases the output is linear as in (5.7) and (5.9). Z_3 and Z_4 can be resistors. Because Z_1 and Z_2 are capacitors (one fixed and the other variable—the sensor), it is necessary to shunt Z_2 by a resistor to provide bias current for the op amp. This resistor should be large enough not to change the output voltage in (5.11).

Figure 5.6*b* shows another solution, which provides a differential output

$$v_o = v_e \left(\frac{Z_2}{Z_1} - \frac{Z_3}{Z_4} \right) \tag{5.12}$$

Z_1 and Z_4 form the differential capacitor, one of whose electrodes can be grounded, which simplifies shielding (Section 5.2.4). If the measurand changes the dielectric or electrode area, the output voltage v_o varies linearly with x. If the measurand changes the plate distance, the (single) sensor can be placed in Z_2 or Z_4. Resistors added to provide bias currents for op amps or required to stabilize them do not affect the output as long as they are matched.

The op amps in Figure 5.6 must have a high enough gain at the working frequency in order for the analysis leading to (5.11) and (5.12) to be valid. Problems 5.2 to 5.5 describe additional pseudobridges.

5.2.3   ac Amplifiers and Power Supply Decoupling

The performance of currently available op amps allows us to easily amplify 10 MHz signals by 10 with a single-stage amplifier. These characteristics are good enough for most ac bridges, so low-cost components suffice in most applications.

Because the central terminal of the ratio arm in ac bridges is usually grounded, one of the output signal terminals is also grounded. Therefore, no differential amplifier is required, in contrast with the case for dc bridges. When an inverting amplifier (Figure 5.7*a*) is used as detector for the circuit in Figure 5.4*d*, we have the advantage that the output signal is independent of any parasitic capacitances Z_p shunting the bridge output because the op amp inverting input is at virtual ground.

On the other hand, if the op amp is assumed to have a finite differential gain

$$A_d = A_{d0} \frac{f_a}{f_a + jf} = \frac{f_T}{f_a + jf} \tag{5.13}$$

the equations to analyze the circuit are

$$\frac{V_o - V_n}{Z} = \frac{V_n - V_s}{Z_s} \tag{5.14}$$

$$V_o = A_d(0 - V_n) \tag{5.15}$$

which lead to

$$\frac{V_o}{V_s} = -\frac{Z/Z_s}{1 + \dfrac{1}{A_d \beta}}$$

(5.16)

where $\beta = Z_s/(Z_s + Z)$. Therefore, the amplified voltage depends on v_s and Z_s, and this can result in a nonlinear dependence on the measurand, even when the source voltage v_s is linear with it.

The noninverting amplifier in Figure 5.7b displays just the opposite characteristics. That is, parasitic impedances affect the amplified signal because the voltage at the noninverting input is

$$V_p = V_s \frac{Z_p}{Z_p + Z_s}$$

(5.17)

and the op amp output is

$$\frac{V_o}{V_s} = \frac{Z_p}{Z_p + Z_s} \frac{Z_2 + Z_1}{Z_1} \frac{1}{1 + \dfrac{1}{A_d \beta}}$$

(5.18)

However, if Z_p is high enough, v_o does not depend on Z_s and therefore if v_s is linear with x, v_o will also be linear.

For a noninverting amplifier based on a current-feedback op amp (CFA) [5, 6], the equations are

$$I_n = \frac{V_s - V_o}{R_2} + \frac{V_s}{R_1}$$

(5.19)

$$V_o = I_n Z_T$$

(5.20)

where Z_T is the open-loop transimpedance and we have used R_1 and R_2 instead of Z_1 and Z_2 because the inverting input of CFAs should see resistive impedance in order to ensure the circuit stability. The output voltage is

$$\frac{V_o}{V_s} = \frac{Z_p}{Z_p + Z_s} \frac{R_2 + R_1}{R_1} \frac{1}{1 + \dfrac{1}{Z_T/R_2}}$$

(5.21)

Therefore, if $Z_p \gg Z_s$, v_o will also be independent of Z_p.

For both inverting and noninverting amplifiers, we can choose impedances to achieve a restricted bandpass appropriate for the signal to be amplified and to reduce noise bandwidth (Section 7.4). For example, if Z_1 is a resistor in series with a capacitor, the gain will decrease at low frequency because of the increase in impedance of the capacitor. If Z_2 is a resistor shunted by a capacitor, the gain will decrease at high frequency because of the decreased impedance of the capacitor.

If the equivalent output signal for the circuit is differential, we can use an instrumentation amplifier such as that in Figure 3.34, whose gain can often be described by (5.13).

Working at ac frequencies requires us to consider several parameters that limit the performance of op amps and instrumentation amplifiers. First, (5.16), (5.18), and (5.21) show that the actual gain of op-amp-based circuits will depend on the open loop gain or transimpedance at the excitation frequency. Because signals from reactance variation sensors are narrow band, the reduced gain can be corrected by calibrating the gain. Note that according to (5.16) and (5.18), voltage feedback amplifiers yield larger gain error for large closed-loop gains. However, (5.21) shows that the gain error in current-feedback amplifiers depends on R_2 but not on the closed-loop gain. Data sheets of instrumentation amplifiers specify their actual gain–bandwidth characteristic.

Ac input impedances in IC amplifiers are far below dc values. This is due to input capacitances that may exceed 3 pF for the component alone. At 1 MHz this implies an impedance of about 50 kΩ. Sockets, circuit layout, and connecting cables further reduce this value.

Other limitations arise from parasitic capacitances in passive components, particularly in resistors. In Figure 5.7b, for example, if $R_2 = 1$ MΩ, then a mere $C = 1$ pF shunting it reduces the -3 dB bandwidth to 160 kHz, even if the op amp has a larger bandwidth. We must therefore avoid high-value resistors and reduce parasitic capacitances.

Stray capacitance C_1 between the inverting terminal and ground in Figure 5.7 may induce oscillation because of the phase shift in the negative feedback loop. To compensate for this delay, R_2 is often deliberately shunted by a capacitor C such as that $R_1 C_1 = R_2 C$. C, however, will limit the signal bandwidth.

Bandwidth limitation due to C can be more restrictive than the one determined by op-amp slew rate (SR). To avoid distortion due to slew rate, the maxi-

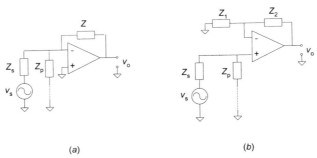

(a)   (b)

Figure 5.7 (a) Inverting and (b) noninverthing amplifier for ac bridges and voltage dividers.

mal frequency for a sine signal having peak amplitude V_p should not exceed

$$f_{max} = \frac{SR}{2\pi V_p} \tag{5.22}$$

Even if the op amp has a high SR, the current from the signal source available to charge C may limit the slew rate.

Example 5.2   Figure E5.2a shows a three-op-amp instrumentation amplifier for the differential capacitive sensor in Figure 4.7c. The sensor plates have radius 2.5 cm and are 0.5 mm apart. If the frequency of the excitation voltage is 10 kHz, calculate R. If the op amps have SR = 13 V/μs, determine the maximal peak amplitude for the excitation voltage.

Figure E5.2b shows the capacitance bridge formed by the four capacitors, and Figure E5.2c shows the equivalent circuit for the sensor connected to the amplifier input. From the sensor geometry we have

$$C_1 = C_3 = \frac{\epsilon_0 \pi r^2}{d\;4}\left(1 - \frac{2\theta}{\pi}\right)$$

$$C_2 = C_4 = \frac{\epsilon_0 \pi r^2}{d\;4}\left(1 + \frac{2\theta}{\pi}\right)$$

Therefore, in the equivalent circuit we have $C_1 + C_2 = C_3 + C_4 = 2C_0$, and from the sensor dimensions we obtain

$$C_0 = (8.85\ \text{pF/m})\frac{\pi \times (0.025\ \text{m})^2}{4 \times 0.0005\ \text{m}} = 8.7\ \text{pF}$$

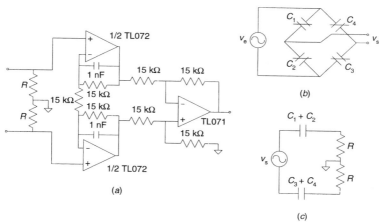

Figure E5.2   Instrumentation amplifier and equivalent circuits for the capacitive sensor in Figure 4.7c.

This capacitance and R form a high-pass filter. If to prevent excessive attenuation we select its corner frequency to be, say, a tenth of the excitation frequency, we need

$$R > \frac{1}{2\pi(10 \text{ kHz})(2 \times 8.7 \text{ pF})} = 9.16 \text{ M}\Omega$$

We can select $R = 10 \text{ M}\Omega$, carbon film, 10% tolerance. This value is high enough to bias an FET-input op amp.

Slew rate limit concerns op amps with the highest output—that is, the output op amp in Figure E5.2c. The open circuit voltage for the bridge is

$$v_s = v_e \left(\frac{C_4}{C_3 + C_4} - \frac{C_1}{C_1 + C_2} \right)$$

The maximal bridge output will correspond to $\theta = \pi/2$ and will be $v_s = v_e$. For the resistors shown, the instrumentation amplifier has a gain of 3. Therefore, the maximal output will be $v_o = 3v_e$; and if $v_e = V_p \sin 2\pi ft$, the condition to fulfill is

$$\left. \frac{dv_o}{dt} \right|_{\max} < 3V_p \times 2\pi f < 13 \text{ V}/\mu s$$

which, when $f = 10$ kHz, leads to $V_p < 138$ V. Op amp power supplies will limit V_p to a value lower than the supply voltage.

Figure 5.8 shows an ac instrumentation amplifier suitable for narrowband signals. The input stage has maximal gain and null phase shift at

$$f_n = \sqrt{f_L f_T} \tag{5.23}$$

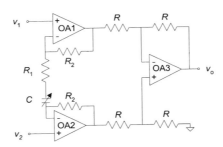

Figure 5.8 Ac instrumentation amplifier for narrowband signals. C permits tuning the working frequency.

where f_T is the gain–bandwidth product of the (matched) op amps, assuming their open-loop response has a dominant pole, and

$$f_L = \frac{1}{2\pi(R_1 + 2R_2)C} \tag{5.24}$$

If $f_L \ll f_T$, the gain at f_n is

$$G = 1 + \frac{2R_2}{R_1} \tag{5.25}$$

Nevertheless, f_T is not accurately known, which requires C to be tunable to set f_n.

Example 5.3   Determine the components in Figure 5.8 in order to obtain $G = 10$ and zero phase shift at 10 kHz when using the OP27 for the input stage. For the OP27, f_T is 5 MHz minimum and 8 MHz maximum. From (5.23),

$$f_L = \frac{f_n^2}{f_T} = \frac{(10 \text{ kHz})^2}{f_T}$$

Therefore, 12.5 Hz $\leq f_L \leq$ 20 Hz. In order to obtain $G = 10$, from (5.25) we infer $2R_2 = 9R_1$. Hence, from (5.24) we have

$$C = \frac{1}{2\pi(10R_1)f_L}$$

If we select $R_1 = 1$ kΩ (metal film, $\pm 1\%$ tolerance), then $R_2 = 4.5$ kΩ and 0.8 μF $\leq C \leq 1.3$ μF. We would select $R_2 = 4.53$ kΩ (metal film, $\pm 1\%$ tolerance). C should be selected for the specific op amp unit because the capacitance of trimmer capacitors does not go beyond tens of picofarads.

When working at high frequency it is convenient to decouple power supplies, as shown in Figure 5.9. The added capacitors provide a low-impedance path to

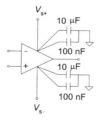

Figure 5.9   Power supply decoupling for an op amp minimizes power supply transients.

the reference terminal for possible transients coupled to power supply lines. The problem to solve is described in Figure 3.40a, where different circuits share common impedances in power supply lines and ground. The drop in voltage across those impedances because of fluctuating supply currents yields voltage fluctuations on the supply terminals of a given component. Op amps have a limited ability to reject these fluctuations, as described by their power supply rejection ratio (PSRR), which is the ratio of power supply change to the equivalent input voltage change that would lead to the observed output voltage change. The PSRR falls off when the frequency increases. For the OPA 605, for example, which is a 200 MHz bandwidth op amp, the PSRR is 100 dB at dc, 70 dB at 1 kHz, and 30 dB at 1 MHz. Some op amps have negative PSRR at high frequency, meaning that power supply fluctuations at those frequencies are in fact amplified. The μA741 can even oscillate if a large transient voltage is applied to one of its supply terminals.

Power supply decoupling aims to reduce the amplitude of transient voltages on op-amp (or IA) supply lines by forming a voltage divider with the impedance of the supply lines and a conveniently placed capacitor. Usually ceramic capacitors of about 100 nF are used because of their low inductance, shunted by 10 μF tantalum capacitors that provide low impedance at frequencies below about 20 kHz. Their mounting leads must be as short as practicable, and they must placed near the amplifier case. The manufacturer may recommend a decoupling circuit different from that in Figure 5.9.

5.2.4 Electrostatic Shields and Driven Shields

Some capacitive sensors have so high an impedance that we cannot neglect the parasitic capacitances between the sensor and its environment. These parasitic impedances change when the sensor moves with respect to nearby conductors, thus leading to interference.

An electric shield has been defined in Section 3.6.1 as a conductive surface enclosing the component or circuit of interest and connected to a fixed potential. Capacitive sensors are shielded to keep the capacitance constant in the presence of any changes in its electric environment.

There are several options for connecting the shield. In Figure 5.10a, the total capacitance will be $C_t = C + C_1$, and it will remain constant, while the capacitance to ground C_G will change depending on the relative position of the conductors. If the ground terminal is part of the measurement circuit, C_G causes an error. This kind of shield is therefore used when we can connect a terminal of C to ground. Otherwise it is better to use a double shield such as that in Figure 5.10b. The added shield keeps C_G constant even if nearby conductors change their positions.

Shielding thus keeps parasitic capacitances constant but does not reduce them. In fact, shielding increases the value for parasitic capacitance, particularly when it is extended to the cables connecting the sensor to the amplifier (coaxial cables) as is usually done. This increase in capacitance decreases the

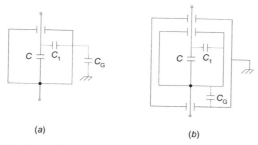

Figure 5.10 Single (*a*) and double (*b*) shield for a capacitive sensor.

sensitivity because the measurand obviously causes only the sensor capacitance to change. Instead of grounding the shield, the parasitic capacitance can be reduced by connecting it to a voltage close to that of the conductors inside it. This technique is known as a driven shield and requires the use of electronic circuits before or included in the ac amplifier. Op amp characteristics influence the efficiency of driven shields.

Figure 5.11 shows a coaxial cable with grounded and driven shield, and the equivalent circuit for driven shield. When the cable shield is grounded (Figure 5.11*a*) the cable capacitance shunts the signal source and the input capacitance of the amplifier. If instead the shield is connected to a voltage close to that of the internal conductor, Figure 5.11*b*, then we have a driven shield. From the equivalent circuit in Figure 5.11*c* we have

$$V_o = A_d(V_p - V_o) = A_d(I_2 Z_c - V_o) \qquad (5.26)$$

$$V_s = I_1(Z + Z_s) - I_2 Z + V_o \qquad (5.27)$$

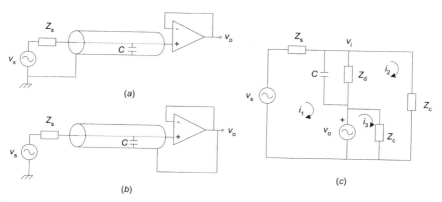

Figure 5.11 (*a*) Common electric shield or guard; (*b*) driven shield; (*c*) equivalent circuit for analyzing a driven shield.

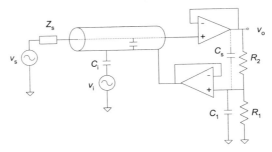

Figure 5.12 Driven shield based on a separate voltage buffer to reduce interference coupled to the shield and a voltage divider to reduce positive feedback.

$$0 = -I_1 Z + I_2(Z + 2Z_c) - I_3 Z_c \qquad (5.28)$$

$$V_o = (I_3 - I_2)Z_c \qquad (5.29)$$

where $Z = Z_d \| (1/j\omega C)$. From these equations we deduce that the input impedance is

$$\frac{V_s}{I_1} = (A_d + 1)Z \| Z_c \qquad (5.30)$$

That is, the capacitive impedance of the cable (and the differential input impedance as well) is multiplied by $A_d + 1$. Therefore, the effective input capacitance for the cable is reduced by a factor somewhat higher than the op-amp open-loop gain. This decreases from values higher than 10^6 at dc, to a value between 10 and 100 at 1 MHz for wide-bandwidth op amps. The higher the value for A_d at the working frequency, the larger the reduction in parasitic capacitance.

Nevertheless, the output impedance of the op amp together with the stray capacitance from the inverting terminal to ground may reduce the net negative feedback. The positive feedback from the output to the cable shield may then lead to oscillation. Hence, it may be convenient to attenuate the signal fed back to the shield. On the other hand, interfering currents along the shield will yield a drop in voltage across the op-amp output impedance. Driving the shield by an op amp different from the signal amplifier solves the problem. Figure 5.12 shows the implementation of both techniques. Usually R_1 is about 100 times R_2.

5.2.5 ac/dc Signal Converters

Applications that do not need to determine the phase of the ac signal can use one of three basic methods to obtain a dc voltage from an ac voltage: rms-to-dc conversion, peak detection, and ac-to-MAV (mean absolute value) conversion (Sections 9.4 and 9.6 in reference 5 and Chapter 5 in reference 6).

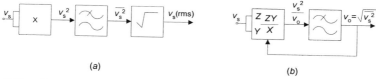

(a)   (b)

Figure 5.13 Computation-based rms-to-dc converters. (*a*) Direct or explicit computation. (*b*) Indirect or implicit computation.

A sine voltage $v_s = V_p \sin(\omega t + \phi)$ has an rms (root mean square) value

$$V_s(\text{rms}) = \sqrt{\frac{1}{T} \int_0^T v_s^2(t)\, dt} = \sqrt{\frac{\omega}{2\pi} \int_0^{2\pi/\omega} V_p^2 \sin^2(\omega t + \phi)} = \frac{V_p}{\sqrt{2}} \qquad (5.31)$$

The peak value divided by the rms value is the *crest factor* (CF). A sine wave has $\text{CF} = \sqrt{2}$. Figure 5.13*a* shows the block diagram of a circuit to compute (5.31) that includes a multiplier (squarer), a low-pass filter (averager), and a square root circuit. Figure 5.13*b* shows how to obtain the rms value by implicit computation based on a multiplier/divider IC and feedback. The squaring operation limits the input dynamic range. There are IC rms-to-dc converters based on each of these methods.

The rms value is alternatively defined as the amplitude of a dc voltage that dissipates the same amount of heat in a resistor as the original signal does. Figure 5.14 shows a thermal rms-to-dc converter that implements this definition. The input signal is applied to a heater, and the output is applied to a similar but thermally isolated heater. The feedback circuit raises the output voltage until both heaters reach the same temperature.

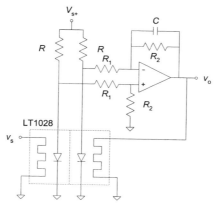

Figure 5.14 In a thermal rms-to-dc converter the output voltage is increased until both heaters reach the same temperature.

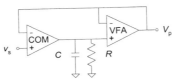

Figure 5.15 Peak detector based on a comparator and a capacitor. Overall feedback reduces op amp errors.

A simple peak detector relies on a comparator and a capacitor acting as a memory (Figure 5.15). The input signal is compared with the stored value, which is updated when it is below the current input. The larger the gain of the comparator, the faster the capacitor is charged. R slowly discharges C to enable tracking fluctuations in the peak value. The op amp should have low drift and bias current. Alternatively, a switch can replace R.

Ac/MAV converters rely on the particular relation between the rms voltage of a sine wave and its mean absolute value after rectification. For a full-wave rectified sine wave we have

$$V_s(\text{MAV}) = \frac{\omega}{\pi} \int_0^{\pi/\omega} V_p \sin \omega t\, dt = \frac{2V_p}{\pi} \tag{5.32}$$

The ratio (rms value)/MAV is termed *form factor* (FF). Therefore, for a sine wave FF $= \pi/(2\sqrt{2}) = 1.11$—that is, 1.11 times the average value after rectification yields the rms value. FF obviously depends on the waveform.

Ac/MAV converters consist of a rectifier and low-pass filter. The circuit in Figure 5.16 merges both functions. Negative inputs turn D1 on and D2 off, so that $v_1 = 0$ V. Ignoring C_1, the output amplifier yields

$$v_o = -\frac{R_5}{R_4} v_s \tag{5.33a}$$

Positive inputs turn D1 off and D2 on, so that $v_1 = -v_s R_2/R_1$, and the output amplifier adds this voltage and v_s to yield

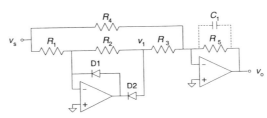

Figure 5.16 Ac/MAV converter based on a full-wave rectifier and a low-pass filter.

$$v_o = -v_s \frac{R_5}{R_4} + v_s \frac{R_2}{R_1} \frac{R_5}{R_3} \qquad (5.33b)$$

If $R_2 R_4 = 2R_1 R_3$, the gain for negative inputs equals that for positive inputs. R_5 does not affect the gain balance and can be selected to provide the desired gain. In addition, R_5 and C_1 form a low-pass filter to obtain a dc output. C_1 can be calculated from the acceptable ripple. The coefficients for the Fourier series for a rectified sinusoid are

$$a_0 = \frac{2}{\pi}$$
$$\qquad (5.34)$$
$$a_n = \frac{\cos(n\pi) + 1}{\pi(n^2 - 1)}$$

and, hence, there are no odd harmonics and the amplitude of even harmonics decreases faster than the slope of the first-order filter formed by R_5 and C_1. Therefore, C_1 can be designed to attenuate the second harmonic as required.

Example 5.4   The circuit in Figure E5.4 is the signal conditioner for a capacitive level sensor that has $C_{\min} = 41.46$ pF, $C_{\max} = 87.07$ pF, and sensitivity 0.19 pF/L (Example 4.1). Design the circuit components to obtain a frequency-independent voltage that is 0 V for the empty tank and 1 V for the full tank.

The circuit is an ac bridge whose output current is converted into a voltage through the op amp. Because the equivalent output impedance for the voltage divider that includes the sensor is capacitive, the transimpedance must also be capacitive. Hence, R is added only to bias the op amp. The output is high-pass filtered by C_4 and R_4, and then rectified and low-pass filtered to obtain the MAV. For an ideal op amp, circuit analysis yields

$$(V_e - V_p)C_1 s = (V_p - V_a)Cs + V_p C_x s$$

$$V_p = V_e \frac{R_3}{R_2 + R_3}$$

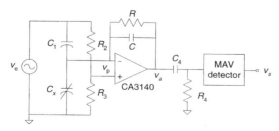

Figure E5.4   Signal conditioner for a capacitive level sensor.

which leads to

$$v_a = v_e \left[\frac{C_1}{C} - \frac{R_3}{R_2 + R_3} \left(\frac{C_1 + C_x}{C} + 1 \right) \right]$$

A null output for an empty tank poses the condition

$$\frac{C_1}{C} = \frac{R_3}{R_2 + R_3} \left(\frac{C_1 + C_{min}}{C} + 1 \right)$$

$$R_2 C_1 = R_3 (C + C_{min})$$

When this condition is fulfilled, the op-amp output is

$$v_a = v_e \frac{R_2 C_1 - R_3 (C + C_x)}{(R_2 + R_3) C} = v_e \frac{R_3}{R_2 + R_3} \frac{C_h}{C}$$

where $C_h = C_x - C_{min}$. In order to obtain 1 V at the output when the tank is full we need

$$k V_e \frac{R_3}{R_2 + R_3} \frac{C_{max}}{C} = 1 \text{ V}$$

where k is the ratio between the MAV and the peak value V_e. For a sine wave we have

$$k = \frac{\text{MAV}}{V_e} = \frac{V_e(\text{rms})}{\text{FF}} \frac{1}{V_e(\text{rms}) \times \text{CF}} = \frac{2\sqrt{2}}{\pi} \frac{1}{\sqrt{2}} = \frac{2}{\pi}$$

Therefore,

$$\frac{2}{\pi} V_e \frac{R_3}{R_2 + R_3} \frac{87.07 \text{ pF}}{C} = 1 \text{ V}$$

If we select $R_2 = R_3$ and $V_e = 10$ V, we need

$$C = \frac{870.7 \text{ pF}}{\pi} = 277 \text{ pF}$$

We can select $C = 270$ pF. The condition for a zero output for an empty tank then leads to

$$C_1 = C + C_{min} = 277 \text{ pF} + 41.46 \text{ pF} = 318.5 \text{ pF}$$

C_1 can be obtained by a 300 pF capacitor shunted by a 20 pF trimmer.

The excitation frequency will be limited by the slew rate of the op amp. The CA3140 has SR $= 7$ V/μs. Because the maximal op-amp output corresponds to the full tank, the condition to fulfill is

$$\left.\frac{dv_a}{dt}\right|_{\text{max}} = 2\pi f_e V_e \frac{R_3}{R_2 + R_3} \frac{C_{\text{max}}}{C} < 7 \text{ V/μs}$$

Using the output condition for the full tank, we obtain

$$2\pi f_e \frac{1 \text{ V}}{k} = 2\pi f_e \frac{\pi}{2}(1 \text{ V}) < 7 \text{ V/μs}$$

That yields $f_e < 709$ kHz. Nevertheless, the CA3140 has an open-loop bandwidth of 10 MHz. Therefore, in order to have a loop gain larger than 100 at the working frequency, this should not exceed 100 kHz.

In order for R not to influence the output, it should be larger than the impedance of C at f_r. Because $R = 10$ MΩ is the maximal common value for carbon film resistors, the condition to fulfill is

$$f_e > \frac{1}{2\pi \times (270 \text{ pF}) \times (10 \text{ MΩ})} = 59 \text{ Hz}$$

This condition is not restrictive at all. Therefore, we can set $f_e = 10$ kHz.

The high-pass filter should not attenuate f_e. If we select its corner frequency to be 1 kHz, and $R_4 = 10$ kΩ to prevent excessive loading at the op-amp output, we need

$$C_4 = \frac{1}{2\pi \times (1 \text{ kHz}) \times (10 \text{ kΩ})} = 16 \text{ nF}$$

5.3 CARRIER AMPLIFIERS AND COHERENT DETECTION

5.3.1 Fundamentals and Structure of Carrier Amplifiers

A carrier amplifier is required for all sensors whose output is an amplitude-modulated (AM) ac signal and that respond to positive and negative values for the measurand. That is the case, for example, for LVDTs, for reactance variation and resistive sensors placed in an ac voltage divider or bridge (particularly differential sensors), for flux-gate sensors, for SQUIDs, and for electromagnetic flowmeters.

A carrier amplifier is a circuit that performs the functions of ac amplification, demodulation, and low-pass filtering, including the necessary oscillator, as shown in Figure 5.17. Carrier amplifiers are available in monolithic form [NE

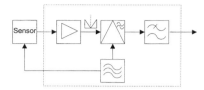

Figure 5.17   A carrier amplifier includes an oscillator that drives the sensor, an ac amplifier, and a coherent demodulator consisting of a multiplier and a low-pass filter.

5521 (Signetics) and AD598 and AD698 (Analog Devices)] but can be built from discrete parts to tailor them to the application in hand. Lock-in or coherent amplifiers are carrier amplifiers whose oscillator drives the measured system rather than the sensor. That is the case, for example, of optical and other radiation-based sensors when the incoming radiation is chopped to feed the sensor a square waveform whose two levels are the unknown level and a reference level. When the signal is to be further processed by a computer, it is possible to implement a digital lock-in amplifier by sampling in synchrony with the reference signal; no analog processing other than amplification is then necessary [7].

The amplitude modulation in ac sensors arises from the product of the supply voltage times the variable to be measured. For example, the output of a voltage divider incorporating a differential sensor is

$$v_o = v_e \frac{1 + x}{2} \tag{5.5}$$

If the driving voltage is sinusoidal with peak value V_e,

$$v_e(t) = V_e \cos 2\pi f_e t \tag{5.35}$$

and, to simplify matters, we assume that the measurand undergoes sinusoidal variations and induces relative changes in impedance that are also sinusoidal with peak value X,

$$x(t) = X \cos(2\pi f_x t + \phi_x) \tag{5.36}$$

from (5.5) we have

$$
\begin{aligned}
v_o(t) &= V_e \cos 2\pi f_e t \frac{1 + X \cos(2\pi f_x t + \phi_x)}{2} \\
&= \frac{V_e}{2} \cos 2\pi f_e t + \frac{V_e X}{4} \{\cos[2\pi(f_e + f_x)t + \phi_x] + \cos[2\pi(f_e - f_x)t - \phi_x]\}
\end{aligned}
\tag{5.37}
$$

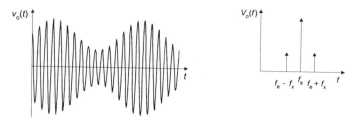

Figure 5.18 Ac voltage dividers yield double-sideband transmitted carrier (DSBTC) AM signals whose spectrum includes that of the signal driving the voltage divider and that of the measurand upwardly translated around the exciting frequency.

which is a double-sideband transmitted carrier (DSBTC) AM signal (Figure 5.18).

Similarly, the output of an ac bridge incorporating a differential sensor is

$$v_0 = v_e \frac{x}{2} \tag{5.7}$$

If the excitation voltage is (5.35) and the measurand induces relative changes in impedance that are sinusoidal (5.36), from (5.7) we have

$$
\begin{aligned}
v_0(t) &= V_e \cos 2\pi f_e t \frac{X \cos(2\pi f_x t + \phi_x)}{2} \\
&= \frac{V_e X}{4} \{\cos[2\pi(f_e + f_x)t + \phi_x] + \cos[2\pi(f_e - f_x)t - \phi_x]\}
\end{aligned} \tag{5.38}
$$

which is a double-sideband suppressed carrier (DSBSC) AM signal (Figure 5.19).

The input ac amplifier may detect voltage or current depending on the sensor type, so that its output is linear according to (5.9) or (5.10). Usually, $f_e \gg f_x$

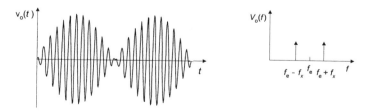

Figure 5.19 Balanced ac bridges yield double-sideband suppressed carrier (DSBSC) AM signals whose spectrum is that of the measurand upwardly translated around the exciting frequency.

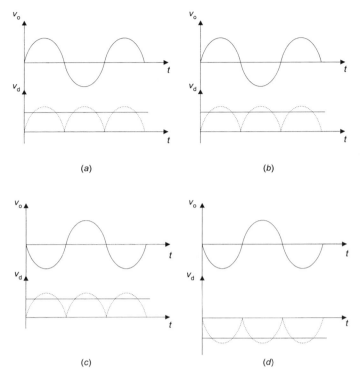

Figure 5.20 Phase recovery problem in amplitude demodulation. In cases (a) and (c) the signal $v_{\mathrm{o}}(t)$ is demodulated by simple rectification and low-pass filtering, and the information about the sign is lost. In cases (b) and (d) the information about the sign is recovered through phase-sensitive demodulation.

and we need a narrow-band amplifier centered on f_{e}. The demodulator must recover X, f_x, and ϕ_x. The demodulation must be synchronous in order to recover both the amplitude and phase of $x(t)$. Using, for example, a simple envelope detector (rectification followed by low-pass filtering) would not obtain information about the sign of $x(t)$. In Figure 5.20, cases a and b, which are based on simple rectification yield the same output in spite of the respective inputs $v_{\mathrm{o}}(t)$ having opposite signs. Cases c and d, which are based on synchronous rectification, yield outputs with the same amplitude but different signs, corresponding to the respective signal $v_{\mathrm{o}}(t)$.

Phase-sensitive (coherent or synchronous) demodulation consists of multiplying the modulated signal $v_{\mathrm{o}}(t)$ by a reference signal $v_{\mathrm{r}}(t)$ in phase with the carrier (excitation) signal $v_{\mathrm{e}}(t)$ and then filtering the resulting signal with a low-pass filter (Figure 5.21). If the reference signal is sinusoidal,

$$v_{\mathrm{r}}(t) = V_{\mathrm{r}} \cos 2\pi f_{\mathrm{r}} t \tag{5.39}$$

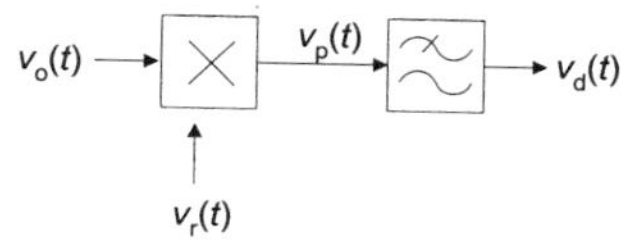

Figure 5.21 Phase-sensitive (synchronous or coherent) demodulation. The modulated signal $v_o(t)$ has been obtained from a carrier in phase with the reference voltage $v_r(t)$.

and the input signal comes from an ac bridge, the output of the multiplier is

$$v_p(t) = v_r(t)v_o(t) = v_r(t)v_e(t)\frac{x(t)}{2}$$

$$= \frac{x(t)}{2}\frac{V_r V_e}{2}[\cos 2\pi(f_e - f_r)t + \cos 2\pi(f_e + f_r)t] \qquad (5.40)$$

If, in addition to the same phase, the excitation and reference signals have the same frequency, $f_e = f_r$, we obtain

$$v_p(t) = \frac{x(t)}{2}\frac{V_r V_e}{2}[1 + \cos 2\pi 2f_e t] \qquad (5.41)$$

The low-pass filter suppresses the high-frequency component ($2f_e$), so that the demodulated output is

$$v_d(t) = \text{lpf}\{v_p(t)\} = \frac{V_r V_e}{2}\frac{x(t)}{2} = \frac{V_r V_e}{4} X \cos(2\pi f_x t + \phi_x) \qquad (5.42)$$

where lpf{ } designates the low-pass filtering function. Therefore, we obtain the amplitude of $x(t)$, scaled by $V_e V_r/4$, and its phase ϕ_x. V_e and V_r must be highly stable because possible fluctuations would be interpreted as produced by x. The AD2S99 from Analog Devices and the SWR300 from Thaler are examples of IC reference oscillators.

Figure 5.22 shows the spectrum changes in amplitude modulation—for example, in an ac bridge [see (5.7)]—and coherent demodulation when the measurand induces a fractional impedance change with spectrum $X(f)$ and maximal frequency f_m. $H(f)$ is the transfer function for the output low-pass filter (LPF), which must pass up to f_m. The system behaves as a narrow-band filter centered on f_e. For a given LPF order, the closer f_m is to f_e (and f_r), the smaller the attenuation of the frequency band to reject (on both sides of $2f_e$) will be. If $f_e > 10f_m$, a first-order LPF may be enough to reject any carrier-induced ripple. The bandwidth of the ac amplifier preceding the multiplier must be at least $0.2f_e$. f_e is selected so that the noise contribution from this amplifier is minimum. Typical excitation frequencies for inductive sensors range from 5 kHz to 10 kHz, and the maximal accepted bandwidth for the modulating signal is in general from 0 Hz to 500 Hz or 1500 Hz. The excitation frequency

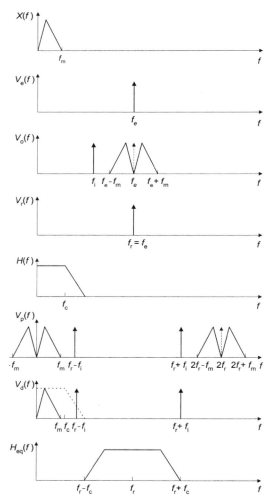

Figure 5.22 Spectrum changes involved in the amplitude modulation of an excitation signal $v_e(t)$ by a measurement signal $x(t)$ to yield $v_o(t)$, and further demodulation consisting of the product by a reference signal $v_r(t)$ to yield $v_p(t)$, followed by low-pass filtering with a filter $H(f)$. The process is equivalent to a bandpass filter centered on the excitation frequency. The rejection of interference of frequency f_i depends on its relative distance to the excitation frequency. The dashed line in $V_d(f)$ is the desirable frequency response of a bandpass amplifier preceding the demodulator.

for capacitive sensors usually ranges from 10 kHz to 500 kHz, with a maximal signal frequency of up to 25 kHz for the highest carrier frequencies.

Figure 5.22 also illustrates the interference-rejection capability of synchronous demodulation. An interference of frequency f_i and peak amplitude V_i added to the modulated signal yields two frequency components $f_r - f_i$ and $f_r + f_i$ at the output of the multiplier. Both components are rejected by the low-pass filter; but depending on $f_r - f_i$, this component may fall inside the filter passband. The amplitude response of a Butterworth LPF with corner frequency f_c and order n is

$$|H(f)| = \frac{1}{\sqrt{1 + \left(\dfrac{f}{f_c}\right)^{2n}}} \tag{5.43}$$

and the output voltage contributed by the interference is

$$|v_d|_i = \frac{V_r V_i}{2\sqrt{1 + \left(\dfrac{f_e - f_i}{f_c}\right)^{2n}}} \tag{5.44}$$

The capability to reject interference added to the input signal is described by the *series* (or *normal*) *mode rejection ratio* (SMRR, NMRR), defined as the quotient between the response to the signals of interest and the response to the interference. The SMRR is usually expressed in decibels. If we apply (5.44) first to a signal with frequency f_e and then to an interference with frequency f_i, we obtain

$$\text{SMRR} = 20 \lg \left| \frac{v_d(f_e)}{v_d(f_i)} \right| = 10 \lg \left[1 + \left(\frac{f_e - f_i}{f_c} \right)^{2n} \right] \approx 20n \lg \frac{|f_e - f_i|}{f_c} \tag{5.45}$$

where the approximation is valid when $f_i \ll f_e$. The inability to reject interference close to f_e restricts the use of the power line frequency and its harmonics for sensor excitation. The SMRR increases with the order of the low-pass filter and also when the input ac amplifier rejects the interference. Hence, it is advisable to use a bandpass amplifier or a combination of preamplifier and bandpass filter.

Example 5.5 Determine the excitation and corner frequencies for a carrier amplifier based on a second-order Butterworth LPF, able to measure a 5 Hz signal with an amplitude error smaller than 1 LSB for a 16 bit ADC and at the same time providing more than a 120 dB attenuation for a 60 Hz interference added at the input of the demodulator.

The amplitude error allowed is $1/2^{16}$. According to (5.43), the condition to fulfill is

$$\frac{1}{\sqrt{1 + \left(\dfrac{5\text{ Hz}}{f_c}\right)^4}} > 1 - \frac{1}{2^{16}}$$

which yields

$$f_c > \frac{5\ \text{Hz}}{\sqrt[4]{\left(\dfrac{2^{16}}{2^{16}-1}\right)^2 - 1}} = 67.27\ \text{Hz}$$

We select $f_c = 68$ Hz to be on the safe side. In order to achieve $\text{SMRR}(60\ \text{Hz}) = 120$ dB, from (5.45) we need

$$120 = 20 \times 2 \times \lg \frac{f_e - 60\ \text{Hz}}{68\ \text{Hz}}$$

Solving for f_e yields $f_e = 68$ kHz. This is a relatively high frequency. If the interference is superimposed on the incoming signal, a bandpass amplifier would contribute to its attenuation and the 120 dB could be shared between the amplifier and the demodulator, thus leading to a lower f_e.

Synchronous demodulation of a double-sideband transmitted carrier (DSBTC) AM signal such as those from ac voltage dividers yields, in addition to the signal of interest $x(t)$ as in (5.42), a large dc output due to the excitation signal (at $f_e = f_r$). If the measurand has very low frequency components, separating them from that dc voltage by an output high-pass filter may be difficult. Therefore, sensors for quantities showing slow variations are better placed in an ac bridge (balanced at a reference condition) than in a voltage divider.

5.3.2 Phase-Sensitive Detectors

The key element in a carrier amplifier is the demodulator, which is called a *phase-sensitive demodulator* because it can detect polarity changes. Equation (5.42) shows that phase-sensitive (synchronous) demodulation can be performed by multiplying the modulated signal by a reference voltage synchronous with the carrier and then filtering with a low-pass circuit. If the reference voltage is sinusoidal as in (5.39), the technique is termed *homodyne detection*. However, precision analog multipliers are expensive.

Simpler synchronous demodulators use a symmetrical square waveform with amplitude $+V_r$ and $-V_r$ as reference. Its Fourier series is

$$v_r(t) = \frac{4V_r}{\pi} \sum_0^\infty (-1)^n \frac{\cos 2\pi(2n+1)f_r t}{2n+1} \tag{5.46}$$

and its spectrum consists of the odd harmonics of f_r with decreasing amplitude. The product $v_r(t) \times v_o(t)$ implies the convolution of $V_r(f)$ and $V_o(f)$ [8]. From Figure 5.22 we infer that this convolution produces a base-band component—from f_e in $v_o(t)$ and $f_r = f_e$ in $v_r(t)$—and intermodulation components at

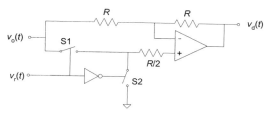

Figure 5.23   Selecting the $+1$ or -1 gain of this switched-gain amplifier according to the carrier frequency of the input signal implements a phase-sensitive demodulator.

$3f_e - f_e = 2f_e$, $5f_e - f_e = 4f_e$ and so on, which are rejected by the output LPF. The output voltage will be

$$v_d(t) = \text{lpf}\{v_p(t)\} = \frac{4V_r}{\pi}\frac{V_e}{2}\frac{x(t)}{2} = \frac{V_r V_e}{\pi} X \cos(2\pi f_x t + \phi_x) \qquad (5.47)$$

which differs from (5.42) only by the scaling factor. Hence, the phase of $x(t)$ is also preserved at the output. But now we have two definite advantages. First, the output still depends on the amplitude of the reference signal V_r, but now this is a square wave and therefore it is easier to keep it constant than for a sine wave. Second, the product can be computed using a simple polarity detector (gain of $+1$ or -1), which is cheaper than an analog multiplier. These phase-sensitive detectors are called *switched-gain detectors*. A shortcoming is that interference or noise in $v_o(t)$ having frequency components at $(2n+1)f_e$ will contribute to the output. Therefore, it is convenient to place a bandpass filter before the demodulator.

The interest of this and other related applications has led to the development of monolithic circuits that integrate the amplifier and the necessary switches for gain commutation. Some examples are OPA675/6 (Burr–Brown), AD630 (Analog Devices), and HA2400/04/05 (Harris).

Figure 5.23 shows a switched-gain amplifier implemented by discrete components. When S1 is off, S2 is on and the output is

$$v_d = -v_o + 2v_o \frac{R_{ON}}{R_{ON} + R_{OFF}} \approx -v_o \qquad (5.48a)$$

When S1 is on, S2 is off and

$$v_d = -v_o + 2v_o \frac{R_{OFF}}{R_{ON} + R_{OFF}} \approx v_o \qquad (5.48b)$$

Therefore, if the control signals for S1 and S2 are obtained from the excitation voltage, we have a synchronous demodulator. Implementing the circuit by an IC

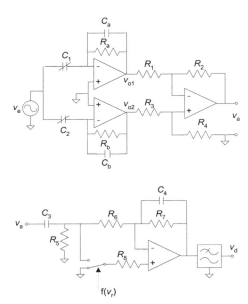

Figure E5.6 Amplifier and switched-gain coherent demodulation for a differential capacitive sensor.

unity-gain integrated difference amplifier (and a SPDT analog switch) improves resistor matching.

Example 5.6 The measurement range for a given displacement differential capacitive sensor is from -1 cm to $+1$ cm. Each capacitor has 10 pF for null displacement and sensitivity 0.1 pF/mm. The sensor is connected to the circuit in Figure E5.6, where the switch is controlled from the excitation voltage supplied to the sensor. Determine the condition to be fulfilled by the circuit components in order for the output voltage of the differential amplifier v_a to be linear with the displacement and independent of the supply frequency. If the excitation voltage has 5 V peak at 20 kHz, design the component values in the first stages to obtain $v_a = 3$ V (peak) at range ends. Design the component values for the HPF, switching-amplifier, and LPF to obtain FSO $= 5$ V.

Each sensor capacitor is connected to a I/V converter whose transimpedance is an RC network. Because the input current to the converter depends on the sensor capacitance, if the resistor has much higher impedance than the feedback capacitor, the converter output will not depend on the signal frequency. The output from the differential amplifier is high-pass filtered and then demodulated by a switched-gain amplifier and LPF. The output of the differential amplifier is

$$V_a = k(V_{o2} - V_{o1}) = kV_r\left(\frac{Z_b}{Z_2} - \frac{Z_a}{Z_1}\right) = kV_r\left(\frac{C_2}{C_b} - \frac{C_1}{C_a}\right)$$

where k is the gain of the differential amplifier and in the last step we have assumed that the impedances of R_a and R_b are much higher than those of the shunting capacitors. If we select $C_a = C_b$ and the sensor is linear with $C_1 = C_0(1 - x)$ and $C_2 = C_0(1 + x)$, we have

$$V_a = kV_r \frac{C_0}{C_a} 2x$$

A displacement of 1 cm produces a 1 pF change in each capacitor, so that $x = 0.1$. In order to obtain 3 V when supplying 5 V we need

$$3 \text{ V} = k(5 \text{ V}) \frac{10 \text{ pF}}{C_a} \times 2 \times 0.1$$

If we select $C_a = 10$ pF, we need $k = 3$. Then, we can select, for example, $R_1 = R_3 = 10$ kΩ and $R_2 = R_4 = 30.1$ kΩ all metal film resistors with 0.1 % tolerance. R_a and R_b must fulfill the condition

$$R_a \gg \frac{1}{2\pi(20 \text{ kHz})(10 \text{ pF})} = 800 \text{ k}\Omega$$

We can select, for example, $R_a = R_b = 10$ MΩ. These resistors need not be accurate. V_d is the MAV obtained by synchronous rectification. Therefore,

$$V_d = \frac{1}{\pi} \int_0^\pi V_{ap} \sin\theta \, d\theta = V_{ap} \frac{2}{\pi}$$

In order to have 5 V at the output, we need a gain

$$G = \frac{5 \text{ V}}{3 \text{ V} \times \frac{2}{\pi}} = 2.6$$

If we select the corner frequency of the high-pass filter to be 2 kHz (one decade below the carrier frequency) and $R_5 = 10$ kΩ to prevent excessive loading to the differential amplifier, we need $C_4 = 8$ nF. In the switched-gain amplifier we need $R_6 = R_7$. For example, $R_6 = R_7 = 10$ kΩ. R_8 permits the matching of the resistance seen from the inverting and noninverting inputs. When the gain is -1, the resistance seen from the inverting input is the parallel combination of R_7 and the series combination of R_5 and R_6. Therefore, we need $R_8 = 6.7$ kΩ. C_4 limits the bandwidth and should not affect the carrier signal. To achieve, for example, 100 kΩ at 20 kHz it should be $C_4 = 80$ pF.

An alternative reference signal for phase-sensitive demodulation is a periodic train of unit-amplitude pulses with frequency $f_r = f_e/k$, where k is any integer.

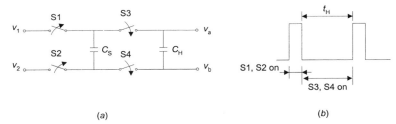

(a)   (b)

Figure 5.24 A floating capacitor clocked by a signal synchronous with the carrier frequency works as a coherent detector.

Multiplying a signal by a pulse train is sampling the signal at time intervals equal to the train period. Figure 5.24 shows this method implemented by the switched capacitor technique [9]. S1 and S2 are on for a short time, and C_s charges to $v_1 - v_2$. Then S1 and S2 turn off and S3 and S4 turn on for a time t_H, so that, in steady state, C_H charges to $v_1 - v_2$ and holds this voltage during t_H. This holding action is equivalent to a low-pass filter. The input voltage can be differential or single-ended. For the differential case, the CMRR is excellent. The voltage across C_H can be measured by a single-ended amplifier, as in Figure 3.31d. Synchronous sampling can also be applied when digitizing signals using an ADC. In any case, because a pulse train has a comb-like spectrum, the signal to be demodulated must be bandpass-filtered to prevent the demodulation of spectral components other than those centered at the carrier frequency.

Applications requiring the detection of the phase shift between two signals of the same frequency but not of their amplitude may use a kind of zero-crossing detectors termed phase comparators. Figure 5.25 shows two possible circuits. First the signals are squared to have only two voltage levels "1" and "0." Squaring can be performed by a comparator—including hysteresis for noisy signals—and does not modify the phase. In Figure 5.25a the flip-flop generates a train of pulses whose width equals the delay between zero crossings of the "set" and "reset" inputs. In Figure 5.25b, the EXCLUSIVE-OR gate yields a rectangular pulse train whose frequency is twice that of the input signals and the pulse width equals the delay between zero crossings. Some phase-locked loops (PLLs) include this kind of phase detector.

To obtain a dc output proportional to the phase shift we can measure the average value of the output pulse train using a low-pass filter. The result is proportional to τ/T_0 in Figure 5.25a and to $2\tau/T_0$ in Figure 5.25b. Because phase shifts close to 0° would yield large errors, we often add 180° or another known phase shift to obtain large dc outputs and then subtract the added value. The circuit in Figure 5.25b has the advantage of having a ripple whose frequency is twice that of the carrier, and hence easier to filter out. An alternative approach to obtain a dc output is to charge a capacitor with a constant current during the time interval when the detector output is at high level.

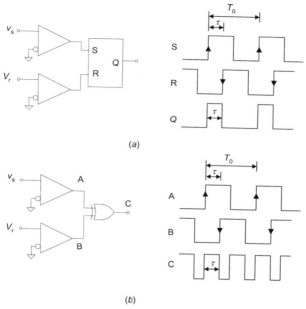

Figure 5.25 Phase detectors based on zero-crossing detectors. (*a*) Using a flip-flop. (*b*) Using an EXCLUSIVE-OR gate.

5.3.3   Application to LVDTs

LVDTs yield an AM voltage whose amplitude is sometimes high enough to allow its demodulation without amplification [10]. Detecting the direction of the core displacement with respect to the central position requires synchronous demodulation.

The simplest solution consists of obtaining a continuous voltage from each secondary winding, then rectifying and subtracting. The sign of the output voltage will indicate the core position. The rectification can be half-wave (Figure 5.26*a*) or full-wave (Figure 5.26*b*). No reference voltage from the primary winding is required, as contrasted to phase detectors based on multiplication. This is due to the particular form for the output signal that is given by three or four terminals while the output of ac bridges comes from two terminals. An additional requirement here is that the output or display device must be differential. A shortcoming is that the diodes must work with voltages larger than their threshold, and this is not always possible at the end of the range of some LVDTs. If we must implement circuits that work as ideal diodes, then we would lose the principal advantage of this method, namely its simplicity.

Carrier amplifiers offer the best solution. But the phase-sensitive detector yields an output dependent also on the phase shift between primary and secondary windings. Therefore, either the LVDT must be supplied at its natural

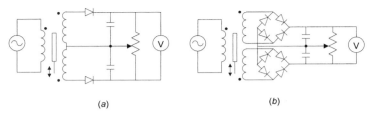

(a)   (b)

Figure 5.26 Phase-sensitive demodulators for LVDTs, based on (a) a half-wave rectifier or (b) a full-wave rectifier.

frequency [equation (4.36)], where the phase shift is zero, or the expected phase shift at the working frequency must be compensated by means of a phase-leading or -lagging network in the primary or secondary winding (Figure 4.21), or the reference signal for demodulation must be phase-shifted. In LVDTs with three or four output terminals, adding the two outputs $(e_{o1} + e_{o2})$ yields a reference for demodulation. The quotient $(e_{o1} - e_{o2})/(e_{o1} + e_{o2})$ yields the desired information. The AD598 (Figure 5.27a) uses this method. An alterna-

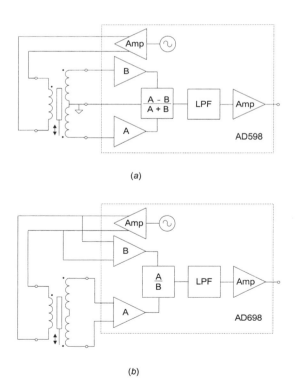

(a)

(b)

Figure 5.27 LVDT signal conditioning based on (a) rectification and subtraction and (b) phase-sensitive demodulation (courtesy of Analog Devices).

tive reference is $2e_{o1} - (e_{o1} - e_{o2})$. The AD698 (Figure 5.27$b$) uses synchronous demodulation to recover amplitude and phase information from $e_{o1} - e_{o2}$ and divides the result by the excitation amplitude to obtain a ratiometric measure. Processors A and B each consist of an absolute value function and a filter. The AD698 can also be applied to other DSBSC AM signals.

5.4   SPECIFIC SIGNAL CONDITIONERS FOR CAPACITIVE SENSORS

Capacitive sensors suit monolithic integration, but bridge circuits with resistors and coils are difficult to integrate. This has led to the development of specific signal conditioners for capacitive sensors suitable for monolithic integration, also amenable to implementation by discrete components. Some conditioners include the capacitive sensor in a variable oscillator (Section 8.3), and others are integrators built from switched capacitors that obtain an output voltage from a difference in electric charge.

The circuit in Figure 5.28 applies the charge redistribution method. There is an autozero phase and a measurement phase. In the autozero phase (Figure 5.28a), a reference dc voltage source charges the sensor C_x at V_r, and the reference capacitor C_r and the integrating capacitor C_i discharge to ground. The op amp output is zero. In the measurement phase, C_x is grounded, C_r is connected to V_r, and C_i closes the op amp feedback loop, hence working as integrator. If $C_x = C_r$, the charge stored in C_x redistributes between them and the op amp output remains at 0 V. But if $C_x \neq C_r$, there is a net charge flow through C_i and the op amp output voltage is proportional to $C_x - C_r$. The output stage is a sample-and-hold amplifier that keeps the last voltage output during the next autozero phase. Reference 11 describes a method to reduce charge injection errors in this circuit.

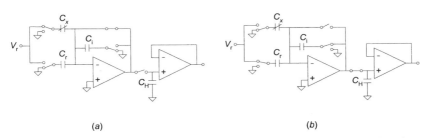

(a)   (b)

Figure 5.28   The charge redistribution method measures capacitance by charging the unknown capacitance to a reference voltage in the autozero phase (a) and then connecting a reference capacitor to the same reference voltage in the measurement phase (b). Any charge not transferred from C_x to C_r charges C_i. The output sample-and-hold amplifier keeps constant the output voltage during the charging phase of C_x.

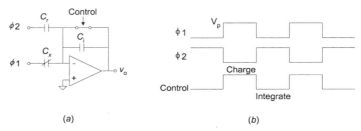

(a)   (b)

Figure 5.29 Capacitance measurement by charge integration and switched capacitances connected to the same voltage level in opposite clock cycles.

Figure 5.29 shows another integrator based on switched capacitors [12]. There are also two capacitors and two phases but instead of charging capacitors at a dc voltage, they are connected to out-of-phase clock signals. The switch resetting C_i is clocked at the same frequency. During the charging phase, C_x charges and the output is zeroed. During the integration phase, the difference in charge between $C_x - C_r$ yields an output voltage

$$v_o = V_p \frac{C_x - C_r}{C_i} \tag{5.49}$$

where V_p is the amplitude of the clock signal. Stray capacitance from the sensor to ground will not interfere as long as the op amp has large dc gain.

The charge transfer method also relies on transferring charge from an unknown capacitor to a known uncharged capacitor. If $C_x \gg C_s$, the voltage across the sampling capacitor in Figure 5.30 will be

$$v_s = V_r \frac{C_x}{C_x + C_s} \approx V_r \frac{C_x}{C_s} \tag{5.50}$$

and the unknown capacitance can be determined from

$$C_x = C_s \frac{v_s}{V_r} \tag{5.51}$$

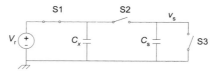

Figure 5.30 Charge transfer method applied to the measurement of a grounded capacitance. First S1 is closed for a short time and C_x charges to V_r. Then S1 opens and S2 is briefly closed to charge C_s to v_s, whose magnitude depends on C_x/C_s.

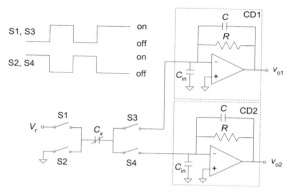

Figure 5.31 Charge transfer method applied to the measurement of an ungrounded capacitance.

In operation, S1 closes momentarily to charge C_x to V_r. Then S1 opens and S2 closes briefly to transfer charge from C_x to the discharged C_s. Later, S2 reopens and v_s is measured. Briefly closing S3 discharges C_s, and the next measurement cycle starts. This circuit has been applied to proximity detection in automatic faucets and humidity measurements [13].

The circuit in Figure 5.31 applies the charge transfer method to an ungrounded capacitance [14]. First, S1 and S3 are closed and S2 and S4 open to charge C_x to a reference voltage V_r. The charging current is converted to a voltage by the current detector CD1. Then S1 and S3 open and S2 and S4 close to discharge C_x to ground potential. This discharging current is converted to a voltage by the current detector CD2. C_{in} (100 nF) ensures that the inverting input of each op amp is kept at virtual ground during the fast charging–discharging cycles. Therefore, stray capacitances do not interfere because C_x has both electrodes connected to low-impedance sources. Feedback capacitors C average the current. The differential output voltage is proportional to the current, hence to C_x. Op amp offset voltages and charge-injection transients are subtracted, hence cancelled provided that they are equal.

The twin-T circuit (Figure 5.32) patented by K. S. Lion in 1964, well before

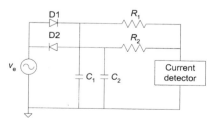

Figure 5.32 Twin-T circuit to measure the difference between two capacitances using voltage excitation and current detection, both grounded.

Figure 5.33 Condenser microphones use a dc bias voltage and yield an output voltage proportional to the displacement of one electrode.

the monolithic integration era, suits sensors with 1 pF to 100 pF capacitance [15]. During the positive half-cycle of the ac source voltage, C_1 charges to the peak value through D1 and C_2 discharges to ground though R_2 and the current detector, which must have slow response. During the negative half-cycle, C_1 discharges to ground through R_1 and the current detector, and C_2 charges to the valley value through D2. The circuit is designed with $R_1 = R_2$. If $C_1 = C_2$ the reading will be zero. But if $C_1 \neq C_2$, the net current through the current detector will be proportional to $C_1 - C_2$. The sensitivity is maximal when the time constants $R_1 C_1$ and $R_2 C_2$ are of the same order of magnitude as the period of the source voltage. Unlike bridge circuits, here the capacitors, the excitation source, and the current detector are all grounded.

Condenser microphones use a dc polarization voltage (above 200 V) rather than ac excitation (Figure 5.33). The capacitor has parallel plates and the acoustic pressure to sense changes the distance between plates. For small displacements relative to the distance between plates, if $R_L C_x \gg 2\pi f$, f being the displacement's frequency, the output voltage is proportional to the displacement and independent of f (reference 16, Chapter 3). Electret microphones apply the same principle but do not require a dc voltage source. Instead, the fixed capacitor electrode has an electret—a plastic (PTFE) or ceramic (CaTiO$_3$) material that traps an electric charge when dc biased at high temperature and retains it for 2 to 10 years when allowed to cool (reference 17, Chapter 14). Commercial microphones integrate an FET transistor to provide high input impedance for voltage detection.

5.5 RESOLVER-TO-DIGITAL AND DIGITAL-TO-RESOLVER CONVERTERS

Resolver-to-digital converters (RDCs) and synchro-to-digital converters (SDCs) give digital signals from input analog-coded angles. Conversely, digital-to-synchro converters (DSCs) and digital-to-resolver converters (DRCs) obtain analog signals from digitally coded angles. All these converters work internally with angles expressed in resolver format, and therefore they implement at their input or output the necessary circuits to convert from one format to another, usually through Scott transformers.

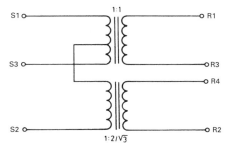

Figure 5.34 Scott transformer to convert angles from synchro (S) to resolver (R) format and conversely.

5.5.1 Synchro-to-Resolver Converters

At its three output terminals, a synchronous transformer or synchro (Section 4.2.4) yields ac voltages that have a fixed frequency and an amplitude that depends on the angle turned by the rotor with respect to the stator. These voltages have the form

$$e_{s13} = K \cos \omega t V \sin \alpha \qquad (5.52)$$

$$e_{s32} = K \cos \omega t V \sin(\alpha + 120°) \qquad (5.53)$$

$$e_{s21} = K \cos \omega t V \sin(\alpha + 240°) \qquad (5.54)$$

where $V \cos \omega t$ is a reference voltage applied to the rotor, and K is a design factor. These three voltages are said to represent the angle α in "synchro format."

At its two output terminals, a resolver yields voltages of the form

$$e_{R13} = KV \sin \alpha \cos \omega t \qquad (5.55)$$

$$e_{R24} = KV \cos \alpha \cos \omega t \qquad (5.56)$$

which represent angle α in "resolver format."

Figure 5.34 shows a Scott transformer that converts one format into another. In addition to voltage-level adaptation, the transformer offers a high isolation between primary and secondary windings. The transformer ratio in Figure 5.34 is 1:1, but different ratios are possible.

To understand the circuit for the Scott transformer, we observe that

$$e_{s32} = KV \cos \omega t V \left[-\frac{1}{2} \sin \alpha + \frac{\sqrt{3}}{2} \cos \alpha \right] \qquad (5.57)$$

$$e_{s21} = KV \cos \omega t V \left[-\frac{1}{2} \sin \alpha - \frac{\sqrt{3}}{2} \cos \alpha \right] \qquad (5.58)$$

and therefore we have

$$e_{R24} = \frac{e_{S32} - e_{S21}}{\sqrt{3}} = \frac{(e_{S32} + e_{S13}/2)2}{\sqrt{3}} \qquad (5.59)$$

The sum indicated in (5.59) is performed in Figure 5.34 by a precision transformer.

In order to calculate the input impedance for each pair of synchro terminals, because the self-inductance is proportional to the square of the number of turns we have

$$L_{S13} = kN^2 \qquad (5.60)$$

$$L_{S12} = k\left(\frac{N}{2}\right)^2 + k\left(\frac{N\sqrt{3}}{2}\right)^2 = kN^2 \qquad (5.61)$$

$$L_{S32} = k\left(\frac{N}{2}\right)^2 + k\left(\frac{N\sqrt{3}}{2}\right)^2 = kN^2 \qquad (5.62)$$

where k is a factor that depends on the material for the transformer core. Therefore, if the resistances are small enough compared with the winding reactances, the three input impedances are equal, provided that the transformers are identical.

When the Scott transformer is used for synchro-to-resolver conversion, the output impedances seen by the following device must be equal because otherwise out-of-phase signals would arise. Figure 5.35 shows this configuration and the equivalent model to describe its output impedance when that of the primary is resistive and well balanced. We have then

$$R_{S13} = R_p N^2 \qquad (5.63)$$

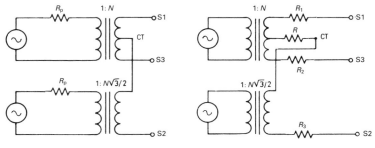

Figure 5.35 Equivalent circuit to calculate the output impedance of a Scott transformer when it is used to convert resolver format in synchro format and input impedances are equal.

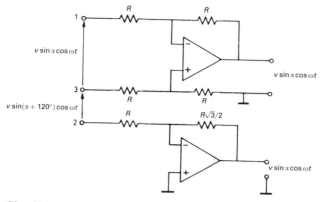

Figure 5.36   Circuit to convert angles in synchro format into angles in resolver format.

$$R_{S1CT} = R_p \left(\frac{N}{2}\right)^2 \tag{5.64}$$

$$R_{S32} = R_p \left(\frac{N}{2}\right)^2 + R_p \left(\frac{N\sqrt{3}}{2}\right)^2 = R_p N^2 \tag{5.65}$$

From these equations we deduce that

$$R_1 = R_2 = \frac{R_p N^2}{2} \tag{5.66}$$

$$R = -\frac{R_p N^2}{4} \tag{5.67}$$

$$R_3 = R_p N^2 + \frac{R_p N^2}{4} - \frac{R_p N^2}{2} = \frac{3}{4} R_p N^2 \tag{5.68}$$

$$R_{S12} = R_1 + R + R_3 = R_p N^2 \tag{5.69}$$

Therefore $R_{S13} = R_{S32} = R_{S12}$. That is, if the primary resistances are balanced, so are those at the secondary windings.

The expression "Scott transformer" is sometimes also used to designate circuits such as that in Figure 5.36 that convert synchro format angles to resolver format angles. But obviously, this circuit cannot do the reciprocal conversion, nor it is possible to step up to high voltage levels, unless special high-voltage op amps are used.

5.5.2   Digital-to-Resolver Converters [18]

Digital-to-resolver converters (DRCs) yield two fixed-frequency sinusoidal voltages whose respective amplitude is proportional to $\sin \alpha$ and $\cos \alpha$, from an input angle α in digital format, usually in natural binary code. Because the maximal angle to represent is 360°, the weight for each bit in that code is

Bit number:	1	2	3	4	5	6	...	n
Degrees:	180	90	45	22.5	11.25	5.625	...	$360/2^n$

Most of these converters rely on sine and cosine multipliers. They are circuits that accept an analog reference signal and a digital signal and yield at their output, in analog form, the product of the first signal times the sine or cosine of the angle represented by the second one. They are thus a special type of non-linear multiplying DAC.

DRCs combine two of these multipliers that accept inputs equivalent to angles from 0° to 90° and are preceded by a quadrant selector that inverts the sign of the reference signal when required. Quadrant selectors are controlled by the two most significant bits of the digital word representing the angle to be converted (Figure 5.37) according to the following rule:

Quadrant	Bit 1	Bit 2	$\sin \alpha$	$\cos \alpha$	Switches Closed (Figure 5.37)
1	0	0	+	+	A, C
2	0	1	+	−	A, D
3	1	0	−	−	B, D
4	1	1	−	+	B, C

The output signal has a high voltage and a low impedance, and therefore it is capable of being transmitted by a long line. Its power ranges from 1 VA to 2 VA, and thus it is able to directly activate the windings of a resolver or a synchro. Output transformers are included, even when the output resolver format is used, because they avoid any circuit damage because of short-circuit.

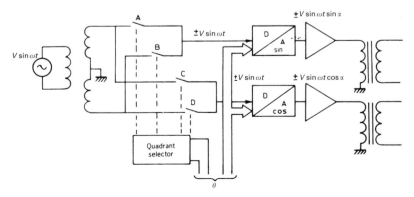

Figure 5.37 Basic circuit for a digital-to-resolver converter [18].

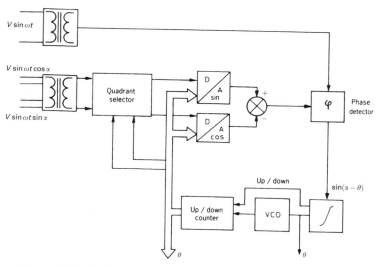

Figure 5.38   Basic circuit for a resolver-to-digital converter [18].

5.5.3   Resolver-to-Digital Converters [18]

Resolver-to-digital converters (RDCs) are also based on sine and cosine multi-pliers that give an angle α in digital format from the analog voltages $\sin \alpha$ and $\cos \alpha$. There are two different configurations: tracking converters and sampling converters.

Tracking converters are the most frequent because of their lower cost and higher noise immunity. Figure 5.38 shows their circuit. They work by internally generating with an up/down counter a digital signal representing an angle θ that is compared with the angle to be converted, α, until they are equal. The comparison is performed with a phase detector whose output controls an oscillator feeding the counter. Its output thus increases or decreases until $\alpha = \theta$. The signal at the input of the synchronous phase detector is

$$V \sin \omega t \cos \alpha \sin \theta - V \sin \omega t \sin \alpha \cos \theta = V \sin \omega t \sin(\alpha - \theta) \quad (5.70)$$

This signal is in phase with the reference signal ($V \sin \omega t$). The output of the phase detector is proportional to the amplitude of each of its inputs and to the cosine of their relative phase. Because $\cos 0° = 1$, in the present case the phase detector output will be proportional to $\sin(\alpha - \theta)$. This can be approximated by $(\alpha - \theta)$ whenever this difference is small. This is the signal sent to the integrator.

The integrator preceding the voltage-controlled oscillator (VCO) makes the system type II (2 integrators; the VCO and the counter form the other integrator). A type II feedback system has neither position nor velocity error, but only acceleration error ([19], Section 7.3). Because most of the control systems where these converters are applied work at constant velocity, type II systems are the most suitable. The acceleration error is $1/k_a$, where k_a is the acceleration constant with values between 1000 rad/s^2 and 600,000 rad/s^2. This means, for example, that for the smaller value, the output angle lags 1° for each 1000 rad/s^2 of input acceleration. The double integration of the error signal provides a high immunity to electromagnetic interference, which usually consists of voltage spikes.

Several variations on this basic structure are commercially available in the form of integrated circuits with resolution ranging from 10 to 22 bits and accuracy about 2 bits less. The AD2S90, for example, has 12 bits and an angular accuracy of 10.6′ ± 1 LSB. Some models incorporate only the two sine and cosine multipliers and accept angles from 0° to 360° in digital form. Other models give the angle in analog form because they incorporate a DAC. Still others with digital output for the angle give an analog velocity signal, thus saving the need for a tachometer. After signal amplification some models can be applied to Inductosyn signal conversion (Section 4.2.4).

Sampling RDCs are usually based on a successive approximation algorithm. They have a faster conversion speed (100 μs to 200 μs, as compared to 1 s for tracking converters). But their cost makes them practical only for multichannel systems with more than six or nine channels (depending on the tracking unit they are compared to).

In sampling models there is a sampling unit for each channel that takes a sample of each of the two inputs in resolver format, at the instant when the reference signal reaches its maximal value. The sampled value is stored in a capacitor and multiplexed to a unit that is shared by all channels and that makes the conversion to digital. This conversion is performed by analyzing the difference between the outputs of two sine and cosine multipliers (as in Figure 5.38). Their input is an angle generated by a successive approximation register that works like those in conventional ADCs. In these models, if the system to be controlled has a constant speed, there is a constant delay between converter output and the present angle position of the system.

5.6   PROBLEMS

5.1   A capacitive pressure sensor gives a 1 kHz output signal that must be amplified by 1000 to obtain an acceptable signal level. The op amp available has a maximal input offset voltage of 3 mV at the working temperature. To prevent it from reducing the dynamic output range, the amplifier in Figure P5.1 is suggested. Give values for the components of the circuit.

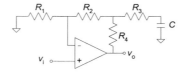

Figure P5.1   High-gain ac amplifier.

5.2 A given differential capacitive sensor is available that is based on the variation in plate distance and whose movable plate is grounded. In order to obtain a ground-referenced output signal proportional to the input displacement, the circuit in Figure P5.2 is proposed. Assume that the op amps are ideal. Determine the conditions to be fulfilled by resistors and capacitors in the circuit in order for the output voltage to be directly proportional to the displacement and independent of oscillator frequency.

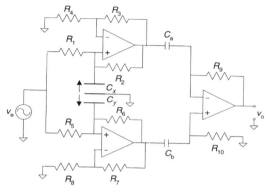

Figure P5.2 Differential capacitive sensor excited by a constant ac current, and differential differentiator.

5.3 Figure P5.3 shows the signal conditioner for a differential capacitive displacement sensor whose movable electrode is grounded. Determine the relation to be fulfilled by passive components in the circuit in order for the output voltage to be independent of the exciting frequency. If the peak output voltage desired is 10 V, determine the maximal excitation frequency allowed by the finite slew rate. If the sensor has parallel plates of

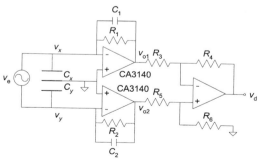

Figure P5.3 Signal conditioner for a differential capacitive sensor with grounded central electrode.

100 cm^2 separated 1 cm, the input displacement range is from -1 mm to $+1$ mm, and the excitation source has 10 V (peak) at 100 kHz, determine the value of resistors and capacitors.

5.4 Figure P5.4 shows a pseudobridge where $Z_1(R_1, C_1)$ is a sensor whose real and reactive impedances are both variable and $Z_3(R_3, C_3)$ is the balancing impedance used to obtain a null output at a selected sensor point. Determine the equation for R_3 and C_3 when this sensor point is (R_{10}, C_{10}). Determine v_1 and v_2 as a function of the components of Z_1. Assume that all op amps are ideal.

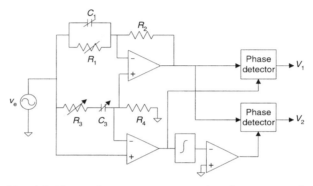

Figure P5.4 Pseudobridge and component separation for sensors whose active and reactive parts change with the measurand.

5.5 The circuit in Figure P5.5 is a pseudobridge used in a precision thermometer based on a 100 Ω platinum probe and supplied by a 72 Hz voltage.

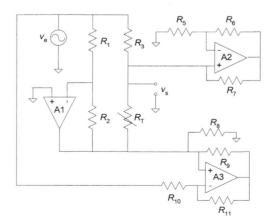

Figure P5.5 Pseudobridge for a resistive sensor excited by a constant current.

Describe the function of each of the op amps and calculate the relationship between resistances in order for the output voltage to be directly proportional to the temperature.

5.6 We wish to measure a temperature close to 100 °C with 0.01 °C resolution. The proposed circuit in Figure P5.6 uses a Pt100 probe that has 100 Ω and $\alpha = 0.004/\text{K}$ at 0 °C. If the bridge is balanced at $T = 100$ °C, determine the output-voltage-to-excitation-voltage ratio at $T = 100.01$ °C. If the probe has $\delta = 100$ mW/K and we wish to keep the self-heating error below 0.001 °C, what is the limit for the peak voltage applied to the bridge? If the output filter is errorless and we wish to obtain a 10 mV output when $T = 100.01$ °C, determine the gain for the amplifier connected to the bridge. If the frequency spectrum of the input temperature is limited to 1 Hz, determine the values for R_1, R_2, R_3, R_4, R_5, C_1, and C_2.

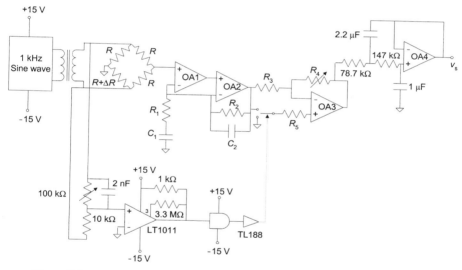

Figure P5.6 Carrier amplifier for high-resolution temperature measurement.

5.7 The LVDT in Figure P5.7 measures displacements up to ± 50 mm and yields a FSO = 250 mV (rms) when excited by 5 V (rms), 2 kHz. At this frequency the primary has 3500 Ω and $+71°$ phase shift. In Figure P5.7 the excitation voltage is 12 V (peak) at 20 kHz. Assume that the primary impedance does not change from 2 kHz to 20 kHz. Determine the value for resistors and capacitors in order to obtain the desired excitation voltage and frequency and a 12 V (peak) FSO. What is the minimal slew rate required for op amps (assumed equal) to avoid distortion?

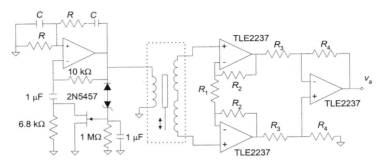

Figure P5.7   Excitation and detection in an LVDT.

REFERENCES

[1] S. M. Huang, A. L. Stott, R. G. Green, and M. S. Beck. Electronic transducers for industrial measurement of low value capacitances. *J. Phys. E: Sci. Instrum.*, **21**, 1988, 242–250.

[2] A. L. Hugill. Displacement transducers based on reactive sensors in transformer ratio bridge circuits. *J. Phys. E: Sci. Instrum.*, **15**, 1982, 597–606.

[3] R. V. Jones and J. C. S. Richards. The design and some applications of sensitive capacitance micrometers. *J. Phys. E: Sci. Instrum.*, **6**, 1973, 589–600.

[4] H. K. P. Neubert. *Instrument Transducers*, 2nd ed. New York: Oxford University Press, 1975.

[5] S. Franco. *Design with Operational Amplifiers and Analog Integrated Circuits*, 2nd ed. New York: McGraw-Hill, 1998.

[6] R. Pallás-Areny and J. G. Webster. *Analog Signal Processing*. New York: John Wiley & Sons, 1999.

[7] X. Wang. Sensitive digital lock-in amplifier using a personal computer. *Rev. Sci. Instrum.*, **61**, 1990, 1999–2001.

[8] A. V. Oppenheim, A. S. Willsky with S. H. Nawab. *Signals and Systems*, 2nd. ed. Upper Saddle River, NJ: Prentice-Hall, 1997.

[9] R. Pallás-Areny and O. Casas. A novel differential synchronous demodulator for ac signals. *IEEE Trans. Instrum. Meas.*, **45**, 1996, 413–416.

[10] E. E. Herceg. *Handbook of Measurement and Control*. Pennsauken, NJ: Schaevitz Engineering, 1976. Fourth printing, 1986.

[11] S. T. Cho and K. D. Wise. A high-performance microflowmeter with built-in test. *Sensors and Actuators A*, **36**, 1993, 47–56.

[12] J. T. Kung, R. N. Mills, and H-S. Lee. Digital cancellation of noise and offset for capacitive sensors. *IEEE Trans. Instrum. Meas.*, **42**, 1993, 939–942.

[13] H. Philipp. The charge transfer sensor. *Sensors*, **13**, November 1996, 36–42.

[14] S. M. Huang. Impedance sensors—dielectric systems. Chapter 4 in: R. A. Williams and M. S. Beck (eds.), *Process Tomography: Principles, Techniques and Applications*. Boston: Butterworth-Heinemann, 1995.

[15] N. M. Patiño and M. E. Valentinuzzi. Lion's twin-T circuit revisited. *IEEE Eng. Med. Biol. Magazine*, **11**, 3, 1992, 61–66.

[16] A. D. Khazan. *Transducers and Their Elements*. Englewood Cliffs, NJ: PTR Prentice-Hall, 1994.

[17] L. K. Baxter. *Capacitive Sensors Design and Applications*. New York: IEEE Press, 1997.

[18] G. S. Boyes (ed.). *Synchro and Resolver Conversion*. Surrey (U.K.): Memory Devices Ltd., 1980.

[19] B. C. Kuo. *Automatic Control Systems*, 6th ed. Englewood Cliffs, NJ: Prentice-Hall, 1991.

6

SELF-GENERATING SENSORS

Self-generating sensors yield an electric signal from a measurand without requiring any electric supply. They offer alternative methods for measuring many common quantities—in particular, temperature, force, pressure, and acceleration. Furthermore, because they are based on reversible effects, these sensors can be used as actuators to obtain nonelectric outputs from electric signals.

This chapter also describes photovoltaic sensors and some sensors for chemical quantities (related to composition). Some effects described in this chapter can happen unexpectedly in circuits, thus becoming a source of interference. That is the case, for example, for thermoelectric voltages, for cable vibrations when they include piezoelectric materials, or for galvanic potentials at soldering points or electric contacts. We will describe the phenomena in sensors, but the same analysis applies to interference minimization.

6.1 THERMOELECTRIC SENSORS: THERMOCOUPLES

6.1.1 Reversible Thermoelectric Effects

Thermoelectric sensors are based on two effects that are reversible as contrasted with the irreversible Joule effect. They are the Peltier effect and the Thomson effect.

Historically, it was Thomas J. Seebeck who first discovered in 1822 that in a circuit with two dissimilar homogeneous metals A and B, having two junctions at different temperatures, an electric current arises (Figure 6.1). That is, there is a conversion from thermal to electric energy. If the circuit is opened, a thermoelectric electromotive force (emf) appears that depends only on the metals

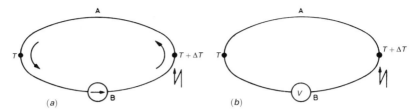

Figure 6.1 Seebeck effect in a thermocouple: (*a*) a current or (*b*) a potential difference appear when there are two metal junctions at different temperatures.

and on the junction temperatures. A pair of different metals with a fixed junction at a point or zone constitutes a *thermocouple*.

The relationship between the emf E_{AB} and the difference in temperature between both junctions T defines the Seebeck coefficient S_{AB},

$$S_{AB} = \frac{dE_{AB}}{dT} = S_A - S_B \tag{6.1}$$

where S_A and S_B are, respectively, the absolute thermoelectric power for A and B. S_{AB} is not in general constant but depends on T, usually increasing with T. It is important to realize that while the current flowing in the circuit depends on conductors' resistances, the emf does not depend on the resistivity, on the conductors' cross sections, or on temperature distribution or gradient. It depends only on the difference in temperature between both junctions and on the metals, provided that they are homogeneous. This emf is due to the Peltier and Thomson effects.

The Peltier effect, named to honor Jean C. A. Peltier, who discovered it in 1834, is the heating or cooling of a junction of two different metals when an electric current flows through it (Figure 6.2). When the current direction reverses, so does the heat flow. That is, if a junction heats (liberates heat), then when the current is reversed, it cools (absorbs heat), and if it cools, then when the current is reversed, it heats. This effect is reversible and does not depend on

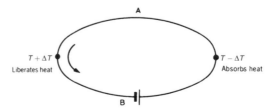

Figure 6.2 Peltier effect: When there is a current along a thermocouple circuit, one junction cools and the other warms.

the contact, namely, on the shape or dimensions of the conductors. It depends only on the junction composition and temperature. Furthermore, this dependence is linear and is described by the Peltier coefficient π_{AB}, sometimes called Peltier voltage because its unit is volts. π_{AB} is defined as the heat generated at the junction between A and B for each unit of (positive charge) flowing from B to A; that is,

$$dQ_P = \pm\pi_{AB}I\,dt \tag{6.2}$$

It can be shown [1] that for a junction at absolute temperature T we have

$$\pi_{AB}(T) = T \times (S_B - S_A) = -\pi_{BA}(T) \tag{6.3}$$

The fact that the amount of heat transferred per unit area at the junction is proportional to the current instead of its square makes this different from the Joule effect. In the Joule effect the heating depends on the square of the current and does not change when current direction reverses.

The Peltier effect is also independent of the origin of the current, which can thus even be thermoelectric as in Figure 6.1a. In this case the junctions reach a temperature different from that of the ambient, and this can be an error source as we will discuss later.

The Thomson effect, discovered by William Thomson (later Lord Kelvin) in 1847–1854, consists of heat absorption or liberation in a homogeneous conductor with a nonhomogeneous temperature when there is a current along it, as shown in Figure 6.3. The heat liberated is proportional to the current, not to its square, and therefore changes its sign for a reversed current. Heat is absorbed when charges flow from the colder to the hotter points, and it is liberated when they flow from the hotter to the colder one. In other words, heat is absorbed when charge and heat flow in opposite directions, and heat is liberated when they flow in the same direction.

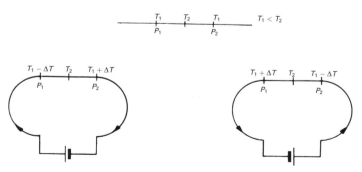

Figure 6.3 Thomson effect: When there is a current along a conductor with non-homogeneous temperature, heat is absorbed or liberated.

The heat flux per unit volume q in a conductor of resistivity r with a longitudinal temperature gradient dT/dx, along which there is a current density i, is

$$q = i^2 r - i\sigma \frac{dT}{dx} \qquad (6.4)$$

where σ is the Thomson coefficient. The first term on the right side describes the irreversible Joule effect, and the second term describes the reversible Thomson effect.

Going back to the circuit in Figure 6.1a, if the current is small enough to make the Joule effect negligible, we can consider only the reversible effects. Then the resulting thermoelectric power $(dE_{AB}/dT)\Delta T$ must equal the net thermal energy converted. In Figure 6.1a where one junction is at temperature $T + \Delta T$ and the other one is at T, the heat absorbed in the hot junction is $\pi_{AB}(T + \Delta T)$, while the heat liberated at the cool junction is $-\pi_{AB}(T)$. By the Thomson effect, there is an amount of heat $-\sigma_A \times \Delta T$ liberated along A while there is an amount of heat $\sigma_B \times \Delta T$ absorbed along B. The power balance is thus

$$\frac{dE_{AB}}{dT}\Delta T = \pi_{AB}(T + \Delta T) - \pi_{AB}(T) + (\sigma_B - \sigma_A) \times \Delta T \qquad (6.5)$$

By dividing both sides by ΔT and taking limits when ΔT goes to zero, we have

$$\frac{dE_{AB}}{dT} = \frac{d\pi_{AB}}{dT} + \sigma_B - \sigma_A \qquad (6.6)$$

This equation constitutes the basic theorem for thermoelectricity and shows that the Seebeck effect results from the Peltier and Thomson effects.

Equations (6.1) and (6.6) allow us to apply thermocouples to temperature measurement. A thermocouple circuit with a junction at constant temperature (reference junction) yields an emf that is a function of the temperature at the other junction, which we call the measuring junction. Tables give the voltages obtained with given thermocouples as a function of the temperature at the measuring junction when the reference junction is kept at $0\,^\circ$C. The equivalent circuit for an ungrounded thermocouple is a voltage source with different output resistance at each terminal (that of the corresponding metal).

The application of thermocouples to temperature measurement is subject to several limitations. First, we must select the type of thermocouple so that it does not melt in our application. We must also be sure that the environment it is placed in does not attack any of the junction metals.

Second, we must keep the current along the thermocouple circuit very small. Otherwise, because the Peltier and Thomson effects are reversible, the temperatures of the conductors and particularly those of the junctions would differ from that of the environment because of the heat flow to and from the circuit. Depending on the intensity of the current, even the Joule effect could be considerable. All this would result in a temperature for the measuring junction different from the one we intend to measure, and also a reference temperature different from the assumed one, thus leading to serious errors. In addition, conductors must be homogeneous, so that caution is needed to prevent any mechanical or thermal stress during installation or operation—for example, because of aging caused by long exposure to large temperature gradients.

Another limitation is that one of the junctions must be kept at a fixed temperature if the temperature at the other junction is to be measured. Any change in that reference junction would result in a serious error because the output voltage is very small, typically from 6 $\mu V/°C$ to 75 $\mu V/°C$. Furthermore, if the reference temperature is not close to the measured temperature, the output signal will have a relatively high constant value undergoing only very small changes due to the temperature changes we are interested in.

When high accuracy is desired, the nonlinearity of the relationship between the emf and the temperature may become important. An approximate formula valid for all thermocouples is

$$E_{AB} \approx C_1(T_1 - T_2) + C_2(T_1^2 - T_2^2) \tag{6.7}$$

where T_1 and T_2 are the respective absolute temperatures for each junction and C_1 and C_2 are constants that depend on materials A and B. From (6.7), we have

$$E_{AB} \approx (T_1 - T_2)[C_1 + C_2(T_1 + T_2)] \tag{6.8}$$

which shows that the emf depends not only on the temperature differences but also on their absolute value. The number of useful thermocouples available is limited because C_2 should be very small, thus reducing the possible choices. For copper-constantan, for example, $C_2 \approx 0.036$ $\mu V/K^2$. This nonlinearity may require a correction to be performed by the signal conditioner. All factors considered, thermocouples seldom achieve errors below 0.5 °C. Tolerance for same-type models can be up to several degrees Celsius.

In spite of the above limitations, thermocouples have many advantages and are by far the most frequently used sensors for temperature measurement. They have a very broad measurement range, as a group from −270 °C to 3000 °C, and each particular model has a broad range. They also display acceptable long-term stability and a high reliability. Furthermore, at low temperatures they have higher accuracy than RTDs. Their small size also yields a fast speed of response, on the order of milliseconds. They are also robust, simple, and easy to use, and very low cost models are available suitable for many applications. Because they do not need excitation, they do not have the self-heating problems suffered by RTDs, particularly in gas measurements. They also accept long connection wires.

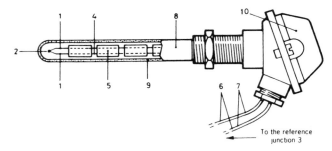

Figure 6.4 Industrial thermocouple with sheath. 1, conductors (different); 2, measurement junction; 3, reference junction; 4, bare thermocouple wires; 5, insulated thermocouple wires; 6, extension leads, of the same wire as that of the thermocouple; 7, compensation leads, different wire from that of the thermocouple but with small emf; 8, probe; 9, protection (external covering); 10, sheath head.

6.1.2 Common Thermocouples

In thermocouple junctions there is a simultaneous requirement for (a) a low-resistivity temperature coefficient, (b) resistance to becoming oxidized at high temperatures, in order to withstand the working environment, and (c) a linearity as high as possible.

Several particular alloys are used that fulfill all these requirements: $Ni_{90}Cr_{10}$ (chromel), $Cu_{57}Ni_{43}$ (constantan), $Ni_{94}Al_2Mn_3Si_1$ (alumel), and so forth. Environmental protection is obtained by a sheath, usually from stainless steel (Figure 6.4). Both speed of response and probe robustness depend on the thickness of the sheath. Both silicon and germanium display thermoelectric properties, but they have found greater application as cooling elements (Peltier elements) than as measurement thermocouples. Table 6.1 gives the characteristics for

TABLE 6.1   Characteristics of Some Common Thermocouples

ANSI Designation	Composition	Usual Range	Full-Range Output (mV)	Error (°C)
B	Pt(6 %)/rhodium–Pt(30 %)/rhodium	38 °C to 1800 °C	13.6	—
C	W(5 %)/rhenium–W-(26 %)/rhenium	0 °C to 2300 °C	37.0	—
E	Chromel–constantan	0 °C to 982 °C	75.0	±1.0
J	Iron–constantan	184 °C to 760 °C	43.0	±2.2
K	Chromel–alumel	−184 °C to 1260 °C	56.0	±2.2
N	Nicrosil (Ni–Cr–Si)–Nisil (Ni–Si–Mg)	−270 °C to 1300 °C	51.8	—
R	Pt(13 %)/rhodium–Pt	0 °C to 1593 °C	18.7	±1.5
S	Pt(10 %)/rhodium–Pt	0 °C to 1538 °C	16.0	±1.5
T	Copper–constantan	−184 °C to 400 °C	26.0	±1.0

some common thermocouples and their ANSI designation. Type C and N are not ANSI standards. There are also thin-film models for surface temperature measurement.

Type J thermocouples are versatile and have low cost. They withstand oxidizing and reducing environments. They are often used in open-air furnaces. Type K thermocouples are used in nonreducing environments and, in their measurement range, are better than types E, J, and T in oxidizing environments. Type T thermocouples resist corrosion; hence they are useful in high-humidity environments. Type E thermocouples have the highest sensitivity, and they withstand corrosion below $0\,°C$ and in oxidizing environments. Type N thermocouples resist oxidation and are stable at high temperature. Thermocouples based on noble metals (types B, R, and S) are highly resistive to oxidation and corrosion.

Standard tables give the output voltage corresponding to different temperatures when the reference junction is at $0.00\,°C$. But this does not mean that a junction placed at $0.00\,°C$ always gives a 0 V output for any thermocouple. This tabulation is only a matter of convenience arising from the fact that in order to measure the voltage generated by a junction, we cannot avoid introducing another junction. Therefore it is more convenient to speak of voltage differences between junctions at different temperatures than to consider the voltage of a single junction for each given temperature. For standardization purposes it has been agreed to take $0.00\,°C$ as the reference temperature for the tables. Table 6.2 shows part of one of these tables [2]. Intermediate voltages or temperatures are obtained by linear interpolation.

Example 6.1 A J-type thermocouple circuit has one junction at $0\,°C$ and the other at $45\,°C$. What is its open circuit emf?

In Table 6.2, at the intersection of the row corresponding to 40 ($°C$) and the column corresponding to 5 ($°C$) we read 2.321 mV.

Example 6.2 A given J-type thermocouple circuit with one junction at $0\,°C$ generates a 5 mV output voltage. What is the temperature at the measuring junction?

At $96°C$ we have 5.050 mV. At $95\,°C$ we have 4.996 mV. Therefore, the sensitivity in this range is 54 $\mu V/°C$, and the junction is at about $95.07\,°C$.

When interpreting this last result it is important to take into account the accuracy of each thermocouple type. For type J it is $\pm 2.2\,°C$ or $0.75\,\%$ (whichever gives the largest error). This means that in the result of the last example the uncertainty would be $\pm 2\,°C$. This does not reduce the usefulness of tables given with $1\,°C$ increments and interpolation because some applications need a high resolution but not necessarily a high accuracy.

Self-calibrating thermocouples [3] have improved accuracy. They include an encapsulated metal located near the junction. When the sensed temperature

TABLE 6.2 Part of the Voltage–Temperature Table for a Type J Thermocouple from 0 °C to 110 °C

Degrees	0	1	2	3	4	5	6	7	8	9	10
0	0.000	0.050	0.101	0.151	0.202	0.253	0.303	0.354	0.405	0.456	0.507
10	0.507	0.558	0.609	0.660	0.711	0.762	0.813	0.865	0.916	0.967	1.019
20	1.019	1.070	1.122	1.174	1.225	1.277	1.329	1.381	1.432	1.484	1.536
30	1.536	1.588	1.640	1.693	1.745	1.797	1.849	1.901	1.954	2.006	2.058
40	2.058	2.111	2.163	2.216	2.268	2.321	2.374	2.426	2.479	2.532	2.585
50	2.585	2.638	2.691	2.743	2.796	2.849	2.902	2.956	3.009	3.062	3.115
60	3.115	3.168	3.221	3.275	3.328	3.381	3.435	3.488	3.542	3.595	3.649
70	3.649	3.702	3.756	3.809	3.863	3.917	3.971	4.024	4.078	4.132	4.186
80	4.186	4.239	4.293	4.347	4.401	4.455	4.509	4.563	4.617	4.671	4.725
90	4.725	4.780	4.834	4.888	4.942	4.996	5.050	5.105	5.159	5.213	5.268
100	5.268	5.322	5.376	5.431	5.485	5.540	5.594	5.649	5.703	5.758	5.812

Note: The reference junction is assumed to be at 0 °C. Voltages are given in millivolts.

transcends the phase transition temperature of the encapsulated metal, the time–temperature record of the thermocouple reaches a plateau. By comparing the plateau temperature with the known phase transition temperature of the encapsulated metal, we perform a single-point calibration.

Systems with computation capability can use polynomials that approximate the values in the tables with accuracy dependent on their order. They all correspond to equations such as

$$T = a_0 + a_1 x + a_2 x^2 + \cdots \tag{6.9}$$

where x is the measured voltage. Table 6.3 gives the polynomial coefficients for different common thermocouples within a specified range and degree of approximation [2]. When the measurement range is very large, instead of using higher order polynomials it is better to divide the whole range into smaller temperature ranges and then use a lower order polynomial for each range.

Figure 6.5 shows different junction types available. Exposed junctions are used for static measurements or in noncorrosive gas flows where a fast response time is required. But they are fragile. Enclosed (ungrounded) junctions are intended for corrosive environments where there is the need for an electrical isolation of the thermocouple. The junction is enclosed by the sheath and is insulated from that by means of a good thermal conductor such as oil, mercury, or metallic powder. When a fast response is needed and a thick sheath is not required, then mineral insulators such as MgO, Al_2O_3, or BeO powders are

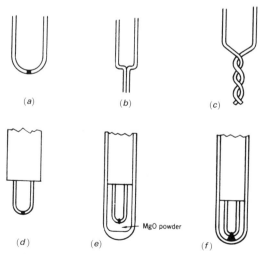

Figure 6.5 Different kinds of thermocouple junctions and their sheaths [4]: (a) butt-welded junction; (b) lap-welded junction; (c) twisted wire; (d) exposed thermocouple for fast response time; (e) enclosed thermocouple—electrical and ambient isolation; (f) grounded thermocouple soldered to the covering—ambient isolation.

TABLE 6.3 Polynomial Coefficients that Give the Approximate Temperature from the Output Voltage for Different Thermocouples According to (6.9)

Polynomial Coefficient	Type E −100°C to 1000°C	Type J 0°C to 760°C	Type K 0°C to 1370°C	Type R 0°C to 1000°C	Type S 0°C to 1750°C	Type T −160°C to 400°C
Accuracy	±0.5°C	±0.1°C	±0.7°C	±0.5°C	±1°C	±0.5°C
a_0	0.104967248	−0.048868252	0.226584602	0.263632917	0.927763167	0.100860910
a_1	17189.45282	19873.14503	24152.10900	179075.491	169526.5150	25727.94369
a_2	−282639.0850	−218614.5353	67233.4248	−48840341.37	−31568363.94	−767345.8295
a_3	12695339.5	11569199.78	2210340.682	1.90002E+10	8990730663	780225595.81
a_4	−448703084.6	−264917531.4	−860963914.9	−4.82704E+12	−1.63565E+12	−9247486589
a_5	1.1086E+10	2018441314	4.83506E+10	7.62091E+14	1.88027E+14	6.97688E+11
a_6	−1.76807E+11		−1.18452E+12	−7.20026E+16	−1.37241E+16	−2.6619E+13
a_7	1.71842E+12		1.38690E+13	3.71496E+18	6.17501E+17	3.94078E+14
a_8	−9.19278E+12		−6.33708E+13	−8.03104E+19	−1.56105E+19	
a_9	2.06132 E+13				1.69535E+20	

used. The final response will depend on the compactness of the insulator, and the maximal allowable temperature will also be different.

Grounded junctions suit the measurement of static temperatures or temperatures in flowing corrosive gases or liquids. They are also used in measurements performed under high pressures. The junction is soldered to the protective sheath so that the thermal response will be faster than when insulated. However, noisy grounds require ungrounded thermocouples.

6.1.3 Practical Thermocouple Laws

In addition to the advantages and disadvantages mentioned above, there are several experimental laws for temperature measurement using thermocouples that greatly simplify the analysis of thermocouple circuits.

6.1.3.1 *Law of Homogeneous Circuits.* It is not possible to maintain a thermoelectric current in a circuit formed by a single homogeneous metal by only applying heat, not even by changing the cross section of the conductor.

Figure 6.6 describes the meaning of this law. In Figure 6.6*a* the temperatures T_3 and T_4 do not alter the emf due to T_1 and T_2. In particular, if $T_1 = T_2$ and A or B are heated, there is no current. In other words, intermediate temperatures along a conductor do not alter the emf produced by a given temperature difference between junctions (Figure 6.6*b*). But this does not mean that if along a conductor there are different temperatures, then long extension wires identical to those of the thermocouple must be used. Instead of these, we can use compensation wires that are made from metals that do not display any appreciable emf and at the same time are cheaper than thermocouple wires. Nevertheless, they are four to five times more expensive than copper wires. Thermocouple wire coverings use standard colors.

6.1.3.2 *Law of Intermediate Metals.* The algebraic sum of all emfs in a circuit composed by several different metals remains zero as long as the entire circuit is at a uniform temperature. This implies that a meter can be inserted into the

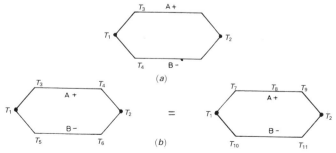

Figure 6.6 Homogeneous circuits law for thermocouples.

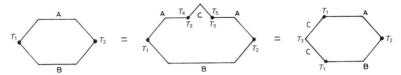

Figure 6.7   Intermediate metals law for thermocouple circuits.

TABLE 6.4   Thermoelectric Sensitivity of Different Metal and Alloy Pairs Common in Electric Circuits

Pair	S_{AB} (μV/K)
Alloy 180–nichrome	42
Au–kovar	25
Cu–Ag	0.3
Cu–Au	0.3
Cu–Cd/Sn	0.3
Cu–Cu	<0.2
Cu–CuO	1000
Cu–kovar	40
Cu–Pb/Sn	1–3

circuit without adding any errors, provided that the new junctions introduced are all at the same temperature, as indicated in Figure 6.7. The measuring instrument can be inserted at a point in a conductor or at a junction. Table 6.4 gives S_{AB} for different metal and alloy pairs common in electric circuits. Alloy 180 is the standard component lead alloy. Nichrome is used in wirewound resistors and strain gages. The Cu–Cu pair refers to copper with different purity grades. The Pb/Sn alloy refers to the common solder alloy, and the Cd/Sn alloy refers to a low-temperature solder alloy. Kovar is an alloy used in some IC pins. Because CuO/Cu yields a large emf, it is advisable to keep electric contacts clean.

A corollary of this law is that if the thermal relationship between each of two materials and a third one is known, then it is possible to deduce the relationship between the two first ones, as shown in Figure 6.8. Therefore it is not necessary to calibrate all the possible metal pairs in order to know the temperature cor-

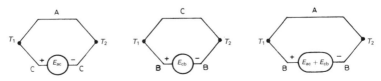

Figure 6.8   Corollary for intermediate metals law in thermocouple circuits.

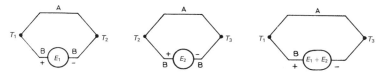

Figure 6.9   Intermediate temperature law for thermocouple circuits.

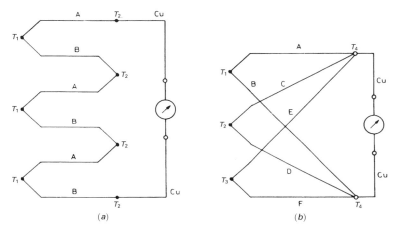

Figure 6.10   (*a*) Series (thermopile) and (*b*) parallel thermocouple connection.

responding to a given emf measured with a given pair. Rather, its behavior with respect a third material is enough. The reference metal is platinum.

6.1.3.3   Law of Successive or Intermediate Temperatures.   If two homogeneous metals yield an emf E_1 when their junctions are at T_1 and T_2, and an emf E_2 when they are at T_2 and T_3, then the emf when the junctions are at T_1 and T_3 will be $E_1 + E_2$ (Figure 6.9). This means, for example, that it is not necessary for the reference junction to be at $0\,°C$. Any other reference temperature is also acceptable.

The preceding laws enable us to analyze circuits such as those in Figure 6.10. Case (*a*) shows several thermocouples connected in series, thus constituting a thermopile. It is straightforward to verify that this increases the sensitivity compared to the case where a single thermocouple is used. Case (*b*) shows a parallel connection, which yields the average temperature if all thermocouples are linear in the measurement range and have the same resistance.

6.1.4   Cold Junction Compensation in Thermocouple Circuits

In order to apply the Seebeck effect to temperature measurement, one junction must remain at a fixed reference temperature. Placing the reference junction

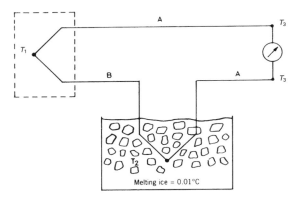

Figure 6.11 Temperature measurement using thermocouples with a junction held at a constant reference temperature.

into melting ice (Figure 6.11) is easy and highly accurate, but it requires frequent maintenance and has a high cost. It is also possible to keep the reference junction at a fixed temperature by means of a Peltier cooler or by means of a constant temperature oven. In any case a long length of one of the thermocouple metal wires must be used, thus increasing cost.

Figure 6.12 shows how to use a cheaper connecting wire (copper), but the need for a constant reference temperature is still expensive. When the expected range of variation for ambient temperature is smaller than the required resolution, we can just leave the reference junction exposed to the ambient. Otherwise, we can use the reference (or cold) junction temperature compensation method.

This consists of leaving the reference junction to undergo the ambient temperature fluctuations but at the same time measuring these by another temperature sensor placed near the reference junction. Then a voltage equal to that generated at the cold junction is subtracted from the one produced by the circuit, as shown in Figure 6.13. The bridge supply voltage must be highly stable and can be provided by a mercury cell or reference voltage generator (Section

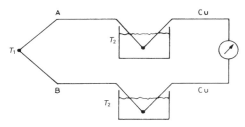

Figure 6.12 Temperature measurement using two junctions at constant temperature and common metal leads.

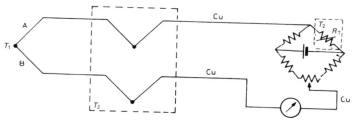

Figure 6.13   Electronic compensation for the reference junction in a thermocouple circuit. Ambient temperature fluctuations are measured by another sensor, and a voltage equal to that generated by the cold junction is subtracted from the output voltage.

3.4.5), for example. There are ICs that measure the ambient temperature and provide the compensation voltage for some specific thermocouples. The LT1025 works with types E, J, K, R, S, and T. The AD594/AD595 is an instrumentation amplifier and thermocouple cold junction compensator (for types J and K, respectively). The AD596/AD597 are monolithic set-point controllers that include the amplifier and cold junction compensation for, respectively, type J and K thermocouples.

Example 6.3   Figure E6.3a shows a circuit to measure a temperature by means of a J-type thermocouple and electronic compensation of the reference junction based on an NTC thermistor at ambient temperature. Design the circuit in

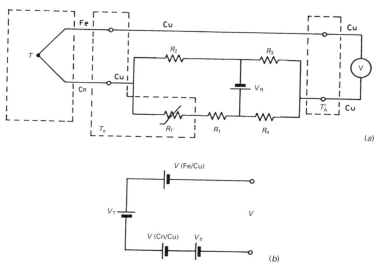

Figure E6.3   (*a*) Proposed circuit for cold junction compensation. (*b*) Equivalent circuit when lead and bridge resistance are much smaller than the input resistance of the voltmeter.

order to have compensation in the range from $10\,°C$ to $40\,°C$ with an NTC thermistor having $B = 3546$ K and resistance 10 kΩ at $25\,°C$.

If we assume that the current along thermocouple wires is very low because of the high input impedance offered by the voltmeter, then the equivalent circuit for the thermocouples is shown in Figure E6.3b. If we call the bridge output V_b, then we have

$$V_T - V(\text{Fe/Cu})|_{T_a} + V(\text{Cn/Cu})|_{T_a} + V_b = V$$

In order for the ambient temperature not to affect the measurement, we need

$$V = V_T$$

By applying the law of intermediate metals, we have

$$-V(\text{Fe/Cu})|_{T_a} + V(\text{Cn/Cu})|_{T_a} = -V(\text{Fe/Cn})|_{T_a} \approx -kT_a$$

where $k \approx 52$ μV/$°C$ is the sensitivity for the J thermocouple in the range from $10\,°C$ to $40\,°C$, assuming that it is constant (Table 6.2).

In principle, we are thus interested in having $V_b = kT_a$; that is,

$$\frac{dV_b}{dT} = k$$

and also $V_b(0\,°C) = 0$ V. The actual bridge output is

$$V_b = -V_R\left(\frac{R_1'}{R_1' + R_2} - \frac{R_4}{R_3 + R_4}\right)$$

where $R_1' = R_1 + R_0 \exp(B/T - B/T_0)$. The bridge sensitivity is

$$\frac{dV_b}{dT} = V_R \frac{\dfrac{BR_2}{T^2} R_0 e^{B(1/T - 1/T_0)}}{(R_1 + R_0 e^{B(1/T - 1/T_0)} + R_2)^2}$$

which, far from being constant, depends on the temperature. This means that the bridge is nonlinear. If as linearization criterion we chose to have the desired slope at the middle of the temperature range to be compensated ($25\,°C$), then we have

$$V_R \frac{\dfrac{BR_2}{(298\text{ K})^2} R_0}{(R_1 + R_0 + R_2)^2} = k$$

Another condition that we can force is that at this same temperature the bridge output equals the voltage at the reference junction, namely about 1.3 mV (Table 6.2). Therefore

$$1.3 \text{ mV} = -V_R\left(\frac{R_1'}{R_1' + R_2} - \frac{R_4}{R_3 + R_4}\right)$$

In order for V_R to be stable, we can choose a mercury cell (1.35 V). In order to calculate the values for the resistors, we can choose R_2 and then the equation for the slope determines R_1 (or conversely) and the output at 25 °C determines the ratio R_3/R_4. In the first case, for example, if $R_2 = 100$ Ω we have

$$R_1 = 22{,}097 \text{ Ω}$$

$$\frac{R_3}{R_4} = 2.15 \times 10^{-3}$$

For example, $R_1 = 22.1$ kΩ, $R_3 = 1$ kΩ, and $R_4 = 46.4$ kΩ.

6.2 PIEZOELECTRIC SENSORS

6.2.1 The Piezoelectric Effect

The piezoelectric effect is the appearance of an electric polarization in a material that strains under stress. It is a reversible effect. Therefore, when applying an electric voltage between two sides of a piezoelectric material, it strains. Both effects were discovered by Jacques and Pierre Curie in 1880–1881.

Piezoelectricity must not be confused with ferroelectricity, which is the property of having a spontaneous or induced electric dipole moment. Ferroelectricity was first discovered by J. Valasek in 1921 in Rochelle salt. All ferroelectric materials are piezoelectric, but the converse is not always true. Piezoelectricity is related to the crystalline (ionic) structure. Ferromagnetism is instead related to electron spin.

Piezoelectric equations describe the relationship between electric and mechanical quantities in a piezoelectric material. In Figure 6.14a, where two metal plates have been placed to form a capacitor, for a dielectric nonpiezoelectric material we have that an applied force F yields a strain S that, according to Hooke's law (Section 2.2), in the elastic range is

$$S = sT \tag{6.10}$$

where s is compliance, $1/s$ is Young's modulus, and T is the stress (F/A).

A potential difference applied between plates creates an electric field E and we have

$$\boldsymbol{D} = \epsilon \boldsymbol{E} = \epsilon_0 \boldsymbol{E} + \boldsymbol{P} \tag{6.11}$$

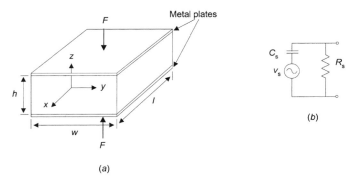

(a)

(b)

Figure 6.14 (a) Parameters used in piezoelectric equations. (b) Equivalent circuit for a piezoelectric sensor.

where D is the displacement vector (or electric flux density), ϵ is the dielectric constant, $\epsilon_0 = 8.85$ pF/m is the permittivity of vacuum, and P is the polarization vector.

For a unidimensional piezoelectric material with field, stress, strain, and polarization in the same direction, according to the principle of energy conservation, at low frequency we have

$$D = dT + \epsilon^T E \tag{6.12}$$

$$S = s^E T + d'E \tag{6.13}$$

where ϵ^T is the permittivity at constant stress and s^E is the compliance at constant electric field. Therefore, when compared to a nonpiezoelectric material, there is also a strain due to the electric field and an electric charge due to the mechanical stress (charges displaced inside the material induce opposite polarity surface charges on the plates).

When the surface area does not change under the applied stress (which is not true in polymers), then $d = d'$ [5]. d is the *piezoelectric charge coefficient* or piezoelectric constant, whose dimensions are coulombs divided by newtons [C/N].

Solving equation (6.12) for E yields

$$E = \frac{D}{\epsilon^T} - \frac{Td}{\epsilon^T} = \frac{D}{\epsilon^T} - gT \tag{6.14}$$

where $g = d/\epsilon^T$ is the *piezoelectric voltage coefficient*.

By solving (6.13) for T, we have

$$T = -\frac{d}{s^E} E + \frac{1}{s^E} S = c^E S - eE \tag{6.15}$$

where $e = d/s^E$ is the *piezoelectric stress coefficient*.

Example 6.4   For lead titanate we have in the principal direction $d = -44$ pC/ N, $\epsilon^T = 600\epsilon_0$, $g = -8$ (mV/m)/(N/m^2), $\epsilon = -4.4$ C/m^2, and $s^E = 1/(100$ GPa). For a cube with 1 cm sides, from (6.14) 1000 N (≈ 100 kg) yields at open circuit $(D = 0)$

$$ E = -\frac{dT}{\epsilon^T} = -82.8 \text{ kV/m} $$

that is, 828 V between two sides.

If 1 kV is applied between two sides, the resulting strain is

$$ S = dE = -44 \times 10^{-7} = -4.4 \text{ } \mu\varepsilon $$

and the elongation is

$$ \Delta l = (1 \text{ cm}) \times 44 \times 10^{-7} = 44 \text{ nm} $$

The *electromechanical coupling coefficient* is the square root of the quotient between the energy available at the output and the stored energy, at frequencies well below that of mechanical resonance. Therefore, it is nondimensional. It can be shown that

$$ k = \sqrt{\frac{d^2}{\epsilon^T s^E}} \tag{6.16} $$

A three-dimensional crystalline solid can experience tension and also compression forces along the three coordinate axes, designated by the subscripts 1, 2, 3, and also torsion forces designated by the subscripts 4, 5, and 6 (Figure 6.15). Using this notation, when there is no piezoelectric effect we have

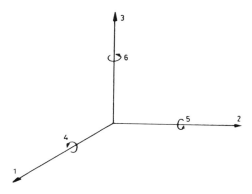

Figure 6.15   Meaning of the indices for the directions in a piezoelectric material.

$$[S_i] = [s_{ij}][T_j] \qquad i = 1, 2, 3$$
$$j = 1, \ldots, 6 \tag{6.17}$$

$$[D_i] = [\epsilon_{ij}][E_j] \qquad i, j = 1, 2, 3 \tag{6.18}$$

When there is a piezoelectric effect, the piezoelectric equations are

$$[S_i] = [s_{ij}][T_j] + [d_{ik}][E_k] \tag{6.19}$$
$$[D_i] = [\epsilon_{lm}][E_m] + [d_{ln}][T_n] \tag{6.20}$$

where $j, n = 1, \ldots, 6$, and $i, k, l, m = 1, 2, 3$.

The coefficients d_{ij} are the piezoelectric constants, which relate the electric field in direction i to the deformation in direction j, and also the surface charge density in the direction perpendicular to i with the stress in direction j. It holds that $d_{ij} = d_{ji}$, and it also holds that $\epsilon_{lm} = 0$ whenever $l \neq m$.

Also we have

$$d_{ij} = \epsilon_i g_{ij} \tag{6.21}$$

The same notation applies to the subscripts of the coupling coefficient k.

Example 6.5 PXE 5 material (Philips) has the following specifications:

Piezoelectric Charge Constants	Piezoelectric Voltage Constants	Coupling Coefficient
$d_{33} = 384$ pC/N	$g_{33} = 24.2 \times 10^{-3}$ V·m/N	$k_{33} = 0.70$
$d_{31} = -169$ pC/N	$g_{33} = -10.7 \times 10^{-3}$ V·m/N	$k_{31} = 0.34$
$d_{15} = 515$ pC/N	$g_{15} = 32.5 \times 10^{-3}$ V·m/N	$k_{15} = 0.66$

This means, for example, that a torsion stress of 1 N/m^2 applied around the axis 2 (direction "5") induces a charge density of 515 pC/m^2 in two metal plates placed on the material in the direction 1.

6.2.2 Piezoelectric Materials

Piezoelectric properties are present in 20 of the 32 crystallographic classes, although only a few of them are used; they are also present in amorphous ferroelectric materials. Of those 20 classes, only 10 display ferroelectric properties.

Whatever the case, all piezoelectric materials are necessarily anisotropic. Figure 6.16 shows why it must be so. In case (*a*) there is central symmetry. An applied force does not yield any electric polarization. In case (*b*), on the contrary, an applied force yields a parallel electric polarization, while in case (*c*) an applied force yields a perpendicular polarization.

The natural piezoelectric materials most frequently used are quartz and

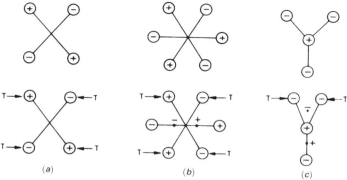

Figure 6.16 Effects of a mechanical stress on different molecules depending on their symmetry [5]: (*a*) When there is central symmetry, no electric polarization arises. (*b*) Polarization parallel to the effort. (*c*) Polarization perpendicular to the effort.

tourmaline. The synthetic materials more extensively used are not crystalline but ceramics. These are formed by many little tightly compacted monocrystals (about 1 μm in size). These ceramics are ferroelectrics, and to align the monocrystals in the same direction (i.e., to polarize them) they are subjected to a strong electric field during their fabrication. The applied field depends on the material thickness, but values of about 10 kV/cm are common at temperatures slightly above the Curie temperature (at higher temperatures they are too conductive). They are cooled in the maintained field. When the field is removed, the monocrystals cannot reorder randomly because of the mechanical stresses accumulated, so that a permanent electric polarization remains.

Piezoelectric ceramics display a high thermal and physical stability and can be manufactured in many different shapes and with a broad range of values for the properties of interest (dielectric constant, piezoelectric coefficient, Curie temperature, etc.). Their main shortcomings are the temperature sensitivity of their parameters and their susceptibility to aging (loss of piezoelectric properties) when they are close to their Curie temperature. The most commonly used ceramics are lead zirconate titanate (PZT), barium titanate, and lead niobate. Bimorphs consist of two ceramic plates glued together and with opposite polarization. If one end is clamped and a mechanical load is applied to the other, one plate elongates and the other shortens, thus generating two voltages of the same amplitude.

Some polymers lacking central symmetry also display piezoelectric properties with a value high enough to consider them for those applications where because of the size and shape required it would be impossible to use other solid materials. The most common is polyvinylidene fluoride (PVF_2 or PVDF), whose piezoelectric voltage coefficient is about four times that of quartz, and its copolymers. Electrodes are screen printed or vacuum deposited.

In order to improve the mechanical properties for piezoelectric sensors, pie-

TABLE 6.5   Some Properties for Common Piezoelectric Materials

Parameter Unit	Density $(kg \cdot m^{-3})$	T_C $(°C)$	$\epsilon^T_{11}/\epsilon_0$	$\epsilon^T_{33}/\epsilon_0$	d (pC/N)	Resistivity $(\Omega \cdot cm)$
Quartz	2649	550	4.52	4.68	d_{11} d_{14} 2.31 0.73	$\approx 10^{14}$
PZT	7500–7900	193–490	—	425–1900	d_{33} 80–593	$\approx 10^{13}$
PVDF (Kynar)	1780	—	—	12	d_{31} 23	$\approx 10^{15}$

zoelectric "composite" materials are used. They are heterogeneous systems consisting of two or more different phases, one of which at least shows piezoelectric properties. Table 6.5 lists the most important properties of some common piezoelectric materials.

6.2.3   Applications

The application of the piezoelectric effect to sense mechanical quantities based on (6.19) and (6.20) is restricted by several limitations. First, the electric resistance for piezoelectric materials is very high but never infinite (Figure 6.14b). Therefore, a constant stress initially generates (or, better, displaces) a charge that will slowly drain off as time passes. Hence, there is no dc response.

Example 6.6   A given piezoelectric sensor uses a PVDF strip measuring 10 cm by 10 cm and 52 μm thick, with deposited electrodes in the vertical direction (direction 3 in Figure 6.15). Calculate the voltage output when applying a 40 kg compression along direction 1. Determine the lowest frequency of a dynamic compression for an allowed amplitude error of 5%.

When stressing the film, electric charge $(Q_3 = D_3 A_3)$ accumulates on the electrodes of area A_3 to yield a voltage $V_3 = Q_3/C_3$. From (6.19), the electric polarization is $D_3 = d_{31} T_1 = d_{31} F_1/A_1$. From Table 6.5, $d_{31} = 23$ pC/N and $\epsilon^T_{33} = 12\epsilon_0$. Therefore,

$$V_3 = d_{31}\frac{F_1}{A_1}A_3\frac{h}{\epsilon^T_{33}A_3} = d_{31}\frac{F_1}{lh}\frac{h}{\epsilon^T_{33}} = d_{31}\frac{F_1}{l\epsilon^T_{33}}$$

$$= \left(23\frac{pC}{N}\right)\frac{40 \times 9.8 \text{ N}}{(0.1 \text{ m})\left(12 \times 8.85\frac{pF}{m}\right)} = 849 \text{ V}$$

From Table 6.5, $\rho = 10$ TΩ·m. Therefore, the leakage resistance between electrodes is

$$R_3 = \rho\frac{h}{A_3} = (10^{13}\ \Omega \cdot m)\frac{52\ \mu m}{0.1 \times 0.1\ m^2} = 52\ G\Omega$$

The capacitance between electrodes and the leakage resistance form a high-pass filter whose transfer function is

$$H(f) = \frac{j2\pi f R_3 C_3}{1 + j2\pi f R_3 C_3} = \frac{1}{1 - j\dfrac{f_c}{f}}$$

where $f_c = 1/(2\pi R_3 C_3)$. C_3 can be estimated as the capacitance of a parallel-plate capacitor,

$$C_3 = 12 \times 8.85 \frac{\text{pF}}{\text{m}} \times \frac{0.1 \times 0.1 \text{ m}^2}{56 \,\mu\text{m}} = 20.4 \text{ nF}$$

An amplitude error below 5% imposes the condition

$$\frac{1}{\sqrt{1 + \left(\dfrac{f_c}{f}\right)^2}} > 0.95$$

$$f > 3.04 f_c = 3.04 \frac{1}{2\pi \times (52 \text{ G}\Omega) \times (20.4 \text{ nF})} = 0.5 \text{ mHz}$$

Piezoelectric sensors show a high resonant peak in their frequency response. This is because when a dynamic force is applied to them, the only damping source is the internal friction in the material. Thus we must always work at frequencies well below the mechanical resonant frequency, and the sensor output must be low-pass filtered to prevent amplifier saturation. Figure 6.17 shows the frequency response of a commercial piezoelectric accelerometer. The gain at the resonant frequency (35 kHz) is 20 times that in the 5 Hz to 7 kHz band, where the frequency response is flat within ±5%. Reference 6 describes a method to increase the useful range up to the resonant frequency. It is based on electromechanical feedback relying on the reversibility of the piezoelectric effect, thus providing damping to the otherwise undamped system.

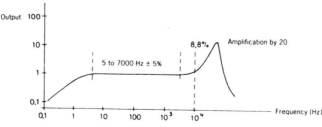

Figure 6.17 Frequency response for a piezoelectric accelerometer displaying a large resonance and lack of dc response.

The piezoelectric coefficients are temperature-sensitive. Furthermore, above the Curie temperature all materials lose their piezoelectric properties. That temperature is different for each material, and in some cases it is even lower than typical temperatures in industrial environments. Quartz is used up to 260 °C, tourmaline up to 700 °C, barium titanate up to 125 °C, and PVDF up to 135 °C. Some materials that display piezoelectric properties are hygroscopic and therefore are inappropriate for sensors.

Piezoelectric materials have a very high output impedance (small capacitance with a high leakage resistance) (Figure 6.14*b*). Therefore, in order to measure the signal generated we must use an electrometer (voltage mode) or charge amplifiers (charge mode) (Sections 7.2 and 7.3). Some sensors include an integrated amplifier, but this limits the temperature of operation to the range acceptable for the electronic components.

Piezoelectric sensors offer high sensitivity (more than one thousand times that of strain gages), usually at a low cost. They undergo deformations smaller than 1 μm, and this high mechanical stiffness makes them suitable for measuring effort variables (force, pressure). Equation (1.31) shows that a high stiffness results in a broad frequency range. Their small size (even less than 1 mm) and the possibility of manufacturing devices with unidirectional sensitivity are also properties of interest in many applications, particularly for vibration monitoring.

Figure 6.18 shows several simplified examples illustrating different possible

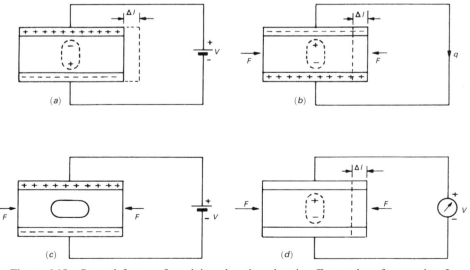

Figure 6.18   Several forms of applying the piezoelectric effect at low frequencies. In each case, one of the quantities is zero. (*a*) Null effort, $T = 0$; (*b*) null electric field, $E = 0$; (*c*) null strain, $S = 0$; (*d*) null charge density, $D = 0$. (From W. Welkowitz and S. Deutsch, *Biomedical Instruments Theory and Design*, copyright 1976. Reprinted by permission of Academic Press, Orlando, FL.)

low-frequency applications for the piezoelectric effect [7]. In case (a) no force is applied but only a voltage V. Therefore, a strain results. Given that $T = 0$, from (6.13) we have

$$S = dE \tag{6.22}$$

With the terminology shown in Figure 6.14, if the strain is in the longitudinal direction (x), we have

$$\frac{\Delta l}{l} = d\,\frac{V}{h} \tag{6.23}$$

From (6.12) we have

$$D = \epsilon^T E \tag{6.24}$$

that is, an electric polarization appears as in any normal capacitor. This arrangement is used for micropositioning—for example, of mirrors in lasers and of samples in scanning tunneling microscopes [8, 9].

In case (b) the metallic plates are short-circuited and a force F is applied. The result is that a polarization appears because electric charges migrate from one plate to the other. Given that $E = 0$, from (6.12) we have

$$D = dT \tag{6.25}$$

The charge obtained will be

$$q = Dlw = lwd\,\frac{F}{hw} = \frac{ld}{h}F \tag{6.26}$$

As in any solid body, a compression strain also results:

$$S = s^E T \tag{6.27}$$

This arrangement is applied to measure vibration, force, pressure, and deformation.

In case (c) the net deformation is zero because a force F is applied just to compensate for the field E due to the applied voltage. Therefore, we have $S = 0$ and from (6.13) we deduce

$$0 = s^E T + dE \tag{6.28}$$

and then

$$F = -\frac{wd}{s^E}V \tag{6.29}$$

The charge induced at each plate can be calculated from (6.12) to be

$$D = \frac{q}{wl} = dT + \epsilon^T E = d\frac{F}{wh} + \epsilon^T \frac{V}{h} \tag{6.30}$$

$$q = V\frac{wl}{h}\left(\epsilon^T - \frac{d^2}{s^E}\right) \tag{6.31}$$

The factor enclosed by the parentheses is designated ϵ^S, and it shows that the dielectric constant decreases because of the piezoelectric effect.

For the open circuit, case (d), it is not possible to transfer any charge from one plate to the other (although there will always be a certain leakage through the voltmeter). Therefore, despite the applied force, we have $D = 0$. From (6.12) we deduce

$$0 = dT + \epsilon^T E \tag{6.32}$$

$$V = -\frac{dhT}{\epsilon^T} = -\frac{dFh}{wh\epsilon^T} = -\frac{dF}{w\epsilon^T} \tag{6.33}$$

The resulting strain will be

$$S = s^E T + dE = s^E \frac{F}{wh} + d\frac{V}{h} \tag{6.34}$$

$$\frac{\Delta l}{l} = \frac{F}{wh}\left(s^E - \frac{d^2}{\epsilon^T}\right) \tag{6.35}$$

The term inside the parentheses is now designated s^D; and it shows that because of the piezoelectric effect, the material stiffness increases. A hammer or cam striking a piezoelectric ceramic generates more than 20 kV. The resulting spark is used for lighting gas ranges or for ignition in small internal combustion engines.

The application of the arrangement in Figure 6.18b to the measurement of forces, pressures, and movements (using a mass–spring system) is straightforward, and it is very similar for the three quantities. Figure 6.19 shows an outline for the three types of sensors. This similarity makes these sensors sensitive to the three quantities, and therefore special designs are required that minimize interference. Figure 6.20 shows a pressure sensor compensated for acceleration by combining signals from the stressed diaphragm and an inertial mass. Table 6.6 gives some characteristics for two quartz pressure sensors. Piezoelectric pressure sensors are used for monitoring internal combustion engines and in hydrophones. Because they lack dc response, they do not suit load cells.

Table 6.7 lists some data for two quartz accelerometers with integral electronics. Piezoelectric sensors with integral electronics are more reliable than sensors with external electronics because the connector is less critical, which is

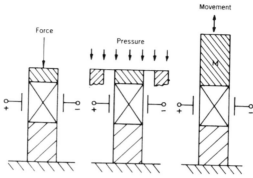

Figure 6.19 Force, pressure, and movement sensors based on piezoelectric elements (courtesy of PCB Piezotronics).

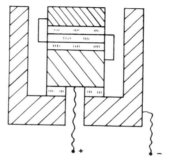

Figure 6.20 Piezoelectric pressure sensor with acceleration compensation by combining the signals from piezoelectric materials sensing both pressure and acceleration with that of one sensing only acceleration (courtesy of PCB Piezotronics).

important in shock and vibration monitoring. Piezoelectric accelerometers offer wider frequency bandwidth (0.1 Hz to 30 kHz), much lower power consumption, and higher shock survivability than micromachined accelerometers. However, they are inferior for static or very low frequency measurements. They are applied to machine monitoring, shock detection in shipment monitoring, impact detection, or drop testing, vehicle dynamics assessment and control, and structural dynamics analysis to detect response to load, fatigue, and resonance.

Polymer-based piezoelectric sensors are applied to microphones, machine monitoring, leak detection in pipes and high-pressure vessels (which produce a characteristic sound), keyboards, coin sensors, occupancy sensing, and vehicle classification and counting in highways. They are becoming relevant in medical applications such as pacemaker rate adjustment according to acceleration, sleep disorder monitoring, blood pressure monitoring, and blood flow and respiratory sounds monitoring in ambulances [10].

TABLE 6.6   Some Characteristics for Quartz Pressure Sensors

Parameter	Unit	6121[a]	112A[b]
Range	MPa	0 to 25	0 to 20.7
Maximal pressure	MPa	35	69
Sensitivity	aC/Pa	0.14	145
Resonant frequency	kHz	60	250
Frequency response	kHz	6	—
Linearity	%FSO	<1.0	1
Hysteresis	%FSO	<0.5	—
Acceleration sensitivity	kPa/g^c		
Axial		0.3	0.014
Transverse		0.05	—
Shock and vibration	g^c	<2000	20000/2000
Temperature coefficient of sensitivity	%/°C	<0.01	0.02
Operating temperature range	°C	−80 to 350	−204 to 240
Insulation at 25 °C/350 °C	TΩ	10/0.01	1/—
Mass	g^d	9.5	5

[a] Kistler.
[b] PCB Piezotronics.
[c] $1\ g = 9.80665\ \mathrm{m\cdot s^{-2}}$.
[d] grams.

TABLE 6.7   Some Characteristics for Quartz Accelerometers with Internal Preamplifier

Parameter	Unit	8702B50[a]	302A09[b]
Range	g^c	±50	±100
Sensitivity	mV/g	100	10
Frequency range			
±5 % limit	Hz	1 to 8000	1 to 5000
±10 % limit	Hz	0.6 to 10,000	0.7 to 10,000
Operating temperature range	°C	−55 to 100	−75 to 120
Mass	g^d	8.6	38
Overload, max.			—
vibration	g^c	±100	
shock (1 ms impulse)	g^c	2000	
Transverse acceleration, max.	g^c	±50	—
Threshold, nominal (rms)	g^c	0.006	0.005
Resonant frequency	kHz	50	20
Transverse sensitivity, max.	%	2.0	<5
Linearity, nominal	%	±1	±1
Temperature coefficient of sensitivity, max.	%/°C	−0.06	0.06
Time constant, nominal	s	1.0	0.5

[a] Kistler.
[b] PCB Piezotronics.
[c] $1\ g = 9.80665\ \mathrm{m\cdot s^{-2}}$.
[d] grams.

6.3 PYROELECTRIC SENSORS

6.3.1 The Pyroelectric Effect

The pyroelectric effect is analogous to the piezoelectric effect, but instead of change in stress displacing electric charge, now it refers to change in temperature causing change in spontaneous polarization and resulting change in electric charge. This effect was named by David Brewster in 1824, but it has been known for more than 2000 years [11].

When the change in temperature ΔT is uniform throughout the material, the pyroelectric effect can be described by means of the *pyroelectric coefficient*, which is a vector p with the equation

$$\Delta P = p\Delta T \tag{6.36}$$

where P is the spontaneous polarization.

This effect is mainly used for thermal radiation detection at ambient temperature (Section 6.3.3). Two metallic electrodes are deposited on faces perpendicular to the direction of the polarization, which yields a capacitor (C_d) acting as thermal sensor. When the detector absorbs radiation, its temperature and hence its polarization changes, thus resulting in a surface charge on the capacitor plates.

If A_d is the area of incident radiation and the detector thickness b is small enough so that the temperature gradient in it is negligible, then the charge induced will be

$$\Delta Q = A_d\Delta P = pA_d\Delta T \tag{6.37}$$

where ΔT is the increment in temperature of the detector. The resulting voltage will be

$$v_o = \frac{pb}{\epsilon}\Delta T \tag{6.38}$$

When the incident radiation is pulsating and has a power P_i, the resulting voltage on the capacitor is

$$v_o = R_v P_i \tag{6.39}$$

where R_v (V/W) is the *responsivity* or voltage sensitivity, given by [11]

$$R_v = \frac{\alpha p}{C_E \epsilon A}\frac{\tau}{\sqrt{1 + (\omega\tau)^2}} \tag{6.40}$$

where

α = the fraction of incident power converted into heat

p = the pyroelectric coefficient for the material

τ = the thermal time constant

C_E = the volumetric specific heat capacity $[\approx Q/(V\Delta T)]$ $[\text{J}/(\text{m}^3 \cdot \text{K})]$

ϵ = the dielectric constant

ω = the pulsating frequency for the incident radiation.

The corresponding short-circuit current is

$$i_{sc} = R_i P_i \tag{6.41}$$

where R_i (A/W) is the *current responsivity*, given by

$$R_i = \omega C_d R_v = \frac{\alpha p}{C_E b} \frac{\omega \tau}{\sqrt{1 + (\omega \tau)^2}} \tag{6.42}$$

which is a high-pass response for frequencies above that determined by the thermal constant of the material. R_v has a bandpass response: (6.40) is a low-pass response because of the thermal time constant, but the device responds only to temperature changes and, hence, there is no dc response. The upper corner frequency for commercial sensors is from 0.1 Hz to above 1 Hz (Figure 6.21). The voltage mode usually yields the best signal-to-noise ratio. The current mode yields a larger output signal and has a flatter frequency response.

As for other radiation detectors, pyroelectric sensors are also sensitive to thermal noise. The *noise equivalent power* (NEP) is the equivalent input power yielding an output response that in a given bandwidth equals that of thermal fluctuations in the detector [12]. The detectivity is $D = 1/\text{NEP}$. The NEP depends on wavelength, operating frequency, temperature, and noise bandwidth (usually 1 Hz). For an ideal detector with area A_d cm^2, at ambient temperature the NEP is about $5.5 \times 10^{-11} \sqrt{A_d}$ $(\text{W}/\sqrt{\text{Hz}})$. The D^* (D-star) parameter normalizes the NEP to a given constant area,

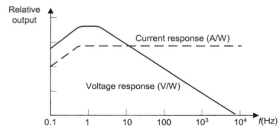

Figure 6.21 Frequency response of pyroelectric sensors in voltage mode (R_v) and current mode (R_i).

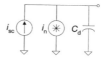

Figure 6.22 Equivalent circuit for a pyroelectric sensor including thermal noise.

$$D^* = \frac{\sqrt{A_d}}{\text{NEP}} \qquad (6.43)$$

Figure 6.22 shows the equivalent circuit for a noisy but otherwise ideal pyroelectric sensor. The star symbol for the current generator modeling thermal noise means that it is a random signal (Section 7.4).

6.3.2 Pyroelectric Materials

Given that pyroelectricity, like piezoelectricity, is also based on crystal anisotropy, many of the piezoelectric materials are also pyroelectric. Ten of the 21 noncentrosymmetrical crystallographic classes have a polar axis of symmetry. All of them display pyroelectric properties.

There are two groups of pyroelectric materials: linear and ferroelectric. The polarization of linear materials cannot be changed by inverting the electric field. This group includes materials such as tourmaline, lithium sulfate, and cadmium and selenium sulfides. Some ferroelectric materials with pyroelectric properties are lithium tantalate, strontium and barium niobate, lead zirconate-titanate, and triglicine sulfate (TGS). Some polymeric materials such as polyvinylidene (PVF$_2$ or PVDF) are also pyroelectric.

Pyroelectric properties disappear at the Curie temperature. For ferroelectric ceramics, the polarization is induced during manufacturing as described in Section 6.2.2.

According to (6.40) the ideal pyroelectric material should simultaneously display a high pyroelectric coefficient, a low volumetric specific heat capacity, and a low permittivity. Table 6.8 lists these parameters for some common pyroelectric materials.

TABLE 6.8 Some Parameters for Common Pyroelectric Materials

Material	Pyroelectric Coefficient (nC/cm$^2\cdot$K)	Relative Permittivity	Specific Heat (J/cm$^3\cdot$K)
Triglicine sulfate, TGS	40	35	2.5
Lithium tantalate, TaO$_3$Li	19	46	3.19
Strontium and barium niobate, SBN	60	400	2.34
PVDF	3	12	2.4

6.3.3   Radiation Laws: Planck, Wien, and Stefan–Boltzmann

We can measure the surface temperature of a heated target by allowing its emitted radiation to be absorbed by a pyroelectric detector, which increases its temperature. Thermocouples (thermopiles), thermistors, RTDs, and photoconductors also suit noncontact temperature measurement.

Any body at a temperature greater than 0 K radiates an amount of electromagnetic energy that depends on its temperature and physical properties. At temperatures above 500 °C, the emitted radiation is visible. Below 500 °C, including ambient temperatures, infrared radiation predominates so that only heat energy is perceived.

In order to study the emission of energy radiating from a body, we consider first its absorption. Of the overall energy received by a body, part is reflected, part is diffused in all directions, part is absorbed, and part is transmitted (i.e., goes through the body). We give the name "blackbody" to a theoretical body that absorbs all the energy incident on it (thereby increasing its temperature). A closed cavity with black walls and controlled temperature, and where only a small aperture is provided, behaves approximately as a blackbody.

At any temperature all bodies emit radiation and absorb that coming from other bodies in their environment. If all bodies are not at the same temperature, the hotter ones will cool and the colder ones will heat, so radiation is enough to reach thermal equilibrium (heat conduction and convection are not required). When equilibrium is reached, all bodies emit as much radiation as they receive. Therefore the bodies emitting more radiation are those that also absorb more, and hence a blackbody is also the best radiation emitter (heat sinks are black). Greenhouses rely on materials that are transparent to visible light but opaque to infrared radiation emitted by bodies heated by that incident light.

The ratio between the energy emitted by a given body per unit area per unit time and that emitted by a blackbody under the same conditions is the *emissivity* of that body ε. For a blackbody, $\varepsilon = 1$. The emissivity depends on the wavelength, the temperature, the physical state, and the chemical characteristics of the surface. For example, unoxidized aluminum has $\varepsilon = 0.02$ at 25 °C and $\varepsilon = 0.06$ at 500 °C; oxidized aluminum has $\varepsilon = 0.11$ at 200 °C and $\varepsilon = 0.19$ at 600 °C.

The energy W_λ emitted by the blackbody per unit time, per unit area, at a given wavelength λ and temperature T, is given by Planck's law,

$$W_\lambda = \frac{c_1}{\lambda^5[\exp(c_2/\lambda T) - 1]} \quad \text{W cm}^2/\mu\text{m} \tag{6.44}$$

where

$c_1 = 2\pi c^2 h = 3.74 \times 10^4 \ \text{W·}\mu\text{m}^4/\text{cm}^2$
$c_2 = hc/k = 1.44 \ \text{cm·K}$
$h = 0.655 \times 10^{-33} \ \text{W·s}^2$ is Planck's constant

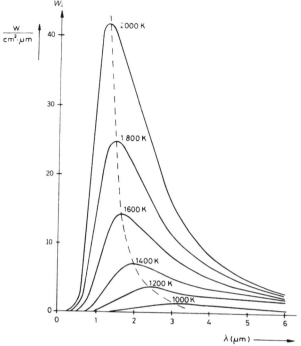

Figure 6.23 Power flux per unit area emitted by the blackbody at different temperatures and at different wavelengths (Planck's law). The dashed line passes through the maximums (Wien's law).

$k = 1.372 \times 10^{-22}$ W·s/K is Boltzmann's constant

$c \approx 300$ Mm/s is the velocity of light.

Figure 6.23 shows the shape of (6.44) for different values of T (in kelvins). The emissivity of real bodies depends on the wavelength, and we have

$$W_{\lambda r} = \varepsilon_{\lambda, T} W_\lambda \tag{6.45}$$

The maximal emitted power for a blackbody is at a wavelength

$$\lambda_{max} = \frac{2896 \ K \cdot \mu m}{T} \tag{6.46}$$

which is the equation for Wien's displacement law (to honor the man who discovered it before Planck's law was discovered). It indicates that the maximum is obtained at a wavelength that decreases for increasing temperatures. For example, the human body, with an assumed surface temperature of 300 K, has its maximal emission at 9.6 μm (region of medium infrared). The sun, whose temperature is 6000 K, has its maximal emission at 483 nm (blue).

The total flux power emitted by the blackbody per unit area is obtained by integrating (6.44) for all wavelengths. In a half-plane (solid angle 2π), the total emitted flux is

$$W = \sigma T^4 \tag{6.47}$$

which shows a dependence on the fourth power of the absolute temperature. $\sigma = 5.67$ pW/cm^2·K^4 is the Stefan–Boltzmann constant.

When this radiation arrives at an object, part of it is absorbed. When the absorption is high, the increase in temperature of the object can be significant. This is the principle of operation for thermal detectors: thermopiles and bolometers (RTD, thermistors, pyroelectric detectors). In contrast, quantum detectors (photoconductors, photovoltaic detectors) are based on the generation of electrons by incident photons, which results in a change in resistance or in the contact voltage of a p–n junction.

In practice, the presence of water vapor, CO_2 and ozone in the air, the dispersion due to dust particles, smoke, and so on, results in a decrease between the radiation emitted and that detected. In addition, the emissivity of the target may be unknown or may change because of varying surface conditions. Two-color or ratio pyrometers solve these problems by measuring two different wavelengths emitted by the target. If those wavelengths are close enough to undergo the same (undesired) variations, the ratio between the signal outputs changes only when the temperature changes at the target. The two signals needed can be obtained by a beam-splitting mirror and matched detectors or by a filter wheel and a single detector. The detailed design of radiometric systems can be found in reference 13.

6.3.4 Applications

The most common application for the pyroelectric effect is the detection of thermal radiation at ambient temperature. It has been applied to pyrometers (noncontact temperature meters in furnaces, melted glass or metal, films, and heat loss assessment in buildings) and radiometers (measurement of power generated by a radiation source). Other applications are IR analyzers (based on the strong absorption of IR by CO_2 and other gases), intruder and position detection, automatic faucet control, fire detection, high-power laser pulse detection, and high-resolution thermometry (6 μK) [14]. Medical thermometers that measure ear temperature detect infrared radiation from the eardrum and surrounding tissue.

Undesirable parasitic charges may neutralize the surface charge induced in the electrodes by desired temperature changes in the detector due to incident radiation. Thus the incident radiation must be modulated, usually by a low-frequency optical chopper that alternatively faces the detector to radiation from

TABLE 6.9   Some Characteristics for Two Pyroelectric Detectors

Characteristic	Unit	P3782[a]	406[b]
Window material		Silicon	Germanium
Sensing area diameter	mm	2	2
Optical bandwidth	μm	2–20	2–15
Voltage responsivity	V/W	1500[c]	275[d]
NEP	pW/$\sqrt{\text{Hz}}$	850[c]	500[d]
D^*	cm$\sqrt{\text{Hz}}$/W	2.2×10^{8c}	1.7×10^{8d}
Operating temperature range	°C	−20 to 60	−55 to 125
Rise time (0–63 %)	ms	100	0.2
Temperature coefficient of sensitivity, max.	%/°C	0.2	0.2

[a] Hamamatsu.
[b] Eltec.
[c] At 1 Hz.
[d] At 10 Hz.

the target and from an object at ambient temperature. A coherent detector synchronous with the chopper eliminates ambient noise. An alternative technique to reduce bulk or common mode effects uses dual-element, series-opposed detectors.

Pyroelectric sensors are faster than other thermal detectors (thermocouples, thermistors) because they are thin, have high sensitivity, and do not need to reach thermal equilibrium with the radiation source because they detect temperature gradients. This makes them appropriate for imaging by scanning the surface to be detected, as used in infrared thermography (thermal imaging—first demonstrated by J. Herschel in 1840) and applied in nondestructive testing, hot spot monitoring, printed circuit testing, and night vision. Placing the sensor at the focus of a parabolic mirror or a Fresnel lens increases its range to tens of meters. Linear arrays can discriminate targets according to the sequence of elements that are activated.

The detector can be suspended, supported by Mylar®, or mounted in a substrate that can be either thermally conductive or insulating. Because all pyroelectric materials are piezoelectric, these detectors have hermetically sealed packages (sometimes even with an internal vacuum), thus reducing the effects of air movements. Table 6.9 lists some parameters for two pyroelectric detectors suitable for the detection of a human body.

6.4   PHOTOVOLTAIC SENSORS

6.4.1   The Photovoltaic Effect

When the internal photoelectric effect discussed for photoconductors (Section 2.6) occurs in a *p–n* junction, it is possible to obtain a voltage that is a function of the incoming radiation intensity. The photovoltaic effect is the generation of an electric potential when the radiation ionizes a region where there is a potential barrier. It was discovered by E. Becquerel in 1839. D. M. Chapin, C. S. Fuller, and G. L. Pearson invented silicon photovoltaic cells in 1954.

When a p-doped semiconductor (doped with acceptors) contacts an n-doped semiconductor (doped with donors), because of the thermal agitation there are electrons that go into the p region and "holes" that move into the n-region. There they recombine with charge carriers of opposite sign. As a result, at both sides of the contact surface there are very few free charge carriers. Also the positive ions in the n region and the negative ions in the p region, fixed in their positions in the crystal structure, produce an intense electric field that opposes the diffusion of additional charge carriers through this potential barrier. This way an equilibrium is attained between the diffusion current and the current induced by this electric field. By placing an external ohmic connection on each semiconductor, no voltage difference is detected because the internal difference in potential at the junction is exactly compensated by contact potentials in the external connections to the semiconductor.

Figure 6.24 shows that radiation whose energy is larger than the semicon-

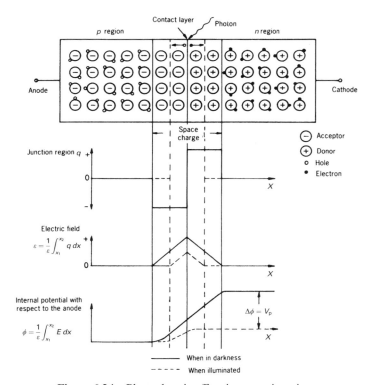

Figure 6.24 Photoelectric effect in a *p–n* junction.

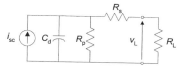

Figure 6.25 Equivalent simplified circuit for a photovoltaic detector. i_{sc} is the short-circuit current, R_p is the parallel resistance, R_s is the output series resistance, and C_d is the junction capacitance. R_L is the load resistance.

ductor band gap generates additional electron–hole pairs that are driven by the electric field in the open circuit p–n junction. The accumulation of electrons in the n region and of holes in the p region results in a change in contact potential V_P that can be measured by means of external connections to a load resistance. This open-circuit voltage increases with the intensity of the incident radiation, until a saturation point is reached (the limit is the band-gap energy). If the contacts are short-circuited, the current is proportional to the irradiation for a broad range of values. Figure 6.25 shows the simplified equivalent circuit.

6.4.2 Materials and Applications

In addition to p–n junctions, there are other methods that produce a potential barrier, but p–n junctions are by far the most common one. If the p–n junction is between semiconductors of the same composition, then it is called a *homojunction*. Otherwise it is called a *heterojunction*.

We select materials for the particular wavelength to be detected, as discussed in Section 2.6 for LDRs. In the visible and near-infrared regions, silicon and selenium are used. Silicon is in the form of homojunctions. Selenium in the form of a selenium layer (p) covering cadmium oxide (n). For silicon sometimes an intrinsic (nondoped) silicon region is added between the p and n regions (p–i–n detectors). This results in a wider depletion region, which yields a better efficiency at large wavelengths, faster speed, and lower noise and dark current. At other wavelengths, germanium, indium antimonide (SbIn), and indium arsenide (AsIn), among others, are used.

Photovoltaic detectors offer better linearity, are faster, and have lower noise than photoconductors, but they require amplification. For large-load resistors, the linearity decreases and the time of response increases. Table 6.10 gives some characteristics for two general purpose photovoltaic cells.

Photovoltaic detectors are used either in applications where light intensity is measured or in applications where light is used to sense a different quantity. They are used, for example, in analytical instruments such as flame photometers and colorimeters, in infrared pyrometers, in pulse laser monitors, in smoke detectors, in exposure meters in photography, and in card readers. Commercial models are available consisting of a matched emitter–detector pair, some of which are already connected to a control relay.

TABLE 6.10   Some Characteristics for Two Photovoltaic Detectors

Characteristic	Unit	S639[a]	J12-18C-R01M[b]
Detector material		Si	AsIn
Sensing area diameter	mm	20	1
Wavelength with maximal sensitivity	μm	0.85	3.6
Responsivity	A/W	0.45	0.7
NEP	pW/$\sqrt{\text{Hz}}$	1	70
D^*	cm$\sqrt{\text{Hz}}$/W	2×10^{12}	—
Shunt resistance[c]	Ω	10^4	20
Junction capacitance	nF	100	0.4
Response time (10–90 %)	μs	200	—
Operating temperature range	°C	−10 to 60	22
Short-circuit current for 100 lx	μA	180	—
Open circuit voltage for 100 lx	V	0.3	—

[a] Hamamatsu.
[b] EG&G Optoelectronics.
[c] Reverse voltage 10 mV.

6.5 ELECTROCHEMICAL SENSORS

Potentiometric electrochemical sensors yield an electric potential in response to a concentration change in a chemical sample. Amperometric sensors use an applied voltage to yield an electric current in response to a concentration change in a chemical sample. They are not self-generating sensors and are described elsewhere [15, 16].

Ion-selective electrodes (ISEs) are potentiometric sensors based on the voltage generated in the interface between phases having different concentrations. This is the same principle for voltaic cells. Assume that there is only one ion species whose concentration changes from one phase to another, or that there are more ions but a selective membrane allows only one specific ion to go through it. Then the tendency for that ion to diffuse from the high-concentration region to the low-concentration region is opposed by an electric potential difference due to the ion electric charge. When we have equilibrium between both forces (diffusion and electric potential), the difference in potential is given by the Nernst equation (first reported in 1899),

$$E = \frac{RT}{zF} \ln \frac{a_{i,1}}{a_{i,2}} \tag{6.48}$$

where $R = 8.31$ J/(mol·K) is the gas constant, T is the temperature in kelvins, z is the valence for the ion, $F = 96,500$ C is Faraday's constant, and a_i is the ion activity. For a liquid solution, activity is defined as

$$a_i = C_i f_i \tag{6.49}$$

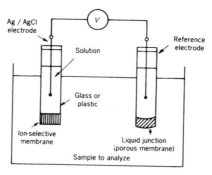

Figure 6.26 Measurement arrangement using an ion-selective electrode (ISE).

where C_i is the concentration for species i, and f_i is the activity coefficient, which describes the extent to which the behavior of species i diverges from the ideal, which assumes that each ion is independent of the others. This is not true at high concentrations and $f_i < 1$. For very diluted concentrations, $f_i \approx 1$.

This measurement principle is applied by using a two electrode arrangement (Figure 6.26). One electrode includes the membrane that is selective to the ion of interest, and it contains a solution having a known concentration for ion species i. The other electrode is a reference, and all ions present in the sample to be measured can freely diffuse through its membrane. This arrangement involves several interfaces, but only one of them generates a variable potential: the one across the ion-selective membrane. From (6.46) we obtain

$$E = E_0 + \frac{RT}{zF} \ln a_i = E_0 + k \lg a_i \qquad (6.50)$$

where a_i is now the activity for the ionic species of interest in the sample, and E_0 and k are constants. At 25 °C, the sensitivity is 59.12 mV for each decade of variation in the change of activity for a monovalent cation, while at 100 °C the sensitivity is 74.00 mV for each decade. It is therefore very important to know the cell temperature in order to correctly interpret the meaning of the measured potential difference. ISEs are bulky (100 mm to 150 mm in length and 10 mm in diameter), fragile, and require maintenance because the electrolyte is volatile.

When the quantity of interest is not the ionic activity but the concentration, from (6.49) and (6.50) we have

$$E = E_0 + k \lg f_i + k \lg C_i \qquad (6.51)$$

If we assume that the activity coefficient is constant, then

$$E = E_0' + k \lg f_i \qquad (6.52)$$

Depending on the material for the membrane, there are different kinds of selective electrodes. Primary electrodes have a single membrane, which may be crystalline. When it is crystalline, it can be homogeneous or heterogeneous. In heterogeneous electrodes the crystalline material is mixed with a matrix of inert material. Crystalline membrane electrodes are applied to concentration measurement for F^-, Cl^-, Br^-, I^-, Cu^{2+}, Pb^{2+}, and Cd^{2+}, among others. The most common electrodes with a noncrystalline membrane are glass electrodes, like those used for pH and Na^+ measurement. Glass composition is chosen depending on the ion to be analyzed. Some metal salts have high electric conductivity and can be deposited on a metal electrode to act as electrolyte. These are termed *solid-state electrodes*. Other electrodes use a membrane (such as PVC or polyethylene) that includes an ion exchanger or a neutral material that transports the ion. K^+, for example, is measured by valinomycin in a PVC membrane.

The most common double-membrane electrodes are gas electrodes. They include a porous membrane through which the gas to be analyzed diffuses and enters into a solution where the presence of the gas produces a change (e.g., in pH), which is the measured variable. This method is applied, for example, to concentration measurement for CO_2, SO_2, and NO_2.

Output impedance for ISEs is very high, normally from 20 MΩ to 1 GΩ, thus requiring electrometer amplifiers with very high input impedance (Section 7.2). Otherwise, current through the cell would imbalance the chemical reaction, leading to variation in its electric potential.

ISEs are used for concentration measurement in multiple applications where they have often replaced flame photometers. They are used, for example, in agriculture to analyze soils and fertilizers, in biomedical sciences and clinical laboratories for blood and urine analysis, in chemical and food industries, and in environmental monitoring to measure ambient pollution.

Solid electrolyte oxygen sensors rely on the influence that oxygen ions adsorbed by a metal oxide have on the concentration of charge carriers and, hence, on conductivity of the oxide—based on ions, hence it is an electrolyte. Oxygen molecules entering interstices in the oxide take two negative charges, so that two "holes" are created in order to keep the charge balance at zero. Oxygen molecules leaving interstices set two electrons free and create a vacancy for another oxygen molecule.

A common solid electrolyte for O_2 detection is yttrium-doped zirconia (ZrO_2–Y_2O_3) placed between two porous thick-film platinum electrodes, inside a temperature-controlled chamber at 600 °C to 850 °C. The open-circuit output voltage is [17]

$$E = E_0 + \frac{RT}{4F} \ln \frac{(p_{O_2})_1}{(p_{O_2})_2} \qquad (6.53)$$

where $(p_{O_2})_1$ and $(p_{O_2})_2$ are the oxygen partial pressures inside and outside the electrolyte. If $(p_{O_2})_2$ is the partial pressure in a reference gas pressure—for

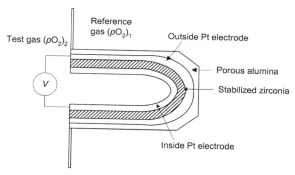

Figure 6.27 A potentiometric oxygen sensor generates a potential difference between electrodes on opposite sides of a stabilized zirconia membrane at high temperature.

example, air $((p_{O_2})_2 = 21 \text{ kPa})$ (Figure 6.27)—from E we can determine $(p_{O_2})_1$ at a given temperature.

These sensors are fast and withstand temperatures from $600\,°C$ to $1200\,°C$, but according to (6.53) they drift with temperature. They can be as small as 1 cm (length) by 2 mm (diameter). Because they consist of solid elements, their sensitivity to acceleration and vibration is minimal. Their main shortcomings are that they need a high temperature to work and that they have a low sensitivity to pressure changes: The voltage in (6.53) is proportional to the logarithm of the ratio between pressures, not to the pressure ratio. For this same reason, however, they can operate over a wide range of oxygen concentration. They are extensively used to determine the air-to-fuel ratio in internal combustion engines—for example, in automobiles, boilers, and furnaces.

6.6 PROBLEMS

6.1 We must repair a temperature-measuring system whose manual is not available. We inspect the measurement circuit and find the result shown in Figure P6.1. From the codes for lead coverings, we deduce that metal A is

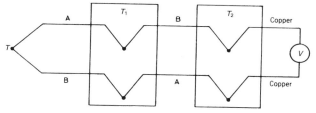

Figure P6.1 Simulation of a $0\,°C$ cold junction by two temperature-controlled chambers.

iron and metal B is constantan. Thus they form a type J thermocouple. Determine temperatures T_1 and T_2 for the temperature-controlled chambers in order to obtain a voltage reading dependent on T but not on T_1 and T_2.

6.2 In order to measure the temperature gradient across a turbine wall, externally covered by a thermal insulator, a temperature probe is available consisting of two iron–constantan thermocouples covered by a material having the same thermal characteristics as the turbine walls. The probe is placed so that one thermocouple is on the inner side of the turbine wall and the other is at its center. Thermocouple wires are long enough to allow for the connections to be made outside of the turbine, and they are made with copper wires. Assume that the temperature distribution in the wall and in the thermal insulator are linear and that the external temperature of the insulator is at ambient temperature, which changes from 10 °C to 30 °C depending on the time of the year.

   a. Design a system to measure the temperature gradient in the wall by means of a voltmeter placed in a rack 50 m distant from the thermocouple. The rack is kept at only 5 °C above the ambient temperature by forced ventilation. Discuss the need for each of the connections made, and discuss their possible influence on the measurement. What is the reading for the voltmeter when the steam temperature inside the turbine is 575 °C and the difference in temperature is 80 °C?

   b. Assume that the temperature at the center of the turbine wall is always the same regardless of the temperature distribution profile. What is the influence of that profile on the measurement?

   c. We use the circuit in Figure P6.2 to measure the steam temperature. If the ambient temperature sensor is a platinum resistance of 100 Ω and $\alpha = 0.00392$ Ω/Ω/K at 0 °C, determine k in order to obtain in the voltmeter a reading corresponding to T_V. What is the error in the compensation of an ambient temperature of 30 °C?

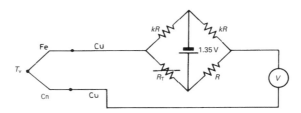

Figure P6.2   Cold junction compensation by an RTD.

6.3 The circuit in Figure P6.3 measures a temperature from 400 °C to 600 °C by a type J thermocouple and cold junction compensation. The LM134 is a current source whose output is I (μA) = (227 Ω) × [273 + T_a(°C)]/R_3.

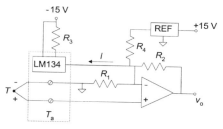

Figure P6.3 Cold junction compensation by a semiconductor temperature sensor.

a. If the op amp is assumed ideal, derive the equation for its output voltage.

b. Determine the gain needed to obtain a -10 V to $+10$ V output range for a 400 °C to 600 °C temperature range.

c. Design the resistances in order to obtain a cold junction compensation for an ambient temperature between 10 °C and 40 °C.

6.4 Figure P6.4 shows a thermometer based on a type T thermocouple and the AD590 for cold junction compensation.

a. Derive the equation for the output voltage as a function of the thermoelectromotive forces and v_c.

b. The compensation circuit includes R_a to prevent any damage to the op amp in case the AD590 is far away. Derive v_c when the op amp is assumed ideal.

c. Design R_1 and R_2 to achieve cold junction compensation when $10\,°C < T_a < 60\,°C$.

d. If the Zener diode yields 6.9 V when its reverse current is from 0.6 mA to 15 mA, design R, R_3, R_4, and R_p to obtain a null output at 0 °C.

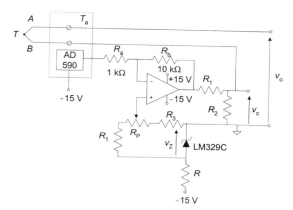

Figure P6.4 Cold junction compensation by a semiconductor temperature sensor.

6.5   We wish to measure a temperature between $-100\,°\text{C}$ and $+100\,°\text{C}$ by using a type K thermocouple and cold junction compensation according to the circuit in Figure P6.5. R_T is a Pt100 with $\alpha = 0.385\,\%/\text{K}$ at $0\,°\text{C}$.

   a. Derive the equation for the output voltage as a function of the sensitivity S_K for the thermocouple and the value for R_T.

   b. Determine the conditions to be fulfilled by the resistors in order to have a null output when $T_a = 0\,°\text{C}$.

   c. If R_2 and R_3 are chosen so that the output voltage of the bridge circuit they constitute with R_T and R_1 can be considered linear, determine their value in order to compensate the cold junction in the entire range for T_a.

   d. Determine the gain for the instrumentation amplifier in order to have an output voltage from -1 V to $+1$ V.

   e. Determine the maximal offset voltage for the instrumentation amplifier in order to limit the error to a maximal $0.1\,°\text{C}$.

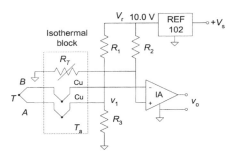

Figure P6.5   Cold junction compensation by an RTD.

6.6   Figure P6.6 shows a thermocouple circuit with a grounded thermocouple wire. The cold (reference) junction is at ambient temperature and includes

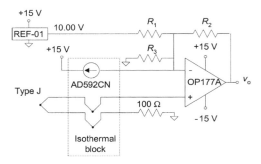

Figure P6.6   Cold junction compensation by a semiconductor temperature sensor.

an AD592CN, which yields 1 μA/K. If the thermocouple is type J, design the resistors to obtain a sensitivity of 10 mV/°C in the range from −25 °C to 105 °C and 0 V at 0 °C.

6.7 We place a quartz crystal piezoelectric sensor of 1 cm² area and 1 mm thickness between two parallel metallic electrodes to measure variable forces perpendicular to the electrodes. Young's modulus for the quartz is 90 GPa, charge sensitivity is 2 pC/N, relative permittivity is 5, and a cubic block with 1 cm sides has a resistance of 100 TΩ between opposite sides. A 100 MΩ resistance and a 20 pF capacitor shunt the electrodes. If a force $F = 0.01 \sin(10^3 t)$ N is applied, calculate the resulting peak-to-peak voltage between electrodes and the maximal deformation for the material.

6.8 A given piezoelectric sensor has a 100 pF capacitance and a charge sensitivity of 4 μC/cm. The connecting cable has a 300 pF capacitance, and the measuring oscilloscope has an input impedance of 1 MΩ in parallel with 50 pF.

   a. Determine the voltage sensitivity (V/cm) for the sensor considered alone.

   b. Determine the voltage sensitivity (V/cm) for the complete measurement system at high frequency.

   c. Determine the minimal frequency of the force we can measure with this system if the maximal allowed amplitude error is 5%.

   d. Determine the capacitance that must be connected in parallel in order to extend the band with 5% error to 10 Hz.

   e. Determine the new voltage sensitivity for the system.

REFERENCES

[1] D. D. Pollock. *Thermoelectricity: Theory, Thermometry, Tool.* ASTM Special Technical Publication 852. Philadelphia, PA: American Society for Testing and Materials, 1985.

[2] Omega Engineering Inc. *The Temperature Handbook.* Stamford, CT, 2000.

[3] F. R. Ruppel. Modeling a self-calibrating thermocouple for use in a smart temperature measurement system. *IEEE Trans. Instrum. Meas.,* **39**, 1990, 898–901.

[4] H. N. Norton. *Sensor and Analyzer Handbook.* Englewood Cliffs, NJ: Prentice-Hall, 1982.

[5] A. J. Pointon. Piezoelectric Devices. *IEE Proc.,* **129**, pt. A, 1982, 285–307.

[6] J. C. L. Van Peppen and K. B. Klaassen. Damping of compression and shear piezoelectric accelerometers by electromechanical feedback. *IEEE Trans. Instrum. Meas.,* **37**, 1988, 572–577.

[7] W. Welkowitz, S. Deutsch, and M. Akay (ed.). *Biomedical Instruments Theory and Design,* 2nd ed. New York: Academic Press, 1991.

[8] D. J. Peters and B. L. Blackfors. Piezoelectric bimorph-based translation device for two-dimensional, remote micropositioning. *Rev. Sci. Instrum.* **60**, 1989, 138–140.

[9] Y. Kuk and P. J. Silverman. Scanning tunneling microscope instrumentation. *Rev. Sci. Instrum.* **60**, 1989, 165–180.

[10] J. V. Chatingny. Health and medical devices that depend on piezopolymer film. *Medical Electronics*, Issue **177**, June 1999, 31–35.

[11] S. T. Liu and D. Long. Pyroelectric detectors and materials. *Proc. IEEE*, **66**, 1978, 14–26.

[12] D. Cima. Introduction to IR pyroelectric detectors. *Sensors*, **10**, March 1993, 70–75.

[13] C. L. Wyatt. *Radiometric System Design*. New York: Macmillan, 1987.

[14] A. Hadni. Applications of the pyroelectric effect. *J. Phys. E: Sci. Instrum.*, 14, 1981, 1233–1240.

[15] M. J. Schöning, O. Glück, and M. Thust. Electrochemical composition measurements. Section 70.1 in: J. G. Webster (ed.), *The Measurement, Instrumentation, and Sensor Handbook*. Boca Raton, FL: CRC Press, 1999.

[16] F. Oehme, Liquid Electrolyte Sensors: Potentiometry, Amperometry, and Conductometry. Chapter 7 in: W. Göpel, T. A. Jones, M. Kleitz, J. Lundstrom, and T. Seiyama (eds.), *Chemical and Biochemical Sensors Part I*, Vol. 2 of *Sensors, A Comprehensive Survey*. W. Göpel, J. Hesse, J. N. Zemel (eds.). New York: VCH Publishers (John Wiley & Sons), 1991.

[17] P. T. Moseley, Solid state gas sensors. *Meas. Sci. Technol.*, **8**, 1997, 223–237.

7

SIGNAL CONDITIONING FOR SELF-GENERATING SENSORS

Self-generating sensors offer a voltage or a current whose amplitude, frequency, and output impedance determine the characteristics required for the signal conditioner.

When the range of the sensor output voltage or current is smaller than the input range of the ADC (or other signal receiver), amplification is required. The amplification may be different from that described in preceding chapters because signals from self-generating sensors are not the output of a bridge circuit or voltage divider.

Besides being very small, the voltages to be amplified sometimes have a very low frequency. This prevents the use of ac-coupled high-gain amplifiers because the capacitors required would not be practical. We must consider dc amplifiers with their offset voltage and bias and offset currents, along with their respective drifts with time and temperature. High-gain dc amplifiers are often based on op amps or instrumentation amplifiers, so we will first discuss the problems and solutions for their drifts.

In other cases the signal to be processed is not small, but it comes from a high output impedance source. Voltage-generating sensors with high output impedance such as pH electrodes need electrometer amplifiers. Current-generating sensors such as piezoelectric sensors may use electrometer, transimpedance, or charge amplifiers—when parasitic capacitance reduces signal bandwidth.

To obtain high resolution, even when drift is not a problem, we must deal with internal noise in amplifiers. Internal noise is inherent to all electronic devices, but we will consider here only op amps and instrumentation amplifiers, which are the most common devices for sensor signal conditioning, and resistors used with them.

7.1 CHOPPER AND LOW-DRIFT AMPLIFIERS

7.1.1 Offset and Drifts in Op Amps

In an ideal op amp the output voltage is zero when both input voltages are zero. The input currents are then also zero. In a real op amp, neither of these conditions holds. In addition to being different from zero when the input voltage is zero, the input currents are not equal to each other. Their difference is called *offset current*. This is due to the imbalance between input transistors [bipolar or field-effect transistors (FETs)]. This imbalance also requires an *offset voltage* between the input terminals for the output voltage to be zero.

We can analyze the effects of these voltages and currents by considering a simple amplifier—for example, the inverting amplifier in Figure 7.1. The output voltage is

$$v_o = \left[-\frac{R_2}{R_1'} v_s + \left(1 + \frac{R_2}{R_1'} \right) V_{io} + I_n R_2 - I_p R_3 \left(1 + \frac{R_2}{R_1'} \right) \right] \frac{1}{1 + \dfrac{1}{A_d \beta}} \qquad (7.1)$$

where $R_1' = R_1 + R_s$, $\beta = R_1'/(R_1' + R_2)$, $A_d = A_{d0}\omega_a/(s + \omega_a) = f_T/(s + \omega_a)$, and we have assumed that $A_d\beta$ is a real number. If, in addition, $A_d\beta \gg 1$, we can approximate

$$v_o = -\frac{R_2}{R_1'} v_s + \left(1 + \frac{R_2}{R_1'} \right) V_{io} + I_n R_2 - I_p R_3 \left(1 + \frac{R_2}{R_1'} \right) \qquad (7.2)$$

The actual sign for V_{io} is unknown and that of I_n and I_p depends on the transistor type (p–n–p, n–p–n) at the op amp input stage. Usually a worst-case condition is assumed and the contribution of offset voltage and current is added.

Equation (7.2) shows that offset voltage and input currents introduce an

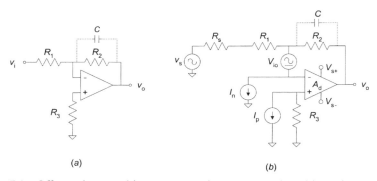

(a)   *(b)*

Figure 7.1   Offset voltage and input currents in an op-amp-based inverting amplifier.

output zero error,

$$\text{OZE} = v_{\text{o}}(0) = \left(1 + \frac{R_2}{R_1'}\right) V_{\text{io}} + I_{\text{n}} R_2 - I_{\text{p}} R_3 \left(1 + \frac{R_2}{R_1'}\right) \tag{7.3}$$

R_3 is not necessary for the amplification function itself, but if its value is chosen such that $R_3 = R_1' \| R_2$, then (7.3) reduces to

$$\text{OZE} = \left(1 + \frac{R_2}{R_1'}\right) V_{\text{io}} + I_{\text{io}} R_2 \tag{7.4}$$

where $I_{\text{io}} = I_{\text{n}} - I_{\text{p}}$ is the offset current. To refer the error to the input we divide the OZE by the gain amplitude $|G| = R_2/R_1'$ to obtain

$$\text{IZE} = \frac{\text{OZE}}{R_2/R_1'} = \left(1 + \frac{R_1'}{R_2}\right) V_{\text{io}} + I_{\text{n}} R_1' - I_{\text{p}} R_3 \left(1 + \frac{R_1'}{R_2}\right) \tag{7.5}$$

which for a matched R_3 reduces to

$$\text{IZE} = \left(1 + \frac{R_1'}{R_2}\right) V_{\text{io}} + I_{\text{io}} R_1' \tag{7.6}$$

This shows that the IZE increases for high input impedance and high gain (R_1 and R_2/R_1 large). If all resistors are reduced by the same factor, the error due to V_{io} does not change while that due to I_{io} decreases. Hence, we must use resistors having a value as small as possible.

The analysis for a noninverting amplifier is similar, but its signal gain is $1 + R_2/R_1$. The resulting IZE is given by (3.38). If R_1 and R_2 are chosen so that their parallel resistance matches the source resistance, $R_{\text{s}} = R_1 \| R_2$, the IZE reduces to (3.39). Alternatively, we can add a resistor R_3 between the inverting terminal and the node connecting R_1 and R_2, such that $R_3 = R_{\text{s}} - R_1' \| R_2$ (Figure 7.3b), where R_1' includes the equivalent output resistance of the voltage level shifting network ($\approx 100\ \Omega$).

The IZE depends on V_{io}, I_{n}, I_{p}, and I_{io}. These parameters depend on the op amp technology and quality. As a rule, bipolar-input op amps display lower offset voltage and drifts, FET-input op amps have lower bias and offset currents, and CMOS op amps have still smaller input currents but larger drifts, particularly at high temperature. Offset voltage and input currents depend on power supply voltages and common mode voltage. Table 7.1 gives some typical values for these parameters. The bias current is $I_{\text{b}} = (I_{\text{n}} + I_{\text{p}})/2$, and therefore $I_{\text{n}} = I_{\text{b}} + I_{\text{io}}/2$ and $I_{\text{p}} = I_{\text{b}} - I_{\text{io}}/2$. I_{b} increases with the common mode voltage. Common bipolar op amps have I_{io} five to ten times smaller than I_{b}. Hence, including R_3 in the circuit reduces the IZE. However, some dc-precision op amps such as the OP07, OP177, AD707, and the like, have compensated

TABLE 7.1 Relevant Data for Some dc Precision Op Amps[a]

Op Amp	A_{d0} Min. ($\times 10^6$)	f_T Min. (MHz)	V_{io} Max. (μV)	$\Delta V_{io}/\Delta T$ Max. (μV/°C)	$\Delta V_{io}/\Delta t$ Avg. (μV/month)	I_b Max. (pA)	I_{io} Max. (pA)	$\Delta I_{io}/\Delta T$ Max. (pA/°C)
Bipolar								
AD705	0.4	0.8	25	0.6	0.3	100	100	0.4
AD707	8.0	0.5	15	0.1	0.2	1000	1000	25
MAX400	0.5	0.4	15	0.3	1.0	2000	2000	25
MAX410	0.6	28.0	250	1	—	1.5×10^5	8×10^4	—
OP07	0.2	0.4	75	1.3	1.0	3000	2800	50
OP77	2.0	0.4	60	0.6	0.2	2800	2800	35
OP97	0.2	0.4	75	2.0	0.3	150	150	7.5
OP177A	5.2	0.4	10	0.3	0.2	2000	1500	25
OPA177G	2.0	0.4	60	1.2	0.2	2800	2800	60
TLE2027C	5.0	7.0	100	1	1	9000	9000	—[b]
Bifet, CMOS								
AD547L	0.1	1.0	250	1	—	25	15	—[c]
AD645C	0.6	2.0	250	1	—	3	0.5	—[d]
AD795K	0.3	1.6	250	3	—	1	0.5	—[d]
LMC6001A	0.4	1.3[e]	350	10	—	0.025	1	—
OP80E	0.1	0.3[e]	1500	—	—	0.25	0.05[e]	—
OPA111SM	0.6	2.0	500	5	—	2	1.5	—[d]
OPA627BM	0.6	16.0	100	0.8	—	5	5	—[d]

Autozero

AD8551	1	1.5^e	5	0.04	—	50	70	—
ICL7650	1.0	2.0	5	0.1	0.1	10	5	—[f]
LTC1052	1.0	1.2	5	0.05	0.1	30	30	—[b]
LTC1150C	10.0	2.5	5	0.05	0.05	100	200	—[c]
LTC1152	0.6	1.0	10	0.1	0.05	100	200	—
LTC1250C	1.8	1.5	10	0.05	0.05	200	400	—[g]
MAX420	0.6	0.5	10	0.05	0.1	100	200	—[g]
MAX430	0.6	0.5	10	0.05	0.1	100	200	—[h]
TLC2652AC	6.0	1.9	1	0.03	0.02	4	2	—[h]
TLC2654C	1.0	1.9	20	0.05	0.02	50	30	—
TSC911B	0.6	1.5	30	0.25	—	120	40	—

Other

LMC2001	0.1	6^e	40	0.015^e	0.006	3^e	6^e	—
TLC4502	0.2	4.7^e	100	1^e	—	1^e	1^e	—

[a] Values are at 25 °C ambient temperature when supplied at ±15 V, except for the MAX410 and some autozero models (±5 V). Other measurement conditions may differ.

[b] I_b remains constant up to 75 °C.

[c] Temperature has a different effect on I_n and I_b.

[d] Doubles every 10 °C above the temperature reached after 5 min operation at 25 °C.

[e] Typical value.

[f] Doubles every 10 °C.

[g] I_b doubles every 10 °C above about 60 °C.

[h] I_b remains almost constant up to 85 °C.

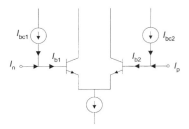

Figure 7.2   Input stage of op amps that have compensated bias current.

input bias current (Figure 7.2). This reduces I_n and I_p, but at the same time, because of the limited accuracy of the added current sources, I_{io} is close to I_n and I_p. Nevertheless, including R_3 still makes sense because it balances the equivalent circuit seen from the op amp input terminals, and balance reduces interference.

The values for offset voltage and input currents in (7.5) and (7.6) are those at the actual temperature of the op amp, which can be calculated as shown in Section 3.2.4,

$$T = T_a + P_d(\theta_{jc} + \theta_{cs} + \theta_{sa}) \tag{7.7}$$

where T_a is the ambient temperature, and θ_{jc}, θ_{js}, and θ_{sa} are the respective thermal resistances between the internal chip and amplifier case (from 100 °C/W to 150 °C/W for plastic packages), between the case and the heat sink, if used, and between the heat sink and the ambient air. P_d includes the quiescent power (P_q), the power supplied to the feedback network, and the power dissipated inside the amplifier by the current delivered to the load. In signal amplifiers, often only P_q is relevant and for split power supplies we have

$$P_q = |V_{s+}||I_{s+}| + |V_{s-}||I_{s-}| \tag{7.8}$$

where I_{s+} and I_{s-} are the respective quiescent currents. Bipolar op amps usually have smaller supply currents than FET-input op amps, and CMOS op amps have very small supply currents.

Manufacturers usually specify offset voltages and input currents from a sample of op amps tested at high speed. This means that the IC does not reach its usual working temperature and therefore the stated parameters correspond to the ambient temperature during the test T_A, usually 25 °C. Data sheets give in addition the average temperature coefficient for V_{io}, I_b, and I_{io}. Therefore, the actual parameter values to use for OZE and IZE in the above equations are

$$V_{io}(T) = V_{io}(T_A) + \frac{\Delta V_{io}}{\Delta T}[T - T_A] \tag{7.9}$$

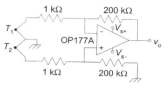

Figure E7.1   Thermocouple amplifier based on a dc precision op amp.

$$I_b(T) = I_b(T_A) + \frac{\Delta I_b}{\Delta T}[T - T_A] \tag{7.10}$$

$$I_{io}(T) = I_{io}(T_A) + \frac{\Delta I_{io}}{\Delta T}[T - T_A] \tag{7.11}$$

Example 7.1   Figure E7.1 shows a differential amplifier connected to a thermocouple circuit whose wires have 10 Ω resistances. Estimate the input zero error when the ambient temperature is 35 °C.

Data sheets for the OP177A yield the following data when working at ±15 V and $T_A = 25$ °C: $V_{io} = 75$ mV (max.); $\Delta V_{io}/\Delta T = 0.1$ mV/°C (max., from −55 °C to +125 °C); $I_b = 1.5$ nA (max.); $I_{io} = 1$ nA (max.); $\Delta I_b/\Delta T = 25$ pA/°C (max., from −55 °C to +125 °C); power consumption, $P_q = 75$ mW (max.); thermal resistance for the "P" package (8-pin plastic DIP), $\theta_{ja} = 103$ °C/W.

The equivalent circuit when considering the offset voltage and input currents reduces to that in Figure 7.1b with $R_3 = (1 \text{ k}\Omega + 10 \text{ }\Omega)\|(200 \text{ k}\Omega) = R_1\|R_2$. Therefore, (7.6) applies with $R_1' = 1 \text{ k}\Omega + 10 \text{ }\Omega$ and $R_2 = 200 \text{ k}\Omega$. We have

$$\text{IZE} = \left(1 + \frac{200}{1.01}\right) V_{io} + I_{io}(1010 \text{ }\Omega)$$

The actual working temperature is calculated from (7.7) by considering that no heat sink is used,

$$T = T_a + P_q\theta_{ja} = 35\,°C + (75 \text{ mW})(103\,°C/W) = 35\,°C + 7.7\,°C \approx 43\,°C$$

Next, the actual values for offset voltage and currents are calculated. For the offset voltage, from (7.9) we have

$$V_{io}(43\,°C) = V_{io}(25\,°C) + \frac{\Delta V_{io}}{\Delta T}[43\,°C - 25\,°C]$$

$$= 10 \text{ }\mu V + (0.1 \text{ }\mu V/°C)(18\,°C) \approx 12 \text{ }\mu V$$

For the input offset current we have

$$I_{io}(43\,°C) = I_{io}(25\,°C) + \frac{\Delta I_{io}}{\Delta T}[43\,°C - 25\,°C]$$

$$= 0.5 \text{ nA} + (25 \text{ pA}/°C)(18\,°C) \approx 1 \text{ nA}$$

Therefore,

$$\text{IZE} = \left(1 + \frac{200}{1.01}\right)(12\ \mu V) + (1\ nA)(1010\ \Omega) = 13\ \mu V$$

For a type J thermocouple (Table 6.2), for example, this yields an error smaller than 0.25 °C.

Offset voltages also change with time. Data sheets specify the average monthly drift $\Delta V_{io}/\Delta t$ after the first 30 days of operation. Because drifts are random, not cumulative, the drift after n months is $\sqrt{n}$ times the monthly drift.

Input zero errors from offset voltages and currents can be cancelled at a given temperature by using the terminal provided in some op amps for internal offset adjustment. This adjustment can also cancel offsets external to the op amp. But in common bipolar op amps, thermal drift increases by 3 to 4 $\mu V/°C$ for each millivolt adjusted. In several precision dc op amps, however, there is no such deleterious effect. There are electrically programmable analog devices (EPAD, Advanced Linear Devices) that provide the user with a programmable offset-voltage adjustment to any desired value. Nevertheless, the adjustment range available either mechanically with a trimmer or electronically in EPADs is quite limited, and some signal conditioners must shift the output signal level by tens of millivolts or even volts. Figure 7.3 shows how to combine amplification and level shifting in a single stage. In order to keep a balanced input, in Figure 7.3a we must select

$$R_3 = \frac{R_1 R_2}{R_1 + R_2} - R_4 \tag{7.12}$$

In Figure 7.3b, where it has been assumed that R_s is not adjustable (e.g., it may be determined by the signal source), R_3 must be

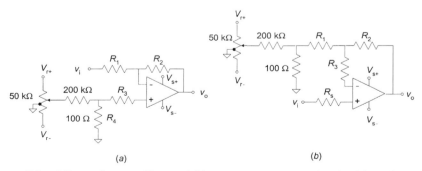

(a)   (b)

Figure 7.3 Offset voltage nulling and bias current compensation in (a) an inverting amplifier and (b) a noninverting amplifier.

$$R_3 = R_s - \frac{R_1' R_2}{R_1' + R_2} \tag{7.13}$$

where $R_1' = R_1 + 100 \ \Omega$. Four-terminal trimming potentiometers that include a center tap, which may be grounded, offer better setability and reduce drifts from V_{r+} and V_{r-}. For better stability, these voltages must come from voltage reference ICs.

Whatever the case, offset nulling or level shifting when the circuit input is held at the reference voltage must be done after the amplifier has reached its operating temperature. Furthermore, temperature gradients in active components must be avoided, and passive components should have a low temperature coefficient. Power supplies must be well-regulated; otherwise their fluctuations would show up at the circuit output. The equivalent input ripple when power supplies change by ΔV_{s+} and ΔV_{s-} is

$$v_r = \frac{\Delta V_{s+}}{\text{PSRR}_+} + \frac{\Delta V_{s-}}{\text{PSRR}_-} \tag{7.14}$$

where PSRR stands for power supply rejection ratio. Some data sheets specify the power supply fluctuation to input voltage error ratio in decibels, whereas others specify the input voltage error to power supply fluctuation ratio in microvolts/volt, but in both cases only for slow voltage fluctuations. Fast transients coupled to power supplies may lead to an erratic behavior. Dc precision op amps have PSRR > 120 dB at dc, but it rolls off 20 dB/decade, in some cases starting at 1 Hz.

7.1.2 Chopper Amplifiers

Before manufacturing techniques made low-drift amplifiers available, some different solutions for drift problems were applied. The most common for many years was to modulate an ac signal with the input voltage, then amplify the ac signal—which makes zero errors irrelevant—and later demodulate. Carrier amplifiers (Section 5.3) process a sine wave modulated by the sensor. But this is not the case for self-generating sensors. Instead, their outputs modulate a square wave generated by a repetitive switch or chopper, hence the name "chopper amplifiers."

In a classic chopper amplifier (Figure 7.4), a repetitive switch alternately connects the input of an ac amplifier to the nearly constant voltage to be measured and to a reference voltage (usually ground). The resulting square wave is high-pass coupled to an ac amplifier whose drifts will not affect the input. The amplified signal is then synchronously demodulated and low-pass filtered in order to reduce any ripple due to oscillator frequency and its harmonics. For the demodulation process to be simple, any possible high-frequency signal at the input is blocked by low-pass filtering (with respect to the oscillator frequency).

Mathematically, the process is similar to amplitude modulation followed by coherent demodulation (Section 5.3). The modulation performed by the input

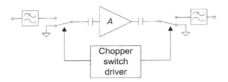

Figure 7.4 Principle of operation for a chopper amplifier.

switch implies the multiplication of the signal to be amplified by the oscillator's signal. The result is a signal whose spectrum consists of the oscillator frequency and two side bands that are symmetrical with respect to the oscillator frequency and whose amplitude depends on that of the input signal. After amplifying the modulated signal, the demodulation performed by the output switch is again a multiplication by the oscillator signal. In this product, all components having the same frequency of both signals will yield a dc component whose amplitude will depend on that of the ac amplifier output. After low-pass filtering, that dc component will be the only component present at the output.

During the amplification process, only the amplifier's high-frequency noise will be superimposed on the signal, thus becoming a source of error. But neither offset voltage drifts nor offset currents drifts will affect the signal that thanks to the modulation is now in the frequency band around the oscillator frequency. The only important dc and low-frequency errors will be those due to the input switch (usually based on FETs), which will not normally exceed those of a conventional op amp. A shortcoming is that signal bandwidth is limited because the maximal frequency for the input signal must be much lower than the switching frequency. A 100 Hz bandwidth, however, is easy to obtain, and this is appropriate for output signal conditioning for many sensors. Nevertheless, current low-drift op amps do better than classic chopper amplifiers in both performance and cost.

7.1.3 Autozero Amplifiers

Some monolithic op amps achieve a very low drift by periodically measuring the offset voltage and then subtracting it from the signal of interest. While the offset voltage is being measured, a hold circuit presents the signal of interest to the output. Figure 7.5 shows the simplified diagram for one of these amplifiers.

During the autozero or nulling phase, switches S1 and S2 are closed, shorting the nulling amplifier to reduce its own input offset voltage by feeding its output back to a subtracting input node. External capacitor C_A stores the nulling potential needed to keep the offset voltage zero during the amplifying phase. During the amplifying phase, switches S3 and S4 are closed to connect the output of the nulling amplifier to a subtracting node of the main amplifier. The offset of the main amplifier is thus nulled, and external capacitor C_B stores the nulling potential needed to keep the offset voltage of the main amplifier nulled during the next nulling phase. If the switching from one phase to the other is fast enough, these commutations do not influence the output waveform.

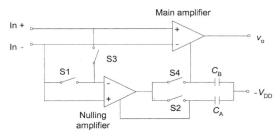

Figure 7.5 Simplified circuit for the TLC2654 (Texas Instruments) autozero op amp. During the nulling phase, S1 and S2 are closed. During the amplifying phase, S3 and S4 are closed. C_A and C_B are external capacitors.

This technique yields dc open-loop gains, PSRR, and CMRR in excess of 120 dB. Table 7.1 includes basic parameters for some models that use autozero or similar methods, such as the self-calibration technique (TLC4502, Texas Instruments). The self-correction method (LMC2001, National Semiconductor) is a dynamic technique that measures and continually corrects the input zero error. Some autozero amplifiers include the capacitors, while others accept an external clock to determine the time duration for each phase. This permits, for example, synchronization of several nearby units to avoid crosstalk between oscillators having close frequencies, which could result in an increased output noise. Their main limitations are clock noise; and because they are sampling systems, the switching frequency must be higher than twice the bandwidth of the input signal. Usual frequencies range from a few hundred hertz to several kilohertz. High clock frequencies increase the noise level. Besides, their supply voltages are below ± 8 V because they are implemented on CMOS technology.

Preserving low offset voltages and drifts requires us to consider thermo-electromotive interference and other pseudonoise sources [1]. Small voltage signals should be connected only through soldered connections. Unavoidable connections through connectors, relays, switches, and so on, should use metal pairs with low thermoelectric sensitivity (Table 6.4). All intervening metal junctions should have matched materials and temperature. Hence, it is advisable to place similar connection pairs close and to introduce redundant components if needed to match the number of junctions [2] (Figure 7.6). Thermal

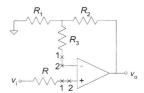

Figure 7.6 Adding R matches the metal junctions 1 and 2 introduced in series with the inverting terminal. Junction 1 is between the (resistor) lead and copper, and junction 2 is between copper and the IC pin (courtesy of Linear Technology).

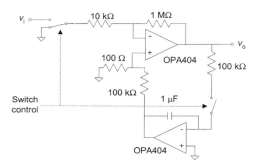

Figure 7.7 Circuit for automatic drift correction in an amplifier using discrete components. Low drift requires a capacitor with a dielectric exhibiting low leakage and low dielectric absorption, such as polypropylene (courtesy of Burr–Brown).

interference increases with power dissipation and when there are thermal gradients that induce airflow. Therefore, place power components far from sensitive circuits and cover these with a thermal insulator. Temperature gradients along resistors yield parasitic voltages: 2 μV/°C in wirewound resistors, 20 μV/°C in metal film resistors, and 450 μV/°C in carbon composition resistors. In addition, each resistor introduces two metal junctions that must be at the same temperature to prevent any net thermoelectromotive force.

Drift correction by repetitive zero subtraction can also be applied to circuits implemented with discrete components if use is made of a sample and hold amplifier or an integrator. Figure 7.7 shows an autozero inverting amplifier that achieves 10 μV, 0.05 μV/°C, and 1 μV/s zero drop, even though the op amp has typical $V_{io} = 260$ μV and $\Delta V_{io}/\Delta T = 3$ μV/°C.

7.1.4 Composite Amplifiers

Dc precision amplifiers usually have quite a restricted bandwidth. Composite amplifiers combine two or more amplifiers to achieve a given overall performance unattainable with a single amplifier. Figure 7.8 shows a feedback com-

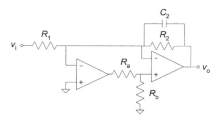

Figure 7.8 Composite amplifier consisting of an input dc precision op amp and an output fast op amp.

posite amplifier that consists of two op amps. The input op amp has low offset errors and the second op amp (VFA or CFA) has good ac performance. In the first case, if we successively consider V_{io1} and V_{io2} and apply superposition, we obtain

$$V_o(0)\left[1 + A_{d2}\frac{R_1}{R_1 + R_2}\left(1 + A_{d1}\frac{R_b}{R_a + R_b}\right)\right] = A_{d2}\left(V_{io2} + V_{io1}A_{d1}\frac{R_b}{R_a + R_b}\right)$$

(7.15)

where A_{d1} and A_{d2} are the respective open loop gains. Because at low frequencies A_{d1} and A_{d2} are very large, we can approximate

$$V_o(0) \approx V_{io1}\left(1 + \frac{R_2}{R_1}\right)$$

(7.16)

Therefore, the first op amp contributes most of the IZE. The equivalent input error from V_{io2} is in fact divided by A_{d1}, and hence becomes negligible. C_2 reduces the gain at high frequencies to prevent oscillation because of the large overall open loop gain. In reference 3 there are additional composite amplifiers.

7.1.5 Offsets and Drifts in Instrumentation Amplifiers

Offset and drift specifications for integrated instrumentation amplifiers are somewhat different from those for op amps. An IA consists of an input section, usually with a selectable gain G, and an output section with fixed gain. Some manufacturers specify errors from the input section as "input errors" and errors from the output section as "output errors." The total error referred to the input (RTI) is then

$$\text{Total error (RTI)} = \text{Input error} + \frac{\text{Output error}}{G}$$

(7.17)

and the total error referred to the output (RTO) is

$$\text{Total error (RTO)} = (\text{Input error}) \times G + \text{Output error}$$

(7.18)

When $G = 1$ the RTI and RTO errors are the same.

Example 7.2 The INA111AP is an IA whose maximal offset error and drift at $25\,°C$ are $(1 + 5/G)$ mV and $(10 + 100/G)$ μV/°C when supplied at ±15 V. Its maximal supply currents are 4.5 mA, and the thermal resistance of the P package is $100\,°C/W$. Calculate the input and output offset errors when $G = 1$ and $G = 1000$, and the ambient temperature is $40\,°C$.

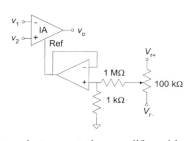

Figure 7.9 Level shift in an instrumentation amplifier without degrading the CMRR.

From (7.7) and (7.8), the IA will reach a temperature

$$T = 40\,°C + 2 \times (15\ V) \times (4.5\ mA) \times (100\,°C/W)$$
$$= 40\,°C + 13.5\,°C = 53.5\,°C$$

From (7.9), when $G = 1$ the RTI offset voltage will be

$$V_{io} = \left(1 + \frac{5}{1}\right) mV + \left(10 + \frac{100}{1}\right)(53.5 - 25)\ \mu V = 9.1\ mV$$

The corresponding RTO offset voltage will be $V_{oo} = V_{io} \times G = 9.1$ mV. When $G = 1000$, the RTI offset voltage will be

$$V_{io} = \left(1 + \frac{5}{1000}\right) mV + \left(10 + \frac{100}{1000}\right)(53.5 - 25)\ \mu V = 1.3\ mV$$

and the corresponding RTO offset voltage will be $V_{oo} = V_{io} \times G = 1.3$ V. Large gains reduce RTI errors.

The OZE in an IA can be nulled out at a given temperature by adding or subtracting a precise voltage at the reference input. Figure 7.9 shows a convenient circuit for such operation that does not affect the CMRR because the (low drift) op amp has very small output resistance. The LTC1100 is an IA with autozero correction. The CS553X series of data acquisition ICs (Cirrus Logic) uses chopper stabilization in its input stage and in the internal programmable-gain instrumentation amplifier.

7.2 ELECTROMETER AND TRANSIMPEDANCE AMPLIFIERS

Signals coming from current sources or from high output impedance voltage sources—for example, semiconductor-junction-based nuclear radiation detectors (e.g., in CT scanners), photoelectric cells, photomultiplier tubes, ionization cells (e.g., for vacuum measurement), photodiodes, ISEs, and piezoelectric sensors—require a measurement system featuring a low input current. When low-frequency signals are of interest, then a voltage amplifier or a current-to-voltage converter (transimpedance amplifier) based on a low-drift op amp is required. Otherwise, as in piezoelectric sensors that do not have dc response or in radiation detectors detecting incoming pulses, either an electrometer or a charge amplifier can be used.

An electrometer is a measuring system having an input resistance larger than 1 TΩ and an input current lower than about 1 pA. There are instruments with 100 TΩ and 50 aA [4]. Electrometer op amps and instrumentation amplifiers also offer high input impedance and low input current. Voltage amplifiers with very large input impedance are useful to interface sensors that give an output voltage with a large series impedance, either resistive (pH electrodes, ISEs) or capacitive (piezoelectric and pyroelectric sensors).

Example 7.3 A piezoelectric hydrophone whose equivalent capacitance is 450 pF is connected to a noninverting amplifier as shown in Figure E7.3. Determine the value for passive components in order to have a gain of 100 and a gain amplitude error smaller than 10% at 5 Hz.

The output voltage from a noninverting amplifier is

$$v_o = v_p \left(1 + \frac{R_2}{R_1} \right)$$

where v_p is the voltage at the noninverting input, which can be obtained by replacing the sensor by its equivalent circuit—a voltage source v_s in series with a capacitance C_s. The circuit equation is

$$j\omega C_s (V_s - V_p) = \frac{V_p}{R_b}$$

From these equations we obtain

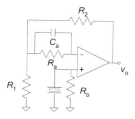

Figure E7.3 Voltage amplifier for a piezoelectric sensor.

$$\frac{V_o(\omega)}{V_s(\omega)} = \frac{j\omega R_b C_s}{1 + j\omega R_b C_s}\left(1 + \frac{R_2}{R_1}\right) = \frac{j\omega\tau}{1 + j\omega\tau}\left(1 + \frac{R_2}{R_1}\right)$$

where $\tau = R_b C_s$. In order to have a gain amplitude error smaller than 10 % at 5 Hz we need

$$\frac{2\pi(5\ \text{Hz})\tau}{\sqrt{1 + [2\pi(5\ \text{Hz})\tau]^2}} > 0.9$$

which yields $\tau > 65.7$ ms. Therefore,

$$R_b > \frac{65.7\ \text{ms}}{450\ \text{pF}} = 146\ \text{M}\Omega$$

To obtain a gain of 100 we need $R_2 = 99R_1$. We can select $R_1 = 1$ kΩ and $R_2 = 98.8$ kΩ, and we can add a trimming resistor if needed for the gain accuracy desired.

R_a balances the resistance seen from op amps inputs, and therefore $R_a = 146$ MΩ. C_a reduces the noise bandwidth for R_a (Section 7.4).

Small currents can be measured with an electrometer amplifier by two different methods: by directly measuring the drop in voltage across a high-value resistor (Figure 7.10a) or by a current-to-voltage conversion based on a transimpedance amplifier (Figure 7.10b) or on a current integrator (Section 7.2.2).

In the first method, when R has a high value it is not possible to measure high-frequency phenomena because the capacitance of the sensor together with that of the cable and amplifier input limit the maximal response. If, for example, $R = 10$ GΩ and $C_p = 100$ pF, the system has a low-pass response with corner frequency $f = 1/(2\pi RC_p) = 0.16$ Hz. The response time is therefore $\tau = 0.35/f = 2.2$ s, which is quite slow. This method is useful for low-frequency current sources only.

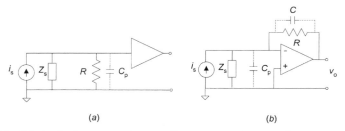

(a)   (b)

Figure 7.10   Methods for measuring small currents using an electrometer amplifier. (a) Detecting the drop in voltage across a resistor. (b) Converting current to voltage.

7.2.1 Transimpedance Amplifiers

The transimpedance amplifier in Figure 7.10b is faster than the voltage meter in Figure 7.10a. The transimpedance is ([1], Section 3.41)

$$\frac{V_o(f)}{I_s(f)} = -\frac{Z}{1 + \dfrac{1}{A_d}\left(1 + \dfrac{Z}{Z_s \| Z_D}\right)} = -\frac{Z}{1 + \dfrac{1}{A_d\beta}} \tag{7.19}$$

where Z is the impedance of R shunted by C, Z_s is the sensor impedance shunted by the impedance of the connecting cable, Z_D is the differential input impedance of the op amp circuit ($Z_D = Z_d \| Z_c$, where Z_d and Z_c are the respective differential and common mode input impedance in op amp data sheets), $A_d = A_{d0}/(1 + jf/f_a)$, and

$$\beta = \frac{1}{1 + \dfrac{Z}{Z_s \| Z_D}} \tag{7.20}$$

If stray capacitances have impedances much larger than those of the resistors shunting them, (7.19) simplifies to

$$\frac{V_o(f)}{I_s(f)} = -\frac{R}{1 + \dfrac{1}{A_d}\left(1 + \dfrac{R}{R_s \| R_D}\right)} = Z_T(f) \tag{7.21}$$

For frequencies larger than f_a we can approximate

$$Z_T(f \gg f_a) \approx -\frac{R}{1 + j\dfrac{f/f_a}{A_{d0}\beta}} = -\frac{R}{1 + j\dfrac{f}{f_H}} \tag{7.22}$$

where

$$f_H = A_{d0}f_a\beta = \frac{A_{d0}f_a}{1 + \dfrac{R}{R_s \| R_D}} = \frac{f_T}{1 + \dfrac{R}{R_s \| R_D}} \tag{7.23}$$

Therefore, the transimpedance is R at low frequencies, and it decreases by 20 dB/decade from f_H up; also, the larger R is, the smaller f_H is. A broad signal bandwidth requires a large f_T. Note that for a given op amp it is not possible to separately set the gain and the bandwidth. Furthermore, if $R \gg R_s$, the input noise is amplified by $1 + R/R_s$.

A resistor T-network yields a large transimpedance without using large

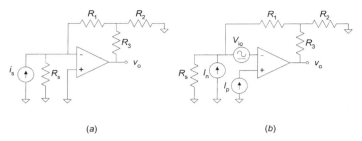

(a)   (b)

Figure 7.11   (a) Use of a T-network to simulate high-value resistances using low-value resistors. (b) Equivalent circuit for offset analysis.

resistors (Figure 7.11a). At low frequencies, the transimpedance is

$$Z_T(f \ll f_H) = R_3 + R_1\left(1 + \frac{R_3}{R_2}\right) \tag{7.24}$$

If $R_3 \gg R_2$ we have $R \approx R_1(1 + R_3/R_2)$. Nevertheless, this circuit has a large OZE (and noise). From Figure 7.11b,

$$OZE = I_n\left[R_3 + R_1\left(1 + \frac{R_3}{R_2}\right)\right] + V_{io}\left[1 + \frac{R_1}{R_s} + \frac{R_3}{R_2}\right] \tag{7.25}$$

Therefore, for $R_s \gg R_1$, for example, the contribution from V_{io} increases by $(1 + R_3/R_2)$.

If the signal source impedances are capacitive—meaning that the impedances of sensor and stray capacitances are smaller than resistors shunting them—(7.19) leads to

$$Z_T = -R\frac{A_{d0}\omega_a\omega_1}{(A_{d0}+1)\omega_a\omega_1 + s\omega_1\left(1 + \frac{\omega_a}{\omega_1} + \frac{A_{d0}\omega_a}{\omega_0}\right) + s^2} \tag{7.26}$$

where $\omega_0 = 1/(RC)$ and $\omega_1 = 2\pi f_1 = 1/[R(C + C_D + C_s)]$. This frequency-dependent transimpedance can be characterized by its natural frequency

$$f_n \approx \sqrt{A_{d0}f_a f_1} = \sqrt{f_T f_1} \tag{7.27a}$$

and its damping factor

$$\zeta = \frac{1}{2}\left(\frac{f_1 + f_a}{f_n} + \frac{f_n}{f_0}\right) \tag{7.27b}$$

Because an underdamped transimpedance may lead to oscillation, circuit design should seek a flat frequency characteristic. If $C_s + C_D \gg C$ and $f_1 \gg f_a$, we can approximate

$$C \cong \frac{1}{2\pi R}\left(\frac{2\zeta}{\sqrt{f_T f_1}} - \frac{1}{f_T}\right) \tag{7.28}$$

Example 7.4 A current signal from a photodiode with $R_s = 20$ GΩ and $C_s = 600$ pF is converted into a voltage by an I/V converter such as that in Figure 7.10b using $R = 100$ kΩ and an op amp with $f_a = 1$ Hz, $f_T = 1$ MHz (typical), differential input capacitance 1 pF, and common mode input capacitance 2 pF. Calculate the capacitance required to prevent gain peaking and the resulting converter bandwidth.

In Figure 7.10b, the input capacitance at the op amp inverting terminal is the parallel combination of the differential input capacitance and the common mode input capacitance from that terminal to ground. In our case, 1 pF + 2 pF = 3 pF. Because $R_s = 20$ GΩ and $C_s = 600$ pF, the sensor impedance is capacitive and the transimpedance is better described by (7.26).

To prevent gain peaking, we will design the circuit to achieve a maximally flat response, which requires $\zeta = \sqrt{2}/2$. Therefore, from (7.28) we wish

$$\frac{\sqrt{2}}{2} = \frac{1}{2}\left(\frac{f_1 + f_a}{f_n} + \frac{f_n}{f_0}\right)$$

where $f_a = 1$ Hz, $f_T = 1$ MHz,

$$f_1 = \frac{1}{2\pi R(C + C_D + C_s)} = \frac{1}{2\pi \times 100 \text{ kΩ} \times (C + 3 \text{ pF} + 600 \text{ pF})}$$

$$f_0 = \frac{1}{2\pi RC} = \frac{1}{2\pi \times 100 \text{ kΩ} \times C}$$

We have two unknowns (C and f_n) and two equations. However, instead of solving the resulting equation system, we can estimate the value for C by assuming that it will be well below 603 pF, so that we can approximate

$$f_1 = \frac{1}{2\pi R(C + C_D + C_s)} \approx \frac{1}{2\pi \times 100 \text{ kΩ} \times (3 \text{ pF} + 600 \text{ pF})} = 2639 \text{ Hz}$$

$$f_n = \sqrt{f_T f_1} = \sqrt{(1 \text{ MHz})(2639 \text{ Hz})} = 51{,}375 \text{ Hz}$$

Because $f_1 \gg f_a$, we can apply (7.28) to obtain

$$C = \frac{1}{2\pi \times 100 \text{ kΩ}}\left(\frac{\sqrt{2}}{51{,}375 \text{ Hz}} - \frac{1}{10^6 \text{ Hz}}\right) = 42 \text{ pF}$$

which is certainly much smaller than 603 pF. By shunting the gain resistor with 42 pF we would obtain a flat frequency response. Because of the uncertainty in f_T we may need some trimming for C. For $\zeta = \sqrt{2}/2$, the -3 dB bandwidth is f_n, hence about 51 kHz. Higher-frequency poles in A_d can affect the actual frequency response, particularly for broad signal bandwidth.

7.2.2   Current Measurement by Integration

An alternative to I/V converters to measure small currents is to integrate them. Figure 7.12a shows the simplified structure of an IC integrator based on an op amp (ACF2101, IVC102). When S1 is closed and S2 is open, the input current charges C, provided that the op amp is ideal and sinks (or sources) negligible current. The voltage across C will be $V_C = Q/C$, where Q is the integrated input current during the time τ considered. Therefore,

$$v_o = -V_C = -\frac{1}{C}\int_0^\tau i_i(t)\,dt \tag{7.29}$$

If the input current is constant during τ, $i_i(t) = I_i$, the output voltage (Figure 7.12b) will be

$$V_o = -\frac{I_i\tau}{C} \tag{7.30}$$

Therefore, we can obtain a large output by increasing τ or reducing C. In both cases, however, the output will eventually saturate. Therefore, after a convenient time τ, S1 opens for a while, the output voltage is read, and S2 closes to discharge C and reset the output at 0 V.

Input offset voltage and bias or leakage current yield an actual output

$$V_o = -\frac{(I_i + I_n)\tau}{C} - V_{io} \tag{7.31}$$

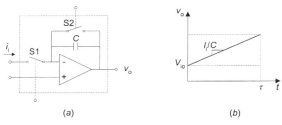

(a)   (b)

Figure 7.12   (a) Current integrator. S1 is closed and S2 is open from $t = 0$ until $t = \tau$. (b) Output voltage.

These error sources can be estimated and later compensated for. In the absence of input signal, closing S2 yields an estimate of V_{io}, and closing S1 with S2 open yields an output voltage because of I_n.

If the input signal is not constant, the output voltage is proportional to the average current. For a sinusoidal current we will have

$$V_o(f) = -\frac{1}{C} \int_{t_1}^{t_1+\tau} I \sin 2\pi f t \, dt = \frac{I}{2\pi f C} 2 \sin(2\pi f t_1 + \pi f \tau) \sin(\pi f \tau) \quad (7.32)$$

This voltage will be maximal when $2\pi f t_1 + \pi f t = \pi/2$. Then,

$$V_o(f) = \frac{I\tau}{C} \frac{\sin \pi f \tau}{\pi f \tau} = V_o(0) \frac{\sin \pi f \tau}{\pi f \tau} \quad (7.33)$$

Therefore, the higher the value of f, the smaller the output. Nevertheless, (7.33) also shows that whenever $f\tau = n$ (where n is any integer), the output is zero. For this reason, choosing τ a multiple of the power line frequency minimizes interference coming from it.

Example 7.5 If the integration time of a current integrator is selected to cancel power line interference for both 60 Hz and 50 Hz systems, and we accept a maximal amplitude error of 0.1%, calculate the maximal frequency of the current to integrate.

Canceling 60 Hz interference requires $\tau = n_1/(60 \text{ Hz})$. Canceling 50 Hz requires $\tau = n_2/(50 \text{ Hz})$. Hence, the minimal τ must be 100 ms ($n_1 = 6$, $n_2 = 5$). The maximal input frequency must fulfill the condition

$$0.999 V_o(0) = V_o(0) \frac{\sin \pi f_m \tau}{\pi f_m \tau}$$

From this,

$$\pi f_m \tau = 0.08$$

$$f_m = \frac{0.08}{\pi \times 100 \text{ ms}} = 0.25 \text{ Hz}$$

7.2.3 Cautions in Designing Electrometer Circuits

High-impedance circuits require us to pay attention to resistors, insulation (dielectrics), and cabling. High-value resistors are from carbon or metal oxide film with ceramic substrate. Glass encapsulation prevents humidity from entering in contact with the resistive element, hence reducing its resistance.

High-quality dielectrics in electrometer circuits are needed because a mere

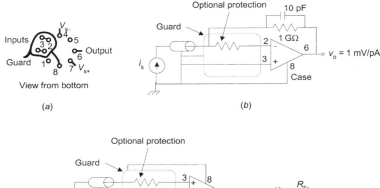

Figure 7.13 Use of guards to reduce parasitic input currents. (*a*) Arrangement on both sides of a printed circuit. (*b*) Guard connection for a current-to-voltage converter. (*c*) Guard connection for a noninverting amplifier.

10 V, such as the op amp voltage supply, induces through a 1 TΩ insulation a 10 pA current in the amplifier input, which may be larger than the current to be measured. Therefore it becomes necessary to use high volumetric resistivity materials such as Teflon™, polypropylene, and polystyrene and then to place guard rings around the sensitive terminals on both sides of printed circuit boards. Boards should be cleaned (e.g., by alcohol) and blow-dried with compressed air. After cleaning, the boards should be coated with epoxy or silicone rubber to prevent contamination.

Guard rings consist of conductive zones encircling the terminal to be protected and connected to a voltage close to that of the terminal, as shown in Figure 7.13. Thus for an inverting amplifier the guard encircles the negative terminal and connects to the reference (common) voltage (Figure 7.13*b*), whereas in a noninverting amplifier it encircles the positive terminal and connects to a voltage divider (Figure 7.13*c*). The connection to pin 8 shown in these three figures allows the guard potential to drive the amplifier metallic can internally connected to that pin.

Cautions with dielectrics also apply to capacitor selection. Their leakage resistance must be high, and their dielectric absorption must be low. Again Teflon™, polypropylene, and polystyrene capacitors are recommended; for less demanding applications, capacitors using mica and polyester are recommended.

Concerning circuit layout, first the amplifier must be placed as close as possible to the signal source. Second, connecting wires must be carefully selected. The best wires are rigid, with a shield connected to the guard, and use high-quality insulating materials free of any piezoelectric effect. The operating temperature should be as low as possible. Supply and common mode voltages must be as low as practical, and excessive output loading must be avoided, because these factors affect bias currents.

When switched gains are desired in a feedback I/V converter, ordinary solid-state switches do not have a high enough OFF resistance for electrometer circuits. Low-cost commercial reed relays, which have an open resistance greater than 10 TΩ, are more appropriate [5]. Diodes for input protection should have low leakage, such as the PAD series (Vishay), or use JFETs (2N4117A) connected as diodes.

Commercially available op amps best suited for building electrometer amplifiers are those whose input stages are based on FET or MOSFET transistors. In general, JFET input stages have a lower resistance but also lower noise and drifts than MOSFET inputs.

7.3 CHARGE AMPLIFIERS

A charge amplifier is a circuit whose equivalent input impedance is a capacitance that provides a very high value of impedance at low frequencies. Thus contrary to what its name may suggest, a charge amplifier does not amplify the electric charge present at its input. Its function is actually to obtain a voltage proportional to that charge and yield a low output impedance. Hence, it is a charge-to-voltage converter.

Figure 7.14a shows its circuit, which was proposed by W. P. Kistler in 1949. It consists of an op amp with a single capacitor as feedback. The circuit is the same proposed in Figure 5.1b for capacitive sensors. Its interest becomes obvious after considering the alternative circuit for measuring the signal of a high output impedance sensor (e.g., a piezoelectric accelerometer), namely a voltage amplifier based on an electrometer amplifier (Figure 7.14b). If the accel-

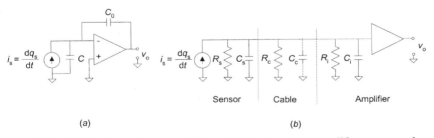

(a) (b)

Figure 7.14 (a) Idealized charge amplifier. (b) Electrometer amplifier connected to a piezoelectric sensor.

erometer has charge sensitivity S_q (C/g) the voltage amplifier yields

$$\frac{V_o(\omega)}{a(\omega)} = \frac{S_q}{C}\frac{j\omega RC}{1 + j\omega RC} \tag{7.34}$$

where $R = R_s \| R_c \| R_i$, $C = C_s + C_c + C_i$, and the subscripts are s for sensor, c for cable, and i for input (amplifier). From (7.34) we deduce that the sensor sensitivity undergoes a reduction that depends on the length of the connecting cable, and that the frequency response is high pass with a corner frequency that depends both on cable length and insulation. This may change with temperature and ambient humidity for some models.

Placing a large capacitor shunting the amplifier input overcomes the dependence on other (variable) capacitances, but at the cost of a reduced sensitivity (see Problem 6.8). The influence of cable capacity can also be reduced with a driven shield (Section 5.2.4). But this adds complexity because then a triaxial cable is necessary (a driven shield and another shield for sensor ground). Therefore, voltage amplifiers suit applications with the sensor close to the amplifier, as in piezoelectric microphones or hydrophones.

The charge amplifier in Figure 7.14a is usually a better solution. It is based on a charge transfer from the sensor (in parallel with the cable and amplifier input) to a fixed capacitor, C_0, and then measuring the voltage across it with an amplifier such as an electrometer. If the open loop gain for the amplifier is A_d, then we have

$$V_o = \frac{Q_s}{C_0 + \dfrac{C + C_0}{A_d}} \approx \frac{Q_s}{C_0} \tag{7.35}$$

where the final approximation assumes $A_d \gg 1$, which is true only at low frequencies. Now the sensitivity does not depend on the cable, except at high frequency, where A_d decreases. Cable capacitance may be important when C_0 is small to have a high sensitivity. Gain accuracy depends on C_0, which consequently must have high stability and low leakage. Stray capacitance must be reduced by shielding if necessary. Nevertheless, capacitors drift more than resistors, so that frequent recalibration may be needed.

To some extent the charge amplifier described is certainly ideal because we have omitted sensor and cable leakage resistances and amplifier input resistances, and also amplifier offset voltage and currents together with C_0 leakage. Figure 7.15 includes all these factors. If we let $R_e = R \| [R_0/(A_d + 1)] = R_s \| R_c \| R_i \| [R_0/(A_d + 1)]$, we have (reference 1, Section 3.4.2)

$$\frac{V_o(\omega)}{Q_s(\omega)} = -\frac{1}{C_0}\frac{j\omega R_e A_d C_0}{1 + j\omega R_e[C + C_0(A_d + 1)]} \tag{7.36}$$

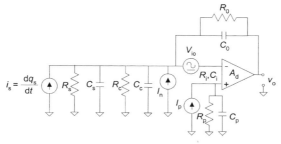

Figure 7.15 Sources of zero error in a real charge amplifier.

This is a high-pass response such as that given by (7.34), and therefore it is not possible to measure static phenomena. Nevertheless, the corner frequency is now lower because C_0 is smaller than sensor and cable capacitance. The value of C_0 is multiplied by A_d; and that for R_0, required to provide a bias path for the op amp, is divided by $1 + A_d$. If, for example, $R_s = 1$ TΩ, which means a quartz sensor, not a ceramic one, the amplifier has a JFET or MOSFET input stage, and $R_0 = 1$ TΩ, for $C_0 = 10$ pF the corner frequency is

$$f_c = \frac{1}{2\pi R_e C_0} = \frac{1}{2\pi \times (0.5 \text{ TΩ}) \times (10^{-11} \text{ pF})} = 0.03 \text{ Hz} \qquad (7.37)$$

Hence, it is possible to measure very slow phenomena. For the voltage amplifier in Figure 7.14*b*, according to (7.34), this measurement would be possible by increasing C, thus resulting in a decreased sensitivity, and by increasing R to the extent determined by sensor leakage. Sensors that withstand high temperatures have higher leakage, hence reduced low-frequency response. Systems whose corner frequency is very small have a long recovery time after saturation.

R_0 also affects the contribution of V_{io} and I_{io} to the output. From Figure 7.15, if $R_p = R \| R_0$, the output voltage due exclusively to offset voltage and currents is

$$v_o(0) = (V_{io} + I_{io}R_p)\left(1 + \frac{R_0}{R}\right) \qquad (7.38)$$

where $I_{io} = I_n - I_p$, and $R = R_s \| R_c$. Ac coupling the charge amplifier to the following stage eliminates this offset. Nevertheless, this error must be kept small enough not to reduce the output dynamic range. C_p reduces the noise contributed by R_p (Section 7.4).

From (7.38) R_0 must be as small as possible for the minimal frequency to be measured. Usually a resistor is placed shunting C_0 rather than relying on the leakage resistance of this capacitor. Given that R is mainly determined by sensor leakage, for quartz sensors R_0 is chosen between 10 GΩ and 10 TΩ, with C_0

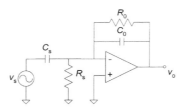

Figure E7.6   Equivalent circuit for a piezoelectric sensor connected to a charge amplifier.

ranging from 10 pF to 100 nF. The sensitivity for ceramics is determined by C_0 ranging from 10 pF to 1 nF, and R_0 values from 100 MΩ to 10 GΩ are used. Larger values for R_0 would be meaningless because of the relatively large sensor leakage. Obviously, ceramic sensors do not permit quasi-static measurements. Instead of a single resistor we can use a T network as in Figure 7.11a, but this increases the offset error and noise.

Example 7.6   Design a charge amplifier for the piezoelectric sensor in Example 6.6 ($R_s = 52$ GΩ, $C_s = 20.4$ nF) so that the output sensitivity is -10 mV/Pa and R_0 does not limit the low-frequency response more than the sensor leakage resistance would do if the sensor were connected to an ideal voltage amplifier.

Figure E7.6 shows the equivalent circuit for the sensor connected to the charge amplifier. If R_0 is first assumed infinite, the output voltage will be

$$v_o = -v_s \frac{C_s}{C_0} = -\frac{q_s}{C_s} \frac{C_s}{C_0} = -\frac{q_s}{C_0} = -d_{31} \frac{F_1}{A_1} A_3 \frac{1}{C_0}$$

To obtain the desired sensitivity we need

$$-10 \text{ mV} = -23 \frac{\text{pC}}{\text{N}} (1 \text{ N/m}^2)(0.1 \times 0.1 \text{ m}^2) \frac{1}{C_0}$$

and, hence, $C_0 = 23$ pF. The closest normalized values (0.1% tolerance) are 22.9 pF and 23.2 pF.

If we now consider R_0, for A_d infinite (7.36) reduces to

$$\frac{V_o(j\omega)}{Q_s(j\omega)} = -\frac{1}{C_0} \frac{j\omega R_0 C_0}{1 + j\omega R_0 C_0}$$

which is a high-pass transfer function. If the sensor were connected to an ideal voltage amplifier with gain G, the transfer function would be

$$\frac{V_o(j\omega)}{V_s(j\omega)} = \frac{j\omega R_s C_s}{1 + j\omega R_s C_s} G$$

which is also a high-pass transfer function. Therefore, the frequency response will be the same when $R_0 C_0 = R_s C_s$. We need

$$R_0 = \frac{R_s C_s}{C_0} = \frac{(52 \text{ G}\Omega)(20.4 \text{ nF})}{23 \text{ pF}} = 46 \text{ T}\Omega$$

We could use either a single (and expensive) resistor or a T network, which would increase the output offset (and noise).

The high-frequency response of charge amplifiers is limited by A_d. At high frequency, $A_d \approx f_T / jf$ and the transfer function is determined by circuit capacitances. The circuit response then reduces to

$$\frac{V_o(f)}{Q_s(f)} = -\frac{1}{C_0} \frac{1}{1 + j\dfrac{f}{f_H}} \tag{7.39}$$

where

$$f_H = \frac{f_T}{1 + \dfrac{C}{C_0}} \tag{7.40}$$

Therefore, a small C_0 increases the sensitivity but also reduces the bandwidth. Adding a resistor (1 kΩ to 10 kΩ) in series with the op amp inverting input improves stability and limits input currents due to accidental contact with a high voltage. This resistor and C_0 may then limit the high-frequency response more than A_d or the sensor resonance (Figure 6.17).

Example 7.7 A given piezoelectric hydrophone has a sensitivity of 0.5 pC/Pa and equivalent capacitance of 10 nF. Design a charge amplifier with -3 dB bandwidth from 10 Hz to 10 kHz and sensitivity -1 mV/Pa, including a protection resistor in series with the op amp inverting input.

Figure E7.7 shows the proposed circuit. From (7.36), the output voltage in the passband is $v_o = -q_s / C_0$. Therefore,

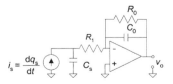

Figure E7.7 Equivalent circuit for a piezoelectric sensor connected to a charge amplifier whose input is protected by a series resistor.

$$C_0 = -\frac{q_s}{v_o} = -\frac{q_s/p}{v_o/p} = -\frac{0.5 \text{ pC/Pa}}{-1 \text{ mV/Pa}} = 500 \text{ pF}$$

We can select $C_0 = 499$ pF, $\pm 1\%$ tolerance, polystyrene.

If we initially assume that R_1 is small enough, (7.36) reduces to

$$\frac{V_o(\omega)}{Q_s(\omega)} = -\frac{1}{C_0}\frac{j\omega R_0 C_0}{1 + j\omega R_0 C_0}$$

In order to obtain -3 dB at 10 Hz we need

$$2\pi \times (10 \text{ Hz}) \times R_0 C_0 = 1$$

From this,

$$R_0 = \frac{1}{2\pi \times (10 \text{ Hz}) \times (499 \text{ pF})} = 32 \text{ M}\Omega$$

R_1 becomes important at high frequencies because its impedance may be comparable to that of the sensor (C_s). For a 10 kHz corner frequency we need

$$2\pi \times (10 \text{ kHz}) \times R_1 C_s = 1$$

which leads to

$$R_1 = \frac{1}{2\pi \times (10 \text{ kHz}) \times (10 \text{ nF})} = 1.6 \text{ k}\Omega$$

If the op amp withstands a 10 mA input current, R_1 protects from contacts with voltage sources up to 16 V.

The comments in Section 7.2.3 on the resistances, insulations, and layout of electrometer amplifiers can also be applied to charge amplifiers built from discrete components.

We wish to evaluate the sensitivity of a charge amplifier before interfacing it with the sensor or to adjust its gain in an assembled system. Then it is useful to have an electronic calibration system without requiring the application of any known mechanical load (acceleration, force, pressure, etc.). Figure 7.16 shows a calibration system, consisting of connecting an adjustable frequency voltage source in series with the sensor [6]. During calibration no mechanical load is applied to the system. If cable resistance is smaller than R_a and R_a is smaller than other circuit impedances at the maximal frequency of interest, the output corresponding to v_c is

$$V_o = V_c \frac{C}{C_0}\frac{A_d}{1 + A_d} \approx V_c \frac{C}{C_0} \qquad (7.41)$$

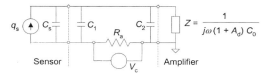

Figure 7.16 In-place calibration of a charge amplifier without exciting the sensor [6].

where $C = C_s + C_1$. In terms of sensor charge sensitivity S_q, the voltage that must be applied for one acceleration unit when the sensor is an accelerometer is

$$\frac{v_c}{a} = \frac{S_q}{C}\,\text{V}/g \qquad (7.42)$$

Therefore the calibration depends on cable length between the voltage connection point and the sensor (C_1). Thus, we can include R_a in the sensor itself. Another conclusion from (7.42) is that if connecting v_c after the amplifier has been calibrated yields an output v_o that does not agree with the expected value, that would mean that C_s has changed. This change is usually associated with a variation in sensor sensitivity, and therefore it is possible to detect sensitivity changes even without exciting the sensor. This system also allows us to detect faults in cables and connectors.

In order for this calibration method to be acceptable, it must not result in any perturbation when the sensor is excited. Thus when disconnecting v_c and R_a remains in series with the circuit, its impedance must be low enough at the maximal frequency of interest. That is,

$$R_a \ll \frac{C_s + C_1 + C_2 + (1 + A_d)C_0}{2\pi f(C_s + C_1)[C_2 + (1 + A_d)C_0]} \approx \frac{1}{2\pi f(C_s + C_1)} \qquad (7.43)$$

We assume this condition in order to deduce (7.42). Values from 10 Ω to 100 Ω usually fulfill this condition.

7.4 NOISE IN AMPLIFIERS

Whenever we process small signals or desire high resolution, particularly when working at low frequencies or with broadband signals, we must consider the intrinsic perturbations or noise associated with the components used to build the circuit. Reference 7 discusses noise and interference in electronic circuits. Reference 8 analyses noise and low-noise design in depth. Chapter 11 in reference 1 summarizes noise analysis in instrumentation electronics.

7.4.1 Noise Fundamentals

7.4.1.1 Noise Description. Noise is a random signal, meaning that we cannot know its actual amplitude at a given moment. However, we can infer some information about it from its statistical description. The *mean-square value*, or intensity, of a signal $x(t)$ is the average of the squares of the instantaneous values of the signal,

$$\Psi_x^2 = \lim_{T \to \infty} \frac{1}{T} \int_0^T x^2(t)\, dt \tag{7.44}$$

This value indicates the power of the signal and can be separated into a time-invariant part—the *signal average* or *mean value* μ_x—and a dynamic part or *signal variance*, which is defined as the mean-square value of $x(t)$ about its mean value. Because electronic noise has zero average, we have

$$\sigma_x^2 = \lim_{T \to \infty} \frac{1}{T} \int_0^T [x(t) - \mu_x]^2\, dt = \lim_{T \to \infty} \frac{1}{T} \int_0^T x^2(t)\, dt \tag{7.45}$$

Hence, $\Psi_x^2 = \sigma_x^2$ and the noise variance equals the noise power. σ_x is the *standard deviation*.

The *probability density function* (PDF), $p(x)$, describes random signals in the amplitude domain as follows:

$$p(x) = \lim_{\Delta x \to 0} \frac{\text{Prob}[x < x(t) < x + \Delta x]}{\Delta x} = \lim_{\Delta x \to 0} \frac{1}{\Delta x} \left[\lim_{T \to \infty} \frac{T_x}{T} \right] \tag{7.46}$$

where T_x is the amount of time in which $x(t)$ is within x and $x + \Delta x$. Electronic noise has a Gaussian, or normal, PDF the same as other processes that result from many independent random events. A Gaussian PDF implies that positive and negative amplitudes with similar values happen the same number of times, and that large amplitudes are less frequent than small amplitudes. The peak value divided by its root-mean-square (rms) value is termed *crest factor* (CF). Table 7.2 lists CF for different probabilities of occurrence. For example,

TABLE 7.2   Crest Factors for Gaussian Signals Depending on the Probability of Actually Having a Larger Peak Amplitude

Probability (%)	Crest Factor
4.6	2
1	2.6
0.37	3
0.1	3.3
0.01	3.9
0.006	4
0.001	4.4
0.0001	4.9

$CF = 3.3$ for a 0.1 % probability means that the peak value exceeds $3.3\sigma_x$—and the peak-to-peak value exceeds $6.6\sigma_x$—less than 0.1 % of the time.

The *power spectral density* (PSD) describes random signals in the frequency domain as follows:

$$G_{xx}(f) = \lim_{\Delta f \to 0} \frac{\Psi_x^2(f, \Delta f)}{\Delta f} = \lim_{\Delta f \to 0} \frac{1}{\Delta f} \left[\lim_{T \to \infty} \frac{1}{T} \int_0^T x^2(t, f, \Delta f)\, dt \right] \quad (7.47)$$

where $\Psi_x^2(f, \Delta f)$ is the signal power in the frequency band from f to $f + \Delta f$, and $x(t, f, \Delta f)$ is that part of $x(t)$ contributing to power in the frequency band from f to $f + \Delta f$. A random signal having the same power density at all frequencies in a given frequency band is said to have a *white spectrum* in that band. Gaussian noise is completely described by its variance and PSD.

7.4.1.2 Noise Sources. The main noise sources in electronic circuits are thermal noise, shot noise, and $1/f$ or low-frequency noise. *Thermal noise* arises in any medium that dissipates energy, such as conductors. It is also called *Johnson noise* and *Nyquist noise* because J. B. Johnson first measured it and H. Nyquist analyzed it in 1927–1928. The thermal noise power $E_t^2 (E_t^2 = \sigma^2 = \Psi^2)$ available from a conductor with resistance R is

$$E_t^2 = 4kTBR \quad (7.48)$$

where $k = 1.38 \times 10^{-23}$ J/K is Boltzmann's constant, T is the absolute temperature (kelvins), and B is the noise bandwidth (Section 7.4.1.3). E_t is the thermal noise voltage (rms value). A 1 kΩ resistance at room temperature yields 4 nV in a 1 Hz bandwidth. The equivalent noise current is

$$I_t^2 = \frac{4kTB}{R} \quad (7.49)$$

Therefore, a 1 kΩ resistance at room temperature yields 4 pA in a 1 Hz bandwidth.

Equation (7.48) shows that the thermal noise depends on the bandwidth but not on the frequency; hence it is white, in addition to Gaussian. The most effective way to reduce the thermal noise is by reducing B, for example by shunting large-value resistors with a capacitor, provided that it does not affect the signal bandwidth.

The PSD for thermal noise is $S_t = e_t^2 = E_t^2/B = 4kTR$; hence it increases with bandwidth. Nevertheless, (7.48) does not apply above 1 THz or near 0 K.

Shot or *Schottky noise* arises from the random fluctuations in the number of electric charges that cross a potential barrier interposed in a charge flow. The rms shot noise current is

$$I_{sh} = \sqrt{2qI_{dc}B} \quad (7.50)$$

where $q = 0.1602$ aC is the electron charge, I_{dc} is the average current through the barrier, and B is the noise bandwidth. The total instantaneous current is $i(t) = I_{dc} + i_{sh}(t)$, where $i_{sh}(t)$ is the random shot noise, for which there is no analytical expression but whose rms value is I_{sh}. Shot noise arises for example in p–n junctions. It is white and Gaussian, and its PSD is $S_{sh} = i_{sh}^2 = I_{sh}^2/B = 2qI_{dc}$.

Low-frequency or $1/f$ *noise* accounts for the excess noise actually measured across resistors or semiconductor junctions when there is current through them. Its PDF is Gaussian and its PSD is inversely proportional to the frequency according to

$$S_f(f) = e_f^2 = \frac{K_f}{f^\alpha} \tag{7.51}$$

Usually $\alpha = 1$, hence the name $1/f$ noise. Therefore, unlike thermal and shot noise, low-frequency noise is not white. The noise power from f_L to f_H is

$$E_f^2(f_L, f_H) = \int_{f_L}^{f_H} S_f \, df = \int_{f_L}^{f_H} \frac{K_f}{f} \, df = K_f \ln \frac{f_H}{f_L} \tag{7.52}$$

Hence, each frequency decade has the same noise, but for a given bandwidth the noise is larger at low frequencies. Nevertheless, because of the logarithmic relation, reducing f_L by extending the circuit operation to several years only increases $1/f$ noise voltage (E_f) by about two or three times. Alternatively, $1/f$ noise can be described by a current I_f.

7.4.1.3 Noise Bandwidth. A random signal $x(t)$ applied to the input of a system whose power gain is $H(f)$ yields a random output $y(t)$ whose PSD is

$$G_{yy}(f) = |H(f)|^2 G_{xx}(f) \tag{7.53}$$

Therefore, the total output power in Figure 7.17a is

$$\Psi_y^2 = \int_0^\infty G_{yy}(f) \, df = \int_0^\infty |H(f)|^2 G_{xx}(f) \, df \tag{7.54}$$

Because $H(f)$ is often unknown but it is proportional to the voltage gain $G(f)$, we normally use $G(f)$ instead of $H(f)$. The *noise bandwidth* B of a system $H(f)$ is the frequency span of a rectangular power–gain curve that yields the same power as the actual system (Figure 7.17b). For a white noise input (constant PSD), the actual system yields

$$\Psi_y^2 = G_{xx} \int_0^\infty |G(f)|^2 \, df \tag{7.55}$$

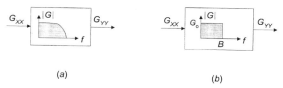

(a)

(b)

Figure 7.17 (a) Input–output relationship for a real system. (b) A theoretical system having the same noise bandwidth B.

whereas the system in Figure 7.17b yields

$$\Psi_y^2 = G_{xx}|G_0|^2 B \tag{7.56}$$

Therefore,

$$B = \frac{1}{|G_0|^2} \int_0^\infty |G(f)|^2 \, df \tag{7.57}$$

which means that the area under the actual power–gain curve equals that of a theoretical rectangular power–gain curve with width B and height of a convenient value $|G_0|^2$.

Systems with steep gain roll-off have B close to the -3 dB (signal) bandwidth. A first-order, low-pass system with corner frequency f_c has $B = \pi f_c/2$. Practical measurement systems cannot have infinite bandwidth but can have a finite frequency band from f_L to f_H. A bandpass system with 20 dB/decade slopes and those corner frequencies has $B = \pi(f_L + f_H)/2$ (reference 1, Section 11.1.5).

7.4.2 Noise in Op Amps

Op amp noise can be modeled with a circuit similar to that in Figure 7.1b for offset voltage and current, as shown in Figure 7.18. The stars inside noise voltage and current generators indicate that the corresponding signals are random.

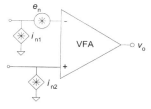

Figure 7.18 The usual noise model for op amps includes a voltage noise generator and two current noise generators added to the ideal op amp.

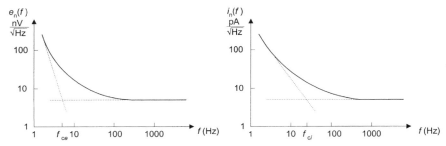

Figure 7.19   Frequency dependence of voltage noise spectral density and current noise spectral density for a low-noise op amp.

These noise generators arise from thermal, shot, and $1/f$ noise inside the op amp. Therefore, they depend on the signal frequency considered. e_n describes the thermal noise and the effect of base and emitter shot currents and excess noise currents (I_f) on internal resistances. This voltage would be the only term to consider if signal source output resistance and other circuit resistances were zero. Whenever this is not the case, we must add the noise from each resistance and also a current noise generator at each input, i_{n1} and i_{n2}. Most op amps have $i_{n1} = i_{n2}$, so that a single i_n is specified.

Figure 7.19 shows the typical frequency dependence of $e_n(f)$ and $i_n(f)$ for common bipolar op amps. Autozero (chopper) amplifiers do not have a noticeable equivalent input $1/f$ noise. They display noise peaks related to clock frequency instead. Manufacturers specify $e_n(f)$ and $i_n(f)$ at a convenient frequency where they are constant, such as 1 kHz $[e_n(1 \text{ kHz}) = e_n$ and $i_n(1 \text{ kHz}) = i_n]$, and also the "corner" frequency [for voltage (f_{ce}) and current (f_{ci})] where $1/f$ noise power equals white noise. Therefore, at any arbitrary frequency we have

$$S_e(f) = e_n^2(f) = e_n^2\left(1 + \frac{f_{ce}}{f}\right) \tag{7.58}$$

$$S_i(f) = i_n^2(f) = i_n^2\left(1 + \frac{f_{ci}}{f}\right) \tag{7.59}$$

To avoid the increased noise at low frequencies, carrier amplifiers use a carrier frequency high enough to translate the information to a high-frequency band where amplification introduces less noise.

Op amps with JFET and MOSFET input normally have f_{ce} much higher than bipolar op amps, and $i_n(f)$ increases by 20 dB/decade above some frequency f_{pi}. Then, instead of (7.59) we have

$$S_i(f) = i_n^2(f) = i_n^2\left(1 + \frac{f}{f_{pi}}\right)^2 \tag{7.60}$$

CFAs have larger e_n, i_n, and f_{ce} than VFAs. Both e_n and i_n increase with temperature. For FET inputs, as the leakage current doubles for every $10\,°C$ increase in temperature, i_n increases by $\sqrt{2}$ for every $10\,°C$ temperature rise.

Alternatively, voltage noise at low frequencies can be specified through the total noise in a given frequency band—for example, from 0.1 Hz to 10 Hz. Because noise in a frequency band is independent from noise in another frequency band, the total noise power is the sum of the respective noise powers. However, noise in a frequency band involving low frequencies is often specified as peak-to-peak value, whereas noise at high frequencies is normally specified as an rms value. Therefore, before adding noise powers we must use the convenient crest factor in Table 7.2 to convert from peak-to-peak values to rms values, or conversely. Normally, peak-to-peak noise is divided by 6 to 7.8 to obtain rms values.

Table 7.3 lists noise parameters for some low-noise op amps. Bipolar op amps yield the lowest noise for signals with low output resistance. FET-input op amps suit high-resistance sources. Autozero op amps yield the lowest noise for low-frequency signals whose bandwidth is below about 0.25 Hz.

The PSD of the output voltage noise depends on the particular circuit implemented by the op amp. Figure 7.20a shows a differential amplifier with the corresponding op amp noise sources and thermal noise for the circuit resistors. One method to determine the circuit gain for each noise source is to replace it by a conventional voltage or current source (Figure 7.20b) and calculate the corresponding gain. The circuit gain for noise sources will be the same as for the conventional source placed in the same position. If in Figure 7.20b we replace R_3 and R_4 by their parallel R_p (Figure 7.20c), circuit analysis when the input terminals are grounded yields

$$V_o = A_d(V_p - V_n) \tag{7.61}$$

$$V_p = V_{tp} + I_{n2}R_p \tag{7.62}$$

$$\frac{V_o + V_{t2} + V_{ni} - V_n}{R_2} + I_{n1} = \frac{V_n - V_{ni} - V_{t1}}{R_1} \tag{7.63}$$

where the subscript t stands for thermal (noise). If we initially assume A_d infinite, solving for V_p and V_n yields

$$V_o = \left(1 + \frac{R_2}{R_1}\right)(-V_{ni} + V_{tp} + I_{n2}R_p) - \frac{R_2}{R_1}V_{t1} - V_{t2} - I_{n1}R_2 \tag{7.64}$$

which shows the respective gain for each voltage and current source. Therefore, the PSD of the output voltage noise will be

$$e_{no}^2 = \left(1 + \frac{R_2}{R_1}\right)^2 (e_n^2 + e_{tp}^2 + i_{n2}^2 R_p^2) + \left(\frac{R_2}{R_1}\right)^2 e_{t1}^2 + e_{t2}^2 + i_{n1}^2 R_2^2 \tag{7.65}$$

TABLE 7.3   Noise Parameters for Some Low-Noise Operational Amplifiers[a]

Op Amp	e_n (1 kHz) $(nV/\sqrt{Hz})$	f_{ce} (Hz)	i_n (1 kHz) $(fA/\sqrt{Hz})$	f_{ci} (Hz)	E_n (pp) (0.1–10 Hz) (μV)
AD745	5	<10	6.9[b]	120	0.38[b]
AD797	1.2	<100	2000[b]	—	0.05[b]
AM427B	3	2.7	600	140	0.18
HA5127A	3.8	<10	600	<10	0.18
LMV751	7[b]	—	10[b]	—	—
LT1001C	11	4	120	70	0.6
LT1007	3.8	2	600	120	0.13
LT1012C	22	2.5	6	120	0.5[b]
LT1028	1.1	3.5	1600	250	0.075
LT1113	6	120	10[b]	—	2.4[b]
LT1169	8	60	1	—	2.4[b]
LT1793	8	30	1[b]	—	2.4[b]
MAX410	2.4	90	1200[b]	220	0.34[b]
NE5534	3.5	100	400[b]	200	—
OP07	11	10	170	50	0.6
OP27A	3.8	2.7	600	140	0.18
OP77A	11	≈2	170	—	0.6
OP297G	17[b]	—	20[b,c]	—	0.5[b]
OPA111AM	15	1000	0.3	—	3.3
TL051	30	100	10	—	4[b]
TLC2201	8[b]	50	0.6[b]	—	0.7[b]
TLE2027AC	3.8	3	600	—	0.13
TLE2662	60	20	1[b]	—	1.1[b]

[a] Voltage and current values are maximal, unless noted, at 25 °C, but not necessarily measured with the same voltage supply and source impedance. Frequency values are typical and some have been estimated from figures.

[b] Typical values.

[c] At 10 Hz.

which for matched resistors $(R_p = R_3\|R_4 = R_1\|R_2)$ reduces to

$$e_{no}^2 = \left(1 + \frac{R_2}{R_1}\right)^2 (e_n^2 + e_{tp}^2) + \left(\frac{R_2}{R_1}\right)^2 e_{t1}^2 + e_{t2}^2 + (i_{n1}^2 + i_{n2}^2)R_2^2 \qquad (7.66)$$

The equivalent input noise power spectral density is

$$e_{ni}^2 = \frac{e_{no}^2}{(R_2/R_1)^2} \qquad (7.67)$$

From (7.54), the output mean-square voltage noise when measured with a device whose bandwidth goes from f_L to f_H is

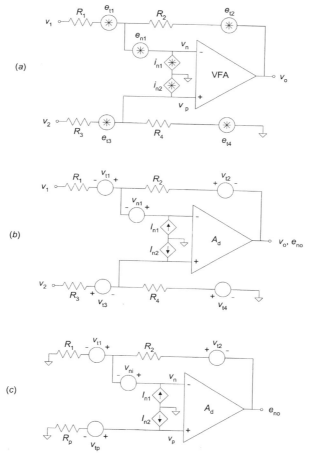

Figure 7.20 (*a*) Noise sources in an op amp and external feedback resistors. (*b*) Equivalent circuit to determine voltage gain for noise sources in (*a*). (*c*) Simplified circuit when R_3 and R_4 in (*b*) are replaced by their parallel equivalent.

$$E_{no}^2 = \int_{f_L}^{f_H} e_{no}^2(f)\, df \tag{7.68}$$

Using (7.66) for e_{no}, and (7.58) and (7.59) for the respective frequency dependence of e_n and i_n for the op amp, if the filters determining f_L to f_H are sharp enough, we finally obtain

$$E_{no}^2 = \left(1 + \frac{R_2}{R_1}\right)^2 \left[e_n^2\left(f_H - f_L + f_{ce}\ln\frac{f_H}{f_L}\right) + e_{tp}^2(f_H - f_L)\right]$$

$$+ \left[\left(\frac{R_2}{R_1}\right)^2 e_{t1}^2 + e_{t2}^2\right](f_H - f_L) + 2i_n^2 R_2^2\left(f_H - f_L + f_{ci}\ln\frac{f_H}{f_L}\right) \tag{7.69}$$

TABLE 7.4   Noise Contribution in the Differential Amplifier in Figure 7.20

Noise Source	PSD	Gain	B	Output Noise Power
R_1, thermal	$e_{t1}{}^2 = 4kTR_1$	$\left(\dfrac{R_2}{R_1}\right)^2$	$f_H - f_L$	$e_{t1}^2\left(\dfrac{R_2}{R_1}\right)^2(f_H - f_L)$
R_2, thermal	$e_{t2}{}^2 = 4kTR_2$	1	$f_H - f_L$	$e_{t2}^2(f_H - f_L)$
$R_p{}^a$, thermal	$e_{tp}{}^2 = 4kTR_p$	$\left(1+\dfrac{R_2}{R_1}\right)^2$	$f_H - f_L$	$e_{tp}^2\left(1+\dfrac{R_2}{R_1}\right)^2(f_H - f_L)$
e_n, white noise	$e_n{}^2$	$\left(1+\dfrac{R_2}{R_1}\right)^2$	$f_H - f_L$	$e_n^2\left(1+\dfrac{R_2}{R_1}\right)^2(f_H - f_L)$
e_n, excess noise	—	$\left(1+\dfrac{R_2}{R_1}\right)^2$	—	$e_n^2\left(1+\dfrac{R_2}{R_1}\right)^2 f_{ce}\ln\dfrac{f_H}{f_L}$
i_{n1}, white noise	$i_{n1}{}^2$	$R_2{}^2$	$f_H - f_L$	$i_{n1}^2 R_2^2(f_H - f_L)$
i_{n1}, excess noise	—	$R_2{}^2$	—	$i_{n1}^2 R_2^2 f_{ci}\ln\dfrac{f_H}{f_L}$
i_{n2}, white noise	$i_{n2}{}^2$	$R_p{}^2\left(1+\dfrac{R_2}{R_1}\right)^2$	$f_H - f_L$	$i_{n2}^2 R_2^2(f_H - f_L)^b$
i_{n2}, excess noise	—	$R_p{}^2\left(1+\dfrac{R_2}{R_1}\right)^2$	—	$i_{n2}^2 R_2^2 f_{ci}\ln\dfrac{f_H}{f_L}{}^b$

$^a R_p = R_3 \| R_4$.
^{b} When $R_1 = R_3$ and $R_2 = R_4$.

where $i_n{}^2 = i_{n1}{}^2 = i_{n2}{}^2$. Table 7.4 summarizes the contribution of each single noise source to the output noise power.

If the op amp has a specified peak-to-peak low-frequency noise voltage E_{lf} from f_L to f_{lf} (and $f_{lf} > f_{ce}$), (7.69) can be replaced by

$$E_{no}^2 = \left(1+\frac{R_2}{R_1}\right)^2\left[\left(\frac{E_{lf}}{2\times CF}\right)^2 + e_n^2(f_H - f_{lf}) + e_{tp}^2(f_H - f_L)\right]$$

$$+ \left[\left(\frac{R_2}{R_1}\right)^2 e_{t1}^2 + e_{t2}^2\right](f_H - f_L) + 2i_n^2 R_2^2\left(f_H - f_L + f_{ci}\ln\frac{f_H}{f_L}\right) \quad (7.70)$$

If the frequency band from f_L to f_H is not determined by sharp filters, the actual noise is somewhat larger than that calculated by (7.69) and (7.70) (reference 1, Section 11.2.3.2). However, because of the uncertainty in e_n and i_n, noise calculations provide an estimate, not an exact result, so that it is not usually worth the trouble to derive the complete expression for the total output noise E_{no} by including the actual transfer function of the filters in (7.68).

If the op amp open-loop gain is considered to be finite, (7.69) and (7.70) still apply if the op amp gain–bandwidth product is much larger than the desired gain–bandwidth product for the signal—that is, if $A_{d0}f_a \gg Gf_H$ (reference 1, Section 11.2.3.1). Because this is the same condition to achieve the desired signal bandwidth, we conclude that A_d does not significantly influence circuit noise.

Example 7.8 Calculate the mean-square output noise for the differential amplifier in Figure 7.20a in the band from 0.1 Hz to 100 Hz when $R_1 = R_3 = 1$ kΩ, $R_2 = R4 = 200$ kΩ, and the op amp is the OP27A.

The gain–bandwidth product for the OP27 is 5 MHz minimum (8 MHz typical), and the desired gain–bandwidth product for the signal is 20 kHz. Hence we can assume A_d infinite. From Table 7.3, for the OP27A we have $f_{ce} = 2.7$ Hz, $e_n = 3.8$ nV/$\sqrt{\text{Hz}}$, $f_{ci} = 140$ Hz, $i_n = 0.6$ pA/$\sqrt{\text{Hz}}$, and $E_{lf}(0.1, 10) = 0.18$ μV (peak-to-peak). Therefore, we can apply (7.70) with $f_{lf} = 10$ Hz. We need first to calculate the thermal noise from resistors R_1, R_2, and $R_3 \| R_4 = R_p$. At 25 °C,

$$e_{t1}^2 = 4kTR_1 = 4 \times 1.38 \times 10^{-23}(\text{J/K}) \times (298 \text{ K}) \times (1 \text{ k}\Omega)$$
$$= 1.64 \times 10^{-17} \text{ V}^2/\text{Hz}$$

$$e_{t2}^2 = 4kTR_2 = 4 \times 1.38 \times 10^{-23}(\text{J/K}) \times (298 \text{ K}) \times (200 \text{ k}\Omega)$$
$$= 3.3 \times 10^{-15} \text{ V}^2/\text{Hz}$$

$$e_{tp}^2 = 4kTR_p = 4 \times 1.38 \times 10^{-23}(\text{J/K}) \times (298 \text{ K}) \times (995 \text{ }\Omega)$$
$$= 1.64 \times 10^{-17} \text{ V}^2/\text{Hz}$$

If from Table 7.2 we take $CF = 3.3$, (7.70) leads to the following respective contributions from each noise source (see also Table 7.4):

Thermal noise from R_1:

$$(1.64 \times 10^{-17} \text{ V}^2/\text{Hz})\left(\frac{200}{1}\right)^2 (100 \text{ Hz} - 0.1 \text{ Hz}) = (8 \text{ μV})^2$$

Thermal noise from R_2:

$$(3.3 \times 10^{-15} \text{ V}^2/\text{Hz})(100 \text{ Hz} - 0.1 \text{ Hz}) = (0.6 \text{ μV})^2$$

Thermal noise from $R_3 \| R_4$:

$$(1.64 \times 10^{-17} \text{ V}^2/\text{Hz})\left(1 + \frac{200}{1}\right)^2 (100 \text{ Hz} - 0.1 \text{ Hz}) = (8 \text{ μV})^2$$

Voltage noise (op amp):

$$\left(1+\frac{200}{1}\right)^2\left[\left(\frac{0.18\ \mu V}{2\times 3.3}\right)^2+\left(3.8\ \frac{nV}{\sqrt{Hz}}\right)^2(100\ Hz-10\ Hz)\right]=(9\ \mu V)^2$$

Current noise (op amp):

$$2\times(0.6\ pA/\sqrt{Hz})^2(200\ k\Omega)^2\left[100\ Hz-0.1\ Hz+(140\ Hz)\ln\frac{100}{0.1}\right]=(5.5\ \mu V)^2$$

Adding all these contributions yields $E_{no}^2(rms)=(16\ \mu V)^2$ and $E_{no}(rms)=16\ \mu V$. Because the signal gain is 200, the equivalent input noise voltage is $16\ \mu V/200=78\ nV$.

Noise for inverting and noninverting amplifiers can also be calculated from the above expressions by replacing resistors in Figure 7.20a as convenient. For inverting amplifiers, matching input resistances by adding R_p does not reduce the effect for noise currents as contrasted with bias currents (Figure 7.1b). Rather, the noise increases because of the contribution from the thermal noise of the added resistor and op amp input noise currents (i_{n2}). Nevertheless, shunting R_p with a large capacitor reduces the noise bandwidth of its thermal noise. The same is true for noninverting amplifiers (see Figure E7.3).

Example 7.9   Calculate the power spectral density of the output noise of a buffer amplifier based on an op amp, as a function of the output resistance of the signal source R_p.
    We can use the circuit in Figure 7.20a with $R_1=\infty$, $R_2=0\ \Omega$, and $R_p=R_s$. Equation (7.65) reduces then to

$$e_{no}^2=e_n^2+e_{ts}^2+i_n^2R_s^2=e_n^2+4kTR_s+i_n^2R_s^2$$

If we place a resistor equal to R_s in the feedback loop in order to compensate for bias currents, we would have $R_1=\infty$, $R_2=R_p=R_s$, and

$$e_{no}^2=e_n^2+e_{ts}^2+e_{ts}^2+(i_{n1}^2+i_{n2}^2)R_s^2=e_n^2+8kTR_s+2i_n^2R_s^2$$

Unless e_n is very large, which would be an odd design choice for a voltage buffer, adding R_s in the feedback loop increases the output noise.

If the feedback network around the op amp includes capacitors in addition to resistors, the noise calculation procedure is the same as above leading from (7.61) to (7.69) but replacing resistor by impedances. This makes the calculation step in (7.68) more involved because of the frequency dependence of the respective gain for each noise source. Nevertheless, in simple circuits such as inverting and noninverting amplifiers, the only relevant additional component to consider is often a capacitor shunting the feedback resistor, as in Figure 3.15a. This capacitor limits the high-frequency gain for noise sources, but the high-pass frequency in the noise bandwidth may be still determined by an external filter (e.g., an antialiasing filter), so that (7.69) and (7.70) still provide an acceptable noise estimate. If that were not the case—that is, $f_2 = 1/(2\pi R_2 C)$ were lower than the corner frequency of the antialiasing filter—we could still apply (7.69) and (7.70) by taking $f_H = f_2$. Certainly, the high-frequency gain for some noise sources is 1 even if $R_2 = 0 \ \Omega$ (see Table 7.4), but the filter following the amplifier will reduce this additional noise.

With regard to f_L, if the feedback network does not limit low-frequency gain, we can always take $f_L = 0.01$ Hz because noise at lower frequencies are attributed to offset drift. If the feedback network limits low-frequency gain below a given f_1, we can apply (7.69) and (7.70) by taking $f_L = f_1$.

Example 7.10   The electrometer amplifier in Example 7.3 is followed by a filter whose noise bandwidth is from 5 Hz to 10 kHz. If the op amp is the LT1113 and $T = 308$ K, estimate the noise voltage at the filter output.

According to Table 7.3, the current noise for the LT1113 is so small that it can be neglected. Figure E7.10 shows the remaining noise sources, with $R_1 = 1$ kΩ, $R_2 = 98.8$ kΩ, $R_a = R_b = 146$ MΩ, $C_b = 450$ pF, and C_a is to be decided. From Table 7.3, $e_n = 6$ nV/$\sqrt{\text{Hz}}$ and $f_{ce} = 120$ Hz. Table 7.4 gives the noise gain for the thermal noise of R_1 and R_2, and the op amp voltage noise. Their respective contributions to the output noise power will be

$$E_{no1}^2 = 4 \times (1.38 \times 10^{-23} \text{ J/K}) \times (1 \text{ k}\Omega) \times (100)^2 (10^4 \text{ Hz} - 5 \text{ Hz})$$
$$= 1.7 \times 10^{-10} \text{ V}^2$$

$$E_{no2}^2 = 4 \times (1.38 \times 10^{-23} \text{ J/K}) \times (98.8 \text{ k}\Omega) \times (10^4 \text{ Hz} - 5 \text{ Hz})$$
$$= 1.7 \times 10^{-11} \text{ V}^2$$

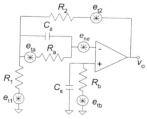

Figure E7.10   Equivalent circuit for an electrometer amplifier when considering its noise sources.

$$E_{noe}^2 = (6 \text{ nV}/\sqrt{\text{Hz}})^2 \times (100)^2 \left(10^4 \text{ Hz} - 5 \text{ Hz} + (120 \text{ Hz}) \ln\frac{10 \text{ kHz}}{5 \text{ Hz}}\right)$$

$$= 4.0 \times 10^{-9} \text{ V}^2$$

The gain for the thermal noise of R_a and R_b includes the effect of the capacitors shunting them. The PSD at the op amp output because of e_{tb} will be

$$e_{nob}^2(f) = \left|\frac{1}{1 + j2\pi f R_b C_s}\right|^2 \left(1 + \frac{R_2}{R_1}\right)^2 e_{tb}^2$$

The corresponding noise power at the filter output will be

$$E_{nb}^2 = \int_{f_L}^{f_H} e_{nob}^2(f)\,df = \left(1 + \frac{R_2}{R_1}\right)^2 e_{tb}^2 \int_{f_L}^{f_H} \frac{1}{1 + (2\pi f R_b C_s)^2}\,df$$

$$= \left(1 + \frac{R_2}{R_1}\right)^2 (4kTR_b)\frac{\arctan 2\pi f_H R_b C_s - \arctan 2\pi f_L R_b C_s}{2\pi R_b C_s}$$

$$= (100)^2 \left(\frac{4 \times (1.38 \times 10^{-23} \text{ J/K}) \times (308 \text{ K})}{2\pi \times 450 \text{ pF}}\right)(1.57 - 1.12)$$

$$= 8.5 \times 10^{-8} \text{ V}^2$$

A large C_s reduces the thermal noise of R_b.

Because e_{ta} is also at the op amp input, the circuit gain for it will be the same as for e_{tb}. Therefore, its contribution to the noise power at the filter output will be

$$E_{na}^2 = \left(1 + \frac{R_2}{R_1}\right)^2 (4kTR_a)\frac{\arctan 2\pi f_H R_a C_a - \arctan 2\pi f_L R_a C_a}{2\pi R_a C_a}$$

Hence, if we select C_a large enough, say 10 nF, this noise power will be negligible. The total noise power will come mostly from e_n and R_b,

$$E_{no}^2 \approx E_{noe}^2 + E_{nob}^2 = (4 \times 10^{-9} + 8.5 \times 10^{-8})\text{V}^2 = 89 \times 10^{-9} \text{ V}^2$$

The corresponding noise voltage is $E_{no}(\text{rms}) = 0.3$ mV. If we use CF $= 3.3$, the peak-to-peak voltage will be $E_{no} = 2 \times 3.3 \times (0.3 \text{ mV}) = 2$ mV.

7.4.3 Noise in Transimpedance Amplifiers

Transimpedance amplifiers use FET-input op amps whose input noise current is better modeled by (7.60). If in Figure 7.10b Z_s is the parallel of R_s and C_s, the

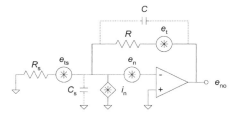

Figure 7.21 Equivalent circuit for a transimpedance amplifier when considering its noise sources.

equivalent circuit for noise analysis is that in Figure 7.21. If we initially disregard C, Table 7.4 gives the gain for each noise source with $R_2 = R$, $R_1 = R_s$, and $R_p = 0\ \Omega$. Therefore, (7.66) reduces to

$$e_{no}^2 = \left(1 + \frac{R}{R_s}\right)^2 e_n^2 + \left(\frac{R}{R_s}\right)^2 e_{ts}^2 + e_t^2 + i_n^2 R^2 \qquad (7.71)$$

Because the transfer impedance in the passband is R, the PSD of the input current noise is

$$i_{ni}^2 = \frac{e_{no}^2}{R^2} = \left(\frac{1}{R} + \frac{1}{R_s}\right)^2 e_n^2 + i_{ts}^2 + i_t^2 + i_n^2 \qquad (7.72)$$

where i_{ts} and i_t are the respective thermal current noise from R_s and R—derived from (7.49). A large R reduces the current noise. For a noise bandwidth from f_L to f_H we finally have

$$E_{no}^2 = \int_{f_L}^{f_H} e_{no}^2\, df = \left(1 + \frac{R}{R_s}\right)^2 e_n^2 \left(f_H - f_L + f_{ce} \ln \frac{f_H}{f_L}\right)$$

$$+ \left[\left(\frac{R}{R_s^2}\right)^2 e_{ts}^2 + e_t^2\right](f_H - f_L)$$

$$+ \frac{i_n^2 R^2 f_{pi}}{3}\left[\left(1 + \frac{f_H}{f_{pi}}\right)^3 - \left(1 + \frac{f_L}{f_{pi}}\right)^3\right] \qquad (7.73)$$

If $f_H < f_{pi}$, noise currents would contribute $i_n^2 R^2 (f_H - f_L)$. If R is large enough to determine the system bandwidth, then (7.23) gives f_H. If C is not negligible so that $f_0 = 1/(2\pi RC) < f_H$, replacing f_H by f_0 in (7.73) still provides a noise estimate. Chapter 5 in reference 9 discusses a more detailed noise analysis method.

7.4.4 Noise in Charge Amplifiers

Charge amplifiers have the same equivalent circuit as transimpedance amplifiers (compare Figure 7.14a to Figure 7.10b) but have a different relative value
for resistors and capacitors, so that the transfer impedance is capacitive.
Therefore, Figure 7.21 applies with $R = R_0$ and $C = C_0$, which is not negligible. If the thermal noise of R_s is modeled by a current source i_{ts} rather than e_{ts},
circuit analysis yields

$$e_{no}^2 = \left|1 + \frac{Z_0}{Z_s}\right|^2 e_n^2 + e_t^2 + (i_n^2 + i_{ts}^2)|Z_0^2| \tag{7.74}$$

where e_t is the thermal noise of the real part of Z_0. The squared gains for noise
sources are

$$|Z_0^2| = \left|\frac{R_0}{1 + j\omega R_0 C_0}\right|^2 = R_0^2 \frac{f_0^2}{f^2 + f_0^2} \tag{7.75}$$

$$\left|1 + \frac{Z_0}{Z_s}\right|^2 = \left(\frac{C_0 + C_s}{C_0}\right)^2 \frac{f^2 + f_p^2}{f^2 + f_0^2} \tag{7.76}$$

where $f_0 = 1/(2\pi R_0 C_0)$ and

$$f_p = \frac{1}{2\pi(R_0\|R_s)(C_0 + C_s)} \tag{7.77}$$

Normally, $R_s \gg R_0$ and $f_p < f_0$.

If the noise bandwidth is from f_H to f_L, and the PSD of the op amp noise
current is white, the output mean-square noise is (reference 1, Section 11.2.7)

$$E_{no}^2 \approx e_n^2 \left(\frac{C_0 + C_s}{C_0}\right)^2 \left(f_H - f_L + f_{ce} \ln\frac{f_H^2}{f_L^2 + f_0^2}\right)$$

$$+ f_0(4kTR_0 + i_{ts}^2 R_0^2 + i_n^2 R_0^2)\left(\arctan\frac{f_H}{f_0} - \arctan\frac{f_L}{f_0}\right) \tag{7.78}$$

A large noise bandwidth increases the contribution from e_n. The reduced
bandwidth of the $R_0 C_0$ network reduces the contribution of the thermal noise
of R_s, the thermal noise of R_0, and the op amp current noise.

Example 7.11   A given piezoelectric accelerometer with sensitivity 1 pC/(m/s^2)
and equivalent capacity 1 nF is connected to a charge amplifier with $C_0 = 100$
pF and $R_0 = 100$ GΩ. If the op amp is the TLC2201, and a filter with noise
bandwidth from 0.1 Hz to 2 kHz follows the amplifier, determine the minimal
acceleration that we can measure because of the noise at 35 °C.

The feedback network determines a corner frequency

$$f_0 = \frac{1}{2\pi R_0 C_0} = \frac{1}{2\pi \times (100 \text{ G}\Omega) \times (100 \text{ pF})} = 16 \text{ mHz}$$

From (7.77), for a very large R_s we have

$$f_p = \frac{1}{2\pi R_0 (C_0 + C_s)} = \frac{1}{2\pi \times (100 \text{ G}\Omega) \times (100 \text{ pF} + 1 \text{ nF})} = 1.5 \text{ mHz}$$

Therefore, $f_p < f_0$. The TLC2201 has $f_T = 1.9$ MHz, so that according to (7.40) it limits the -3 dB bandwidth to $(1.9 \text{ MHz})/11 = 173$ kHz. Therefore, it does not influence the noise bandwidth. From Table 7.3, $e_n = 8 \text{ nV}/\sqrt{\text{Hz}}$, $f_{ce} = 50$ Hz, and $i_n = 0.6 \text{ fA}/\sqrt{\text{Hz}}$. We can apply (7.78),

$$E_{no}^2 \approx (8 \text{ nV}/\sqrt{\text{Hz}})^2 \left(\frac{100 \text{ pF} + 1 \text{ nF}}{100 \text{ pF}}\right)^2$$

$$\times \left(2 \text{ kHz} - 0.1 \text{ Hz} + (50 \text{ Hz}) \ln \frac{(2 \text{ kHz})^2}{(0.1 \text{ Hz})^2 + (16 \text{ mHz})^2}\right)$$

$$+ (16 \text{ mHz})[4 \times (1.38 \times 10^{-23} \text{ J/K}) \times (308 \text{ K}) \times (10^{11} \text{ }\Omega)$$

$$+ (0.6 \text{ fA})^2 \times (10^{11} \text{ }\Omega)^2](1.54 - 1.41) \text{ V}^2$$

$$= 2.3 \times 10^{-11} \text{ V}^2 + 1.8 \times 10^{-11} \text{ V}^2 = 4 \times 10^{-11} \text{ V}^2$$

Therefore, $E_{no}(\text{rms}) = 6.4 \text{ }\mu\text{V}$. The peak-to-peak noise voltage for CF $= 3.3$ will be $E_{no} = 42 \text{ }\mu\text{V}$. The voltage sensitivity will be

$$S = \frac{q/a}{C_0} = \frac{1 \text{ pC}/(\text{m/s}^2)}{100 \text{ pF}} = 10 \text{ mV}/(\text{m/s}^2)$$

In an analog display where noise peaks mask small accelerations, the resolution would be $4.2 \times 10^{-3} \text{ m/s}^2$.

7.4.5 Noise in Instrumentation Amplifiers

The transfer function of instrumentation amplifiers does not depend on external feedback networks. Figure 7.22 shows the circuit model for noise analysis, where e_{ta} and e_{tb} are the thermal noise voltages of the source impedances—that is, the thermal noise of their respective real parts. If the gain is G, the PSD of the output noise is

$$e_{no}^2 = (e_n^2 + i_n^2 |Z_o|^2 + i_n^2 |Z_o'|^2 + e_{ta}^2 + e_{tb}^2)|G|^2 \qquad (7.79)$$

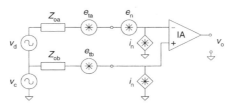

Figure 7.22 Noise model for an instrumentation amplifier including the thermal noise from the signal source output impedance.

Similarly to offset voltages and their drifts (Section 7.1.5), noise in instrumentation amplifiers also consists of one component due to the input stage and another component due to the output stage. This way the input-referred noise voltage depends on the gain G,

$$e_n^2 = e_{n1}^2 + \frac{e_{n2}^2}{G^2} \qquad (7.80)$$

where e_{n1} is the spectral density of the noise voltage due to the input stage and e_{n2} is that due to the output stage. IAs also have $1/f$ noise, and e_{n1} and e_{n2} may have different corner frequencies, f_{ce_1} and f_{ce_2}. Table 7.5 summarizes noise parameters of some instrumentation amplifiers.

The mean-square voltage noise in a frequency band from f_H to f_L is obtained by integrating (7.79) according to (7.68). If the source impedance is resistive and balanced, the result is

$$E_{no}^2 = e_{n1}^2 \left(f_H - f_L + f_{ce1} \ln \frac{f_H}{f_L} \right) G^2 + e_{n2}^2 \left(f_H - f_L + f_{ce2} \ln \frac{f_H}{f_L} \right)$$

$$+ 2i_n^2 R_0^2 \left(f_H - f_L + f_{ci} \ln \frac{f_H}{f_L} \right) G^2 + 2 \times 4kTR_0 \times G^2 \qquad (7.81)$$

If the IA data sheet specifies the peak-to-peak excess noise voltage in a frequency band from f_L to f_{lf} ($f_{lf} < f_{c1}, f_{c2}$) for a gain G, the corresponding noise power can be added to the remaining noise power after converting it to rms noise by using a convenient crest factor (Table 7.2). The mean-square output noise is then

$$E_{no}^2 = \left(\frac{E_{n1G}}{2 \times CF} \right) + e_{n1}^2 \left(f_H - f_{lf} + f_{ce1} \ln \frac{f_H}{f_{lf}} \right) G^2$$

$$+ \left(\frac{E_{n2G}}{2 \times CF} \right) + e_{n2}^2 \left(f_H - f_{lf} + f_{ce2} \ln \frac{f_H}{f_{lf}} \right)$$

$$+ 2i_n^2 R_0^2 \left(f_H - f_L + f_{ci} \ln \frac{f_H}{f_L} \right) G^2 + 2 \times 4kTR_0 \times G^2 \qquad (7.82)$$

TABLE 7.5   Noise Parameters for Some Instrumentation Amplifiers[a]

IA	E_n (pp) (μV) 0.1–10 Hz	e_n (nV/$\sqrt{Hz}$)			I_n (pp) (pA) 0.1–10 Hz	i_n (pA/$\sqrt{Hz}$)	
		10 Hz	100 Hz	1 kHz		10 Hz	100 Hz
AD620A	0.28	—	—	9	10	—	—
AD623A	1.5^b	—	—	35	1.5	—	—
AD624A	0.3	—	—	4	60	—	—
AD625A	0.3	—	—	4	60	—	—
AD626A	2	—	—	250	—	—	—
AMP02F	0.5	—	—	10	—	—	0.4^b
INA101HP	0.8^c	18^b	15^b	13^b	50^b	0.8	0.35
INA103KP	—	2	1.2	1	—	—	2
INA110AP	1	—	—	10^d	—	—	0.0018^d
INA111AP^b	1	—	13	10	—	—	0.0008^d
INA114AP	0.4	15	11	11	18	0.4	0.2
INA131AP	0.4	16	12	12	18^e	0.4	0.2
LTC1100C	2	100	97	90	—	—	—
LT1167C	0.28	—	—	7.5	10	0.124	—
PGA202KP	1.7	—	—	12^d	—	—	—

[a] All values are typical, at 25 °C ambient temperature when supplied at $\pm$15 V or $\pm$5 V and for $G = 100$, unless otherwise noted.
[b] For $G = 1000$.
[c] From 0.01 Hz to 10 Hz.
[d] At 10 kHz.
[e] From 0.1 Hz to 100 Hz.

If the voltage gain provided by a single instrumentation amplifier is not enough and more stages are used, to obtain the minimal output noise we should concentrate most of the gain in the first stage. For better interference rejection, the IA should be preceded by an amplifier with differential input and output, such as the input stage in Figure 3.34.

7.5 NOISE AND DRIFT IN RESISTORS

7.5.1 Drifts in Fixed Resistors

Design calculations for electronic circuits usually give nonstandard values. We select the closest standard value with the appropriate tolerance and a material suitable for the application in hand (metal film, carbon film, chip, wire wound, or other) [10]. However, the actual resistance value changes with time, depending on the material and power dissipated.

Resistors dissipate power according to Joule's law, and the resulting temperature increase depends on heat transmission by conduction, convection, and radiation. The maximum body temperature is called *hot-spot temperature* and usually occurs in the middle of the resistor. The temperature rise at the hot spot in the normal operating temperature range of resistors is proportional to the power dissipated,

$$\Delta T = \theta \times P \qquad (7.83)$$

where θ is the thermal resistance (K/W), which depends on the dimensions of the resistor, the heat conductivity of the materials used, and, to a lesser degree, the way of mounting. If the ambient temperature is T_a, the hot spot will reach a temperature

$$T = T_a + \Delta T \qquad (7.84)$$

The stability of a resistor depends on T and the materials it is made from. Besides, resistor connections will reach a higher temperature for a higher T so that we must prevent connections from reaching the melting point for solder. Common film resistors have a specified maximal hot-spot temperature of 155 °C.

From (7.83) and (7.84) we deduce that the power–temperature characteristic for resistors is a straight line,

$$P = \frac{T - T_a}{\theta} \qquad (7.85)$$

whose slope is $1/\theta$. Different values of ambient temperature result in a set of parallel straight lines.

Manufacturers determine the stability $\Delta R/R$ for given time periods (usually 1000 h—about 6 weeks), as a function of the hot-spot temperature with the resistance value R as a parameter. Because the resistance changes exponentially with temperature, $\lg(\Delta R/R)$ plotted against T yields a straight line. Figure 7.23 is a nomogram that combines the plots of P and $\Delta R/R$ against T and permits the calculation of several variables for a given resistor value under different operating conditions. The specified $\Delta R/R$ value is not the exact percent change in R but is an interval that has a given probability—usually 95%—of including the actual change. $\Delta R/R$ values corresponding to $P = 0$ W for a given T_a indicate the intrinsic long-term stability for that particular resistor type at the given ambient temperature.

Example 7.12   A 10 kΩ resistor-type CR25 dissipates 0.35 W in a given circuit in an ambient temperature of 40 °C. Determine if this power dissipation is allowed and estimate the resistance value after 1000 h.

In Figure 7.23, the horizontal line corresponding to $P = 0.35$ W intersects the straight line corresponding to $T_a = 40$ °C at the vertical of $T = 125$ °C. Because the maximal temperature allowed is 155 °C, the resistor is in the safe zone.

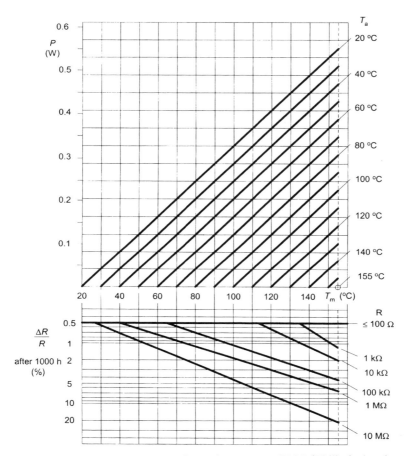

Figure 7.23 Performance nomogram for resistors type CR25 (Philips) showing relationship between power dissipation (P), ambient temperature (T_a), hot-spot temperature (T_m), resistor value (R), and maximum resistance drift $\Delta R/R$—95% probability—after 1000 h of operation.

Extending the vertical line corresponding to $T = 125\,°C$ into the lower half of the nomogram, it intersects the line corresponding to 10 kΩ at a point placed on the horizontal line for $\Delta R/R = 0.7\%$. That is, $\Delta R = 70\ \Omega$. Therefore, the interval 9930 $\Omega < R < 10070\ \Omega$ has a 95% probability of including the actual value for R after 1000 h of continuous operation.

This example shows that we cannot guarantee a resistor value by merely selecting a close tolerance if the resistor must dissipate an appreciable amount of power. In the example above, we just need 6 mA to get 0.35 W. Nevertheless, the drop in voltage across the 10 kΩ resistor would be 60 V, which is not common in signal circuits.

If the operation time t_x is higher than 1000 h, we can estimate the resistor drift by multiplying that reading on the nomogram for 1000 h by $(t_x/1000$ h$)^{1/2}$. Resistors intended for circuits working uninterruptedly for long time periods are specified for 10,000 h (as in cars) and even 225,000 h (as in telephone exchanges). In the last case the maximum hot-spot temperature allowed is usually limited to 125 °C.

Nomograms for resistors such as that in Figure 7.23 also permit us to estimate the maximal power dissipation allowed for a given acceptable drift under particular operating conditions.

Example 7.13   A given 1 kΩ resistor-type CR25 works in a circuit operating at 70 °C. Determine the maximal power dissipation allowed if its drift after 4000 h must be less than 1 %.

The nomogram in Figure 7.23 gives drifts for 1000 h of operation. In order to have 1 % drift after 4000 h, we need about $(1\,\%)/\sqrt{4} = 0.5\,\%$ drift in 1000 h. From the lower part of the nomogram, for 0.5 % drift the temperature of a 1 kΩ resistor should not exceed 135 °C. In the upper part of the nomogram, for $T_a = 70\,°C$ and $T = 135\,°C$ the power dissipated should not exceed about 0.32 W. Hence, the current through the resistor should be less than 17 mA, and the drop in voltage across it should be less than 17 V.

An alternative method to specify resistor drift is by the power derating curve. This curve shows the maximal power dissipation allowed to guarantee, within a given probability, that a resistor does not achieve a hot-spot temperature higher than, say 155 °C, or a drift exceeding a given value, say 1.5 %. Common derating curves are flat for ambient temperature up to about 70 °C and then progressively decrease up to 0 W for the maximal temperature (155 °C).

Resistors in precision circuits do not dissipate much power, so that their stability depends mostly on the effect of ambient temperature changes because of their temperature coefficient of resistance (TCR), which strongly depends on resistor material and on resistor value for some materials. It may exceed $1200 \times 10^{-6}/K$ for carbon composition resistors, and it is about $200 \times 10^{-6}/K$ for carbon-film and chip resistors, from $25 \times 10^{-6}/K$ to $50 \times 10^{-6}/K$ for metal-film resistors, and less than $3 \times 10^{-6}/K$ for wire wound resistors.

7.5.2   Drifts in Adjustable Resistors (Potentiometers)

Electronic circuits use two main types of potentiometers: trimming (or preset) potentiometers and control potentiometers. Trimming potentiometers are designed for a limited number of wiper movements—for example, to eliminate circuit tolerances or for circuit readjustment. Control potentiometers are intended to control gain, volume, tone, and so on. The stability in the set value is very important in trimming potentiometers but not in control potentiometers.

Potentiometers are selected according to their nominal resistance R_n, which is the nominal value of the resistance between the end terminals when the wiper is at the end stop. Tolerance for the nominal value is usually $\pm 10\%$ to $\pm 20\%$. The TCR is above $\pm 200 \times 10^{-6}/\text{K}$ for cermet, carbon composition, and carbon-film resistance elements, and it is about $\pm 50 \times 10^{-6}/\text{K}$ for wire wound and metal film resistance elements. The total resistance R_T is the resistance measured between the end terminals, and their drifts are similar to those of resistors built from the same material. The maximal voltage that can be applied to a potentiometer able to dissipate $P_{\max}$ is $(P_{\max} \times R_n)^{1/2}$. The maximal current that may be passed between the resistance element and the wiper is $(P_{\max}/R_n)^{1/2}$, provided that the load connected to the wiper is at least $10R_n$ (see Problem 2.1). To prevent the development of dry circuit problems—oxide film—in the metallic electric contact between the wiper and the resistance element, some manufacturers recommend a minimal wiper current above $100\ \mu\text{A}$.

Trimming potentiometers are also selected according to their resolution (adjustability). The resolution of wire wound potentiometers is limited by the number of wire turns used in their manufacture. Common single-turn trimmers can be set to within 0.05% of the applied voltage (voltage-mode operation) or 0.1% of R_T. Adding resistance in series with the trimmer allows finer setting but reduces the overall adjustment range. Multiturn trimmers (3, 10, 20, 40 turns) yield better adjustability. Setting stability is the ability of a potentiometer to maintain its initial setting during mechanical and environmental stresses (shock, vibration). It is normally expressed as a percentage change in output voltage with respect to the total applied voltage, and it ranges from 0.5% for wire wound models to 2% for carbon models.

7.5.3 Noise in Resistors

Noise in metal- or carbon-film resistors is larger than the thermal noise given by (7.48) for a conductor with resistance R. Because a relative large proportion of electrons in film resistors flow along the surface rather than inside the bulk (thin) material, there is an excess noise whose voltage spectral density is given by

$$e_{\text{ex}} = \frac{mI_{\text{dc}}R}{\sqrt{f}} = \frac{mV_{\text{dc}}}{\sqrt{f}} \tag{7.86}$$

where m is a parameter that depends on the manufacturing process and V_{dc} is the voltage drop across the resistor. The mean square voltage noise from f_H to f_L will be

$$E_{\text{ex}}^2 = \int_{f_L}^{f_H} e_{\text{ex}}^2\, df = \int_{f_L}^{f_H} \left(\frac{mV_{\text{dc}}}{\sqrt{f}} \right) df = m^2 V_{\text{dc}}^2 \ln \frac{f_H}{f_L} \tag{7.87}$$

Therefore, excess noise is not white and, the same as $1/f$ noise in op amps, there is the same noise power in each decade.

Resistor manufacturers specify the excess noise by the *noise index* (NI), defined as the ratio between the rms excess noise voltage, in microvolts, in a frequency decade (E_{ex}) and the dc voltage, in volts, across the resistor (V_{dc}). NI is sometimes expressed in decibels,

$$\text{NI} = 20 \lg \frac{E_{ex}}{V_{dc}} \qquad (7.88)$$

and NI = 0 dB when $E_{ex}/V_{dc} = 1 \ \mu V/V$.

NI depends on the resistor material and value and, for a given resistor type, it has a dispersion of up to 20 dB. Some manufacturers, however, specify NI with a 95% confidence level. Carbon composition resistors are the noisiest, followed by carbon-film and metal-film resistors. Resistors with larger power rating are quieter because of the increased relative amount of volume conduction in them. Resistors having noise larger than usual are unreliable and may be on the verge of failing.

To calculate the mean-square voltage of excess noise from the specified NI in decibels, we apply (7.87) in a frequency decade, substitute in (7.88), and solve for m:

$$m = \frac{10^{[(\text{NI}/20)-6]}}{\sqrt{10}} = 0.66 \times 10^{[(\text{NI}/20)-6]} \qquad (7.89)$$

From (7.87), the excess voltage from f_H to f_L will be

$$E_{ex}(\text{rms}) = 0.66 \times 10^{[(\text{NI}/20)-6]} \times V_{dc} \sqrt{\ln \frac{f_H}{f_L}} \qquad (7.90)$$

If NI is specified in microvolts per volt instead of decibels, we have

$$E_{ex}(\text{rms}) = 0.66 \times \text{NI} \times V_{dc} \sqrt{\ln \frac{f_H}{f_L}} \qquad (7.91)$$

The total noise power will be the sum of the thermal noise power in (7.48) and the power of excess noise above:

$$E_R^2 = E_t^2 + E_{ex}^2 \qquad (7.92)$$

Excess noise is important in broadband circuits when there is a large drop in voltage across resistors, particularly below 1 kHz.

Example 7.14 A given carbon-film resistor has 100 kΩ and has 1.6 $\mu V/V$ excess noise, and it works in a circuit at 40 °C ambient temperature. Calculate its voltage noise from 0.1 Hz to 10 kHz.

The power dissipated by the resistor will be $P = (10\text{ V})^2/(100\text{ k}\Omega) = 1\text{ mW}$. Therefore, there is negligible heating and the resistor temperature will be $40\,^\circ\text{C}$ (313 K). From (7.91), the excess noise voltage will be

$$E_{\text{ex}}(\text{rms}) = 0.66 \times (1\ \mu\text{V/V}) \times (10\text{ V})\sqrt{\ln\frac{10\text{ kHz}}{0.1\text{ Hz}}} = 36\ \mu\text{V}$$

and the excess noise power will be $E_{\text{ex}}{}^2 = 1.3 \times 10^{-9}\text{ V}^2$.

From (7.48), the thermal noise power will be

$$E_t^2 = 4kTBR = 4 \times 1.38 \times 10^{-23}(\text{J/K}) \times (313\text{ K})$$

$$\times (10\text{ kHz} - 0.1\text{ Hz})(100\text{ k}\Omega) = 1.7 \times 10^{-11}\text{ V}^2$$

which is much smaller than excess noise. Hence, the resistor noise will be about $36\ \mu\text{V}$.

7.6 PROBLEMS

7.1 Estimate the IZE, gain error, and drifts of a noninverting amplifier intended for a sensor whose output resistance is 1 kΩ when implemented by the OP77GP with $R_1 = 100\ \Omega$, $R_2 = 100\text{ k}\Omega$ (metal film, 1% tolerance, TCR $= \pm 50 \times 10^{-6}/^\circ\text{C}$), the ambient temperature is $40\,^\circ\text{C}$, and the power supplies ($+15$ V, -15 V) have a maximal $\pm 1\%$ ripple.

7.2 The electrometer amplifier in Figure P7.2 is connected to a sensor that yields a current from 0.1 Hz to 10 kHz with an output impedance of 1 GΩ‖5 pF. The circuit includes a potentiometer in order to obtain a balancing voltage so that a zero output can be obtained when the sensor current is zero.

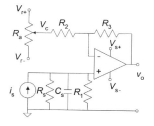

Figure P7.2 Electrometer amplifier to amplify the voltage drop across a resistor R_1 connected to a sensor with current output.

a. Design R_2, R_3, and R_a to obtain a voltage gain of 400 and to compensate bias currents so that only offset currents interfere at the output.

b. If the op amp has an open loop gain of 2×10^5 with a corner frequency at 20 Hz, what is the error due to this finite gain?

7.3   Figure P7.3 shows a piezoelectric pressure sensor whose sensitivity is 50 mV/Pa and capacitance about 50 pF connected to a voltage amplifier. Capacitor C is for sensitivity adjustment, and the 1 kΩ resistor protects the op amp.

a. Determine the gain, R_1, and R_2 to obtain a 5 V output margin corresponding to a pressure range from 26 dB to 110 dB (0 dB = 20 µPa).

b. If the desired −3 dB bandwidth is from 15 Hz to 16 kHz, calculate the value for the components that determine it. Disregard C.

c. Calculate R_0 and C_0.

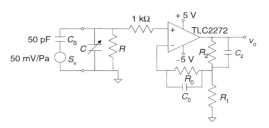

Figure P7.3   Electrometer amplifier connected to a piezoelectric sensor.

7.4   Portable computers are sensitive to shocks that may displace magnetic heads in hard disks outside their target zone. Figure P7.4 shows a circuit able to obtain a control signal from a piezoelectric accelerometer whose sensitivity is 4.7 mV/g and whose equivalent capacitance is 150 pF.

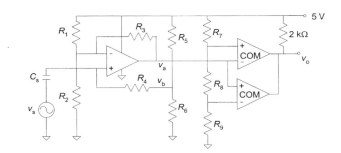

Figure P7.4   Piezoelectric accelerometer connected to a voltage amplifier and window comparator for shock detection.

a. Determine the amplifier gain needed to obtain 1 V for a 20 g shock.

b. Determine resistors R_1 to R_6 to obtain the desired gain, a corner frequency of 20 Hz, and the output voltage centered around midrange of the power supply voltage. Assume that the op amp is ideal.

c. Determine R_7, R_8, and R_9 to obtain an output pulse when the acceleration falls outside the range -20 g to $+20$ g. The comparators have open collector output stages.

7.5 A given sensor offers an output current from 0 pA to 100 pA in the band from 1 Hz to 100 Hz, with a 10 GΩ internal resistance. In order to obtain a corresponding voltage from 0 V to 10 V, a two-stage amplifier is considered. The first stage would consist of the I/V converter in Figure P7.5, and the second stage would consist of a voltage amplifier with a gain of -1000.

a. If the largest resistor available is 10 MΩ, calculate the values for R, R_1, and R_2.

b. If the op amp has $V_{io} = 15$ mV, with drifts of 10 μV/°C, $I_b = 50$ pA that doubles each 10 °C, and it is assumed to be ideal in other aspects, what is the effect at the output of the first stage of V_{io} and I_b?

c. If the input impedance and open-loop gain for the op amp are assumed finite, what is the value for the input impedance for the circuit?

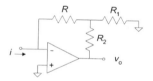

Figure P7.5   Current-to-voltage converter based on a resistor T network.

7.6 Determine the condition to be fulfilled for the wideband I/V converter in Figure P7.6 in order to have a transimpedance R.

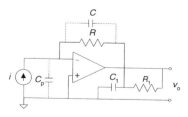

Figure P7.6   Wideband current-to-voltage converter.

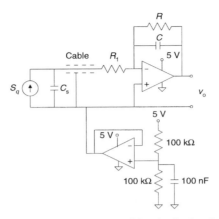

Figure P7.7   Single-supply charge amplifier for hydrophone interfacing.

7.7 A given hydrophone uses PVDF with charge sensitivity 0.5 pC/Pa and equivalent capacitance C_s. Figure P7.7 shows its connection to a single-supply charge amplifier, including 1 kΩ for protection.

   a. If the sheet of PVDF has $\epsilon_r = 11$, 20 cm^2 area, and is 12 μm thick, determine the voltage sensitivity of the sensor.

   b. Determine C to obtain a sensitivity of -250 μV/Pa at the output. Assume that R_1 is negligible.

   c. If the amplitude response at 10 Hz should not differ from that in the passband by more than 5%, determine the minimal value for R.

   d. Quantify the effect of R_1 in the frequency response.

7.8 A given piezoelectric accelerometer has a sensitivity of 1 pC/(m/s^2), 1 nF capacitance, and a very large leakage resistance. In order to measure accelerations from 0.1 Hz to 1 kHz, we connect it to the circuit in Figure P7.8, where op amps are assumed ideal. Determine the values for the components in the circuit in order to have an output sensitivity tunable

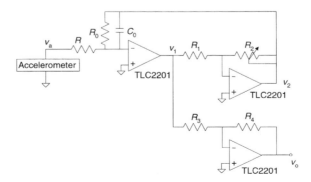

Figure P7.8   Charge amplifier with gain adjustable by a resistor.

between 1 mV/(m/s^2) and 100 mV/(m/s^2). The passband is determined by an external filter.

7.9 A given piezoelectric accelerometer is available whose sensitivity is 10 pC/g and whose internal impedance is 10 G$\Omega$$\|$$100$ pF. The cable available to connect it to a charge amplifier has a capacitance of 30 pF/m and an insulation resistance of 100 GΩ·m. The charge amplifier is based on an op amp whose open-loop gain is 120 dB at dc and 60 dB at 1 kHz, and it is placed 2 m from the sensor. The output voltage to obtain is 1 V/g, and the bandwidth is from 1 Hz to 1 kHz. Calculate the values for the feedback capacitor and resistance in the charge amplifier in order to obtain the desired response.

7.10 Figure P7.10 shows a charge amplifier that includes a capacitor C_1 to reduce the effect of bias currents and a series resistor R to limit damaging currents.

    a. If R, R_1, and C_1 are initially disregarded, determine C_0 to obtain an output sensitivity of -10 mV/(m/s^2) when the accelerometer has a sensitivity of 1 pC/(m/s^2).

    b. Calculate R, R_1, and C_1 to obtain a passband from 1 Hz to 10 kHz and to obtain an error from bias currents smaller than 10% of the dynamic range.

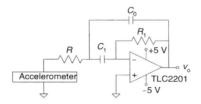

Figure P7.10   Modified charge amplifier with reduced error from op amp bias currents.

7.11 Determine the 95% confidence interval for the value of a 100 kΩ resistor-type CR25 after operating for 5 years with 0.25 W dissipation in an environment at $40\,°C$.

7.12 In order to measure a variable force in the band from 0.1 Hz to 1000 Hz, a load cell is available that consists of four strain gages undergoing opposite deformations, bonded on a material whose Young's modulus is 200 GPa, and arranged in a Wheatstone bridge. Their nominal value is $120\ \Omega$, the gage factor is 2, and the maximal current 25 mA. The bridge output is connected to a model AD 624 instrumentation amplifier. Calculate the limit posed by amplifier noise on the resolution in the measured mechanical stress.

7.13 The circuit in Figure P7.13 is a low-noise preamplifier used to observe the output signal of a pair of face-to-face biopotential electrodes on a

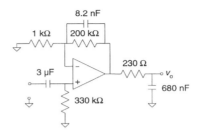

Figure P7.13   Ac low-noise preamplifier.

storage oscilloscope. Calculate the equivalent input peak-to-peak voltage noise when the resistance of the electrode pair is 1 kΩ and the op amp is the OP-07.

7.14   A given photodiode with sensitivity 0.43 μA/μW, stray capacitance 20 pF, and leakage current 100 MΩ is connected to a transimpedance amplifier such as that in Figure 7.10*b* with $R = 10$ MΩ and $C = 2$ pF. The op amp is the TLC2201B supplied between 0 V and 5 V, and its output is connected to a sharp filter with bandpass from 10 Hz to 10 kHz. Determine the output noise voltage.

REFERENCES

[1] R. Pallás-Areny and J. G. Webster. *Analog Signal Processing*. New York: John Wiley & Sons, 1999.

[2] J. Williams. *Application Considerations and New Circuits for a New Chopper-Stabilized Op Amp*. Application Note 9. Milpitas, CA: Linear Technology, 1991.

[3] J. Williams. *High Speed Amplifier Techniques*. Application Note 47. Milpitas, CA: Linear Technology, 1990.

[4] J. F. Keithley, J. R. Yeager, and R. J. Erdman. *Low Level Measurements: For Effective Low Current, Low Voltage, and High Impedance Measurements*, 4th ed. Cleveland, OH: Keithley Instruments, 1992.

[5] G. F. V. Vanderchmidt. Inexpensive, automatic range-switching electrometer. *Rev. Sci. Instrum.* **61**, 1990, 1988–1989.

[6] J. E. Rhodes. *Piezoelectric Transducer Calibration Simulation Method Using Series Voltage Insertion*. Technical Publication 216. San Juan Capistrano, CA: Endevco Corp.

[7] H. W. Ott. *Noise Reduction Techniques in Electronic Systems*, 2nd ed. New York: John Wiley & Sons, 1988.

[8] C. D. Motchenbacher and J. A. Connelly. *Low-Noise Electronic Design*. New York: John Wiley & Sons, 1993.

[9] J. Graeme. *Photodiode Amplifiers*. New York: McGraw-Hill, 1996.

[10] D. D. Dunlap. Resistors. Chapter 1 in: C. A. Harper (ed.), *Passive Electronic Component Handbook*. New York: McGraw-Hill, 1997.

8

DIGITAL AND INTELLIGENT SENSORS

The overwhelming presence of digital systems for information processing and display in measurement and control systems makes digital sensors very attractive. Because their output is directly in digital form, they require only very simple signal conditioning and are often less susceptible to electromagnetic interference than analog sensors.

We distinguish here three classes of digital sensors. The first yields a digital version of the measurand. This group includes position encoders. The second group relies on some physical oscillatory phenomenon that is later sensed by a conventional modulating or generating sensor. Sensors in this group are sometimes designated as *self-resonant, variable-frequency*, or *quasi-digital sensors*, and they require an electronic circuit (a digital counter) in order to yield the desired digital output signal. The third group of digital sensors use modulating sensors included in variable electronic oscillators. Because we can digitally measure the oscillation frequency, these sensors do not need any ADC either.

Digital computation has evolved from large mainframes to personal computers and microprocessors that offer powerful resources at low cost. Process control has similarly evolved from centralized to distributed control. Silicon technology has achieved circuit densities that permit the fabrication of sensors that integrate computation and communication capabilities, termed *intelligent* or *smart sensors*.

8.1 POSITION ENCODERS

Linear and angular position sensors are the only type of digital output sensors that are available in several different commercial models. Incremental encoders

433

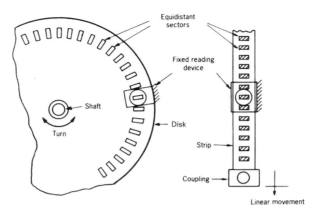

Figure 8.1 Principle of linear and rotary incremental position encoders. (From N. Norton, *Sensor and Analyzer Handbook*, copyright 1982, p. 105. Reprinted by permission of Prentice-Hall, Englewood Cliffs, NJ.)

are in fact quasi-digital, but we discuss them here because they are related. Each September issue of *Measurements & Control* lists the manufacturers and types of encoders.

8.1.1 Incremental Position Encoders

An incremental position encoder consists of a linear rule or a low-inertia disk driven by the part whose position is to be determined. That element includes two types of regions or sectors having a property that differentiates them, and these regions are arranged in a repetitive pattern (Figure 8.1). Sensing that property by a fixed head or reading device yields a definite output change when there is an increment in position equal to twice the pitch p. A disk with diameter d gives

$$m = \frac{\pi d}{2p} \qquad (8.1)$$

pulses for each turn.

This sensing method is simple and economic but has some shortcomings. First, the information about the position is lost whenever power fails, or just after switch-on, and also under strong interference. Second, in order to obtain a digital output compatible with input–output peripherals in a computer, an up–down counter is necessary. Third, they do not detect the movement direction unless additional elements are added to those in Figure 8.1. Physical properties used for sector differentiation can be magnetic, electric, or optic. The basic output is a pulse train with 50% duty cycle.

A toothed wheel or etched metal tape scale of ferromagnetic material yields

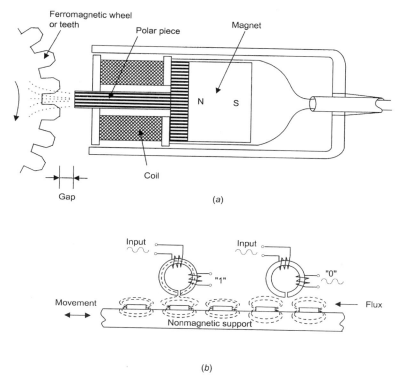

Figure 8.2 Different sensors for magnetic incremental position encoders. (*a*) Coil and magnet (courtesy of Orbit Controls). (*b*) Toroidal core.

a voltage impulse each time it passes by a fixed coil placed in a constant magnetic field, as shown in Figure 8.2*a*. The resulting signal is almost sinusoidal, but it is easy to convert it into a square wave or just to determine its zero crossings. There is a minimal and maximal velocity that determines the application range for this method, which is used, for example, in antilock braking systems (ABS) in cars. Alternatively, an AMR or GMR sensor (Section 2.5) can replace the coil to obtain a change in resistance whose amplitude does not depend on the turning speed.

Figure 8.2*b* shows another inductive system but this time based on a toroid with two windings. One winding is used for exciting, using currents between 20 kHz and 200 kHz, and the other is used for detection. There are two output states: "1" when no voltage is detected and "0" when a voltage with the exciting frequency is detected. The moving element includes regions with magnetized material. Each time one of these regions passes in front of the reading head, the core saturates because the flux emanating from the material adds to the flux created by the exciting signal. When the core saturates, the secondary winding does not detect any voltage ($e = -d\Phi/dt$, with Φ constant—saturation

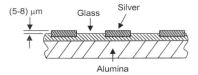

Figure 8.3 Silver-in-glass technology for contacting incremental position encoder (courtesy of Spectrol).

flux): state "1." When there is a region with no magnetization in front of the reading head, the secondary winding detects a voltage induced by the primary, state "0." A Hall effect sensor (Section 4.3.2), magnetoresistor (Section 2.5), or Wiegand sensor (Section 4.2.6) can replace the toroidal core. Inductive encoders are sensitive to stray magnetic fields.

Electric encoders can be capacitive or contacting encoders. Capacitive encoders use a patterned strip such as that in the Inductosyn™ (Section 4.2.4.3) but without shielding between the scale and the ruler. This yields a cyclic change in capacitance with a period equal to the pitch, which can be as low as 0.4 mm. The contacting encoder in Figure 8.3 has a moving element formed by an alumina substrate with a fused glass layer and a conductive palladium–silver pattern screen printed on top of the glass. When kiln-firing during the fabrication process, the conductive pattern sinks into the glass to yield a 5 µm to 8 µm step-height differential between the conductor and the insulator surface. The wiper is from precious metal. Former encoders with copper foil etched on printed circuit board had 25 µm to 35 µm steps, which increased wear and contact bouncing. This silver-in-glass technology offers low cost, ruggedness, high-resistance to corrosion, and life expectancy of up to fifteen million cycles, far above the 100,000 cycles of former PCB designs.

Optical encoders can be based on opaque and transparent regions, on reflective and nonreflective regions, and also on interference fringes. Whatever the case, the fixed reading head includes a light source (infrared LED), and a photodetector (phototransistor or photodiode). Main problems result from dust-particle buildup, time and temperature drifts for optoelectronic components, and vibration effects on focusing elements. High-performance sensors have either a lens or an aperture to provide collimated light output and minimal spurious reflections.

When opaque and transparent regions—chromium on glass, slotted metal, and so forth (Figure 8.4a)—are used, the emitter and the receiver must be placed on each side of the moving element. In contrast, when relying on reflective and nonreflective zones—for example, polished steel with an engraved surface (Figure 8.4b)—the emitter and the detector must be on the same side of the coding element. Glass disks are more stable, rigid, hard, and flat than metal disks, but are less resistant to vibration and shock. Reference 1 discusses the design of encoders based on integrated optical subsystems.

Interference fringe encoders are based on moiré patterns. To create them

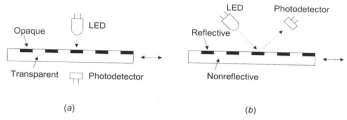

(a)   (b)

Figure 8.4 Incremental optical encoder. (*a*) With opaque and transparent sectors. (*b*) With reflective and nonreflective sectors.

from a linear movement, we can use a fixed and a movable rule having lines inclined with respect to each other (Figure 8.5). If the inclination α is such that $\tan \alpha = p/d$, a relative displacement p (line pitch) produces a vertical displacement d of a dark horizontal fringe. If the relative inclination is n times higher, n dark horizontal fringes appear. Interference fringes from a rotary movement can be obtained from two superimposed disks, one fixed and the other one movable, one with N radial lines and the other one with $N + 1$. Interference fringes are also obtained if both disks have N lines but one is off center, or one has N lines with a different inclination. A light emitter–detector pair yields an almost sinusoidal signal with N cycles/turn for a rotary encoder.

Incremental encoders have resolution ranges from 100 counts/turn to 81,000 counts/turn and accuracy up to 30″. When the detector offers two sinusoidal outputs with different phase shifts, interpolation using various phase-shifting methods can increase the resolution by a factor of up to 256. One interpolation method [2] digitally measures the phase from the quadrature outputs.

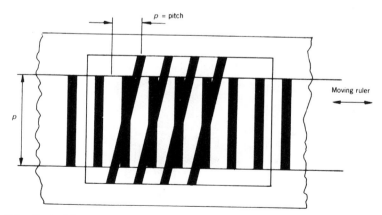

Figure 8.5 Optical incremental encoder based on interference fringes (moiré patterns). The horizontal dark fringe moves in the vertical direction when the sliding rule moves horizontally.

$$V_a = V_p \cos N\theta \tag{8.2}$$

$$V_a = V_p \sin N\theta \tag{8.3}$$

where V_p is the amplitude of the output voltage, N is the number of steps (pitches) in one turn, and θ is the current shaft angle, which can be calculated from

$$\theta = \frac{\arctan(V_b/V_a)}{N} \tag{8.4}$$

This is a periodic quantity that gives 2π rad each $360°/N$ angle increment, so that we also need an incremental counter to determine the actual angle. To measure the phase, the 2π rad phase plane is divided in several sectors and the sector corresponding to each V_a, V_b pair is stored in a ROM. V_a and V_b are each digitized by an ADC, and the system looks in the ROM for the corresponding phase. Reference 3 describes a similar approach using software techniques.

Usual diameters for encoder disks are from 25 mm to 90 mm. Linear incremental encoders can measure position with resolution of up to 0.5 μm/period and accuracy up to 50 μm. They are used for position control in general—for example, reading/writing head positioning in disk and magnetic tape drives, paper positioning in printers, copiers, and fax machines, tool positioning in automatic machines, and dimensional metrology. Small rotational units replace control potentiometers in instrument panels because of their longer life.

Optical encoders yield the highest resolution. The limiting factor is the photodetector size. Resolution increases by using one or several fixed grids or masks with opaque and transparent regions, placed between the movable element and the detector, and having the same pitch as the encoded element (Figure 8.6). The detector receives the maximal light when all the grids and the

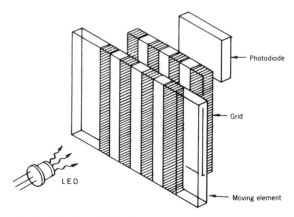

Figure 8.6   Use of a fixed grid to limit the field for a photodetector, thus increasing its resolution (courtesy of TRW Electronics Components Group).

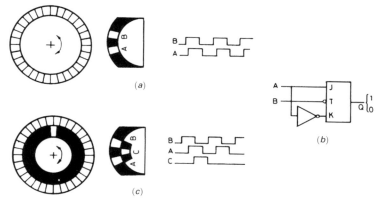

Figure 8.7 Detection of movement direction in incremental encoders. (*a*) By means of two outputs with 90° phase shift. (*b*) Output electronic circuit. (*c*) Additional marker for absolute positioning.

movable encoded element are perfectly aligned. As the encoded element moves from that position, the light received will decrease until reaching a minimum. The photodetector averages the signal from more than one slot, thus compensating for any possible differences between them. The output is a continuous (not discrete) signal between maxima that can be interpolated.

In order to determine the movement direction, we require another reading element, and sometimes another encoded element, together with some appropriate electronic circuits. In inductive encoders, another sensing coil is placed to obtain a 90° out-of-phase signal—quadrature encoding (Figure 8.7*a*). In one turning direction, signal A leads signal B, whereas in the opposite direction, signal B leads signal A. Then a phase detector indicates whether the turning direction is clockwise (CW) or counterclockwise (CCW) (Figure 8.7*b*). In optical and contacting (electric) encoders, another encoded track having a small phase shift with respect to the first one is added, with its corresponding read head. In interference fringe encoders and in high-resolution optical encoders, two optical units are used, which yield two signals with a 90° relative phase shift. Some encoders even use two additional units at 180° with respect to the other two, in order to further increase the resolution.

Pulse multiplication increases the disk resolution by two or four times. An EX-OR gate fed by two out-of-phase signals duplicates the number of pulses (Figure 8.8*a*). Differentiating a single signal yields an impulse for each rising or falling edge, hence duplicating the number of pulses if those impulses are further rectified and stretched (Figure 8.8*b*). Differentiating two pulse channels improves the resolution by four.

In order to detect the absolute position of the movable part, we need an up–down counter fed by pulses from the detector. The direction of counting is determined by the signal that gives the movement direction, and resetting is done

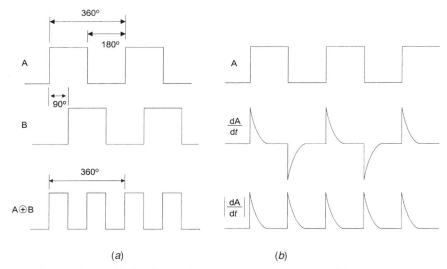

(a)   *(b)*

Figure 8.8 Pulse multiplication to increase encoder resolution. (*a*) From two channel outputs through an EX-OR. (*b*) By pulse differentiation and rectification.

through a third encoder output signal of one pulse each turn (when it is a rotary encoder), termed marker or zero index (Figure 8.7*c*), which also defines a home position. Alternatively, one output controls the direction of counting while the other output is counted (Figure 8.9). Also, the three output signals can be interfaced to I/O lines of a microprocessor or microcontroller, which should poll them at a rate faster than the maximal rate the outputs are expected to change.

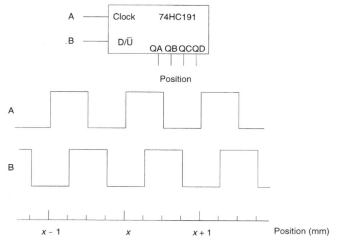

Figure 8.9 An up–down counter offers absolute positioning from an incremental encoder with two out-of-phase outputs.

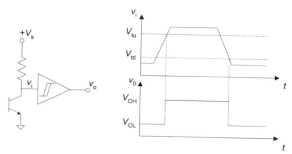

Figure 8.10 Hysteresis in Schmitt triggers avoids erratic transitions between logic levels when input signals do not have steep edges.

When the aim is to measure a rotating speed, incremental encoders are limited by the maximal frequency accepted by electronic circuits if the maximal rotating speed is very high. Digital tachometers based on the same principles yet having only one track or a few tracks yield a lower number of pulses at each turn.

Encoder signals are often open collector and processed by a Schmitt trigger before feeding them to further digital circuits. Schmitt triggers have a lower threshold trigger level when they are in the on state than when they are in the off state (Figure 8.10). This hysteresis minimizes any electric noise switching that could otherwise yield erratic transitions when the output level is close to the threshold of an ordinary digital IC.

Output signals from contacting encoders may display contact bouncing (erratic transition from low to high level with the wiper is at a transition zone). Debouncing circuits, such as the MC14490, introduce an appropriate delay, say 5 ms, so that the output signal does not reflect any transition at its input within that delay period.

8.1.2 Absolute Position Encoders

Absolute position encoders yield a unique digital output corresponding to each resolvable position of a movable element, rule, or disk, with respect to an internal reference. The movable element is formed by regions having a distinguishing property, and designated with the binary values 0 or 1. But unlike incremental encoders, their tracks are so arranged that the reading system directly yields the coded number corresponding to each position (Figure 8.11). Each track corresponds to an output bit, with the innermost track yielding the most significant bit. The most common sensors for these encoders are optical sensors, which are far ahead of contact sensors. Integrated photosensor arrays such as the TSL214 simplify system design.

Absolute encoders have intrinsic immunity to interruptions and electromagnetic interference. They do not accumulate errors. Feedback systems relying on them do not require a homing sequence at start up, or when power fails, or at

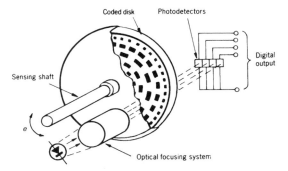

Figure 8.11 Principle of absolute position encoders for linear and rotary movements.

routine intervals during operation. The price to be paid is a more complex reading head as compared to incremental encoders. This is due to the need for as many reading elements as encoded tracks. Also they must be perfectly aligned; otherwise the output code may be ambiguous when changing from one position to a contiguous position. The resulting code can then be very different from the one for the actual position. If, for example, the natural binary code is used, in an 8 bit system the positions 3 and 4 are given by

$$\begin{array}{llllllllll}
\text{Position 3:} & 0 & 0 & 0 & 0 & 0 & 0 & 1 & 1 \\
\text{Position 4:} & 0 & 0 & 0 & 0 & 0 & 1 & 0 & 0
\end{array}$$

Suppose that the reading heads are slightly misaligned—for example, the first two heads are slightly advanced. Then the output reading when moving from position 3 to position 4 would be 0000 0000.

Binary codes with unit distance in all positions including the first one and the last one (cyclic continuous codes) are unambiguous—that is, codes where two contiguous positions differ only in one bit. In the natural binary code there are for N positions $N/2$ transitions where more than one bit changes.

Table 8.1 shows the weight for each bit and the pattern of the coded regions corresponding to different codes. The Gray code is the commonest continuous code and has the same resolution as the natural binary code. A shortcoming is that if the output information is to be sent to a computer, it must be first converted to binary code. The algorithm to obtain the i binary bit is to add modulo-2 (EX-OR function) the $i + 1$ binary bit to the i Gray bit. The most significant binary bit equals the most significant bit in Gray code. That is,

$$\begin{aligned}
B_i &= B_{i+1} \oplus G_i, && 0 \leq i < n \\
B_n &= G_n
\end{aligned} \tag{8.5}$$

Figure 8.12 shows the corresponding circuit. The Gray code does not permit error correction—for example, when transmitting signals in a noisy environ-

TABLE 8.1   Common Codes in Absolute Position Encoders

Decimal number	Natural binary Code (32 16 8 4 2 1)	NB Pattern	BCD Code Tens (8 4 2 1)	BCD Code Units (8 4 2 1)	BCD Pattern Tens	BCD Pattern Units	Gray Code (31 15 7 3 1)	Gray Pattern
0	0 0 0 0 0 0		0 0 0 0	0 0 0 0			0 0 0 0 0	
1	0 0 0 0 0 1		0 0 0 0	0 0 0 1			0 0 0 0 1	
2	0 0 0 0 1 0		0 0 0 0	0 0 1 0			0 0 0 1 1	
3	0 0 0 0 1 1		0 0 0 0	0 0 1 1			0 0 0 1 0	
4	0 0 0 1 0 0		0 0 0 0	0 1 0 0			0 0 1 1 0	
5	0 0 0 1 0 1		0 0 0 0	0 1 0 1			0 0 1 1 1	
6	0 0 0 1 1 0		0 0 0 0	0 1 1 0			0 0 1 0 1	
7	0 0 0 1 1 1		0 0 0 0	0 1 1 1			0 0 1 0 0	
8	0 0 1 0 0 0		0 0 0 0	1 0 0 0			0 1 1 0 0	
9	0 0 1 0 0 1		0 0 0 0	1 0 0 1			0 1 1 0 1	
10	0 0 1 0 1 0		0 0 0 1	0 0 0 0			0 1 1 1 1	
11	0 0 1 0 1 1		0 0 0 1	0 0 0 1			0 1 1 1 0	
12	0 0 1 1 0 0		0 0 0 1	0 0 1 0			0 1 0 1 0	
13	0 0 1 1 0 1		0 0 0 1	0 0 1 1			0 1 0 1 1	
14	0 0 1 1 1 0		0 0 0 1	0 1 0 0			0 1 0 0 1	
15	0 0 1 1 1 1		0 0 0 1	0 1 0 1			1 1 0 0 1	
16	0 1 0 0 0 0		0 0 0 1	0 1 1 0			0 1 0 0 0	
17	0 1 0 0 0 1		0 0 0 1	0 1 1 1			1 1 0 0 0	
18	0 1 0 0 1 0		0 0 0 1	1 0 0 0			1 1 0 0 1	
19	0 1 0 0 1 1		0 0 0 1	1 0 0 1			1 1 0 1 1	
20	0 1 0 1 0 0		0 0 1 0	0 0 0 0			1 1 1 1 0	
21	0 1 0 1 0 1		0 0 1 0	0 0 0 1			1 1 1 1 1	
22	0 1 0 1 1 0		0 0 1 0	0 0 1 0			1 1 1 0 1	
23	0 1 0 1 1 1		0 0 1 0	0 0 1 1			1 1 1 0 0	
24	0 1 1 0 0 0		0 0 1 0	0 1 0 0			1 0 1 0 0	
25	0 1 1 0 0 1		0 0 1 0	0 1 0 1			1 0 1 0 1	

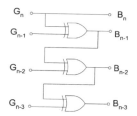

Figure 8.12 Gray-to-binary code converter.

ment. Reference 4 proposes a cyclic code that permits detection of single bit errors, with the exception of those contiguous to the actual position.

If the objective for the measurement is only to numerically display the position, a conversion to BCD code is required. Disks directly coded in the code finally used do not require any conversion, but have the ambiguity problem.

Another method to solve the ambiguity problem is to use a double set of reading heads displaced by a given distance between them, and then to apply some decision rule in order to accept only the reading from one of the two sensors for each track. Also, we can place a marker in the center of each track, then accept the read head signal only when it signals a nontransition zone between positions. A memory stores the last position read, and it is updated when there is a valid change.

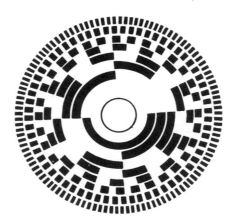

Figure 8.13 Digital encoder disk with an added track (along the external perimeter) to increase the resolution using a system with a fixed grid.

The resolution of absolute encoders is from 6 bits to 21 bits in Gray code (8 bits to 12 bits is common), with diameters from 50 mm to 175 mm for rotary encoders and accuracy up to 20″. The diameter is designed by a number that is ten times the diameter in inches. For example, size 40 means a diameter of 4 inches. The resolution improves by increasing the number of encoded tracks, but the resulting increase in diameter and inertia limits this solution. An alternative is to use a gear and another encoder, but the final resolution remains always limited by that of the first disk. There are multiturn models up to 26 bits. The electric output signal is usually in the form of open collector TTL.

Another method to increase resolution is to add a fixed grid to produce sinusoidal outputs and then interpolate as discussed for incremental encoders (Figure 8.6). Thus the movable disk includes an additional radial track along the disk periphery, as shown in the model in Figure 8.13.

Reference 5 describes a more efficient absolute encoder that needs a single encoded track, which for a disk is placed along the periphery (not radially as in Figure 8.13). The reading head is placed at a distance that depends on the pitch and the desired resolution. For a 0.1 mm pitch (using photodetectors), in order to achieve a 10 bit resolution we need a circumference of 102.4 mm, hence a radius of 16.3 mm. The track code is a pseudorandom binary sequence (PRBS), where only one bit changes from any code to the next. A PRBS with $2^n - 1$ terms can be generated by an n bit register with an appropriate modulo-2 feedback. In Figure 8.14, for example, the 4 bit shift register with the feedback equation $R(5) = R(1) \oplus R(2)$ generates the 15-term sequence 000100110101111—written from $R(0)$ to $R(14)$. Any 4 bit window sliding over this sequence is unique. Figure 8.14 shows the code corresponding to the positions $x = 0$ and $x = 7$. Computer interfacing requires code conversion.

Common applications for absolute position encoders are high-resolution

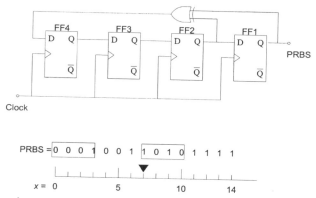

Figure 8.14 A 4 bit shift register with feedback generates a 15 bit pseudorandom binary sequence in which a 4 bit sequence uniquely identifies the absolute position.

measurement and control of linear and angular positions. They suit applications involving slow movements or where the movable element remains inactive for long time periods, such as parabolic antennas. They are used, for example, in robotics, plotters, machine tools, read head positioning in magnetic storage disks, radiation source positioning in radiotherapy, radar, telescope orientation, overhead cranes, and valve control. They can also sense any quantity that we convert to a displacement by means of an appropriate primary sensor—for example, in liquid level measurements using a float.

8.2 RESONANT SENSORS

Sensors based on a resonant physical phenomenon yield an output frequency that depends on a measurand affecting the oscillation frequency. They require a frequency-counter in order to measure either the frequency or the oscillation period based on an accurate and stable clock. The choice of method depends on the desired resolution and also on the available time for the measurement (see Problem 8.1). Resonant structures of single-crystal silicon are particularly suited to IC integration [6, 7].

Sensors use either harmonic oscillators or relaxation oscillators. Harmonic oscillators store energy that changes from one form of storage to another, for example from kinetic energy in a moving mass to potential energy in a stressed spring. In relaxation oscillators there is a single energy storage form, and the stored energy is periodically dissipated through some reset mechanism.

Quartz clocks are accurate enough to derive a time base for most sensor applications, but they drift with time and temperature. Time drifts arise from structural changes due to defects in crystal lattices, mechanical stress from supporting elements (that decrease as time passes and change after thermal

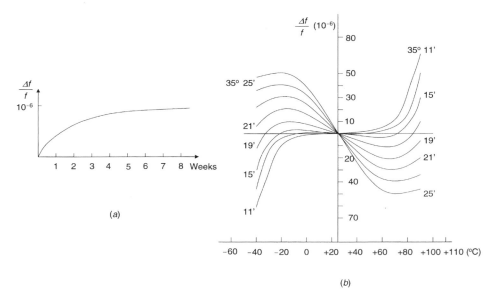

(a)

(b)

Figure 8.15   (a) A quartz crystal oscillator achieves long-term stability a few weeks after turn-on. (b) The temperature stability of quartz oscillators is nonlinear and depends on the angle at which the crystal is cut with respect to the z-axis.

cycling), and mass changes because of absorption and desorption of contaminant gases inside the crystal package. Aging curves describing $\Delta f/f$ are first exponential but become stable after a few weeks or months (Figure 8.15a). Accurate crystals are pre-aged at the factory before shipping.

Thermal drifts are the basis of quartz thermometers (Section 8.2.1) and have the form of a recumbent "S." Their value depends on the particular crystal cut considered. Figure 8.15b shows a family of temperature stability curves for a 10 MHz fundamental quartz crystal with the angle of cut with respect to the z-axis as parameter; the z-axis is the optical axis, which permits light to pass readily. For quartz clocks, the 35°13' cut yields the best stability over a short temperature range about ambient temperature. The 35°15' cut (the AT cut, preferred above 1 MHz) has a better stability from 0 °C to 50 °C. Oven-controlled oscillators eliminate thermal drift by maintaining the crystal at constant temperature, usually at the upper turnover point in Figure 8.15b (about 70 °C for the AT cut). This yields a frequency stability $\Delta f/f$ about 10^{-8} in the range from 0 °C to 50 °C, as compared to a drift higher than 2.5×10^{-6} for a common crystal in the same range (an RTXO—room temperature crystal oscillator). Temperature-compensated crystal oscillators (TCXOs) incorporate a temperature-dependent network with characteristics approximately equal to and opposite from that of the crystal. They achieve frequency stability better than 10^{-6} in the range from −20 °C to +70 °C.

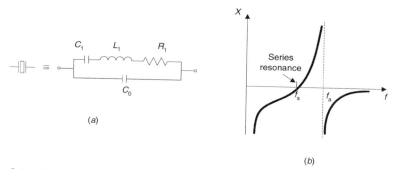

Figure 8.16 (*a*) High-frequency equivalent circuit for a piezoelectric material such as quartz with metal electrodes deposited on two faces. (*b*) The reactance of a quartz crystal varies with the operating frequency near resonance.

8.2.1 Sensors Based on Quartz Resonators

Quartz is piezoelectric and therefore an applied voltage stresses the crystal (Section 6.2). If the voltage alternates at a proper rate, the crystal begins vibrating and yields a steady signal. The equivalent circuit in Figure 8.16*a* then replaces that in Figure 6.14*b*. L_1 is associated with the mass of the crystal, C_1 with its elasticity or mechanical compliance, and R_1 with its internal friction (resulting in heat dissipation) when oscillating. C_0 is the electrostatic capacitance of the crystal between the electrodes plus the holder and the leads. For a crystal used in a 32.768 kHz oscillator, for example, $L_1 = 4451$ H, $C_1 = 5$ fF, $R_1 = 11.2$ kΩ, and $C_0 = 1.84$ pF.

The presence of a resonant circuit in Figure 8.16*a* permits the crystal to be used in an oscillator. At series resonance

$$f_s = \frac{1}{2\pi\sqrt{L_1 C_1}} \tag{8.6}$$

the crystal's reactive elements cancel and provide an effective impedance consisting only of R_1 (Figure 8.16*b*). As the frequency increases, the crystal behaves as a positive reactance in series with a resistance. At the antiresonant-frequency f_a, the crystal's reactance is maximal. The range from f_s to f_a is referred to as the *crystal's bandwidth*.

The series resonant circuit (Figure 8.17*a*) operates above f_s, where the reactance is slightly inductive. Series capacitance is then added to tune the circuit. Figure 8.17*b* shows a basic oscillator based on CMOS inverters. C_{L1} and C_{L2} are loading capacitances (larger for low oscillating frequency than for high oscillating frequency), $R_f = 1$ MΩ, and R_d is a damping resistor that depends on the gate type and oscillation frequency.

Because quartz is inert, using a high-purity single crystal yields a mechanical resonance with large long-term stability. Short-term stability depends on the

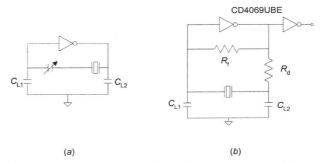

(a)   (b)

Figure 8.17   (*a*) Series resonator oscillator based on a quartz crystal. (*b*) Crystal (series) oscillator based on CMOS inverters.

quality factor Q (stiffness and low hysteresis) and equivalent inductance, which are very large. Short-term stability permits the design of high-resolution sensors. Long-term stability implies a longer time interval between calibrations.

8.2.1.1   *Digital Quartz Thermometers.* The values for the elements in the equivalent circuit for a quartz crystal depend on the temperature, and therefore the oscillation frequency displays a thermal drift. If precision-cut quartz crystals are used, the relationship between temperature and frequency is very stable and repeatable. Then, from the measurement of the oscillation frequency we can infer the temperature of the element. The general equation is

$$f = f_0[1 + a(T - T_0) + b(T - T_0)^2 + c(T - T_0)^3] \qquad (8.7)$$

where T_0 is an arbitrary reference temperature (usually 25 °C), and f_0, a, b, and c depend on the cutting orientation. Ideally we would seek $b = c = 0$, but this is not easy. An alternative is to seek high sensitivity and repeatability instead of linearity and obtain T from $f - f_0$ by a look-up table (Figure 8.18). Some temperature probes using this principle include the electronic circuitry to output a pulse frequency signal enabling remote sensing with low interference susceptibility as compared to systems with analog voltage output.

Oscillator frequencies range from about 256 kHz to 28 MHz with temperature coefficients (a) from $19 \times 10^{-6}/°C$ to $90 \times 10^{-6}/°C$. Sensitivities are up to

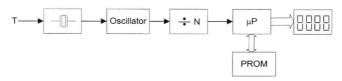

Figure 8.18   Simplified diagram for a quartz digital thermometer.

about 1000 Hz/°C in the range from $-50\,°C$ to $150\,°C$. The resolution can be as high as $0.0001\,°C$, but the better the resolution, the slower the measurement speed (see Problem 8.2). Some probes reach $-40\,°C$ to $300\,°C$, but with reduced linearity unless corrected by a look-up table. Low-mass probes can be applied to infrared radiation intensity measurement.

8.2.1.2   Quartz Microbalances. The oscillation frequency of a crystal resonator decreases when the crystal mass increases. If the initial oscillation frequency is f_0, a deposition of a small mass Δm on a crystal with surface area A and density ρ yields an approximate frequency shift given by the Sauerbrey equation [8]

$$\Delta f \approx -f_0^2 \frac{\Delta m / A}{N \rho} \tag{8.8}$$

where N is a constant and it is assumed that the mass added does not experience any shear deformation during oscillation—that is, it is rigid. For an AT-cut crystal, for example, $\Delta f \approx -2.3 \times 10^{-6} f_0^2 (\Delta m / A)$. Disks with 10 to 15 mm diameter and 0.1 to 0.2 mm thickness yield resonant frequencies from 5 MHz to 20 MHz and have a sensitivity of about 189 ng/(cm²·Hz). Including a resonant temperature sensor compensates temperature interference.

This sensing method is applied to humidity measurement by covering the crystal with a hygroscopic material exposed to the environment where humidity is to be measured. Water absorbed increases the mass and reduces the crystal oscillation frequency [9]. Crystals coated with specific organic nonvolatile materials instead of a hygroscopic material can detect specific volatile compounds in a gas phase with resolution of nanograms per square centimeter [10].

Quartz crystals oscillators are widely used as thin-film thickness monitors to control deposition rates and in situ measurement of coating film thickness in the semiconductor and optical industries. Equation (8.8) is not accurate enough for these monitors. They rely instead on a sensor function that considers the influence of the different acoustic impedances of the deposition materials upon the resonant frequency. The mass sensitivity of these *Z-matching devices* depends on the material deposited [8].

8.2.1.3   Quartz Resonators for Force and Pressure Sensing. Quartz crystals, the same as other single-crystal materials, have highly stable elastic properties with very low creep and hysteresis. Hence, they suit mechanical resonators whose resonance frequency depends on the applied stress. For a string-type resonator, for example, the natural mechanical resonant frequency is [6, 8]

$$f_n = \frac{n}{2l} \sqrt{\frac{T}{\rho}} \tag{8.9}$$

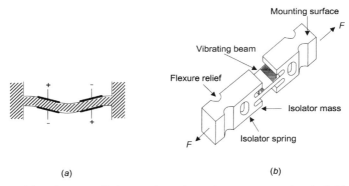

(a)   (b)

Figure 8.19   (*a*) Voltage applied to surface electrodes creates an electric field that flexes a quartz beam. (*b*) Force sensor based on a quartz vibrating beam with electrodes on both surfaces.

where n is the harmonic number considered, l is the length of the "string" (e.g., a long slender quartz beam), T is the stress applied, and ρ is the density of the crystal material. Quartz has the added advantage of being piezoelectric, so that the vibration can be excited by a driving alternating voltage and the oscillation frequency is that of the voltage detected by electrodes deposited on the crystal (Figure 8.19*a*).

The high stiffness of quartz makes it suitable for force, torque, and pressure measurement. A matched crystal placed nearby but not subjected to mechanical stress yields a signal to compensate temperature interference. There is a variety of mechanical structures obtained by micromachining able to sense those measurands. Figure 8.19*b* shows a sensor for tensile and compressive force based on a single quartz beam. Other models have two or three beams. Load cells use a push rod to transmit the input force to the quartz sensor through a lever mechanism. Pressure sensors rely on either the force exerted by a primary sensor (diaphragm, bellows) or the changing resonant frequency because of the pressure directly applied to a quartz diaphragm. Using two diaphragms but exposing only one to the pressure to measure permits differential measurements to compensate for temperature and acceleration interference.

8.2.1.4 Quartz Angular Rate Sensors. A vibrating quartz tuning fork can sense angular velocity because of the Coriolis effect. The sensor typically consists of a double-ended quartz tuning fork micromachined from a single quartz crystal (Figure 8.20) rotating at the angular velocity to measure, Ω [11]. An oscillator at precise amplitude excites the drive tines so that they move toward and away from one another at a high frequency. Because of the Coriolis effect (Section 1.7.7) there is a force acting on each tine,

$$\boldsymbol{F} = 2m\boldsymbol{\Omega} \times \boldsymbol{v}_{\mathrm{r}} \tag{8.10}$$

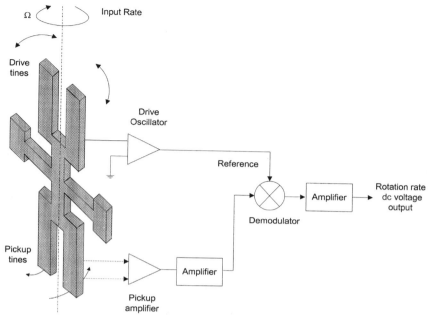

Figure 8.20 Angular rate sensor based on the Coriolis effect on a micromachined quartz tuning fork (courtesy of BEI Sensors and Systems).

where m is the tine mass and v_r is its instantaneous radial velocity. F is perpendicular to both Ω and v_r, hence to the plane of the fork assembly. Because the tines move in opposite directions, the resultant forces also have opposite directions. This produces a torque proportional to Ω. Because v_r is sinusoidal, the torque is also sinusoidal at the same frequency of the drive tines. The pickup tines respond to the oscillating torque by moving in and out of the plane, producing a signal that can be amplified and demodulated to yield a dc voltage proportional to the rate of rotation Ω.

Quartz angular rate sensors replace spinning-wheel gyroscopes because of their lower cost, increased reliability (there are no moving mechanical parts), and light weight. The GyroChip™, for example, weighs less than 60 g, has a sensitivity of 2.5 mV/(°/s) to 50 mV/(°/s), and has an operating life exceeding 5 years. It has been used to control angular velocity in aircraft, robots, and hydraulic equipment, to instrument automobile motions during crash tests, to evaluate rider quality in high-speed trains, to navigate autonomous underwater vehicles, to stabilize infrared cameras on helicopters, and in other applications.

8.2.2 SAW Sensors

A perturbation produces waves on the surface of a liquid, as we all know from our experience on ponds. Similar waves also travel in the surface of a solid.

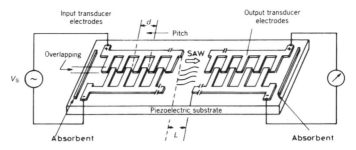

Figure 8.21   Principle of surface acoustic wave (SAW) filters.

Lord Rayleigh analyzed these waves in 1885 and applied the results to seismographic interpretation. In spite of the differences between solid and liquid waves, in both cases they attenuate with depth.

A method to produce a surface perturbation, certainly less convulsive than earthquakes, is to place two interleaved metallic electrodes (e.g., aluminum) on the surface of a piezoelectric material as shown in Figure 8.21. For a distance d between electrodes, a voltage of frequency f applied to the electrodes produces a surface deformation that propagates in both directions as a surface wave with velocity v, depending on the material, provided that $v = 2fd$. A similar electrode pair yields an alternating output voltage when the deformation wave arrives at it. These devices, designated by the acronym SAW (surface acoustic wave), were patented by J. H. Rowen and E. K. Sittig in 1963 and are extensively used in filters and oscillators above 100 MHz.

The velocity v for the surface wave depends on the deformation state for the surface and also on the temperature because they both influence the density and elastic properties for the material, in addition to altering the distance between electrodes. This is the principle for the use of these devices in sensors according to a direct action on the surface or on a particular coating.

SAW sensors are typically constructed as delay lines and placed in the feedback loop of an amplifier, resulting in an oscillator whose oscillation frequency depends on the surface deformation (Figure 8.22). The total phase shift in the feedback loop is

$$\phi_T = \phi_0 \pm \delta\phi_0 \pm \phi_{ex} \tag{8.11}$$

where $\phi_0 = 2\pi fL/v$ is the phase shift due to the wave transit time from one electrode pair to the other; $\delta\phi_0$ is the phase increment due to the substrate deformation and temperature change, if any; and ϕ_{ex} is the phase shift due to the amplifier and to the external impedance matching network. The system oscillates when $\phi_T = 2n\pi$ and the amplifier gain exceeds the total loss in the system. The oscillation frequency conveys information about the measurand.

An alternative is to measure the delay time in a delay line such as that in

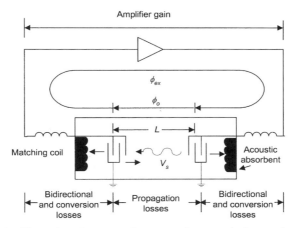

Figure 8.22 Oscillator based on a surface acoustic wave device used as a delay line.

Figure 8.22 with an emitter and a receiver. Any change located in the propagation zone (L) that affects either the velocity v or the length will be detected. The emitter sends a wave packet that propagates on the surface with a velocity that depends on the boundary conditions.

The preferred piezoelectric materials for SAW sensors are quartz and $LiNdO_3$. Interference from temperature or other quantities is cancelled by placing another electrode pair nearby in a region where the measurand does not produce any stress but the interfering quantity does. This yields a reference oscillator whose frequency depends on the interfering quantity in the same way as the measuring oscillator.

SAW sensors are applied to measure temperature, force, torque, pressure, acceleration, and gas concentration by mass adsorption (chemosensors) [10]. Gas flow has been measured by detecting its cooling effect on a SAW delay line oscillator. Detectors for CO, HCl, H_2, H_2S, NH_3, NO_2, SO_2, hydrocarbons, and organophosphorous compounds rely on a selective binding agent into a film deposited on the crystal surface. The coating can be optimized for sensitivity (up to femtograms), selectivity, or fast response. An uncoated SAW sensor working on the selective condensation of vapors depending on the surface temperature can detect the equivalent to 500 chemical species [12].

SAW sensors are very small, because v is of the order of 300 m/s to 4000 m/s and d (spatial periodicity) is about 1 μm. They are simple and manufactured by photolithographic methods compatible with planar technologies, hence relatively inexpensive.

8.2.3 Vibrating Wire Strain Gages

According to (8.9), the lower transverse oscillation frequency for a vibrating taut string or wire of length l is

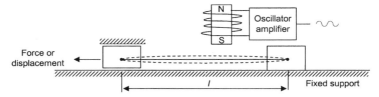

Figure 8.23   Vibrating wire gage. The transverse vibration is excited by a current pulse applied to the coil, which then is used to detect the vibration frequency.

$$f = \frac{1}{2l}\sqrt{\frac{F}{\mu}} \tag{8.12}$$

where F is the mechanical force applied to it and μ is the longitudinal mass density (mass/length). If the position of one of the ends changes because it is mounted on a movable support, then the oscillation period is directly proportional to the displacement. If a force is applied, the resulting oscillation frequency is directly proportional. For strain measurement, equation (8.12) yields

$$\varepsilon = \frac{4l^2\mu}{EA}f^2 \tag{8.13}$$

where E is Young's modulus and A is the wire's cross section.

The oscillation frequency is measured with a variable reluctance sensor (Section 4.2.1) and is in the audible range. Hence it is also called an *acoustic gage*. Usually a self-oscillating system is arranged where the detected signal is amplified and fed back to an electromagnetic driver. Some units use the driver alternately as the detector (Figure 8.23). In order for the oscillation frequency to be independent of the driver electric characteristics, the quality factor for the mechanical resonator must be at least 1000 or higher. It is convenient to use a thin wire, which is enclosed in a sealed chamber to avoid chemical corrosion and dust deposition, which would change its mass.

This principle can sense any physical quantity resulting in a change in l, F, or m. A common application is strain or tension measurement [13]. In contrast with resistive strain gages, vibrating-wire strain gages can detect nonplane deformation. In addition, they are insensitive to resistance changes in connecting wires—for example, those due to temperature. Temperature interferes because it affects the sensing wire length l. In order to compensate for temperature, we can measure the change in resistance of the driving coil wire, as in RTDs (see Problem 8.3).

Other reported applications are the measurement of mass, displacement, pressure (using a diaphragm with an attached magnet as primary sensor), force and weight (using a cantilever as primary sensor).

Vibrating strips are a variation on this same measurement principle. Their lower natural longitudinal oscillation frequency is

$$f = \frac{1}{2l}\sqrt{\frac{E}{\rho}} \tag{8.14}$$

where l is the length, E is Young's modulus, and ρ is the density. Vibrating strips are used for dust deposition measurement of exhaust gases and also to measure viscosity.

8.2.4   Vibrating Cylinder Sensors

If instead of a vibrating wire or strip we use a thin (75 μm)-walled cylinder with a closed end, the oscillation frequency depends on the dimensions and material for the cylinder and on any mass vibrating together with its walls. By using an electromagnetic driver as in the previous case in order to keep the system oscillating, it is possible to measure the difference in pressure between both cylinder sides because it results in mechanical stresses in its walls. We can also apply this system to gas density measurement because the gas near the walls vibrates when the walls do. For corrosive liquids it is better to use a glass or ceramic cylinder and a piezoelectric driver, thus avoiding corrosive-prone elements in electromagnetic drivers.

The most common application for this measurement principle is the measurement of the density of flowing liquids, using an arrangement like that in Figure 8.24. It consists of two parallel conduits through which the liquid flows; the two tubes are clamped at their ends and coupled to the main conduit by a flexible joint. Because the volume is known and the oscillation frequency for

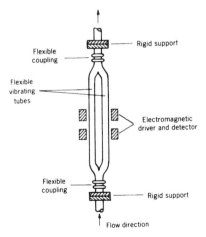

Figure 8.24   Vibrating tube method to measure liquid density.

both conduits, which behave as a tuning fork, depends on the mass, the frequency depends on the density in the form [6]

$$f = \frac{f_0}{\sqrt{1 + \dfrac{\rho}{\rho_0}}} \tag{8.15}$$

where f_0 is the conduit oscillation frequency when there is no liquid, and ρ_0 is a constant that depends on system geometry. The output frequency can be measured, for example, with a PLL whose voltage-controlled oscillator (VCO) drives the vibrating tube.

8.2.5 Digital Flowmeters

8.2.5.1 Vortex Shedding Flowmeters. The detection of oscillations in a flowing fluid allows us to obtain a variable frequency signal, which depends on fluid velocity. Those oscillations can be natural or forced.

The method of forced oscillations is mostly used for gases. It consists of placing inside the pipe a grooved conduit so that the outcoming flow is helical and has its maximal velocity (and minimal pressure according to Bernoulli's theorem, Section 1.7.3) at a point that shifts back and forth. The frequency at which this low-pressure point passes by a fixed detector is proportional to fluid velocity, and therefore to volumetric flow. Fluctuations associated with the shifting point are sensed by a piezoelectric pressure sensor or a thermistor (for temperature changes). Signal frequency ranges from 10 Hz to 1000 Hz.

For liquids it is more common to place inside the conduit a blunt (non-aerodynamic) object (vortex shedder), which is nonaligned with the flow lines (Figure 8.25). This also works for gas and steam. When the fluid layer in contact with the object separates from it, Karman vortices are shed from the object, the same as a flagpole creates vortices in the blowing wind, which wave the flag. Those vortices are usually detected by ultrasound whose intensity undergoes a

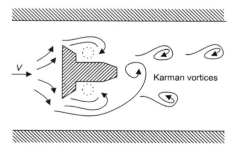

Figure 8.25 A blunt object inside a fluid stream produces downstream vortices whose frequency is proportional to flow velocity.

variable attenuation, and also by temperature fluctuations or by the drag force on the blunt object. Shaping the blunt object with a particular profile yields a frequency for the vortices proportional to the average flow velocity [14]. However, there is always a minimal velocity below which vortex frequency is irregular. Also, the larger the pipe diameter, the lower the output frequency, which limits the diameter to about 350 mm.

This method is fairly accurate (about 0.5%) and independent of fluid viscosity, density, pressure, and temperature. It is particularly indicated for flow measurements at high temperature and high pressure. Its main shortcomings are that it introduces a large drop in pressure and that it is unsuitable for dirty, abrasive, or corroding fluids.

8.2.5.2 Coriolis Effect Mass Flowmeters. A common type of mass flowmeter relies on the Coriolis effect on a U-shaped flow tube (Figure 8.26) vibrated at its natural frequency (about 80 Hz) by an electromagnetic device located at the bend of the tube [15]. As the liquid flows into the tube, the Coriolis force it experiences because of the vertical movements of the tube has opposite sign to that experienced when it leaves the tube because according to (8.10) opposite velocities yield opposite forces. That is, when the liquid enters the tube, it resists being moved upward (or downward) and reacts by pushing down (respectively, up); when the fluid leaves the tube after having been forced upward (or down-

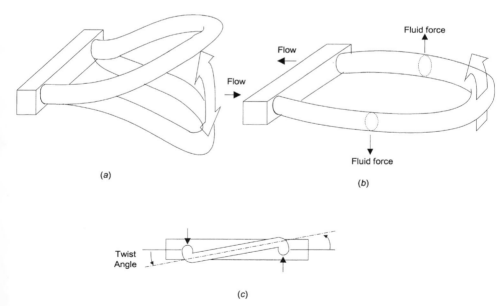

Figure 8.26 (*a*) Coriolis flowmeter based on a vibrating U tube. (*b*) When the tube moves upward, the fluid exerts a downward force at the inlet and an upward force at the outlet that results in (*c*) a tube twist.

ward), it resists having its vertical movement decreased and pushes up (respectively, down). The result is a tube twisting whose amplitude is proportional to the liquid mass flow rate.

Coriolis flowmeters measure mass directly, not through volume or velocity, and can measure corrosive fluids and difficult fluids such as slurries, mud, and mixtures. First marketed in 1978, there are models with two tubes, and different forms (S, Ω, loop). They are not affected by changes in fluid pressure, density, temperature, or viscosity and can achieve an uncertainty of about 0.3%. However, they are not useful for low-pressure gas because of the low forces they develop. Each September issue of *Measurements & Control* lists manufacturers of Coriolis flowmeters in the section on Mass Flowmeters.

8.2.5.3 Turbine Flowmeters. Turbine flowmeters consist of a multiple-blade rotor placed inside a pipe with its rotation axis perpendicular (for low/medium flow rates) or coaxial (for high flow rates) to the fluid flow. The rotor is suspended in the fluid by ball or sleeve bearings. As the fluid passes through the blades, the rotor spins at a velocity that is proportional to the average flow rate. The rotational speed is sensed by a magnetic pickup placed outside the pipe. Each time a turbine blade with an attached magnet or Wiegand wire passes the base of the pickup, it generates an electric pulse, as in incremental encoders. Alternatively, the magnet can be mounted in the pickup and the vane rotation changes the reluctance. Some models use electro-optical sensing. The total number of pulses in a given time interval is proportional to the total volume displaced.

Turbines are intrusive and therefore produce a pressure loss. Also, the immersed materials, including bearings, must be chemically compatible with the fluid. Bearings wear out, particularly at high flow rates. One major advantage is their wide measurement range, usually 20:1 and even 30:1 in some models. Typical uncertainty is about 0.5% for liquids and 1% for gases. Temperature and viscosity (for liquids) or pressure (for gases) corrections improve accuracy. Turbine flowmeters are used, for example, for fuel flow measurement in aircraft, in water service monitoring, and in monitoring spirometers. Each February issue of *Measurements & Control* lists manufacturers of turbine flowmeters.

8.3   VARIABLE OSCILLATORS

When the measurand produces a variable frequency signal, frequency measurement yields a digital output without requiring a voltage-based ADC. Variable-frequency signals have a large dynamic range because voltage saturation or voltage noise do not limit it, particularly in low-voltage systems supplied at 5 V or 3.3 V. Also, they withstand larger interference than voltage signals in short-range telemetry. Therefore, incorporating a modulating sensor in a variable oscillator is an interesting interface option. Often, however, the relationship between the variable frequency and the measurand is not linear (Section 8.3.4).

Furthermore, reactance-variation sensors in variable oscillators do not work at fixed frequency. Oscillators create more interference than circuits processing dc voltage or current, and nearby oscillators may interfere each other. Circuit layout should avoid these problems.

Modulating sensors use both harmonic oscillators (sine wave output) and relaxation oscillators (square wave output). Section 8.2.1 discusses oscillators that include a self-resonant self-generating sensor.

8.3.1   Sinusoidal Oscillators

Sensors can be placed in either RC or LC harmonic oscillators. RC oscillators rely on either RC phase-shift networks or the Wien bridge, which is preferred here because it is more stable. Figure 8.27 shows a basic Wien bridge (reference 16, Section 10.1) and the block diagram to analyze it. If the op amp has $A_d \gg A_c$, the (transform of the) output voltage is

$$V_o = A_d V_o \left(\frac{Z_2}{Z_1 + Z_2} - \frac{R_3}{R_3 + R_4} \right) \tag{8.16}$$

The circuit will oscillate when

$$\frac{R_3}{R_4} = \frac{Z_2}{Z_1} = \frac{R_2}{1 + j\omega R_2 C_2} \frac{j\omega C_1}{1 + j\omega R_1 C_1} \tag{8.17}$$

This condition is fulfilled when

$$\frac{R_4}{R_3} = \frac{R_1}{R_2} + \frac{C_2}{C_1} \tag{8.18}$$

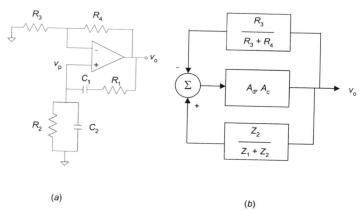

(a)   (b)

Figure 8.27   (a) Basic Wien bridge oscillator and (b) its equivalent block diagram.

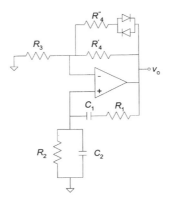

Figure 8.28 Actual Wien bridge with variable diode resistance, which decreases with amplitude to control oscillation amplitude.

and the oscillation frequency is

$$f_o = \frac{1}{2\pi\sqrt{R_1 R_2 C_1 C_2}} \tag{8.19}$$

The slew rate of common op amps limits the maximal output frequency to about 100 kHz.

The sensor can be any of the elements (resistance, capacitance) of Z_1 or Z_2. In order to ensure oscillation at start up, R_3 or R_4 is made dependent on v_o. When v_o is small we need a large gain in order to amplify any noise at frequency f_o at the op amp input. Once v_o has reached large enough amplitude, we wish to diminish the gain to prevent output saturation. Figure 8.28 shows an actual oscillator. When v_o is small, $R_4 = R_4'$, but when v_o is large, the diodes are on, each during one half-cycle, and $R_4 = R_4' \| R_4''$. We can select, for example, $R_4' = 2.1 R_3$ and $R_4'' = 10 R_4'$ (see Problems 8.4 and 5.7). Alternatively, we can use a fixed R_4 and place a filament lamp—which has positive temperature coefficient—in series with R_3.

Higher-frequency oscillators use the circuits proposed by Hartley (capacitive sensors) or Colpitts (inductive sensors) [17]. Nevertheless, all harmonic oscillators have common shortcomings: They cannot directly accommodate differential sensors, and connection leads influence the actual output frequency.

8.3.2 Relaxation Oscillators

Relaxation oscillators are easier to implement than harmonic oscillators, particularly for resistive and capacitive sensors. Figure 8.29 shows an astable multivibrator and its output waveform after an initial transient. The voltage divider formed by R_1 and R_2 determines the voltage V_p at the noninverting in-

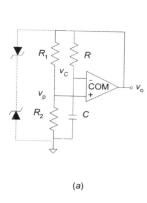 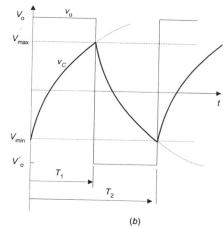

(a)

(b)

Figure 8.29 (a) Astable multivibrator based on a voltage comparator and (b) associated voltage waveforms.

put of the comparator. During the time interval T_1 when the output is at high level (V_o), C charges through R and the voltage across it is

$$v_C(t) = V_{min} + (V_o - V_{min})(1 - e^{-t/RC}) = V_o(1 - e^{-t/RC}) + V_{min}e^{-t/RC} \quad (8.20)$$

where $V_{min} = V_o'R_2/(R_1 + R_2)$. At $t = T_1$, v_C reaches its maximal value (larger than V_p) and the comparator switches its output to V_o'. C then discharges through R according to

$$v_C(t) = V_{max} + (V_o' - V_{max})(1 - e^{-(t-T_1)/RC})$$
$$= V_o'(1 - e^{-(t-T_1)/RC}) + V_{max}e^{-(t-T_1)/RC} \quad (8.21)$$

where $V_{max} = V_oR_2/(R_1 + R_2)$. At $t = T_2$, v_C reaches its minimal value (smaller than V_p) and the comparator toggles back to V_o. A comparator with symmetrical output levels ($V_o' = V_o$) yields $V_{max} = -V_{min} = V_p$ and $T_2 = 2T_1$. Connecting two series-opposition Zener diodes as shown in Figure 8.29 (or another suitable voltage-clamping network) yields symmetrical output levels with matched temperature coefficients. The period of oscillation is $T = 2T_1$. T_1 follows from the condition $v_C = V_{max}$, which leads to

$$T_1 = RC \ln(1 + 2R_2/R_1) \quad (8.22)$$
$$T = 2RC \ln(1 + 2R_2/R_1) \quad (8.23)$$

The sensor can be either R or C (provided its losses are small enough). If $R_2/R_1 = (e - 1)/2 = 0.859$, we have $T = 2RC$. The speed of common com-

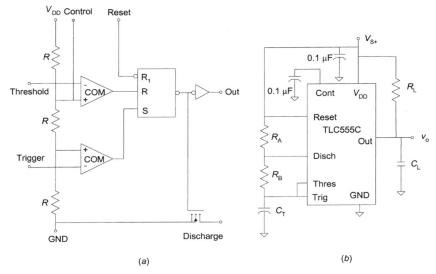

Figure 8.30 (*a*) Functional block diagram of the TLC555C timer and (*b*) connections to implement an astable multivibrator (courtesy of Texas Instruments).

parators determines a maximum operating frequency of about 10 kHz. The exponential charge and discharge of C through a resistor can also be applied to linearize sensors that depend exponentially on the measurand, such as NTC thermistors for temperature measurement (see Problem 8.5).

The 555 family of IC timers includes the comparator and voltage divider network to implement astable multivibrators whose oscillation frequency is determined by an external RC network. Figure 8.30 shows the functional block diagram of a particular model and its connection as an astable multivibrator. The Reset can override trigger (Trig), which can override Threshold (Thres). C_T charges through R_A and R_B to the trigger voltage level (about $2V_{DD}/3$) and then discharges through R_B to the value of the threshold voltage level (about $V_{DD}/3$). The output is high during the charging cycle (t_H) and low during the discharging cycle (t_L), whose respective values are

$$t_H \approx C_T(R_A + R_B) \ln 2 \tag{8.24}$$

$$t_L \approx C_T R_B \ln 2 \tag{8.25}$$

The period of the output waveform is

$$T = t_H + t_L \approx C_T(R_A + 2R_B) \ln 2 \tag{8.26}$$

and its duty cycle is

$$\alpha = \frac{t_H}{t_H + t_L} \approx \frac{R_B}{R_A + 2R_B} \tag{8.27}$$

The 0.1 μF capacitor at the control terminal decreases T by about 10%. A capacitor sensor replacing C_T or resistive sensors replacing R_A or R_B yields a period that will depend on the measurand. Also, a resistive differential sensor can replace R_A and R_B (see Problem 8.6).

Including the internal on-state resistance during discharge through R_B and propagation delay times from the trigger and threshold inputs to the discharge input (t_{PHL} and t_{PLH})) yields a better estimation of the actual charge and discharge time at high frequencies when R_B is close to R_{ON}:

$$t_H \approx C_T(R_A + R_B) \ln\left[3 - \exp\left(\frac{-t_{PLH}}{C_T(R_B + R_{ON})}\right)\right] + t_{PHL} \qquad (8.28)$$

$$t_L \approx C_T(R_B + R_{on}) \ln\left[3 - \exp\left(\frac{-t_{PHL}}{C_T(R_A + R_B)}\right)\right] + t_{PLH} \qquad (8.29)$$

The output period then does not depend linearly on C_T, R_A, or R_B. Furthermore, at 25 °C and 5 V supply, the threshold voltage level for the TLC555C ranges from 2.8 V to 3.8 V and the trigger level ranges from 1.36 V to 1.96 V. The time interval is also sensitive to power supply (0.1 %/V, typical). These factors reduce the accuracy.

8.3.3 Variable CMOS Oscillators

CMOS oscillators that depend on external RC networks can accommodate sensors to yield an output period or frequency proportional to the measurand. Figure 8.31a shows a simple oscillator based on a Schmitt trigger. It relies on

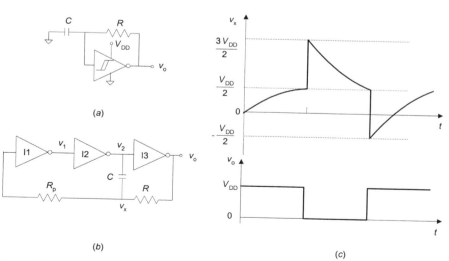

(a)

(b)

(c)

Figure 8.31 Relaxation CMOS oscillators. (a) Based on a Schmitt trigger. (b) Based on three inverters. (c) Waveforms in the three-inverter oscillator.

the charge and discharge of a capacitor, but, unlike the circuit in Figure 8.29a, switching thresholds (V_{TL}, V_{TH}) are set internally. Equations (8.20) and (8.21) apply here too with $V_o \approx V_{DD}$, $V_o' = 0$ V, $V_{max} = V_{TH}$, and $V_{min} = V_{TL}$. Therefore, for $0 < t < T_1$,

$$v_C(t) = V_{DD}(1 - e^{-t/RC}) + V_{TL}e^{-t/RC} \tag{8.30}$$

and for $T_1 < t < T_2$,

$$v_C(t) = V_{TH}e^{-(t-T_1)/RC} \tag{8.31}$$

At $t = T_1$, v_C reaches its maximum (V_{TH}) and the output switches to low level. Therefore, at $t = T_1$ we have

$$V_{TH} = V_{DD}(1 - e^{-T_1/RC}) + V_{TL}e^{-T_1/RC} \tag{8.32}$$

which leads to

$$T_1 = RC\ln\frac{V_{DD} - V_{TL}}{V_{DD} - V_{TH}} \tag{8.33}$$

Similarly, at $t = T_2$, v_C reaches its minimum (V_{TL}) and the output switches to high level. Therefore, at $t = T_2$ we have

$$V_{TL} = V_{TH}e^{-(T_2-T_1)/RC} \tag{8.34}$$

which yields the period of the output waveform,

$$T_2 = RC\ln\frac{V_{TH}}{V_{TL}}\frac{V_{DD} - V_{TL}}{V_{DD} - V_{TH}} \tag{8.35}$$

The approximate output frequency is $f_o \approx 1/T_2$, provided that T_2 is much larger than propagation delays (tens of nanoseconds). The sensor can be either R or C.

This circuit has the shortcoming that V_{TL} and V_{TH} change from unit to unit and also with the supply voltage V_{DD}. The MM74HC14, for example, typically has $V_{TL} = 1.8$ V and $V_{TH} = 2.7$ V when supplied at $V_{DD} = 5$ V. However, the guaranteed thresholds from $-40\,°C$ to $85\,°C$ are 0.9 V $< V_{TL} < 2.2$ V and 2.0 V $< V_{TH} < 3.15$ V.

RC oscillators based on three inverters (Figure 8.31b) are more stable because switching thresholds track closely to 50% of the supply voltage [18]. Figure 8.31c shows the approximate output waveform and the voltage v_x at the charging node. If we initially have $v_2 = 0$ V, then $v_o = v_1 = 1(V_{DD})$. Because v_o is at high level, C charges through R and v_x increases. When v_x reaches $V_T = V_{DD}/2$, I1 switches to 0 and therefore v_2 switches to 1 (V_{DD}) and v_o

switches to 0. C transmits the step in voltage, so that v_x goes from V_T to $V_T + V_{DD} \approx 3V_{DD}/2$. Because $v_o = 0$ V, C discharges through R (and to a lesser degree through R_p) until $v_x = V_T$. At this moment, I1 switches its output to 1, v_2 switches to 0, and v_o switches to 1. The V_{DD} voltage step in v_2 makes v_x change to $V_T - V_{DD} \approx -V_{DD}/2$. C then starts to charge through R and the cycle repeats. Current through R_p is due to the antiparallel diodes at the inverter input that clamp input voltages larger than V_{DD} or smaller than 0 V. The oscillation frequency is

$$f_o = \frac{1}{2RC\left(\dfrac{0.405R_p}{R + R_p} + 0.693\right)} \tag{8.36}$$

The lower the frequency, the higher the stability because propagation delays and effects of threshold shifts comprise a smaller part of the output period. The input capacitance of inverter gates influences the actual oscillation frequency (see Problem 8.7).

The DS1620 (Dallas Semiconductor) is a 9 bit temperature sensor that uses two oscillators whose frequency is determined by a resistor. The resistor in each oscillator has different temperature characteristics, so that the temperature is determined by comparing the two oscillation frequencies.

8.3.4   Linearity in Variable Oscillators

The frequency of oscillators that incorporate a sensor is not proportional to the measurand. For example, for relaxation oscillators we have

$$f = \frac{k}{X} \tag{8.37}$$

where X is the resistance or capacitance that changes in response to the measurand. X can change in a linear form,

$$X = X_0(1 \pm \alpha) \tag{8.38}$$

or in a nonlinear form (e.g., in a parallel plate capacitor whose plate distance changes with the measurand),

$$X = \frac{X_0}{1 \pm \alpha} \tag{8.39}$$

In this latter case we would have a linear output frequency, but not in the former case.

By developing (8.37) in a Taylor's series, we have

$$f = f_0 + (X - X_0)\frac{df}{dX}\bigg|_{X_0} + \frac{(X - X_0)^2}{2}\frac{d^2f}{dX^2}\bigg|_{X_0} + \cdots \qquad (8.40)$$

The first and second derivatives of (8.37) are

$$\frac{df}{dX} = -\frac{k}{X^2} = -\frac{f}{X} \qquad (8.41)$$

$$\frac{d^2f}{dX^2} = \frac{2k}{X^3} = \frac{2f}{X^2} \qquad (8.42)$$

Substituting these equations for $X = X_0$ in (8.40) and neglecting terms with order higher than two yields

$$f \approx f_0 - f_0\frac{X - X_0}{X_0} + \frac{(X - X_0)^2}{2}\frac{2f_0}{X_0^2} = f_0(1 - \alpha + \alpha^2) \qquad (8.43)$$

Solving for α yields

$$\alpha = \frac{1 - \sqrt{1 - 4(1 - f/f_0)}}{2} \qquad (8.44)$$

If f were assumed to depend linearly on α, the result would be

$$\alpha = 1 - \frac{f}{f_0} \qquad (8.45)$$

which is different from that in (8.44).

Example 8.1   A given sensor whose resistance changes by up to $\pm 20\%$ is placed in a relaxation oscillator whose output frequency is measured with negligible error. Determine the relative error for α when the oscillation frequency is assumed linear and when it is calculated from (8.44).

From (8.43), the error because of the nonlinear relationship between f and α will increase for large positive or negative α. When $\alpha = 0.2$, from (8.37) $f = f_0/(1 + 0.2) = f_0/1.2$. If the relationship between f and α is assumed linear, from (8.45) $\alpha = 0.17$, which implies a relative error of 15%. If we use (8.44) instead, then

$$\alpha = \frac{1 - \sqrt{1 - 4(1 - 1/1.2)}}{2} = 0.21$$

and the relative error is 5%.

When $\alpha = -0.2$, from (8.37) $f = f_0/(1 - 0.2) = f_0/0.8$. If the relationship between f and α is assumed linear, from (8.45) $\alpha = -0.25$, which implies a relative error of 25%. If we use (8.44) instead, then

$$\alpha = \frac{1 - \sqrt{1 - 4(1 - 1/0.8)}}{2} = -0.207$$

and the relative error is 4%. If the range for α were larger, the difference in error would increase.

If (8.44) yields an error that exceeds the target, we can use a look-up table to store the coefficients of a polynomial that approximates the actual f–α relationship.

The output frequency of harmonic oscillators has the form

$$f = \frac{k}{X^2} \tag{8.46}$$

so that it is not linear for sensors whose impedance changes according to (8.38) or (8.39). Using the same approach above with a truncated Taylor's series yields

$$f \approx f_0 - f_0 \frac{X - X_0}{2X_0} + \frac{(X - X_0)^2}{8} \frac{3f_0}{X_0^2} \tag{8.47}$$

For sensors modeled by (8.38) we obtain

$$f \approx f_0 \left(1 - \frac{\alpha}{2} + \frac{3\alpha^2}{8} \right) \tag{8.48}$$

whereas for sensors modeled by (8.39) we have

$$f \approx f_0 \left(1 - \frac{\alpha}{2} + \frac{7\alpha^2}{8} \right) \tag{8.49}$$

Solving for α in each case we obtain a better estimate than that resulting from a linear model.

8.4 CONVERSION TO FREQUENCY, PERIOD, OR TIME DURATION

Resonators or variable oscillators cannot directly accommodate self-generating sensors such as thermocouples or pH electrodes that do not yield a variable impedance. One option to obtain a quasi-digital signal from those sensors is to use a common signal conditioner and apply its voltage or current output

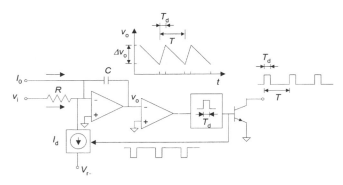

Figure 8.32   Simplified circuit of a voltage-to-frequency converter based on the charge-balancing technique.

to a voltage-to-frequency converter. An alternative is to include the sensor in a voltage-to-frequency (or period) converter, or in a voltage-to-pulse width duration converter (pulse-width modulation, PWM).

8.4.1   Voltage-to-Frequency Conversion

Voltage-to-frequency converters (VFCs) yield from an input voltage or current a pulse train or a square signal, or both, compatible with common logic levels (TTL, CMOS), whose repetition frequency is linearly proportional to the analog input quantity. A voltage-controlled oscillator (VCO) is also a VFC, but has more limited variation range (100 to 1 at best) and poorer linearity. They can work, however, at frequencies well above the common 10 MHz limit for VFCs. Monolithic VFCs yield full-scale output frequencies from 100 kHz to 10 MHz, with a frequency range variation of 1 to 10,000, which is equivalent to 13 bit resolution for an ADC.

Many VFCs work according to the charge balancing technique (Figure 8.32). The circuit consists of an integrator, a comparator, a precision monostable, an output stage, and a switched current source that exhibits high time and temperature stability. When the input voltage is positive, C charges at a rate proportional to the input quantity and yields a negative slope ramp at the output of the integrator v_o. A comparator detects the time when this voltage reaches a preset level and starts a monostable, which outputs a pulse having fixed amplitude and duration T_d. A digital buffer, represented by a simple open collector n–p–n transistor, leads that pulse to the circuit output. This pulse also controls the discharge of the integrator through a constant intensity current source I_d (1 mA is common). The amount of charge drained from the capacitor will be $I_d T_d$; and if the input voltage is still present, it will compensate for this charge after a time that will depend on its magnitude. The process will repeat after a time T such that

$$IT = I_\mathrm{d} T_\mathrm{d} \tag{8.50}$$

$$f = \frac{1}{T} = \frac{I}{I_\mathrm{d} T_\mathrm{d}} \tag{8.51}$$

Note that neither C nor the comparator threshold influences f. Critical parameters are the duration of the monostable pulse and the value for the discharge current, which must be very stable.

Models such as that in Figure 8.32, where the summing point of the input amplifier is accessible, allow us to shift the output range (0 Hz corresponds to 0 V) by adding a current at that point. We can reduce the range so that the maximal frequency is lower than that offered by the circuit, by dividing the input voltage with a resistive attenuator. Alternatively, we can divide the output frequency using a digital counter, which is usually preferred because the attenuator is sensitive to resistor temperature coefficient.

A VFC whose output is fed into a digital counter implements an ADC. For this reason, specifications and advantages for VFCs are usually given in terms relative to other ADC circuits. VFCs are highly linear and have a high resolution and noise immunity, but are relatively slow converters.

Nonlinearity, expressed as percent of full-scale frequency (FS), ranges from 0.002 % FS to 0.05 % FS, depending on the models, and decreases at high output frequencies because of the relative importance of dead times for the comparator and monostable. The linearity of ADCs is usually ±1/2 least significant bit (LSB). Therefore, a 0.002 % FS nonlinearity is equivalent to that of a 14 bit ADC.

Resolution depends on the duration during which the output frequency is being counted and on the maximal value for this frequency. It increases with both factors, provided that the counter does not overflow. By counting a 10 kHz output during 1 s we have a 1 part in 10,000 resolution, better than that of a 13 bit ADC. Models are available with up to 24 bit equivalent resolution (DYMEC 2824).

Noise-rejection ability stems from counting the output pulse train during a given duration, thus averaging any possible input fluctuations that may have resulted in output pulse frequency jitter. That ability, expressed in decibels, can be estimated by the series mode rejection ratio (SMRR)

$$SMRR = 20 \lg \frac{\sin \pi f_\mathrm{noise} t_\mathrm{counter}}{\pi f_\mathrm{noise} t_\mathrm{counter}} \tag{8.52}$$

For 60 Hz voltage interference, for example, if the counting lasts an exact multiple of the period of noise, the SMRR is infinite.

Dynamic range is another important feature. The usual input range is 0 V to 10 V (or 0 mA to 1 mA), with a 1 mV threshold (due to offset voltage and other nonlinear factors). This means a 4 decade range, with some units offering even 6 decades. Still other high input impedance units accept low-level input signals (such as those from thermocouples and strain-gage bridges), so we do not need an amplifier. If the dynamic range required is smaller than that offered by the circuit, it is possible to trade it for response time. For example, suppose that the output for a 10 V input is 100 kHz. When the input is 1 V, the output will be 10 kHz, which implies a 100 μs response time. Therefore when the converter is provided with range adjustment, it is possible to arrange for 100 kHz to correspond to a 1 V input, thus to reduce the response time to 10 μs.

Performance for the different commercially available VFC units depends on the number of external components to be provided by the user in order to have a properly working circuit. Some models available are AD537, AD652, AD654, ADVFC32, LM331, TC9400/1/2, VFC32, and VFC100. Low-power units can be used in a two-wire 4 to 20 mA current loop without requiring any additional supply wires (Section 8.6.1). Using an optocoupler instead of the line transmitter yields a high-linearity isolated measurement system.

Some VFCs based on the charge-balance principle can be connected to perform a frequency-to-voltage conversion instead, for example to obtain an output voltage from sensors whose output is a variable frequency or pulse train. They can also be applied in acoustics and in vibration measurements to obtain signals proportional to frequency. Conversely, they can be used to obtain analog control signals from digital systems that provide output pulses.

8.4.2   Direct Quantity-to-Frequency Conversion

Modulating sensors incorporated into voltage-to-frequency conversion circuits can yield an output frequency or period proportional to the measurand. Figure 8.33 shows a strain-gage bridge connected to a differential integrator (reference 19, Section 4.4.2) whose output is connected to a voltage comparator that excites the bridge. R_a and R_b form a voltage divider that imbalances the bridge at rest $(x = 0)$. Depending on whether the comparator output is at high (V_o) or low (V_o') level, the bridge output will be

$$v_s = V_o x \tag{8.53}$$

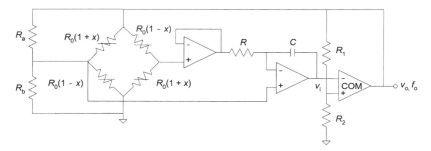

Figure 8.33   Strain-to-frequency converter based on a differential integrator and a voltage comparator.

or

$$v_s = V_o'x \tag{8.54}$$

In the first case, the output of the integrator will be

$$v_i(t) = \frac{V_o x}{RC}t + V_i(0) \tag{8.55}$$

where $V_i(0)$ is its output at $t = 0$ s. At $t = T_1$, v_i reaches the level needed to switch the comparator output to V_o'. That is,

$$v_i(T_1) = \frac{V_o x}{RC}T_1 + V_i(0) = V_o \frac{R_2}{R_1 + R_2} \tag{8.56}$$

From T_1 on we will have

$$v_i(t) = \frac{V_o'x}{RC}(t - T_1) + V_o \frac{R_2}{R_1 + R_2} \tag{8.57}$$

At $t = T_2$, v_i will reach the level needed to switch the comparator back to V_o. Therefore,

$$v_i(T_2) = \frac{V_o'x}{RC}(T_2 - T_1) + V_o \frac{R_2}{R_1 + R_2} = V_o'\frac{R_2}{R_1 + R_2} \tag{8.58}$$

This means that in (8.55) we have $V_i(0) = V_o'R_2/(R_1 + R_2)$. Hence, we can solve (8.56) for T_1 to obtain

$$T_1 = (V_o - V_o')\frac{R_2}{R_1 + R_2}\frac{RC}{V_o x} \tag{8.59}$$

Consequently, from (8.58) we obtain

$$T_2 = (V_o - V_o')\frac{R_2}{R_1 + R_2}\frac{RC}{x}\left(\frac{1}{V_o} - \frac{1}{V_o'}\right) \tag{8.60}$$

The frequency of the output signal is the reciprocal of T_2,

$$f_o = \frac{x}{RC}\frac{R_1 + R_2}{R_2}\frac{-V_o V_o'}{(V_o - V_o')^2} \tag{8.61}$$

hence proportional to the strain. If the comparator has symmetrical output levels ($V_o = -V_o'$), the result reduces to

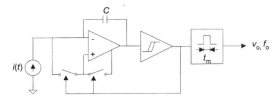

Figure 8.34   Simplified diagram of the TSL220 light-to-frequency converter (courtesy of Texas Instruments).

$$f_{\mathrm{o}} = \frac{x}{RC} \frac{R_1 + R_2}{4R_2} \tag{8.62}$$

so that it does not depend on voltage levels.

Sensors with current output can readily yield a frequency output by integrating that current until reaching a level able to switch a comparator or Schmitt trigger. Figure 8.34 shows the simplified internal diagram of the TSL220 light-to-frequency converter, which integrates a photodiode, an op amp integrator, transistor reset switches, a level detector with hysteresis, and a one-shot pulse generator. The photodiode current charges C so that the output of the integrator is a ramp whose slope is proportional to the current. When the ramp reaches the upper threshold voltage, the level detector switches its output, resets the integrator, and triggers the monostable to produce a fixed-width output pulse. When the integrator output decreases below the lower threshold voltage, the output of the level detector switches again, opening the switches to start a new cycle. Because the discharge is quite fast as compared to the integration time, the integration time determines the output frequency, which will therefore be proportional to the photodiode current. The TSL230, TSL235, and TSL245 use the same technique.

The TMP03 is a monolithic temperature sensor that includes a charge-balance (or sigma–delta) modulator (Figure 8.35) and yields an output signal with a mark–space ratio that encodes the temperature as follows:

$$T(^{\circ}\mathrm{C}) = 235 - \frac{400 \times T_1}{T_2} \tag{8.63}$$

The TMP04 is a similar sensor for temperatures in Fahrenheit degrees encoded as

$$T(^{\circ}\mathrm{F}) = 455 - \frac{720 \times T_1}{T_2} \tag{8.64}$$

Because the result depends on a quotient, using the same clock for measuring T_1 and T_2 makes the result independent of the particular clock frequency used. T_1 is about 10 ms and relatively insensitive to the temperature as compared to T_2. The temperature decode error depends on clock rate and counter resolution.

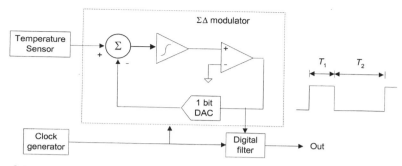

Figure 8.35   The TMP03/TMP04 temperature sensor includes a sigma–delta modulator to yield a serial digital output (courtesy of Analog Devices).

Example 8.2   We wish to measure a temperature from $-25\,°C$ to $100\,°C$ using the TMP03 and a 12 bit counter. Determine the maximal clock frequency that does not overflow the counter. If we use a clock of 100 kHz, determine the temperature resolution when measuring $0\,°C$ and $100\,°C$.

The counter may overflow when measuring the longest time. The TMP03 has $T_1 = 10$ ms. We will estimate T_2 for end-range temperature. From (8.63), at $100\,°C$,

$$T_2 = \frac{400\,T_1}{235 - T(°C)} = \frac{400 \times 10 \text{ ms}}{235 - 100} = 29.6 \text{ ms}$$

The maximal reading from a 12 bit counter is 4095. Therefore, the clock frequency must be

$$f_c < \frac{4095}{29.6 \text{ ms}} = 138 \text{ kHz}$$

If we use $f_c = 100$ kHz, when measuring T_1 we will obtain

$$N_1 = (100 \text{ kHz}) \times (10 \text{ ms}) = 1000$$

At $0\,°C$, the counter reading for T_2 will be

$$N_2 = (100 \text{ kHz}) \times \frac{400 \times (10 \text{ ms})}{235 - 0} = 1702$$

The result will be obtained from the quotient of these two readings according to (8.63). That is,

$$T(°C) = 235 - \frac{400 \times N_1}{N_2} = 234 - \frac{400 \times 1000}{1702} = 0.02$$

Nevertheless, each counting operation has a ± 1 count uncertainty. Therefore, the worst-case situation is

$$\frac{N_1 - 1}{N_2 - 1} = \frac{999}{1703} = 0.02$$

$$T(^\circ C) = 235 - 400 \times 0.5866 = 0.35$$

which implies a $0.35\,^\circ C$ error. The actual error depends on the number of digits used in calculations.

Similarly, at $100\,^\circ C$ we have

$$N_2 = (100 \text{ kHz}) \times \frac{400 \times (10 \text{ ms})}{235 - 100} = 2963$$

$$T(^\circ C) = 235 - \frac{400 \times N_1}{N_2} = 234 - \frac{400 \times 1000}{2963} = 100$$

When considering the ± 1 count uncertainty, the worst-case situation is

$$\frac{N_1 - 1}{N_2 - 1} = \frac{999}{2964} = 0.337$$

$$T(^\circ C) = 235 - 400 \times 0.337 = 100.18$$

which implies a $0.18\,^\circ C$ error. Therefore, the error increases for small counter outputs.

Other temperature sensors with digital output are the AD7814, DS1720, LM74, LM75, LM76, LM77, LM84, MAX6576, and MAX6577. Williams [20] describes direct converters to frequency for thermocouples, piezoelectric accelerometers, temperature sensors with current output, strain gages, photodiodes, capacitive hygrometers, and bubble-based tilt sensors.

8.4.3 Direct Quantity-to-Time Duration Conversion

Daugherty [21] describes a simple technique to measure an unknown resistance, which can be applied to resistive sensor interfacing. The resistor is determined by measuring the ratio of time to either charge or discharge a capacitor using the unknown resistor and a known one. Figure 8.36a shows a microcontroller-based circuit to implement this technique. There are three connections to the microcontroller: one to charge C through the unknown R_x, one to charge C through the reference R_r, and one to monitor the capacitor voltage and discharge it. The time to charge C through R_x to a threshold voltage V_{TH} from a voltage V_{OH} (with the I/O port connected to R_r in high-impedance state) is

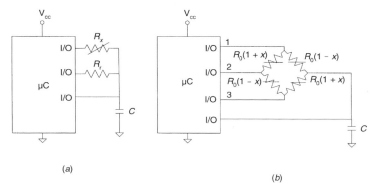

Figure 8.36 (*a*) Basic circuit to determine an unknown resistor from the time ratio needed to charge a capacitor using the unknown and a reference resistor. (*b*) The ratio $(t_1 - t_3)/t_2$—time needed to charge a capacitor through different ports—is proportional to x (patent applied for).

$$t_x = -R_x C \ln\left(1 - \frac{V_{TH}}{V_{OH}}\right) \tag{8.65}$$

Upon reaching V_{TH}, C is discharged through the I/O port it is connected to. Then the I/O port connected to R_x is placed in high-impedance state and the I/O port connected to R_r is set at high level (V_{OH}). The time needed to charge C is now

$$t_r = -R_r C \ln\left(1 - \frac{V_{TH}}{V_{OH}}\right) \tag{8.66}$$

Solving for the unknown resistance yields

$$R_x = R_r \frac{t_x}{t_r} \tag{8.67}$$

which does not depend on C or on V_{OH} or V_{TH}. C must be selected so that the ± 1 count uncertainty in the time intervals to measure is compatible with the desired resolution, and the counting device does not overflow.

Nevertheless, the output resistance of I/O ports is not zero when in the high-level (or low-level) state, the port impedance in the high-impedance state is finite, and the input current at the port monitoring the voltage across C may not be negligible. These factors result in a nonlinear relationship between R_x and t_x. Using n reference resistors permits us to determine n calibration points and interpolate between them to determine R_x from t_x.

If $R_x = R_0(1 + x)$ and instead of R_r in Figure 8.36a we place another linear sensor $R_0(1 - x)$, we can subtract the respective times to charge C and divide by the total charging time to obtain

$$\frac{t_2 - t_1}{t_2 + t_1} = \frac{R_0(1+x) - R_0(1-x)}{2R_0} = x \qquad (8.68)$$

If we measure the time difference only, the result depends on C (see Problem 8.8).

A sensor bridge with four active arms can be connected as shown in Figure 8.36b. When charging C by activating I/O port 1, the charging resistance is $R_0(1+x)(3-x)/4$. When I/O port 2 is activated, the charging resistance is R_0. And when activating I/O port 3, that resistance is $R_0(1-x)(3+x)/4$. Next we compute

$$\frac{t_1 - t_3}{t_2} = \frac{R_0(1+x)(3-x) - R_0(1-x)(3+x)}{4R_0} = x \qquad (8.69)$$

Here too, the finite input and output impedances of microcontroller ports limit the linearity.

8.5   DIRECT SENSOR–MICROCONTROLLER INTERFACING

Interfacing sensors to microcontrollers without an intervening ADC requires a method to measure frequency, period, or time interval, which encodes the information about the measurand. Signals to be interfaced to microcontrollers must have the appropriate logical levels and fast voltage transitions (edges) to prevent false triggering by interference and metastable conditions [22].

8.5.1   Frequency Measurement

Frequency is usually measured by counting cycles during a known time interval. Figure 8.37 shows that this interval (gate time) is usually obtained from a precision clock by frequency dividers. The result is

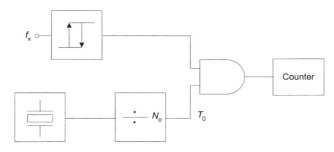

Figure 8.37   Basic block diagram of a frequency meter.

$$N = f_x \times T_0 \tag{8.70}$$

Because the input signal and the internal clock are asynchronous, we may stop counting just before the next input transition arrived or just after it arrived. This implies an uncertainty of 1 count, usually described by saying that the actual result is $N \pm 1$ counts.

Because the resolution of this measuring method is 1 count, the relative resolution is $1/N$, which improves for large N. However, a large N implies a long measurement time, particularly for low frequencies. For example, measuring 10 kHz with uncertainty below 0.1% requires $N = 1000$ and, because each input cycle lasts 100 μs, the measurement time will be 100 ms.

Example 8.3   A common sensor for traffic control is a buried flat coil supplied by a current from 20 kHz to 150 kHz. As a car enters the coil, its inductance decreases because of eddy currents induced in the car, and when the car leaves the coil the inductance returns to its original value. If the coil is included in a harmonic oscillator, the change in oscillation frequency signals the presence of a car. If the oscillation frequency changes by 10% because of a car and we measure it with an 8 bit counter, determine the maximal measurement time before the counter overflows.

The oscillation frequency of an LC harmonic oscillator is

$$f_0 = \frac{K}{\sqrt{LC}}$$

Therefore, if only the inductance changes

$$\frac{df_0}{f_0} = \frac{1}{2}\frac{dL}{L}$$

For a 10% change in inductance we can approximate

$$\Delta f_0 = -0.5 f_0 \frac{\Delta L}{L} = -0.5 f_0(-0.1) = 0.05 f_0$$

The maximal frequency will therefore be

$$f_{\max} = f_0(1 + 0.05) = 1.05 \times 20 \text{ kHz} = 21 \text{ kHz}$$

Because the maximal reading of an 8 bit counter is 255, the counting time is limited to

$$T \le \frac{255}{21 \text{ kHz}} = 12 \text{ ms}$$

Microcontrollers do not include a frequency divider able to provide the time base for frequency measurements according to Figure 8.37. An alternative method is to use two programmable counters, one to count the elapsed time and one to count the input pulses. For example, the 8051 includes two 16 bit timers/counters [23]. When working as timers, the register is automatically incremented every machine cycle—which consists of 12 clock cycles for this particular model. When working as counters, the register is incremented in response to a 1-to-0 transition at its corresponding external input pin. Because recognizing such a transition takes two machine cycles, the maximal count rate is 1/24 the clock frequency. To measure frequency, one timer is preset at the measurement time T_0 and the input pulses are counted until the timer arrives at 0, at which moment it generates an interrupt that stops counting. Then the counter is read and the timer is reloaded at the preset time. If the frequency range goes from f_{min} to f_{max}, the time needed to obtain a resolution of n bits ($N = 2^n$) is

$$T_0 = \frac{2^n}{f_{max} - f_{min}} \tag{8.71}$$

For a given resolution, the narrower the frequency range, the larger the value of T_0.

Example 8.4   A given sensor has an output frequency of 9 kHz to 11 kHz. Determine the measurement time and the number of counts needed to measure the frequency with a 12 bit resolution.
From (8.71),

$$T_0 = \frac{2^{12}}{11 \text{ kHz} - 9 \text{ kHz}} = 2.048 \text{ s}$$

A 9 kHz input yields $N_{min} = (9 \times 10^3) \times (2.048 \text{ s}) = 18,342$. An 11 kHz input yields $N_{max} = (11 \times 10^3) \times (2.048 \text{ s}) = 22,528$. Note that $N_{max} - N_{min} = 4096 = 2^{12}$ as desired, but each counter must have more than 14 bits.

8.5.2   Period and Time-Interval Measurement

Achieving high resolution when measuring low frequencies or narrow frequency ranges, which are common in sensors, takes a long time. This restricts dynamic measurements because the frequency signal must be constant while it is being counted. Measuring the signal period instead is faster, even when including the time needed to calculate the frequency from the measured period.
    The period of a signal can be measured by counting pulses of a reference signal (clock) over a number of M periods (Figure 8.38). The result is

$$N = f_c M T_x \tag{8.72}$$

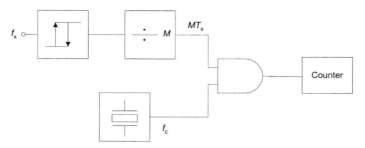

Figure 8.38   Basic block diagram of a period meter.

The measurement takes MT_x and, because the resolution is $1/N$, from (8.72) we infer that the resolution–measurement time product is constant $(1/f_c)$. For example, because a 10 kHz signal has 100 μs period, measuring it with uncertainty below 0.1 % requires $N = 1000$. If $f_c = 1$ MHz, we need $MT_x = 1$ ms; that is, $M = 10$. The input signal must be stable for 10 periods, far less than the 1000 periods needed to measure frequency with the same resolution. A faster clock would permit a shorter measurement time for the same resolution.

Some microcontrollers cannot directly implement the method in Figure 8.38. Instead they can measure the total elapsed time over k input pulses. The result is $kT_{x(m)}$, where $T_{x(m)}$ is the period of the input signal in machine cycles. That is,

$$T_{x(m)} = \frac{1/f_x}{T_m} = \frac{f_c}{pf_x} \tag{8.73}$$

For the 8051, $p = 12$. To implement the measurement, one timer is used to measure the elapsed time $kT_{x(m)}$ (i.e., counting to k), and the sensor signal is connected to an external interrupt pin. Each 1 to 0 transition generates an interrupt. The interrupt routine counts sensor pulses; and when it gets to the predetermined k, it reads and clears the timer.

k, which must be an integer, depends on the desired resolution. To obtain m bit resolution we need

$$k \geq \frac{2^m}{T_{max(m)} - T_{min(m)}} \tag{8.74}$$

where $T_{max(m)}$ and $T_{min(m)}$ are measured in machine cycles.

Example 8.5   A given sensor has an output frequency of 9 kHz to 11 kHz. Determine how many signal cycles must elapse to measure its period with 12 bit resolution by using an 8051 microcontroller whose clock frequency is 12 MHz.

In order to apply (8.74) we must first determine the maximal and minimal signal periods in machine cycles. From (8.73), with $p = 12$,

$$T_{\max(m)} = \frac{12\text{ MHz}}{12 \times (9\text{ kHz})} = 111.1\text{ machine cycles}$$

$$T_{\min(m)} = \frac{12\text{ MHz}}{12 \times (11\text{ kHz})} = 90.9\text{ machine cycles}$$

Applying now (8.74),

$$k = \frac{2^{12}}{111.1 - 90.9} = 202.8$$

Therefore, we should measure for 203 signal cycles.

If the information about the measurand is encoded in the frequency rather than in the period, the result is

$$f_x = \frac{k}{k T_{x(m)}} \tag{8.75}$$

Hence, in order to obtain an n bit resolution in the frequency we need to know the resolution required when measuring the time $k T_{x(m)}$. By taking the derivative of (8.75), we obtain

$$\mathrm{d}f_x = -\frac{k}{(k T_{x(m)})^2}\, \mathrm{d}(k T_{x(m)}) = -\frac{f_x^2}{k}\, \mathrm{d}(k T_{x(m)}) \tag{8.76}$$

which means that the resolution in f_x is (f_x^2/k) times the resolution in $k T_{x(m)}$. Therefore, to have n bit resolution in the calculated frequency, we must measure the period with a number of bits m such that

$$\frac{f_{\max} - f_{\min}}{2^n} \geq \frac{f_{\max}^2}{k}\, \frac{k(1/f_{\min} - 1/f_{\max})}{2^m} \tag{8.77}$$

which leads to

$$m \geq n + \frac{\ln(f_{\max}/f_{\min})}{\ln 2} \tag{8.78}$$

From this equation and (8.74), the predetermined number of cycles during which the microcontroller must count is

$$k \geq 2^n p \frac{f_{\max}}{f_c} \frac{f_{\max}}{f_{\max} - f_{\min}} \tag{8.79}$$

where f_c is the clock frequency (see Problems 8.9 to 8.11).

Example 8.6   Consider the same sensor in Examples 8.4 and 8.5 whose output frequency is from 9 kHz to 11 kHz. How many signal cycles must elapse to determine its frequency with a 12 bit resolution from the measurement of its period using an 8051 microcontroller whose clock frequency is 12 MHz? What is the resolution needed in the period measurement? How long does the measurement take?

From (8.79), with $p = 12$ and $f_c = 12$ MHz, we have

$$k \geq 2^{12} \times 12 \times \frac{11 \text{ kHz}}{12 \text{ MHz}} \times \frac{11}{11 - 9} = 247.8 \approx 248$$

From (8.78) we obtain

$$m \geq 12 + \frac{\ln(11/9)}{\ln 2} = 12.289 \approx 13 \text{ bit}$$

The measurement will be longer when the signal period is short. Because the period of a 9 kHz signal is 0.11 ms, 248 cycles will last 27.5 ms. This is shorter than the 2.048 s needed to achieve the same resolution from a frequency measurement (Example 8.4).

The overall measurement time should include the time needed to determine the frequency by either computing the reciprocal of the period measured or by searching a look-up table. The size of this table will depend on $f_{max} - f_{min}$ and on the number of bits needed to represent each frequency value. Alternatively, the look-up table may store the values of the measurand whose information is encoded in the frequency.

Some microcontrollers include timers that are turned on by a 0-to-1 transition in an external interrupt pin and that are turned off by a 1-to-0 transition in the same pin. This last transition also generates an interrupt that permits the interrupt routine to read and reset the timer. This operation mode enables the direct measurement of pulse width—for example, when the sensor produces a PWM modulation (Section 8.4.3).

Example 8.7   We wish to measure a temperature from $-20\,°$C to $60\,°$C with $0.1\,°$C resolution using a linearized silicon PTC 2000 Ω thermistor with $\alpha = 0.79\,\%/$K at $25\,°$C. The interface is a microcontroller that measures the time interval needed to charge an external capacitor C through the sensor up to the transition level needed to generate an interrupt. Determine C when that level is half the supply voltage applied to the resistor and the effective clock frequency is 1 MHz.

The voltage across the capacitor will rise exponentially according to

$$v_C(t) = V_{CC}(1 - e^{-t/R_T C})$$

The time needed to reach $V_{CC}/2$ is

$$t_{th} = R_T C \ln 2 = 0.693 R_T C$$

The model for the sensor is

$$R_T = R_0[1 + \alpha(T - T_0)] = (2000\ \Omega)[1 + (0.0079/\text{K})(T - 25\,^\circ\text{C})]$$

Therefore, the sensor has 1289 Ω at $-20\,^\circ\text{C}$ and 2553 Ω at $60\,^\circ\text{C}$. The corresponding times to reach the threshold voltage are

$$t_{th}(-20) = 0.693(1289\ \Omega)C$$
$$t_{th}(60) = 0.693(2553\ \Omega)C$$

The time difference is

$$\Delta t_{th} = 0.693(1264\ \Omega)C = (876\ \Omega)C$$

The dynamic range needed is

$$\text{DR} = \frac{60\,^\circ\text{C} - (-20\,^\circ\text{C})}{0.1\,^\circ\text{C}} = 800$$

Because the resolution when counting is 1 (count), we need

$$\Delta N = 800 = f_c \Delta t_{th} = (1\ \text{MHz})(876\ \Omega)C$$

From this, $C = 0.9$ µF. We would select $C = 1$ µF. The maximal number of counts obtained would be 1769 at $60\,^\circ\text{C}$ and the minimal 893 at $-20\,^\circ\text{C}$. Because encoding the larger number requires 12 bits, and we need to store 800 different values, a look-up table should have about 1.2 kbyte.

8.5.3   Calculation and Compensations

Measurement bridges compensate multiplicative interference and readily perform difference measurements. This section explores these properties in frequency measurements.

Figure 8.39 shows how to measure the ratio (quotient) between two frequencies. The counter will read

$$N = \frac{f_x}{M} \frac{Q}{f_y} \tag{8.80}$$

In order to have a large resolution (large N), if $f_x > f_y$ we select Q large and M small ($M = 1$). Multiplicative errors common to f_x and f_y will cancel.

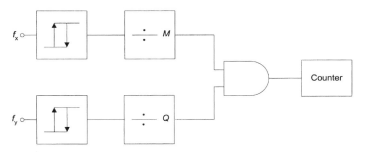

Figure 8.39   Basic block diagram of a frequency ratio meter.

To add or subtract two frequencies measured simultaneously we can perform the corresponding function with the digital outputs of the respective counters. If the frequencies are measured successively instead, to add them we can load the counter with the result of the first measurement and then measure the second frequency. To subtract, we load the counter with the complement to 2 of the first reading.

Frequency measurement applied to differential sensors also yields some advantages. If each sensor of a differential pair is placed in a relaxation oscillator, from (8.37) to (8.39) the respective output frequency can be either directly proportional to the measurand,

$$f_1 = kX_0(1 + \alpha) \tag{8.81}$$

$$f_2 = kX_0(1 - \alpha) \tag{8.82}$$

or inversely proportional to it:

$$f_1 = \frac{k}{X_0(1 + \alpha)} \tag{8.83}$$

$$f_2 = \frac{k}{X_0(1 - \alpha)} \tag{8.84}$$

Instead of measuring each oscillation frequency separately for a given time, we add the respective counter outputs until the total number of counts is a predetermined value N:

$$N = (f_1 + f_2)T_N \tag{8.85}$$

If during this time we measure the difference between oscillator frequencies, the result is

$$N_1 - N_2 = (f_1 - f_2)T_N = \frac{f_1 - f_2}{f_1 + f_2}N \tag{8.86}$$

Therefore, from (8.81) and (8.82) we have

$$N_1 - N_2 = \alpha N \tag{8.87}$$

and from (8.82) and (8.83) we obtain

$$N_1 - N_2 = -\alpha N \tag{8.88}$$

That is, the output is proportional to the measurand for both sensor types. This means that if instead of measuring the difference in frequency we measure the difference in period, which may be faster, the output will also be linear. Also, from (8.85) the measurement time T_N is $T_N = N/(2kX_0)$—independent of α— for a sensor described by (8.81) and (8.82) but depends on α for a sensor described by (8.83) and (8.84).

8.5.4   Velocity Measurement. Digital Tachometers

Speed measurement from encoder pulses (Section 8.1.1) should be fast enough to accommodate fluctuating speeds but also provide a high resolution, which requires a high number of counts. If the encoder yields m pulses each turn and we measure its frequency using a counter as in Figure 8.37, the rotational speed in turns per second is

$$n = \frac{N_c}{T_0} \frac{1}{m} \tag{8.89}$$

where N_c is the number of pulses counted during the time interval T_0. If this interval is well known, since the relative uncertainty in N_c is $1/N_c$, the relative uncertainty in n is

$$\frac{dn}{n} = \frac{dN_c}{N_c} = \frac{1}{N_c} = \frac{1}{nT_0} \frac{1}{m} \tag{8.90}$$

Hence, if T_0 is short (for dynamic speed measurement) the uncertainty increases for small n and m.

If we measure the period of the input pulses instead and count N_p pulses from a clock f_c, the rotational speed is

$$n = \frac{f_c}{N_p} \frac{1}{m} \tag{8.91}$$

If f_c is constant, the relative uncertainty in the measured speed is

$$\frac{dn}{n} = -\frac{dN_p}{N_p} = \frac{-1}{N_p} = \frac{n \times m}{f_c} \tag{8.92}$$

The uncertainty now increases with n and m. The measurement lasts

$$T_p = \frac{N_p}{f_c} = \frac{1}{n \times m} \tag{8.93}$$

Hence, it is longer for small n and m.

The constant elapsed time (CET) method [24] solves the trade-off between resolution and measurement time. It combines encoder pulse counting and time measurement. The time is measured by counting pulses from a precision clock of period T_c. The measured time interval is selected so that it is larger than or equal to the desired elapsed time T_e and contains an integer number of encoder pulses. The pulse counter and the time counter start at the rising edge of the encoder pulse and stop at the first rising edge of the encoder pulse occurring after the interval T_e. Hence, there is no uncertainty in N_p. If the respective counter for pulses and time read N_p and N_c, the rotational speed is

$$n = \frac{N_p}{N_c T_c} \frac{1}{m} = \frac{N_p f_c}{N_c m} \tag{8.94}$$

At very low speed, during T_e we will count a single encoder pulse ($N_p = 1$) and the method is equivalent to period measurement. The relative uncertainty in n will be

$$\frac{dn}{n} = -\frac{dN_c}{N_c} = \frac{-1}{N_c} = \frac{n \times m}{N_p f_c} \tag{8.95}$$

At high speed, we obtain a small uncertainty by increasing N_p. If the maximal measurement time accepted is T_{max} and we select N_p so that

$$T_{max} \leq \frac{N_p}{n \times m} \tag{8.96}$$

then the relative uncertainty is constant and smaller than $1/(f_c T_{max})$. If the encoder pulse length exceeds T_{max}, the speed signal is set equal to zero. The minimal measurable speed is

$$n_{min} = \frac{1}{T_{max} \times m} \tag{8.97}$$

which obviously decreases with the maximal response time.

8.6   COMMUNICATION SYSTEMS FOR SENSORS

The output of signal conditioners must be communicated to a receiver or display. Whenever the measurement and data display points are separated (as in remote process control) or are inaccessible (as in flight tests, biomedical applications, and wildlife management), we need an installation able to transmit/receive measurement data or commands, or both.

Measurements and commands can be transmitted by mechanical or electric means. Mechanical means, fluidic, or pneumatic systems suit only short distances (<300 m) but can easily achieve high-power actions if a compressed air installation is available. Then the measured quantity is converted to a pneumatic pressure in the standard pressure range from 20 kPa to 100 kPa (0.2 to 1.0 bar). They are used in those situations where electronics is unsuitable—for example, when there is ionizing radiation or strong electromagnetic fields, or for reasons such as safety (explosive media) or reliability, in case the electric supply fails and a (nonelectric) compressor unit is available. But they are unsuitable when the information is to be sent to a computer or any other electronic data-processing system.

Short-distance communication using wire transmission (twisted pair, coaxial cable, telephone wire) is relatively inexpensive. Wire transmission is also used in large area installations that include an appropriate infrastructure such as power distribution networks and pipelines. However, wire transmission has restricted bandwidth and speed. Optical fibers overcome these limitations and withstand strong electromagnetic interference but are more expensive. Communication involving long distance or inaccessible emitter or receivers use radio-frequency (RF) telemetering. Some special short-range applications use ultrasound, infrared radiation, or simple capacitive or inductive coupling.

Whatever the communication means, sensor signals must be conditioned in order to be adapted to its characteristics. This conditioning process may require more than one stage, as indicated in Figure 8.40. The sensor signal, once conditioned, is modulated (if analog) or encoded (if digital), in order to combine it with other signals sharing the same transmission means, or in order to make it acceptable to the transmitter modulator, which couples the signals to the communication channel. Because duplex communication is normally desired to check the transmitted measurements and to transmit commands, the same modulating unit usually includes a demodulator, thus constituting a modem. The receiver performs the converse functions.

For short distance the information is directly sent in baseband under the form of a voltage, current, or frequency coming from the modulator or encoder next to the sensor. In voltage telemetry we convert the sensor output signal into a proportional voltage, connect this voltage to a line consisting of two wires, and measure the voltage at the receiving end. Standardized line voltage ranges (low end of the scale to the upper end) are 1 V to 5 V, 0 to ± 1 V, 0 to ± 10 V, both dc, and ac from 0 V to 5 V. Excluding 0 V of the range enables the receiver to detect short circuits. The maximal distance depends on wire resistance be-

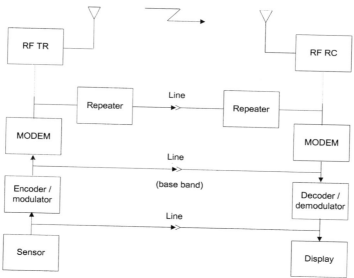

Figure 8.40 General structure for a telemetry system indicating the possible levels to establish the link between transmitter and receiver.

cause of the drop in voltage through it, and on the current in the line, which depends on the input resistance of the detector.

The main shortcoming of voltage telemetering is the induction of parasitic voltages in the loop formed by the connecting wires, which superimpose on the transmitted signal. When the transmitted signal has much lower bandwidth than the interference, a simple low-pass filter may cancel these. But because 60 Hz power lines are one of the most important interference sources, the filter solution is of limited use. Twisting the wires to minimize the effective loop area reduces magnetic interference. Capacitive interference decreases when the circuit has small equivalent impedance, but this is not the case for signal sources with high output impedance (Section 3.6.1). A grounded conductive shield reduces capacitive interference (Section 3.6.3) but at the added cost of shielded cables.

8.6.1 Current Telemetry: 4 to 20 mA Loop

In current telemetry we convert the sensor signal to a proportional current, which is sent to the connecting lines. The receiver detects this current by measuring the drop in voltage across a known resistor. Standard current values are 4 to 20 mA (by far the most common), 0 to 5 mA, 0 to 20 mA, 10 to 50 mA, 1 to 5 mA, and 2 to 10 mA. Using 4 mA (or other nonzero value) for the 0 level makes the detection of an open-circuit condition easy (0 mA).

Current telemetry is insensitive to parasitic thermocouples and drops in voltage along cable resistances, as long as the transmitter is capable of maintaining the value for the current in the circuit. This permits us to use thinner wires, which are less expensive. Voltages induced by magnetic coupling into the wire loop have no significant influence, provided that the output resistance for the current source is high enough, because then the current due to the interference will be very low. Using a twisted pair further reduces magnetic interference.

Capacitive interference results in an error dependent on receiver resistance, which is usually 250 Ω for the 4 to 20 mA system. This resistor is small enough to yield insignificant errors, and therefore the allowed cable length is longer than for a voltage telemetering system. Current telemetering has the added advantage that a single reading or recording device can switch to different channels having different cable lengths without this resulting in a different accuracy for each length.

When the transmitter is floating, sometimes it is possible to complete the circuit using only two wires, shared by the supply and the signal. Figure 8.41a shows the general circuit using four wires, two for the power supply and two for signal transmission. It is usually possible to share a return wire as indicated in Figure 8.41b. In Figure 8.41c, the power supply is connected in series with the reading device or devices and any other resistances inside the loop including the

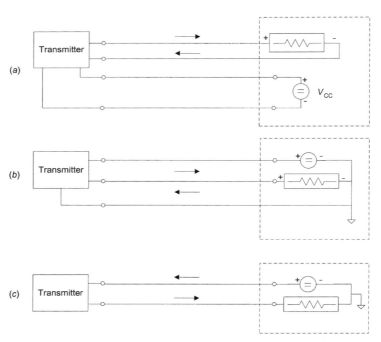

Figure 8.41   Current telemetering using: (a) four wires; (b) three wires; (c) two wires.

sensor. The standard supply voltage is 24 V: 12 V for the drop in voltage in the transmitter, 5 V for the drop in voltage in the receiver (250 Ω), 2 V for the drop in voltage along the line (100 Ω), and 5 V for the drop in voltage across a series resistor added to provide intrinsic safety in case of short circuit. The supply voltage must increase 1 V for each 50 Ω increase in load resistance.

The 4 to 20 mA loop has extended use in process control to transmit measurement data and also to control some actuators that readily accept commands in this standard. The AD693, the AD 694 (Analog Devices), and the XTR series (Burr–Brown) are monolithic signal conditioning circuits that accept low-level signals from different sensors (RTDs, resistance bridges, thermocouples) to control a standard 4 to 20 mA output. Some models include excitation voltage or current for modulating sensors.

8.6.2 Simultaneous Analog and Digital Communication

Communication systems using a current loop are point to point and one way. Because they are point to point, adding a new sensor implies modification of the system cabling. Because they are one way, the controller cannot query the transmitter. Digital communication permits, for example, that devices incorporate all the information needed about them: manufacturer, model, serial number, calibration factors, configuration, process variables, measurement ranges, diagnostics, and so on. This information does not improve process control by itself but simplifies installation and maintenance. Status information can be obtained at any time. Nevertheless, the whole replacement of existing analog communication systems by bus-type digital communication systems would be extremely expensive.

The HART® (Highway Addressable Remote Transducer) Field Communications Protocol preserves the 4 to 20 mA signal and enables two-way digital communications to occur without disturbing the integrity of the 4 to 20 mA signal. Several vendors support HART®, originally developed by Rosemount in 1988 (http://www.hartcomm.org). The HART® protocol uses the Bell 202 frequency-shift-keying (FSK) standard to superimpose a digital signal on top of a 4 to 20 mA analog signal (Figure 8.42). 1200 Hz represents a logical 1 (mark),

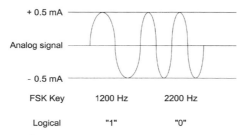

Figure 8.42 The HART® field communication system uses FSK to encode information on top of the 4 to 20 mA analog signal.

and 2200 Hz represents a logical 0 (space). Because the digital FSK signal is phase continuous, there is no interference with the 4 to 20 mA signal. Each byte has one start bit, eight data bits, one odd-parity bit, and one stop bit. The transmission speed is 1200 bit/s. If the entire transmitter consumes less than 3.5 mA (4 mA − 0.5 mA), a single twisted pair supplies the power and transmits analog and digital signals. HART® can be installed in any analog installation without having to run new cabling. The minimal wire size is 24AWG (about 0.51 mm in diameter). It can be used for point-to-point communication and also to communicate up to 15 devices in a loop (multidrop mode), with a maximal length of 3000 m (shielded wire pair). Each device has a particular direction in the loop. HART® implements a master/slave protocol; a remote "slave" device responds only when addressed by the master. The poll/response rate is 2/s. The integrity of HART® communication is very secure because status information is included with every reply message and extensive error checking occurs with each transaction. Up to four process variables can be communicated in one HART message, and each device may have up to 256 variables. There are HART®-compatible intrinsic safety barriers and isolators that pass the HART® signals for use in hazardous areas.

8.6.3 Sensor Buses: Fieldbus

Analog communication systems imply a double signal conversion. Output signals from analog sensors must be digitized in order to be processed by a microprocessor, for example for linearization, limit detection, or any other operation. If the resulting information is to be transmitted by, say, a 4 to 20 mA loop, digital signals must be converted back to analog, and then digitized again in the central processor that controls the system. Sensors able to communicate digitally skip that double conversion. Furthermore, a bus-type communication permits a single communication channel to be shared by different devices in two-way communication. This simplifies cabling, particularly when using a serial bus (Figure 8.43). A fieldbus is a digital, two-way, multidrop communication link among intelligent measurement and control devices with user interface. It serves as a local area network (LAN) for advanced process control, remote input/output, and high-speed factory automation applications.

The functions performed in a bus-based measurement and control system are the same as in Figure 1.1, but the system is configured around the bus rather than cabled according to each device function. Adding new sensors or actuators is fast because only the system software must be updated, not the cabling, and there is no need to stop the process. Nevertheless, all devices connected to a given bus must be compatible, and there is not yet a universally accepted standard. Consequently there are many fieldbuses.

In 1985, the International Electrotechnical Commission (IEC) initiated an effort, led by the committee SP-50 of the ISA (then Instrument Society of America, now International Society for Measurement and Control), to create a fieldbus standard. The standard would follow the seven-layer OSI (Open

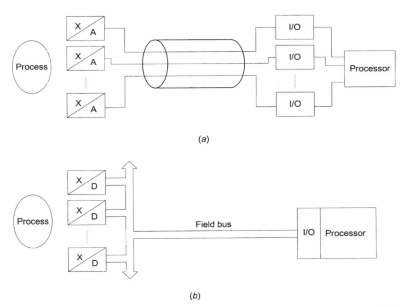

Figure 8.43 Basic structure of a system with (*a*) analog communication and (*b*) bus-based digital communication.

Systems Interconnection) model approved by the IEC and the ISO (International Standards Organization) [25]. The years-long effort led to a simplified model with three layers or levels (physical, data link, and application), with an added layer for user interface. Each layer includes specific rules termed protocols, relative to data format and timing.

The physical layer is concerned with the access to, and control of, the physical medium in order to ensure the transfer of information over physical distances. The physical layer receives messages, converts them into physical signals on the transmission medium, and vice versa. Hence, it deals with electrical and mechanical compatibility and with the functioning and protocol requirements posed by the physical medium (wire, RF, fiber optic) in order to emit and receive information through it. A transmitter is an active device containing circuitry that applies a digital signal on the bus. The code selected for wire communication is Manchester II.

Link layer protocols operate between the ends of the transmission medium to overcome the deficiencies of the physical layer and to manage the link resources (e.g., bandwidth). The link layer determines, for example, which device can "talk" at a given moment, and it detects and corrects errors.

Application layer protocols are strictly not part of the communications system, but the user of it. They define how to write, read, understand, and execute a message. The user layer is the interface between the user and the communication system that allows the host system to communicate with devices without the need for custom programming.

It has turned out, however, that because of the large demand for field buses, several manufacturers have marketed products using proprietary protocols before reaching unanimous consensus about an international standard. What should have been a single standard set (IEC61158), now known as Type 1 (or H1), led to an eight-part proposal, including: ControlNet (Type 2), Profibus (Type 3), P-Net (Type 4), Fieldbus Foundation high-speed Ethernet (HSE) (Type 5), SwiftNet (Type 6), WorldFIP (World Factory Implementation Protocol) (Type 7), and Interbus-S (Type 8). Developments in this field can be found at the Fieldbus Foundation website (http://www.fieldbus.org). *InTech* magazine includes a monthly section on Fieldbus news.

8.7 INTELLIGENT SENSORS

An intelligent (or smart) sensor has a built-in microprocessor for automatic operation, for processing data, or for achieving greater versatility. There is a wide variety of available intelligent sensors ranging from those that merely include digital "trimming" helped by an onboard microprocessor, to devices that combine data processing and communication, sometimes in a monolithic chip (e.g., MicroConverter™ from Analog Devices). Honeywell introduced the first commercial intelligent sensor in 1983. It included two pressure sensors (differential and static) and one temperature sensor (for compensation), multiplexed into an ADC and microprocessor. Input signals were processed to yield a digital output that was converted back into analog by a DAC connected to a 4 to 20 mA loop.

The low cost of digital processors enables the production of affordable intelligent sensors with digital output capable of self-identification, self-testing, adaptive calibration, noisy data filtering, sending and receiving data, making logical decisions, and so on. These advantages, however, are somewhat obscured by the forest of incompatible industrial networks or fieldbuses that do not permit sensors to plug directly into them. The IEEE 1451 family of "Standards for Smart Transducer Interface for Sensors and Actuators" aims to cover network-capable smart transducers from the interface to the transducer itself up to a high-level, object-model representation of behavior, attributes, and data communications, in order to ensure transducer-to-network interoperability and interchangeability [26].

Figure 8.44 shows the basic components of a network independent sensor compatible with IEEE Std1451.2-1997 (Transducer-to-Microprocessor Communications Protocols and Transducer Electronic Data Sheet (TEDS) Formats). The smart sensor (or actuator) itself is referred to as STIM (Smart Transducer Interface Module). It integrates one or more sensors (and/or actuators), signal conditioners, and converters to connect them with the logic (microcontroller) to implement the interface and the TEDS. The TEDS is a data sheet describing a

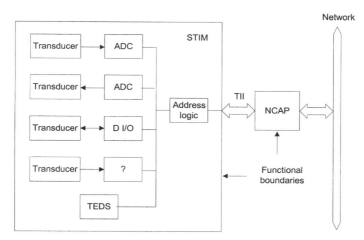

Figure 8.44 The IEEE 1451.2 standard defines the Smart Transducer Interface Module (STIM), the Transducer Electronic Data Sheet (TEDS), the Network Capable Applications Processor (NCAP), and the Transducer Independent Interface (TII).

transducer (parameters of operation and conditions of usage) and stored in some form of electrically readable memory. The NCAP (Network Capable Application Processor) is a device between the STIM and the network that performs network communications, STIM communication, and data conversion functions. The TII (Transducer Independent Interface) is a 10 wire serial digital interface that connects a STIM to an NCAP. It is based on the SPI (Serial Peripheral Interface) synchronous serial communication protocol and allows any STIM to plug into any industrial network through a NCAP node. The standard covers the STIM, the TII, and the portion on the NCAP transducer block that communicates with the STIM. It also defines the various formats for the TEDS, the transducer functional type (e.g., sensor, actuator, buffered sensor, and event sensor), and a general-purpose calibration and correction engine.

The IEEE P1451.1 proposed standard (NCAP Information Model) defines an object-oriented model of the components of a networked smart transducer using a high level of abstraction. The model is a virtual backplane into which are plugged a series of functional hardware and software blocks to provide the desired functionality. It provides the interface to the transducer block, so that its hardware looks like an I/O driver, and the interface to the NCAP including different network protocol implementations.

The IEEE 1451.3 proposed standard (Digital Communications and TEDS Formats for Distributed Multidrop Systems) defines a high-speed, multidrop digital bus for short-run connection of multiple transducers to a STIM and NCAP. That is, it defines the local bus inside the STIM in Figure 8.44 that connects the address logic to the ADC and DAC. The standard suits data acquisition systems that use many transducers distributed over a limited area and where it is not practical to run the fieldbus cables or control network to each transducer.

The IEEE P1451.4 proposed standard (Mixed-Mode Communication Protocols and TEDS Formats) describes a two-way communication protocol for mixed mode sensors (and actuators) that connects to smart transducers via conventional analog wiring.

8.8 PROBLEMS

8.1 Determine the rotating speed above which it is better to measure pulse frequency than pulse interval when the available clock for the counter is 10 MHz and the maximal counting time available is 1 s.

8.2 A given quartz crystal that has a linear drift of $35.4 \times 10^{-6}/°C$ is used to design a digital thermometer. Calculate the frequency for the oscillator and the gate time for the counter in order to have a sensitivity of 1000 Hz/°C and a resolution of 0.0001 °C.

8.3 A vibrating wire gage is mounted on concrete which has an expansion coefficient of $8 \times 10^{-6}/°C$. The steel wire has an expansion coefficient of $14 \times 10^{-6}/°C$. In order to compensate for temperature interference, we sense the resistance of the driving system coil of 150 Ω and with $\alpha = 4.3$ (mΩ/Ω)/K at 25 °C. Calculate the correction factor for the deformation due to the temperature indicated by the gage when the coil resistance is 141 Ω.

8.4 A temperature is to be measured from −40 °C to 85 °C by using an NTC thermistor that has 4700 Ω at 25 °C and $B = 3500$ K in the measurement range. We wish to obtain a frequency output yielding 1 kHz at 0 °C. Because we do not need a linear frequency–temperature relationship, we consider a Wien bridge such as that in Figure 8.28. Design the circuit components and determine the range for the output frequency.

8.5 Figure P8.5 shows a circuit that over a reduced temperature range

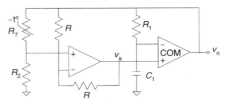

Figure P8.5 Relaxation oscillator whose period is proportional to a temperature increment over a reduced temperature range.

yields a period that is proportional to the temperature increment with respect to T_0. If the NTC thermistor is modeled by $R_T = a \times b^{-\Delta T}$ ($\Delta T = T - T_0$) and has 12,400 Ω at 25 °C and $B = 2734$ K in the measurement range, design the circuit to obtain 50 μs/K from 35 °C to 45 °C.

8.6 The astable multivibrator in Figure P8.6 has an output signal that depends on two strain gages that have 120 Ω and gage factor 2, bonded so that when one increases its value the other decreases by the same amount. The up–down counter increments or decrements its output depending on which of its two inputs receives the incoming pulse. Determine C and the clock frequency in order to have a resolution of 1 με. Use typical values in the data sheet of the TLC555C.

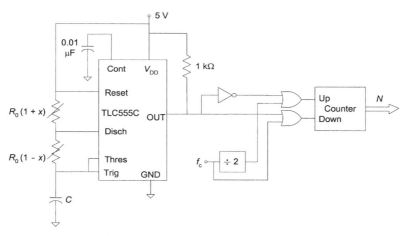

Figure P8.6 Deformation measurement using two strain gages in an astable multivibrator.

8.7 The oscillation frequency in the capacitive hygrometer in Figure P8.7a is $f_H = 0.559/(RC)$, where $C = C_H + C_G$. C_H is the sensor capacitance, which has a temperature drift of 0.05 %RH/K, and C_G is the input capacitance of the inverter gate, which depends on the supply voltage V_{cc}. This last dependence can be applied to correct the thermal drift of the sensor by making V_{cc} variable according to the output of a circuit that senses the temperature by an AD590 that yields 1 μA/K (Figure P8.7b). In order to determine the dependence of C_G on V_{cc}, we keep the sensor at 20 °C and 50 %RH and measure the output frequency (Figure P8.7c). We obtain 8129 counts for $V_{cc} = 6$ V, 7986 counts for $V_{cc} = 9$ V, and a linear relationship. If the humidity sensor has 107 pF, 110 pF, and 122 pF

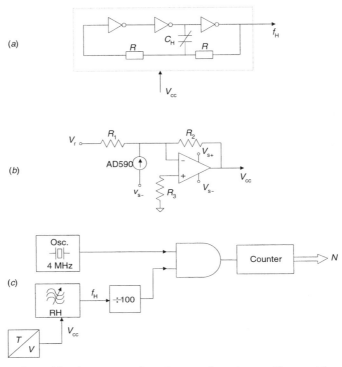

(a)

(b)

(c)

Figure P8.7 Capacitive hygrometer based on a relaxation oscillator with temperature compensation of the thermal drift of the input capacitance of inverter gates.

at 20 °C and, respectively, 0%RH, 50%RH, and 100%RH, design the circuit for temperature compensation.

8.8 Figure P8.8 shows a circuit to measure a temperature difference. Each temperature sensor is an RTD that has 1000 Ω and $\alpha = 3.912 \times 10^{-3}$

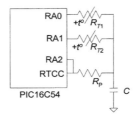

Figure P8.8 Direct sensor–microcontroller interface for differential temperature measurement based on measuring a time difference.

$\Omega/\Omega/K$ at $0\,°C$. The circuit works by measuring the difference in time to charge C through either R_{T1} or R_{T2}. R_p limits the discharge current through RA2 and RTCC. The microcontroller is supplied at 5 V, has an 8 MHz clock, and needs four clock cycles to perform a machine cycle ($p = 4$). An internal 8 bit counter determines how long it takes to charge C to the high threshold voltage since RA0 and RA1 are successively set to high level. If the temperature range is from $-50\,°C$ to $500\,°C$, and the maximal temperature difference between probes is $50\,°C$, determine C in order to obtain the maximal possible resolution available for the temperature difference. Consider that the output high level is 5 V and that the input trigger level is the minimum in data sheets. What would happen if the high levels at RA0 and RA1 were the minimum in data sheets?

8.9 A given sensor yields a frequency output from 5 kHz to 10 kHz. If we wish to determine that frequency with 10 bit resolution from the measurement of its period, what is the resolution needed? If we use an 8051 microcontroller with a 1 MHz clock, how long does the measurement take in the worst case?

8.10 A given sensor yields a frequency output from 9 kHz to 11 kHz. If we wish to measure it with 10 Hz resolution by measuring its period with an 8051 microcontroller that has a 1 MHz clock, determine the number of bits needed to measure the period and the number of cycles of the input signal to count.

8.11 The H1 capacitive humidity sensor (Philips) has a sensitivity of about 0.4 pF/% RH and $C = 122$ pF when RH $= 43\%$. In order to measure the relative humidity in a room, we place the sensor in a relaxation oscillator whose frequency is determined from the period measured by an 8051 microcontroller that counts how much time it takes to count a given number of input cycles.

  a. If the clock is 10 MHz and the measurement time must be shorter than 10 ms, determine the oscillation frequency for the relaxation oscillator and the number of periods to count in order to have a resolution of 2% RH.

  b. If the relaxation oscillator oscillates at 100 kHz when RH $= 43\%$, determine the error if we assumed that the oscillation frequency were proportional to the relative humidity.

8.12 We measure the phase shift between two 10 kHz sine waves by counting the time between zero crossings.

  a. If the maximal phase shift to measure is $90°$, determine the clock frequency needed to obtain a $0.1°$ resolution.

  b. If the clock available is 10 MHz, determine the resolution we can achieve when measuring a phase shift of $45°$ and calculate the time it takes to measure it.

REFERENCES

[1] Anonymous. *Design and Operational Considerations for the HEDS-5000 Incremental Shaft Encoder.* Application Note 1011. Palo Alto: Hewlett Packard, 1981.

[2] J. R. R. Mayer. Optical encoder displacement sensors. Section 6.8 in: J. G. Webster (ed.). *The Measurement, Instrumentation, and Sensor Handbook.* Boca Raton, FL: CRC Press, 1999.

[3] J. R. R. Mayer. High-resolution of rotary encoder analog quadrature signals. *IEEE Trans. Instrum. Meas.*, **43**, 1994, 494–498.

[4] M. Heiss. Error-detection unit-distance code. *IEEE Trans. Instrum. Meas.*, **39**, 1990, 730–734.

[5] E. M. Petriu. Absolute position measurement using a pseudorandom binary encoding. *IEEE Instrum. Meas. Magazine*, **1**, September 1998, 19–23.

[6] R. A. Busser. Resonant sensors. Chapter 7 in: H. H. Bau, N. F. de Rooij, and B. Kloeck (eds.) *Mechanical Sensors*, Vol. 7 of *Sensors, A Comprehensive Survey*, W. Göpel, J. Hesse and J. N. Zemel (eds.). New York: VCH Publishers (John Wiley & Sons), 1994.

[7] G. Stemme. Resonant silicon sensors. *J. Micromech. Microeng.*, **1**, 1991, 113–125.

[8] E. Benes, M. Gröschl, W. Burger, and M. Schmid. Sensors based on piezoelectric resonators. *Sensors and Actuators A*, **48**, 1995, 1–21.

[9] H. Ito. Balanced adsorption quartz hygrometer. *IEEE Trans. Ultrason. Ferroelectr. Freq. Control*, **34**, 1987, 136–141.

[10] G. Fischerauer, A. Mauder, and R. Müller. Acoustic wave devices. Chapter 5 in: H. Meixner and R. Jones (eds.) Micro- and Nanosensor Technology/Trends in Sensor markets, Vol. 8 of *Sensors, A Comprehensive Survey*, W. Göpel, J. Hesse, and J. N. Zemel (eds.). New York: VCH Publishers (John Wiley & Sons), 1995.

[11] A. M. Madni and R. D. Geddes. A micromachined quartz angular rate sensor for automotive and advanced inertial applications. *Sensors*, **16**, August 1999, 26–34.

[12] E. J. Staples. A new electronic nose. *Sensors*, **16**, May 1999, 33–40.

[13] W. S. Arnold and W. I. Norman. Noncontact tension measurement. *Measurements & Control*, Issue **175**, 1996, 121–127.

[14] W. M. Mattar and J. H. Vignos. Vortex shedding flowmeters. Section 28.8 in: J. G. Webster (ed.). *The Measurement, Instrumentation, and Sensor Handbook.* Boca Raton, FL: CRC Press, 1999.

[15] J. Yoder. Coriolis effect mass flowmeters. Section 28.10 in: J. G. Webster (ed.). *The Measurement, Instrumentation, and Sensor Handbook.* Boca Raton, FL: CRC Press, 1999.

[16] S. Franco. *Design with Operational Amplifiers and Analog Integrated Circuits*, 2nd ed. New York: McGraw-Hill, 1998.

[17] B. Parzen. *Design of Crystal and Other Harmonic Oscillators.* New York: John Wiley & Sons, 1983.

[18] M. Watts. *CMOS Oscillators.* Application Note 118. Santa Clara, CA: National Semiconductor.

[19] R. Pallàs-Areny and J. G. Webster. *Analog Signal Processing.* New York: John Wiley & Sons, 1999.

[20] J. Williams. *Some Techniques for Direct Digitization of Transducer Outputs.* Application Note 7. Milpitas, CA: Linear Technology, 1985.

[21] K. M. Daugherty. *Analog-to-Digital Conversion, A Practical Approach.* New York: McGraw-Hill, 1995.

[22] J. E. Buchanan. *Signal and Power Integrity in Digital Systems.* New York: McGraw-Hill, 1996.

[23] T. Williamson. Using the 8051 microcontroller with resonant transducers. *IEEE Trans. Ind. Electr.*, **32**, 1985, 308–312.

[24] R. Bonnert. Design of a high performance digital tachometer with a microcontroller. *IEEE Trans. Instrum. Meas.*, **38**, 1989, 1104–1108.

[25] L. W. Thompson. *Industrial Data Communications, Fundamental and Applications,* 2nd ed., Research Triangle Park, NC: Instrument Society of America, 1997.

[26] R. N. Johnson. Building plug-and-play networked smart transducers. *Sensors*, **14**, October 1997, 40–61.

9

OTHER SENSING METHODS

The very wide variety of devices and methods used to measure different physical quantities makes it difficult for any sensor classification criterion to be exhaustive. The criterion followed in this book is not an exception. This chapter discusses some additional sensors and measurement methods that are not based on any of the measurement principles described in previous chapters. They rely on semiconductor devices (not just semiconductor materials) or on some radiation modified by the measurand.

9.1 SENSORS BASED ON SEMICONDUCTOR JUNCTIONS

Semiconductor junctions are the basis of self-generating sensors such as some photoelectric cells (Section 6.4) and of several modulating sensors. The latter, however, need a current or voltage bias in order to provide a useful output, in the same way that modulating sensors based on resistance or reactance variation need voltage or current excitation.

Sensors based on semiconductor junctions are of twofold interest. First, the large yield of microfabrication processes results in very competitive prices for them. Second, it is possible to include sensor, signal conditioning, signal processing, and communication circuits to produce intelligent sensors (Section 8.7). The journal *Sensors and Actuators* reports on scientific research in sensors (and actuators) based on semiconductors (in general). The magazine *Sensors* covers industrial developments. Reference 1 authoritatively reviews fundamentals and applications of semiconductor sensors.

501

9.1.1   Thermometers Based on Semiconductor Junctions

The forward characteristic for a diode is temperature-dependent (about -2 mV/°C for silicon diodes), which is usually considered a shortcoming. However, we can use that dependence to measure temperature or any other quantity related to a change in temperature (see Problem 9.1). But this dependence is nonlinear and not repetitive enough for accurate measurements. It is therefore better to use the temperature dependence of the base–emitter voltage v_{BE} of a transistor supplied with a constant collector current.

According to the Ebers–Moll model, the collector current for an ideal transistor is

$$i_C = \alpha_F I_{ES}(e^{qv_{BE}/kT} - 1) - I_{CS}(e^{-qv_{CB}/kT} - 1) \tag{9.1}$$

where

α_F = the forward current transfer ratio
I_{ES} = the emitter saturation current
$q = 0.160$ aC is the electron charge
v_{BE} = the base–emitter voltage
$k = 1.3807 \times 10^{-23}$ J/K is Boltzmann's constant
T = the absolute temperature
I_{CS} = the collector saturation current
v_{CB} = the collector–base voltage

The product $a_F I_{ES}$ is sometimes designated I_S. In the active zone, $i_C \gg I_S$. If in addition we make the collector–base voltage zero, from (9.1) we deduce

$$v_{BE} = \frac{kT}{q} \ln \frac{i_C}{I_S} \tag{9.2}$$

which shows that v_{BE} depends on the temperature, but I_S is also temperature-dependent according to [2]

$$I_S = BT^3 e^{(-qV_{g0})/kT} \tag{9.3}$$

where B is a constant that depends on doping level and on the geometry but does not depend on the temperature, and V_{g0} is the band-gap voltage (1.12 V at 300 K for silicon).

By combining (9.2) and (9.3), we obtain

$$v_{BE} = \frac{kT}{q} \ln \frac{i_C}{BT^3} + V_{g0} \tag{9.4}$$

If we designate V_{BE0} the base–emitter voltage corresponding to a constant collector current I_{C0} at a given temperature T_0, then we have

$$v_{BE} = \frac{kT}{q} \ln \frac{i_C}{I_{C0}} \left(\frac{T_0}{T}\right)^3 + (V_{BE0} - V_{g0})\frac{T}{T_0} + V_{g0} \tag{9.5}$$

The relation between v_{BE} and T is therefore nonlinear and depends on the collector current. To quantify the nonlinearity, we take the derivative with respect to the temperature at a given constant collector current. For $i_C = I_{C0}$ we have

$$\left.\frac{dv_{BE}}{dT}\right|_{i_C=I_{C0}} = \frac{V_{BE0} - V_{g0}}{T_0} - \frac{3k}{q}\left(1 + \ln \frac{T}{T_0}\right) \tag{9.6}$$

The first term on the right-hand side is the sensitivity, while the second term describes the nonlinearity. Their respective values for silicon are about -2.2 mV/°C and 0.34 mV/°C.

Example 9.1 The thermometer in Figure E9.1 uses a diode-connected transistor that at 25 °C has $v_{BE} = 0.595$ V and -2.265 mV/°C temperature coefficient when the collector current is 100 μA. If $I_0 = 100$ μA, design the circuit to obtain an output range from 0 V to 10 V for a temperature range from 0 °C to 100 °C. Determine the temperature error at 0 °C because of the op amps' offset voltages when the op amps are at ambient temperature of 30 °C. If the resistors have 1% tolerance, determine the error due to their standardized value and their tolerance.

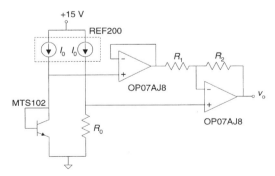

Figure E9.1 Thermometer based on the temperature coefficient of the base–emitter junction of a diode-connected transistor.

The output voltage will be

$$v_{\mathrm{o}} = I_0 R_0 \left(1 + \frac{R_2}{R_1}\right) - v_{\mathrm{BE}} \frac{R_2}{R_1}$$

where the base–emitter voltage is

$$v_{\mathrm{BE}}(T) = 0.595 \text{ V} - (2.265 \text{ mV/}^\circ\text{C})(T - 25\,^\circ\text{C})$$

The conditions to fulfill at $0\,^\circ$C and $100\,^\circ$C are

$$0 \text{ V} = I_0 R_0 \left(1 + \frac{R_2}{R_1}\right) - 0.595 \text{ V} - (2.265 \text{ mV/}^\circ\text{C})(0\,^\circ\text{C} - 25\,^\circ\text{C})\frac{R_2}{R_1}$$

$$10 \text{ V} = I_0 R_0 \left(1 + \frac{R_2}{R_1}\right) - 0.595 \text{ V} - (2.265 \text{ mV/}^\circ\text{C})(100\,^\circ\text{C} - 25\,^\circ\text{C})\frac{R_2}{R_1}$$

which lead to the equation system

$$0 \text{ V} = (10^{-4} \text{ A}) R_0 \left(1 + \frac{R_2}{R_1}\right) - (0.6516 \text{ V})\frac{R_2}{R_1}$$

$$10 \text{ V} = (10^{-4} \text{ A}) R_0 \left(1 + \frac{R_2}{R_1}\right) - (0.4521 \text{ V})\frac{R_2}{R_1}$$

We obtain $R_2/R_1 = 44.15$. If $R_1 = 1$ kΩ, we need $R_2 = 44.1$ kΩ and $R_0 = 6371\ \Omega$.

The output voltage because of offset voltages is

$$v_{\mathrm{o}}(0) = V_{\mathrm{io2}}\left(1 + \frac{R_2}{R_1}\right) - V_{\mathrm{io1}}\frac{R_2}{R_1} = 45.15 V_{\mathrm{io2}} - 44.15 V_{\mathrm{io1}}$$

Because of power dissipation, op amps will raise their temperature above $30\,^\circ$C. Nevertheless, the OP07A has its offset voltage specified after warm-up. Therefore, we need to consider only the temperature difference from $25\,^\circ$C ambient temperature in data sheets to $30\,^\circ$C actual ambient temperature. In a worst-case condition, for the output op amp we have

$$V_{\mathrm{io2}} = 25\ \mu\text{V} + (0.6\ \mu\text{V/}^\circ\text{C})(30\,^\circ\text{C} - 25\,^\circ\text{C}) = 28\ \mu\text{V}$$

For the first op amp, the worst-case condition is to have an equal but opposite initial voltage $(-25\ \mu\text{V})$ and the typical drift (instead of the maximal drift as supposed for the output op amp). Hence,

$$V_{\mathrm{io2}} = -25\ \mu\text{V} + (0.2\ \mu\text{V/}^\circ\text{C})(30\,^\circ\text{C} - 25\,^\circ\text{C}) = -24\ \mu\text{V}$$

and

$$v_o(0) = 45.15 \times (28\ \mu V) - 44.15(-24\ \mu V) = 2.3\ mV$$

which implies an error of about $0.02\,°C$.

If we do not trim each resistor, we have errors because calculated values may be different from standard resistor values, and also because of resistor tolerance. If $R_1 = 1\ k\Omega$ the closest standard values for R_2 and R_0 with $1\,\%$ tolerance are $R_2 = 44.2\ k\Omega$ and $R_0 = 6.34\ k\Omega$. We will therefore have zero and sensitivity error. The worst-case situation because of tolerance will happen when R_0 and R_2 have their minimal value and R_1 is maximal. At $0\,°C$ we will have

$$v_o(0) = (10^{-4}\ A) \times (6.34\ k\Omega) \times 0.99 \times \left(1 + \frac{44.2 \times 0.99}{1 \times 1.01}\right)$$

$$- (0.6516\ V)\frac{44.2 \times 0.99}{1 \times 1.01} = -0.4\ V$$

which implies an error of $-4\,°C$. At $100\,°C$, the same resistors would yield

$$v_o(100) = (10^{-4}\ A) \times (6.34\ k\Omega) \times 0.99 \times \left(1 + \frac{44.2 \times 0.99}{1 \times 1.01}\right)$$

$$- (0.4251\ V)\frac{44.2 \times 0.99}{1 \times 1.01} = 9.4\ V$$

Therefore, the sensitivity would be $98\ mV/°C$ instead of $100\ mV/°C$.

The nonlinearity of the base–emitter voltage and the requirement for a collector current that must be kept constant with time and temperature make this solution unattractive. The usual alternative consists of using two bipolar transistors whose emitter current densities have a constant ratio.

A method for that uses two identical transistors supplied by different collector currents (Figure 9.1a). If both sensors are at the same temperature, the difference between the respective base–emitter currents is

$$v_d = v_{BE1} - v_{BE2} = \frac{kT}{q} \ln \frac{I_{C1}}{I_{S1}} - \frac{kT}{q} \ln \frac{I_{C2}}{I_{S2}} \tag{9.7}$$

If both transistors are assumed identical, we have $I_{S1} \approx I_{S2}$ and

$$v_d = \frac{kT}{q} \ln \frac{I_{C1}}{I_{C2}} \tag{9.8}$$

Therefore, if I_{C1}/I_{C2} is constant, v_d will be proportional to T, without requiring any current source to be kept constant. It is sufficient to have this ratio between both current sources constant. In Figure 9.1a, $I_{C1}/I_{C2} = 2$, so that

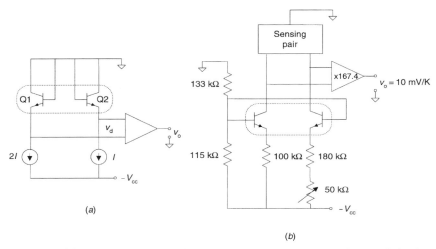

Figure 9.1 (*a*) Thermometer based upon the temperature dependence of the base–emitter voltage in a bipolar transistor. The use of two current sources with a given ratio makes it unnecessary to have a stable reference current and provides an increased linearity. (*b*) Circuit for current sources of 5 μA and 10 μA.

$v_d/T = 59.73$ μV/K. A following differential amplifier with a gain of 167.4 yields an output of 10 mV/K (Figure 9.1*b*) (see also Problem 9.2).

An alternative thermometer uses two transistors with different emitter areas but with the same collector current. Figure 9.2 shows the simplified diagram for a widely used sensor of this type that behaves as a temperature-to-current converter [3]. Its equivalent circuit is a two-terminal current source that passes a current equal (in microamperes) to the absolute temperature.

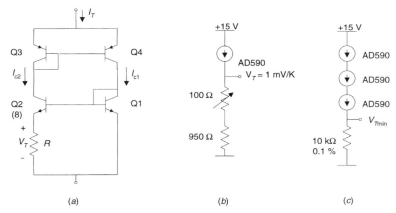

Figure 9.2 (*a*) Simplified circuit for a temperature-to-current converter. (*b*) Thermometer with 1 mV/K output. (*c*) Circuit to detect the minimal temperature. (Courtesy of Analog Devices.)

Transistors Q3 and Q4 are equal and form a current mirror so that

$$I_{C1} = I_{C2} = \frac{I_T}{2} \qquad (9.9)$$

Q2 consists of 8 transistors in parallel, equal to each other and to Q1. Then the emitter current density is 8 times larger in Q1 than in Q2. By designating I_1 the collector current for transistor Q1 and designating I_2 the collector current for each of the Q2 transistors, the output voltage will be

$$v_T = \frac{kT}{q} \ln \frac{I_1}{I_2} = \frac{k}{q} (\ln 8) T = (179 \ \mu V/K) \times T \qquad (9.10)$$

The input current will be

$$I_T = 2I_{C2} = \frac{2v_T}{R} \qquad (9.11)$$

If R is adjusted to 358 Ω, we will have, independently of the applied voltage (over a given range),

$$\frac{I_T}{T} = 1 \ \mu A/K \qquad (9.12)$$

Alternatively, we can use a single transistor, switch two known currents, and subtract the respective voltages.

Having a current output is an advantage for remote measurements because cable length and interfering voltages due to capacitive interference do not noticeably affect the measurement thanks to the low impedance for the detecting circuit. Figure 9.2b shows how to obtain a voltage output. Figure 9.2c shows several sensors connected in series, so that the output voltage is proportional to the minimal temperature.

Table 9.1 gives some characteristics for these and similar temperature sen-

TABLE 9.1 Characteristics of Several Temperature Sensors with Analog Output Based on Semiconductor Junctions

Model	Sensitivity	Range	Accuracy
AD592CN	1 μA/K	$-25\,°C$ to $+105\,°C$	$\pm 0.5\,°C$
ADT43	20.0 mV/°C	$5\,°C$ to $+100\,°C$	$\pm 1.0\,°C$
AD22100K	22.5 mV/°C	$-50\,°C$ to $+150\,°C$	$\pm 2.0\,°C$
LM35A	10.0 mV/°C	$-55\,°C$ to $+150\,°C$	$\pm 1.0\,°C$
LM35D	10.0 mV/°C	$0\,°C$ to $+100\,°C$	$\pm 2.0\,°C$
LM45B	10.0 mV/°C	$-30\,°C$ to $+100\,°C$	$\pm 2.0\,°C$
LM62	15.6 mV/°C	$-10\,°C$ to $+125\,°C$	$\pm 2.0\,°C$
TC1046	6.25 mV/°C	$-40\,°C$ to $+125\,°C$	$\pm 2.0\,°C$
TMP01	5 mV/K	$-55\,°C$ to $+125\,°C$	$\pm 1.0\,°C$
TMP17F	1 μA/K	$-40\,°C$ to $+105\,°C$	$\pm 2.5\,°C$

sors, some with voltage proportional to the absolute temperature (VPTAT). In general, they are less expensive than RTDs, are more linear than thermistors and thermocouples, and offer a higher-level output than RTDs or thermo-couples. However, they have a reduced operating range ($-55\,°C$ to $150\,°C$ maximum, are less accurate and linear than RTDs, and have slower response than bare thermocouples. They are commonly used in temperature controllers, thermostats, HVAC systems, thermal protection (e.g., in PCs), industrial pro-cess control, and temperature probes for digital multimeters. Because of their relative low mass, they are fast (1.5 to 10 s response time for $50\,°C$ changes). Furthermore, if the probe is electrically insulated, they permit measurement of the temperature of operating active components. Because their measurement range includes ambient temperature, they are also used for temperature coeffi-cient compensation and for cold junction compensation in thermocouple cir-cuits (see Problems 9.3 to 9.5 and 6.3, 6.4, and 6.6). They are also the basis of IC sensors with an integrated window comparator for temperature control. Some models even integrate an ADC (e.g., AD7816, BU9817FV, LM75, LTC1392, THMC50, TMP03, TMP04). The AD22103 has an output that is ratiometric with its supply voltage V_s according to

$$v_o = \frac{V_s}{3.3\ \text{V}} \left(0.25\ \text{V} + \frac{28\ \text{mV}}{°C} \times T_A \right)$$

If V_s is also used as the reference for an ADC, the digital output depends only on the temperature.

9.1.2   Magnetodiodes and Magnetotransistors

I–V characteristics for a diode change in a magnetic field perpendicular to the direction of travel of charge carriers because the Lorentz force deviates those carriers from their trajectories. If a diode is designed so that the carriers are deviated to a high recombination region, then the I–V characteristic of the re-sulting magnetodiode depends on the magnetic field intensity. The sensitivity to the magnetic field increases when recombination characteristics for the high and low recombination regions differ substantially. This sensitivity is about ten times higher than that of a silicon Hall-effect device. However, magnetodiodes need unconventional IC processes that are expensive.

This same principle can be applied to magnetotransistor design, but another structure is preferred, which consists of a base, an emitter, and two collectors (reference 1, Section 5.4). When no magnetic field is present, both collector currents are equal. When a magnetic field is applied, one collector current in-creases and the other decreases. The difference between them is a measure of the applied field intensity.

An alternative sensor uses a Hall element and two transistors. The Hall ele-ment is the base common to both transistors and has two contacts, one at each base. When a magnetic field generates a Hall voltage between both contacts, the base voltage for one transistor is larger than that for the other, thus result-ing in collector current imbalance, which is a measure of the applied field. It is also possible to arrange the Hall element so that it controls the gate voltage of a field-effect transistor.

None of these devices has yet found broad commercial use, mostly because of their poor repeatability, low sensitivity, and offset problems. In addition, some of the best devices are incompatible with standard IC processes [4].

9.1.3 Photodiodes

Section 6.4.1 discusses how the internal photoelectric effect in a p–n junction results in a change in the contact potential or in the short-circuit current, which depends on the intensity of the incident radiation. Photodiodes are based on the same principle; but instead of using them as self-generating sensors, we apply an inverse bias voltage, usually from 5 V to 30 V. This increases the width of the depletion region and yields a faster response time and a current propor-tional to the radiation intensity.

Figure 9.3 shows the structure for a photodiode. Because nondepleted p and n regions are conductive, any applied voltage is applied to the depletion region, where it creates an electric field. Any incident radiation absorbed produces electron–hole pairs, which accumulate in the p and n regions because of the electric field, thus resulting in a voltage (photovoltaic effect). In order to collect the output current, charges have to migrate to the diode surface, which slows the response time. This results in a higher recombination probability, which reduces the responsivity (sensitivity).

Figure 9.4 shows the response of a photodiode to a square pulse of radiation. When there is no polarization (0 V), the response is slow because of the slow charge migration toward the surface. But when a small inverse voltage is ap-plied (5 V), charges generated in the depletion region are quickly collected and are responsible for the fast initial response. Charges produced outside the de-

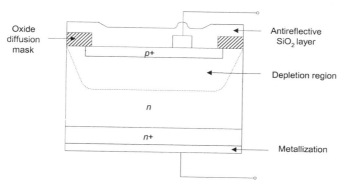

Figure 9.3 Photodiode structure showing the depletion region and the thin p-layer.

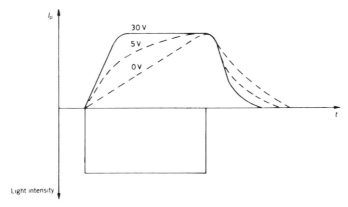

Figure 9.4 Speed of response for a photodiode as a function of reverse bias voltage amplitude (courtesy of Centronic).

pletion region migrate very slowly and are responsible for the slow part of the response. For a larger applied voltage (30 V) the depletion region extends to the entire device depth, which results in a single fast rising edge.

A method to increase the sensitivity and spectral bandwidth for photodiodes consists of placing a region of intrinsic semiconductor between the p and n regions, thus forming a p–i–n diode. Then most of the incident photons are absorbed in this intrinsic region where there is a lower recombination rate. The increased separation between doped zones also results in a reduced internal capacitance.

A bias voltage large enough to bring the photodiode near breakdown yields a chain reaction—termed avalanche multiplication—which amplifies the basic photodiode current by up to 100. This permits low-light measurement and also high-speed measurement. However, avalanche photodiodes (APDs) are so sensitive to tolerances in bias voltage and diode characteristics that they may need individual circuit adjustment [5].

Spectral response for a photodiode depends on the absorption of its window and also on the detecting material itself. Silicon, for example, is transparent to radiation with a wavelength longer than 1100 nm. Hence this radiation is neither absorbed nor detected. Also, very short wavelengths hardly penetrate into the material and are absorbed only in a very thin surface layer. Therefore surface finishing is critical and the p-doped zone is made very thin. There is also a loss due to the absorption in antireflective coatings, because these improve the response to some wavelengths but reduce it at those wavelengths where they are somewhat reflective. Detector input windows are selected to improve the response to the desired wavelengths in the intended application. For example, to detect infrared radiation we can use either (a) a plastic window that filters visible light ($\lambda < 800$ nm) and transmits radiation with 850 nm $< \lambda < 1000$ nm or (b) germanium that is transparent in the range from 800 nm to 1800 nm. Photodiodes are available for the wavelength range from 0.2 μm to 2 μm.

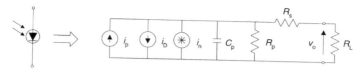

Figure 9.5 Photodiode equivalent circuit. i_p is the signal current; i_D is the leakage current (dark current); i_n is the total noise current spectral density; R_p is the dynamic resistance; C_p is diode capacitance; R_s is the series resistance; R_L is the load resistance.

Some color sensors use a red, a blue, and a green filter preceding the photodiode. The color is determined by measuring the current generated by the light transmitted through each filter. A shortcoming is that filters also attenuate the intensity of light of the desired wavelength. An alternative method uses a transparent window and two stacked (back-to-back) p–i–n photodiodes. The photodiodes' spectral response depends on the voltage difference applied to them. By sequentially applying three different bias voltages we can detect the intensity of the three basic colors.

Figure 9.5 shows the equivalent circuit for a photodiode connected to a load resistance R_L. It is similar to that in Figure 6.25 for a photoelectric cell, but now leakage (i_D, dark current) and noise currents (i_n) have been added. Table 9.2 gives some specifications for two particular photodiodes.

If we neglect the noise current in Figure 9.5, the output voltage is

$$v_o = (i_p - i_D) \frac{R_L R_p}{R_s + R_L + R_p} \tag{9.13}$$

The current sources are

$$i_p = \alpha q \Phi A = S \times P \tag{9.14}$$

$$i_D = I_s(e^{qv_d/kT} - 1) \tag{9.15}$$

TABLE 9.2 Some Specifications of Two Commercial Photodiodes

Parameter	Hewlett Packard 5082-4203 (p–i–n)	Hamamatsu G1115 (GaAsP)
Active area	0.2 mm^2	1.7 mm^2
Sensitivity (responsivity)	0.5 A/W at 770 nm	0.29 A/W at 560 nm
Leakage current at 25 °C	2.0 nA at −10 V	10 pA at −10 mV
NEP/$\sqrt{B}$	51 fW/$\sqrt{\text{Hz}}$	0.9 fW/$\sqrt{\text{Hz}}$
C_p	1.5 pF at −25 V	600 pF
R_s	50 Ω	—
R_p	100 GΩ	20 GΩ
Rise and fall time without bias	300 ns with 50 Ω load	1.5 µs with 1 kΩ load
Rise and fall time at −20 V	1 ns with 50 Ω load	—

where

α = the quantum yield for the detector (electron–hole pairs generated per second divided by number of photons incident per second)

q = the charge of an electron

Φ = the incident flux density

A = the detector area

S = sensitivity (responsivity)

P = incident power

I_S = the reverse saturation current

v_d = the voltage applied to the diode

k = Boltzmann's constant

T = the absolute temperature

Figure 9.6 shows the current through the photodiode, as a function of the applied voltage. For a given load resistance, from these curves and (9.13) we can calculate the output voltage. Because the sensitivity depends on temperature, to achieve a constant sensitivity the bias voltage must track changes in temperature.

As in other electronic devices, noise limits the minimal detectable signal. If noise is considered as a signal due to an incident radiation, the power for the radiation necessary to yield that signal is called *noise equivalent power* (NEP). For a biased diode, the major noise source is the shot current (Section 7.4.1) associated with the average leakage or dark current I_D (as low as 100 pA). From (7.50), this noise current is

$$I_{sh} = \sqrt{2qI_D B} \qquad (9.16)$$

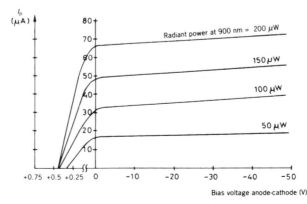

Figure 9.6   *I–V* characteristic for a photodiode (courtesy of Hewlett Packard).

where $q = 0.169$ aC is the electron charge and B the noise bandwidth. If the detector flux responsivity at the working wavelength is S [A/W], we have

$$\text{NEP} = \frac{I_{sh}}{S} \tag{9.17}$$

When the diode is unbiased (photovoltaic mode), the leakage current is very small (a few picoamperes). From (9.15), if v_d is very small, the leakage current tends to 0 A ($I_s - I_s$). However, the shot noise of these currents do not cancel each other; rather their power (intensity) adds. The resulting noise is equal to the thermal noise associated with the dynamic resistance R_p,

$$I_{tp} = \sqrt{\frac{4kTB}{R_p}} \tag{9.18}$$

If the load resistance in Figure 9.5 is not very large, in (9.18) we must replace R_p by $R_p \| R_L$. A large bias voltage increases I_D and R_p. Therefore for high reverse bias the shot noise predominates while at low reverse bias the thermal noise predominates. At frequencies below about 20 Hz or 30 Hz there is an additional $1/f$ noise current.

Temperature affects the noise due to the increase in leakage current because of thermal electron–hole pair generation. These currents double each $10\,°C$ of temperature increase. In an unbiased photodiode, we can calculate the temperature effects from (9.18). There is also a reduction in R_p due to thermal electron–hole pair generation.

Noise also depends on sensing area. A large area increases noise. However, whereas noise increases according to the square root of the electric capacitance of the diode, the output current increases proportionally to the area. Therefore, large-area photodiodes (up to 1 cm^2) have increased signal-to-noise ratio. Nevertheless, since capacitance increases linearly with area, large-area photodiodes have slow response.

Example 9.2 Calculate the NEP for a photodiode biased so that $I_D = 10$ nA, $R_p = 100$ MΩ, $S = 0.5$ A/W, when operating at 45 °C and the noise bandwidth is from 10 kHz to 100 kHz.

From (9.16), the shot noise from the dark current is

$$I_{sh} = \sqrt{2 \times (0.169\ \text{aC}) \times (10\ \text{nA}) \times (100\ \text{kHz} - 10\ \text{kHz})} = 17\ \text{pA}$$

From (9.18), the thermal noise from the parallel resistance is

$$I_{tp} = \sqrt{\frac{4 \times (1.38 \times 10^{-23}\ \text{J/K}) \times (318\ \text{K})(100\ \text{kHz} - 10\ \text{kHz})}{100\ \text{MΩ}}} = 4\ \text{pA}$$

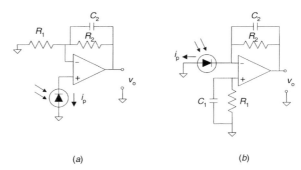

(a)   (b)

Figure 9.7 Amplifiers for unbiased photodiodes. (*a*) Photovoltaic mode. (*b*) Current detection mode.

We assume that both current sources are independent and, hence, add their powers. Because 17 pA is more than three times 4 pA, the overall current power is about $(17 \text{ pA})^2$. Therefore, from (9.17) we have

$$\text{NEP} = \frac{17 \text{ pA}}{0.5 \text{ A/W}} = 34 \text{ pW}$$

Figure 9.7 shows circuits for amplifying the output of an unbiased photo-diode [6]. In the photovoltaic mode we do not apply any bias and measure the open-circuit voltage (Figure 9.7*a*) or the short-circuit current by a transimpe-dance amplifier (Figure 9.7*b*). The absence of leakage current results in a very low noise, but the relatively high value for C_p due to the lack of applied reverse voltage reduces bandwidth. Since the op amp input current is negligible, the open-circuit output voltage v_o when we measure voltage (Figure 9.7*a*) can be obtained from

$$0 = i_p - I_s(e^{qv_d/kT} - 1) \tag{9.19}$$

which gives

$$v_d = \frac{kT}{q} \ln\left(1 + \frac{i_p}{I_s}\right) \tag{9.20}$$

$$v_o = \left(1 + \frac{R_2}{R_1}\right)\frac{kT}{q} \ln\left(1 + \frac{i_p}{I_s}\right) \tag{9.21}$$

The response is therefore logarithmic.

If we measure the short-circuit current instead (Figure 9.7*b*), $v_d \approx 0$ V and the output voltage is

$$v_o = i_p R_2 \tag{9.22}$$

Choosing $R_1 = R_2$ reduces the effect of op amp input currents to those of its offset current. R_2 can be replaced by a resistor T-network (Figure 7.11) in both circuits. C_2 avoids gain peaking (by compensating C_p—see Problem 9.6), and C_1 reduces noise bandwidth. Problem 9.7 proposes an alternative circuit.

Example 9.3   The HP 5082-4204 *p–i–n* photodiode has a sensitivity of 0.5 μA/μW at 770 nm, a leakage resistance of 100 GΩ, and a capacitance of 6.5 pF (when unbiased). When connected to the transimpedance amplifier in Figure E9.3*a*, the op amp and circuit layout add 3.5 pF to ground. Determine R to obtain an output sensitivity of 1 V/μW at low frequencies. If the incident light is pulsed at 10 kHz, determine the error because of the finite op amp gain and input resistance.

At low frequencies we assume the op amp ideal, so that we have

$$R = \frac{v_o}{i_p} = \frac{1 \text{ V/μW}}{0.5 \text{ μA/μW}} = 2 \text{ M}\Omega$$

Figure E6.3b shows the equivalent circuit when considering stray capacitance and leakage resistance. The circuit equations are

$$\frac{V_o - V_n}{Z} = I_p + \frac{V_n}{Z_p}$$

$$V_o = -A_d \times V_n$$

where Z is the parallel equivalent of R and C while Z_p is the parallel equivalent of R_p and C_p. From this, the transfer function is

$$\frac{V_o}{I_p} = \frac{Z}{1 + \dfrac{1}{A_d}\left(1 + \dfrac{Z}{Z_d}\right)}$$

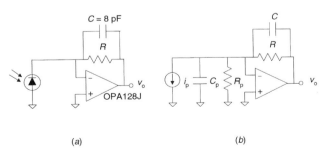

(a)   (b)

Figure E9.3   (*a*) Transimpedance amplifier for an unbiased photodiode. (*b*) Equivalent circuit.

At 10 kHz we have

$$Z = \frac{R}{1 + j2\pi f RC} = \frac{2 \text{ M}\Omega}{1 + j2\pi(10 \text{ kHz})(2 \text{ M}\Omega)(8 \text{ pF})} = \frac{2 \text{ M}\Omega}{1 + j0.32\pi}$$

$$Z_p = \frac{R_p}{1 + j2\pi f R_p C_p} = \frac{100 \text{ G}\Omega}{1 + j2\pi(10 \text{ kHz})(100 \text{ G}\Omega)(10 \text{ pF})}$$

$$= \frac{100 \text{ G}\Omega}{1 + j2\pi \times 10^4} \cong \frac{10 \text{ M}\Omega}{j2\pi}$$

$$A_d(10 \text{ kHz}) = j10^{45/20} = j178$$

A_d has been obtained from the data sheets for the OPA128. The transfer function is then

$$\frac{V_o}{I_p}(10 \text{ kHz}) = \frac{\dfrac{2 \text{ M}\Omega}{1 + j0.32\pi}}{1 + \dfrac{1}{j178}\left(1 + \dfrac{2 \text{ M}\Omega}{1 + j0.32\pi}\dfrac{j2\pi}{10 \text{ M}\Omega}\right)} \cong \frac{2 \text{ M}\Omega}{1 + j}$$

whose amplitude is about 141 kΩ and whose phase is about $-45°$. The gain at 10 kHz is about 30% below that at dc.

The OPA128 has a differential input resistance of 10 TΩ shunted by 1 pF and common mode input impedance of 10 PΩ shunted by 2 pF. Therefore, the input impedance from the inverting input to ground in Figure E6.3 is about 10 TΩ shunted by 3 pF. At 10 kHz the sensor impedance is capacitive and well below 10 TΩ. Therefore, the op amp impedance does not significantly influence the result.

When the diode is (reverse) biased, the model in Figure 9.5 suggests that we measure the output current. Figure 9.8a shows a possible circuit. Biased photodiodes yield a higher noise but also a higher bandwidth than unbiased photodiodes, with rise times as short as 10 ps. The output voltage is now

$$v_o = -(i_p + i_D)R_1 \tag{9.23}$$

Input bias current (I_b) adds to the photodiode current, and op amp input offset voltage (V_{io}) adds to the output voltage. Adding a matching dummy photodiode and a matching resistor as in Figure 9.8b reduces errors from leakage currents.

Photodiodes are commercially available that include amplification, temperature compensation, and stabilization on the same chip [e.g., OPT202, OPT209, OPT301 (Burr–Brown) and TSL25X and TSL26X series (Texas Instruments)]. The TSL220 and TSL230 yield an output voltage whose frequency is proportional to the incident light intensity. Integration reduces leakage currents, inter-

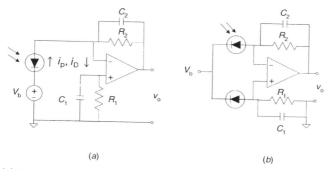

(a)   (b)

Figure 9.8 (a) Amplifier for a reverse-biased photodiode. (b) Adding a matched dummy photodiode and a matched resistor reduces the effect of leakage currents.

ference, and gain-peaking because of stray capacitance. In addition to single elements, arrays are also available, formed by placing several photodiodes 1 mm apart or by diffusing several photodiodes in the same wafer. The TSL213 has 64 pixels with 125 μm center-to-center spacing. The TSL215 has two 64-pixel arrays. Photodiode arrays need fewer supply voltages and clock signals than CCDs (Section 9.3) and can be read pixel by pixel, but they are slower and have poorer resolution.

Example 9.4 The TSL251 integrates a photodiode and an I/V converter, has a sensitivity of 45 mV/(μW/cm²), and has a dark voltage of 3 mV. Its output is connected to an ADC through an intervening voltage amplifier. If the ADC has an input range of 5 V and we wish to obtain null digital output when in dark and maximal digital output when the input irradiance is 50 μW/cm², determine the amplifier gain and the number of bits of the ADC. If the temperature coefficient of the sensor is 1 mV/K, determine the maximal temperature change it may experience without affecting the ADC output.

The gain is obtained from the quotient between the output voltage range and the input voltage range:

$$G = \frac{v_o(\text{max}) - v_o(\text{min})}{v_i(\text{max}) - v_i(\text{min})}$$

The minimal input and output voltages are 0 V. The maximal input voltage will be obtained for the maximal irradiance:

$$v_i(\text{max}) = 45\frac{\text{mV}}{\text{μW/cm}^2} \times 50 \text{ μW/cm}^2 = 2.250 \text{ V}$$

Therefore, $G = (5 \text{ V})/(2.25 \text{ V}) = 2.22$.

In order for the digital output be zero when the sensor is in dark, the output voltage must be less than 1 LSB. Therefore, we need

$$q = \frac{5 \text{ V}}{2^n} > (3 \text{ mV}) \times 2.22 = 6.7 \text{ mV}$$

That is, $n > 9.55$ bit. We would select $n = 10$ bit. If the thermal drift should not induce a bit transition, we need

$$(1 \text{ mV/K}) \times \Delta T < \frac{q}{2} = \frac{4.88 \text{ mV}}{2}$$

which leads to $\Delta T < 2.4\,^\circ\text{C}$.

Photodiodes are used in optical communications, photometers, illumination and brightness control, infrared remote control, distance meters, thickness gages, and optical absorbance measurement. They are also applied to count bills in sorting machines by triggering when the amount of light passing through the bills is less than a set threshold. Ultraviolet (UV) photodiodes are used in spectroscopy, photometry, fluorescence analysis, UV intensity and exposure metering, and for gas and oil flame monitoring. Color sensors are used in product inspection and quality control. Photodiode arrays are used in linear and rotary encoders, bar-code readers, edge detection and positioning, imaging, and paper handling (optical character recognition, document classification, copying machines, facsimile), as well as for automatic focus and exposure control in cameras.

9.1.4   Position-Sensitive Detectors (PSDs)

If a nonuniform radiation strikes a p–n junction, the maximal photocurrent will appear in the region receiving the maximal intensity and a lateral voltage difference will arise different from the transversal voltage difference in Figure 9.3. This effect was discovered by Wallmark in 1957. If the junction is reverse-biased and one layer (say the p-layer) works as a resistor by placing two contacts on it (Figure 9.9a), we have a lateral photodiode or position sensitive detector (PSD). The device works as an optoelectronic potentiometer. If we apply the same voltage to those two contacts, the photocurrent i_p splits in two components i_1 and i_2 such that $i_1 R_1 = i_2 R_2$. That is, i_p divides inversely proportional to the distance from the incident spot to the electrodes. If we designate x the relative position of the light spot (Figure 9.9b) ($-1 \leq x \leq 1$), we have

$$x = \frac{R_2 - R_1}{R_2 + R_1} = \frac{i_1 - i_2}{i_1 + i_2} \tag{9.24}$$

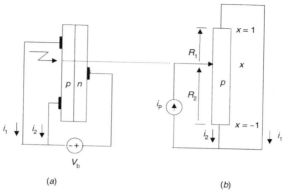

Figure 9.9 (*a*) Lateral photodiode and (*b*) simplified model.

Therefore, the above current ratio is proportional to the relative position of the light spot on the sensor surface. The resolution is larger than 10 μm. The detecting material is chosen to match the wavelength of the incident light. Duo-lateral photodiodes detect the 2-D position of the light spot [7].

Silicon PSDs often have a built-in amplifier with a bandpass characteristic that suppresses (a) constant light and low-frequency alternating light on the one hand and (b) high-frequency interference on the other hand. Applications include autofocus mechanisms, activation of ATMs (automatic teller machines) or vending machines when the user approaches, range finders, and precise location sensing in industrial equipment.

9.1.5 Phototransistors

A phototransistor is an integrated combination of photodiode and n–p–n transistor where the optical radiation illuminates the base. The collector current is

$$i_C = (\beta + 1)(i_p + i_D) \qquad (9.25)$$

where i_p and i_D are given by (9.14) and (9.15), and β is the current gain for the transistor in the common emitter configuration (from 100 to 1000). This gain is not constant but depends on the current and therefore on the illumination level. This lack of linearity and also their lower bandwidth as compared to photodiodes (due to the large base–collector capacitance) makes them less suitable for measurement. Phototransistors, however, suit switching applications because of their current gain. For applications requiring gains up to 100,000, photo-Darlingtons are available that consist of a phototransistor feeding the base of a second transistor. Phototransistors work in the wavelength range from 0.4 μm to 1.1 μm.

Phototransistors are extensively used in photoelectric sensing for industrial

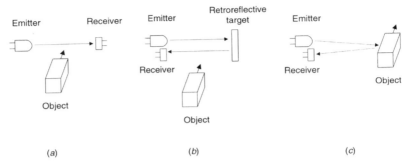

Figure 9.10   Photoelectric sensing modes: (*a*) opposed, (*b*) reflex, and (*c*) proximity.

applications [8]. Photoelectric sensing uses a light source (LED or incandescent lamp—for color applications) and a photoelectric sensor—phototransistor—that responds to a change in the intensity of the light falling upon it. Applications requiring high sensitivity to visible light use photoconductors. Applications requiring a very fast response or a linear response over a broad range of irradiance use photodiodes. The emitted light is modulated by turning the LED on and off at several kilohertz. The amplifier following the phototransistor is tuned to the modulation frequency so that interfering light signals—that is, those not pulsing at the same frequency—are rejected. Intense ambient light, however, may saturate the detector. Mechanical shielding from ambient light reduces interference. Phototransistors have maximal sensitivity to infrared LEDs, but red, green, yellow, and blue LEDs are also used. Sensing ranges are from about 20 cm to more than 50 m.

There are three basic photoelectric sensing modes: opposed, retroreflective, and proximity. In the *opposed sensing mode*, also referred to as through-beam, beam-break, or direct scanning, the light from the emitter is directly aimed at the receiver (Figure 9.10*a*). Any object interrupting the light path is detected. To detect small parts or inspect small profiles, or for very accurate position sensing, the diameter of the light beam can be reduced by placing an aperture at the emitter or receiver side, or at both.

In the *retroreflective sensing mode*, also termed *reflex*, or retro for short, the emitter aims the light beam at the retroreflective target, which bounces it back to the receiver (Figure 9.10*b*). An object is sensed when it interrupts the beam, provided that it is not highly reflective.

In *proximity mode sensing* the object establishes a light beam rather than interrupts it. The emitter and the receiver are on the same side of the object. The emitter sends energy and the receiver detects the energy reflected from the object (Figure 9.10*c*). Bright white surfaces are sensed at a greater range than dull black surfaces. Also, large objects return more energy than small objects. Lens systems focus the emitted light to an exact point in front of the sensor, and they also focus the receiver at the same point. This determines a well-defined sensing area. Shiny objects, however, are not detected unless they remain parallel to the sensor lens.

Phototransistors are also common in chilled-mirror hygrometers. Because warm air can hold more water vapor than cold air, when air is cooled a point is reached where water condenses to form dew, mist, or frost. In chilled-mirror hygrometers, a gold- or rhodium-plated copper surface is thermoelectrically cooled and the dew layer formed is detected by sensing with a phototransistor the reduced infrared light from an LED reflected from the mirror. The mirror temperature is the dew point, with typical accuracy $\pm 0.2\,^{\circ}\mathrm{C}$. A separate LED and phototransistor pair are used to compensate temperature interference. If the mirror is continually cooled, airborne contaminants may adhere to it, thus affecting its reflectivity. Cycling chilled mirrors retain dew on the mirror only a fraction of the time, thus avoiding the contamination problem [9].

9.1.6 Semiconductor-Junction Nuclear Radiation Detectors

Nuclear radiation consists of either subatomic particles (α, β^{+}, β^{-}, protons, neutrons) or photons (electromagnetic radiation) whose energy is high enough to remove electrons from atoms (i.e., ionize them). X rays are also high-energy photons, but they come from the electron shell of atoms instead. Nuclear radiation passing through a semiconductor produces many electron–hole pairs along its track. If those electric charges are produced in the presence of an intense electric field, they can be collected to yield an electric current before they recombine. A reverse-biased diode has a depletion zone (Figure 9.3) suitable for radiation detection. The base silicon must be highly pure to reduce stray currents.

Depending on the fabrication method, there are different semiconductor diode detectors [10]. Diffuse-junction detectors start with a homogeneous crystal of p-type material. One surface is exposed to a high-temperature vapor ($900\,^{\circ}\mathrm{C}$ to $1100\,^{\circ}\mathrm{C}$) of n-type impurity (phosphorous), which converts a region of the crystal close to the surface from p-type to n-type material. At about $0.1\,\mu\mathrm{m}$ to $2\,\mu\mathrm{m}$ from the surface the n-type and p-type impurities reverse their relative concentration and a junction is formed. Because the n-type surface layer is heavily doped compared with the p-type original crystal, the depletion zone extends (from $50\,\mu\mathrm{m}$ to $500\,\mu\mathrm{m}$) primarily on the p side of the junction. The surface layer remaining outside the depletion region constitutes a window through which the incident radiation must pass, hence absorbing some radiation.

Surface barrier detectors start from n-type or p-type silicon. A gold (for n-type) or aluminum (for p-type) layer is deposited by evaporation in conditions allowing slight surface oxidation. The deposited layer is thin enough not to absorb any significant amount of radiation. This window, however, is so thin that it may be attacked by some gases and also permits light transmission, thus resulting in interference.

In ion implanted detectors, the doping impurities at the surface of the semiconductor are introduced in the crystal by bombarding it with ions from a particle accelerator. The acceleration voltage determines the penetration depth. Annealing after bombardment reduces effects from radiation damage. Because the annealing temperature ($500\,^\circ$C) is smaller than that required for diffusing impurities, the structure of the crystal is less disturbed than in a diffused junction detector. Also, entrance windows can be as thin as 34 nm, yet stronger than in surface barrier detectors.

Regardless of the fabrication method, the junction has an approximate capacitance $C_x \approx \epsilon A/x$, where A is the detector area and x is the depth of the depletion layer. If the starting material is p-type silicon ($\epsilon_r \approx 12$), if $x = 1$ mm, then $C_x \approx 10$ pF/cm^2. Because the energy needed to create an electron–hole pair in silicon is 3.62 eV, an incident radiation of 1 MeV would yield $N =$ (1 MeV)/(3.62 eV) = 276,000 pairs, with an electric charge $Q = (276,000) \times$ (0.16 aC) = 40 fC. A detector with $A = 1$ cm^2 would therefore yield $v_\text{o} =$ (0.04 pC)/(10 pF) = 4 mV. In order to obtain an output voltage unaffected by the sensor and the connecting cable capacitance, we can measure this voltage by a charge amplifier. Alternatively, we can measure the current using a transimpedance amplifier.

Gas-filled detectors need 30 eV to produce an electron–hole pair. Scintillation detectors need 500 eV [10]. Therefore, silicon detectors that need just 3.62 eV have better resolution. They are also faster and more linear. Because their density can be up to one thousand times that of gases, they need less volume to capture radiation, hence they have better spatial resolution. Semiconductor diode detectors are used in nuclear medicine, for particle detection in high-energy physics, and in radiography. X-ray detectors are used in tomography, angiography, digital radiography, and nondestructive testing in industry.

9.2   SENSORS BASED ON MOSFET TRANSISTORS

The drain current I_D for an n-channel MOSFET transistor in the linear region ($V_\text{GS} > V_\text{T}$, $V_\text{DS} < V_\text{GS} - V_\text{T}$) is [11]

$$I_\text{D} = \beta V_\text{DS}\left(V_\text{GS} - V_\text{T} - \frac{V_\text{DS}}{2}\right) \tag{9.26}$$

The corresponding equation when in the saturation region ($V_\text{GS} > V_\text{T}$, $V_\text{DS} > V_\text{GS} - V_\text{T}$) is

$$I_\text{D} = \frac{\beta(V_\text{GS} - V_\text{T})^2}{2} \tag{9.27}$$

where β is the sensitivity parameter determined by the gate's dimensions,

$$\beta = \mu C_\text{ox}\frac{W}{L} \tag{9.28}$$

and

μ = the electron mobility in the channel
C_{ox} = the gate capacitance (oxide) per unit area
W = the channel's width
L = the channel's length
V_{GS} = the voltage applied between gate and source
V_{DS} = the voltage applied between drain and source
V_T = the threshold voltage above which an inversion channel is formed

The value of V_T ranges from 1 V to 6 V and is given by

$$V_T = \frac{\Phi_M - \Phi_S}{q} - \frac{Q_{SS}}{C_{ox}} + 2\phi_F - \frac{Q_B}{C_{ox}} \qquad (9.29)$$

where $\Phi_M - \Phi_S$ is the difference in work functions for metal and semiconductor (silicon; the work function is the minimal energy required to move an electron from the Fermi level to infinity), ϕ_F is the Fermi level for the substrate (semiconductor), Q_{SS} is the surface charge density, Q_B is the charge density in the depletion region, and q is the charge of an electron.

When MOSFET transistors are used as electronic components, we assume that all parameters in equations (9.26) through (9.29) are constant. By so doing, we obtain a well-defined input–output relationship between V_G and I_D.

In sensors, on the contrary, we are interested in detecting quantities that modulate some of these parameters, thus changing the input–output relationship. Temperature and radiation, for example, affect μ and ϕ_F. However, this fact is common to many electronic devices, and MOSFET transistors are not frequently used for those applications. The important advantage for MOSFET transistors is the dependence between V_T and the work function (or chemical potential) for the metal, Φ_M [12]. FETs sensitive to chemicals are termed ChemFETs.

If the work function of the metal gate is controlled by an external parameter, the MOS transistor can sense that parameter. If, for example, palladium is used as a gate material instead of the normal aluminum, the palladium adsorbs hydrogen that diffuses into the palladium–oxide interface and forms a dipole layer that changes V_T. Then the threshold voltage gives a measure of hydrogen concentration. The device is made to work at high temperature (50 °C to 150 °C) to promote the palladium catalytic action. This method is sensitive to all gases able to dissociate at the palladium surface, such as H_2O, H_2S, NH_3, and a number of hydrocarbons. Other gases can be detected with a porous gate, where pores permit the gas to reach the palladium–oxide interface. These devices are generically named gasFETs (Figure 9.11a).

Other MOSFET-based sensors are based on gate modifications in a conventional MOSFET that are compatible with usual production technologies.

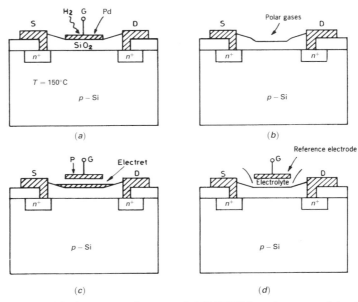

Figure 9.11 Simplified structure for several MOSFET-based sensors: (*a*) with palladium instead of aluminum in the gate; (*b*) ADFET without gate and only a 5 nm oxide layer; (*c*) PRESSFET using an electret on the oxide and separated gate; (*d*) ISFET with electrolyte in contact with the insulator in the gate region and metal reference electrode.

The simplest modification is just to omit the gate. The best-known one consists of interposing a given material between gate and oxide, thus rendering the transistor sensitive to any measurand resulting in changes in that material.

An OGFET (open gate FET) is a MOSFET transistor without a gate that is exposed to a gas. Drain current is then a function of gas partial pressure. An improved version is the ADFET (adsorption FET), whose oxide has a thickness smaller than 5 nm (Figure 9.11*b*). This device senses the concentration of gases having a permanent dipole moment, such as H_2O, NH_3, HCl, CO, NO, NO_2, and SO_2. Its sensitivity is due to the dipolar molecules adsorbed in the oxide layer that create an electric field that controls the drain current. By chemically etching this layer, it seems possible to achieve a selective response. Other variations on the same device are intended to minimize electric interference due to the lack of gate electrode (which when present acts as an electric shield).

If in a conventional MOSFET the gate electrode is separated from the oxide in the vertical direction, equations (9.26) through (9.29) still are valid, but in (9.28) the equivalent capacitance C_{eq} must replace the oxide capacitance C_{ox}. This device will then detect any measurand that changes C_{eq}. Figure 9.11*c* shows the diagram for a pressure sensor based on this principle (PRESSFET), where the pressure applied changes the separation between electrode and oxide.

An electrically polarized material (*electret*) is deposited on the oxide, so it is not necessary to apply any external voltage. An alternative approach is to use a piezoelectric material instead of an electret and an air chamber.

An ISFET (ion-sensitive FET) is a MOSFET transistor that instead of gate electrode has in that region a chemically selective covering or membrane (Section 6.5). ISFETs were first proposed by P. Bergvelt in 1970. When it is immersed into an electrolyte, the potential in the insulation (oxide) depends on the detected ion concentration. That potential depends on the threshold voltage and therefore on the drain current. The reference metal electrode immersed in the same electrolyte can be considered as the equivalent to the gate electrode (which is not present in the device). This way the ISFET can be considered a MOSFET that has an oxide–electrolyte system instead of gate (Figure 9.11*d*). Some problems derive from the poor selectivity and adherence of the membrane, as well as from the stability and photosensitivity of the gate material. The package must allow the contact with the electrolyte but prevent this from entering the electronic circuits. There are commercial ISFETs for pH and glucose, among others. Some biosensors (Section 9.6) rely on ISFETs.

9.3 CHARGE-COUPLED AND CMOS IMAGE SENSORS

9.3.1 Fundamentals

Bell Laboratories patented the charge-coupled device (CCD) in 1970. Several historical papers on CCDs are compiled in reference 13. A CCD is a monolithic array of closely spaced MOS capacitors that transfers an analog signal charge (a "packet") from one capacitor to the next, working as an analog shift register. The charge is stored and transferred between potential wells at or near an $Si–SiO_2$ interface (Figure 9.12). These wells are formed by MOS capacitors (an array of metal electrodes is deposited on the SiO_2) pulsed by a multiphase clock voltage. The charges transferred are electrons or holes, respectively, in *n*-channel and *p*-channel devices and can be introduced either electrically or optically. For optical input the CCD acts as optical sensor. The channels are formed by selectively altering the electric conductivity of the silicon underlying the insulator by diffusing or by growing an *n*-epitaxial layer on the *p* substrate. A complete theoretical analysis can be found in reference 14, while applications are discussed in reference 15.

If in Figure 9.12 a positive step voltage is applied to a gate electrode while adjacent electrodes remain at a lower voltage, a potential-energy well is set up into the *p* substrate, and charge (electrons) is stored under that electrode. If now one of the adjacent electrodes is biased with a voltage higher than that of the previous electrode, a deeper potential well is established and the stored charge travels to it along the surface of the silicon to seek the lowest potential. By clocking electrode voltages, we can thus move, in a given direction, the charge initially stored. However, thermal electron–hole pairs are being continuously

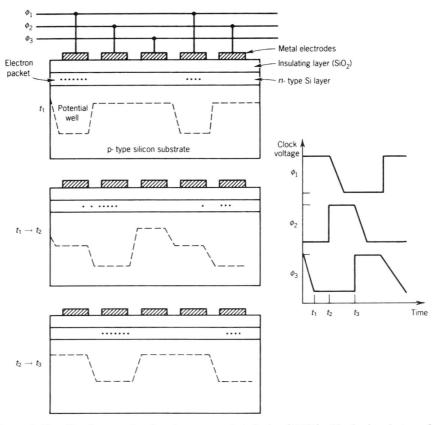

Figure 9.12. Fundamentals of a charge coupled device (CCD). Clock signals transfer electrons accumulated in a potential well.

produced, and electrons eventually fill the well. Therefore, a CCD is a dynamic device in which we can store a charge signal only for a time much shorter than thermal relaxation times for the MOS capacitors. At room temperature, this time is from 1 s to several minutes, depending on the structure and fabrication processes.

Depending on how many clock phases a CCD needs, there are two- and three-phase structures. A three-phase system is like that in Figure 9.12, where every third electrode is connected to the same clock voltage, therefore requiring three separate clock signals. These signals overlap and exhibit a steep leading edge and a linearly falling edge, which increases the efficiency in charge transfer. The charge stored into the wells under the ϕ_1 electrodes at $t = t_1$, not necessarily the same in each well, spreads out to adjacent wells created under electrodes ϕ_2 when a positive voltage step is applied to these electrodes. By t_2, the charge has spread out, and then the voltage ϕ_1 is linearly reduced so that the

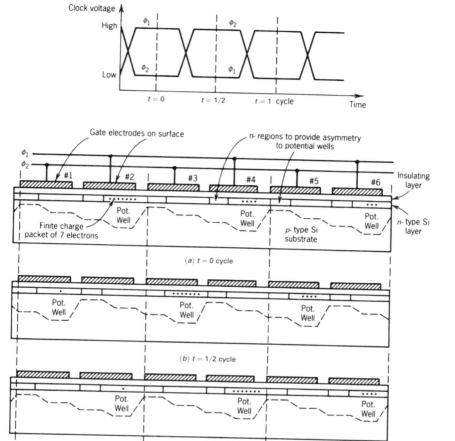

Figure 9.13.   A two-phase CCD based on *n*-type regions of different conductivity (courtesy of Fairchild Wescon Systems).

potential at the wells under the ϕ_1 electrodes rises slowly rather than abruptly. This eases the transfer of charge to the wells under electrodes 2, 5, and others connected to the same clock line. By t_3 the transfer has been completed. A low voltage at ϕ_3 during the interval t_1 to t_3 ensures a transfer to the right with no charge moving to the left of the wells. By repeating the same procedure with ϕ_2 and ϕ_3, we can move the charge another step, and then another one from ϕ_3 to ϕ_1, and so on. By inverting the sequence order, the charge would be moved to the left.

In a two-phase system, the oxide thickness is stepped, or different *n*-type conductivity regions are created (Figure 9.13), so the potential wells beneath each individual electrode are asymmetrical. Now the charge transport is always to the right. Clock pulses do not need to overlap as in a three-phase system, and the layout is more economical. The virtual phase design (Texas Instruments) utilizes a junction–gate region at the substrate dc potential and is placed between the clocked electrodes. This accomplishes the same gating and transport function as a separate gate electrode, thus reducing to one the clock signals required.

In the so-called buried channel CCD, which is the commonest, instead of storing and transferring the charge at the silicon surface under the silicon dioxide, those processes happen in a buried channel away from the Si–SiO$_2$ interface. This yields an increased carrier mobility, hence speed, and independence of surface phenomena (which result in noise and losses).

In a CCD image sensor, light from the object illuminates a CCD, either from the electrode or from the substrate side, and electron–hole pairs are created in the silicon by the photoelectric effect. (Silicon is sensitive to photons with wavelengths from 300 nm to 1100 nm.) By applying the appropriate clock signals, potential wells are created that collect the photon-generated minority carriers for a time called the *optical integration time*, while majority carriers are swept into the substrate. The collected charge packets are shifted down the CCD register and converted into voltage or current at the output terminal. The amount of charge accumulated in each well is a linear function of the illumination intensity and the exposure (or integration) time. The output signal charge is a stepped dc voltage that will linearly vary from a thermally generated background level (noise) at zero illumination (dark condition) to a maximum at saturation under bright illumination.

CMOS image sensors also rely on the photoelectric effect in silicon but do not serially shift charge packets to the common output amplifier. Instead, the charge at each *pixel* (i.e., picture element) is directly sensed and the respective signals are multiplexed toward the signal processing circuits. CMOS image sensors can have passive or active pixels. Sensors with passive pixels (Figure 9.14a) date from about 1967. Each pixel consists of a photodiode and an access transistor to the column bus. After photocharge integration, the array controller turns on the access transistor to transfer the charge to the capacitance of the

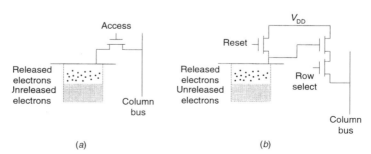

Figure 9.14 (a) Passive photodiode in a CMOS image sensor. (b) Active pixel sensors (APSs) buffer the photodiode with a transistor.

column bus. This bus is connected to a charge-integrating amplifier that senses the voltage and resets the photodiode. The controller then turns off the access transistor. Because each pixel contains only one transistor, the fill factor is high. However, the read noise is high, particularly for far-off pixels that may be unable to charge the distributed capacitance of the bus under low-level illumination. Also, differences in turn-on thresholds for the access transistors yield nonuniform response to identical light levels, thus resulting in fixed-pattern noise (FPN).

Active-pixel sensors (APSs) include an amplifier in each pixel. In active photodiodes (Figure 9.14*b*) there is a source–follower transistor that provides current to charge and discharge the bus capacitance quickly. This permits an increased bus length, hence a larger array size. They also include a reset transistor that controls integration time, hence providing electronic shutter control, and a row-select transistor to coordinate pixel reading. These additional transistors reduce the fill factor and increase FPN because their thresholds are difficult to match. To counter the low fill factor, some CMOS sensors use a microlens for each pixel, but this increases cost because microlens deposition is not a standard CMOS process.

9.3.2   Types of CCD and CMOS Imaging Sensors and Applications

There are two types of CCD imagers available: the line imager and the area imager. The number of pixels ranges from 128 to 16 million. Common linear models have from 1800 to 5400 pixels. Custom models with contiguous lines are limited by the maximal wafer width, but it is possible to stagger several devices to achieve the desired number of pixels. Typical pixel sizes are 27 μm^2 for 512×512 arrays and 12 μm^2 for 1024×1024 arrays.

Line imagers (Figure 9.15) have a line of sensor elements, also called *photo-*

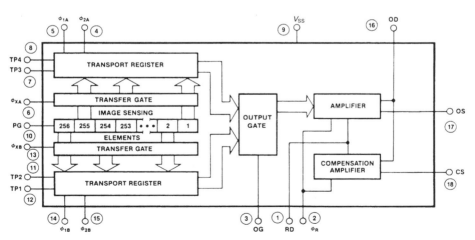

Figure 9.15.   Block diagram of a CCD line image sensor (courtesy of Fairchild Weston Systems).

sites or pixels, with a common electrode called a *photogate* and an individual optical microlens. Pixels are electrically separated from one another by a highly concentrated *p*-type region called a *channel stop*. Some models have some additional sensing elements shielded from external light by an opaque metallization (aluminum) that provide a dark reference level. Subtracting this dummy signal from the required signal eliminates the 6 V to 9 V bias signal. An alternative for offset elimination is correlated double sampling that uses two sample and hold amplifiers (SHAs). The first SHA samples the offset signal during the reference phase and the second SHA samples the complete (offset plus video) signal. Subtracting their outputs yields the video signal. Still other models include white reference cells consisting of input diodes that output a signal of approximately 70 % or 80 % of the saturation level. These signals are useful to external dc restoration and automatic gain control circuits.

Adjacent to the line of sensor elements, there is a structure called a *charge transfer gate*, arranged in two lines, one at each side of the row of sensors. When the transfer gate voltage goes high and the photogate goes low, the photon-generated charge packets, which have accumulated at each pixel when the photogate voltage was high, are alternately transferred into the transfer gate storage well, with odd-numbered pixels to one side, even to the other side. From the transfer gates the charge packets are transferred to the transport registers and moved serially to the output amplifier. The transfer gate also controls the integration time. Transfer gates prevent image smear, which is a fundamental limitation of the basic CCD image sensor; it is due to the pickup of spurious charge by charge packets while they are transferred and light is still incident on the array.

The complementary phase relationship of the clocks for reset and transport, along with the geometrical layout of the transport registers, provides for the alternate delivery of charge packets into the output amplifier thus recreating the original sequence of the line image data. Some models include two outer CCD shift registers or a peripheral diode, to reduce peripheral dark current noise in the inner shift registers.

Output charge packets are transported to a precharged diode whose potential changes linearly in response to the amount of signal charge delivered. This potential is applied to the gate of an output MOS transistor that provides an output pulse train voltage signal. A reset transistor driven by a reset clock recharges the charge detector diode capacitance (0.1 pF to 0.5 pF) before the arrival of each new signal charge packet from the transport registers. By gating the output amplifier, some models provide a sample-and-hold output.

A line imager requires clock signals at least for transfer, transport, and reset. The transfer clock is applied to the transfer gate to move the accumulated charge from the pixels to the transport registers, and it controls the exposure time. Two sets of transport clocks, usually two-phase, are required to move the signal charge packets down the transport registers into the gated charge detector/amplifier. The reset clock returns the charge detector potential to a fixed level. It operates at twice the transport clock frequency, and it determines the output data rate. Clock signals fed into a CCD imaging sensor must carefully follow the manufacturer's directions relative to amplitudes, phases, rise and fall times, and input interfacing. There are several chip sets that provide the necessary signal processing electronics with common pixel resolution from 10 bit to 14 bit.

The main parameters of line image sensors are spatial resolution, responsivity, spectral response, dynamic range, and output data rate. The *spatial resolution* characterizes the ability to discriminate between closely spaced points in the image. It is expressed by the *modulation transfer function* (MTF) of the output, which depends on the spatial frequency of an image. An image consists of periodic intensity variations that can be Fourier analyzed as spatial frequency components. The spatial frequency f is expressed in line pairs per millimeter, where a line pair is the spacing between maxima of intensity. The spacing of electrodes can also be expressed as a spatial frequency f_0 (elements per millimeter), and the frequency image of the object may be normalized as the ratio f/f_0. According to the sampling theorem, the resolution is ultimately limited to $f/f_0 = 0.5$ (Nyquist limit), which means we need at least two sensing elements to image a line pair. The intensity modulation of an image with a given spatial frequency focused on the image sensor yields a modulation depth of the output, which decreases at high spatial frequencies. The MTF is the output modulation depth normalized to unity at zero spatial frequency.

The *responsivity* is the output signal voltage per unit exposure for a specified spectral type of radiation. It equals output voltage divided by exposure ($V/\mu J/cm^2$). The exposure is the product of the irradiance times the exposure (integration) time, which is the time interval between the falling edges of any two transfer clock pulses. The variation in responsivity with wavelength is the spectral response, which is broader than that of the human eye response and whose maximum is around 800 nm. The difference of the response levels of the most and least sensitive elements under uniform illumination gives the so-called photoresponse uniformity.

The *dynamic range* is the quotient between the saturation exposure and the rms noise equivalent exposure, which is the exposure level that gives an output signal equal to the rms noise level at the output in the dark. The best models have a dynamic range of 75,000:1. The dark signal is a thermally generated (random) voltage, and it depends on integration time and on temperature, doubling each 5 °C. This limits the operating temperature to about 70 °C or even 55 °C in some models. The minimal operating temperature is about −25 °C.

The output data rate depends on the number of output terminals. For more than 10 MHz, two outputs are provided: one for even-numbered pixels and one for odd. The higher the sample rate, the lower the resolution in bits. For 12 bit resolution the data rate is 10 MHz. The maximal sample rate in line sensors is about 25 MHz.

For those applications where the dynamic range in parts of the image is larger than the dynamic range of the CCD, there are sensors that include antiblooming and integration control. The antiblooming feature limits the amount of charge that can be collected in any photosite to the value that the shift register can transfer without excessive loss. This prevents the oversaturated photosites from smearing into adjacent areas with consequent loss of resolution (excess charge would remain in the well and combine with the charge packet from the $n + 2$ pixel). The price paid is a loss of signal amplitude information in the oversaturated pixels. Blooming can be eliminated by adding an electrode next to the photogate and biasing both at approximately the same voltage (5 V to 7 V). Excess charge generated in the photosites spills across the barrier formed by that electrode into a nearby electrode "sink."

Integration control consists of reducing the responsivity of all pixels, like an electronically variable shutter activated during each scan time. This way it is possible to prevent any pixel from saturating and overflowing (blooming). Integration control can be implemented by feeding a clock signal to the antiblooming electrode; whenever this electrode is raised 3 V above the photogate voltage, the charge generated in the photosites is dumped into the electrode sink. By controlling the duty cycle of that clock, the effective charge collection time can be adjusted to the desired value. Setting the low level of that clock to the voltage required for antiblooming implements both antiblooming and integration control.

An area image sensor consists of an array of photosensors precisely positioned in rows and columns each with an optical microlens to focus the incoming photons. Models are available with matrix sizes ranging from 192 (H) × 165 (V) to 2048 (H) × 2048 (V); several models conform to the different TV standards (NSTC, PAL/SECAM) and to the common intermediate format (CIF) used in PC-attachable cameras. The first color CCD camera was sold by Sony in 1980.

There are different methods for transfering the sensed image. In the interline-transfer method, each column of sensors is connected to a vertical transport register, all of which feed into a horizontal transport register. After the exposure time, first the charge packets of odd-numbered pixels at each row are transferred to the vertical transport register, and from there to the horizontal transport register, on a line-by-line basis. There they move serially into the output amplifier. After readout of the odd field, the process is repeated for the even-numbered pixels.

In the frame-transfer method there is an image area and a storage area. At the end of the integration time, the signal charge is transferred into the storage area, and from there into a readout serial register that feeds into the output amplifier. The next frame of information is collected while the preceding frame is being read out.

Some features found in line sensors, such as dark reference level output and antiblooming control, are also available. There are also chip sets that provide the necessary support functions for CCD imagers. Color sensitivity is obtained in some area sensors by laminating a color stripe filter on top of the image-sensing area and aligning it precisely with the columns of sensing elements. This separates the columns in three groups corresponding to the red, green, and blue colors used in the filter.

CCD imaging sensors are used in solid-state TV cameras and other imaging devices working with visible or infrared light, where an optical system focuses the image on the sensor. The precise knowledge of photosensor locations with respect to one another gives them a definite advantage as compared to vacuum tube TV cameras. Its low-power and low-voltage requirements make them even more attractive, particularly for hand-held devices such as camcorders. They are used, for example, in inspection, measurement, surveillance, telecine, facsimile, spectrometers, and optical character recognition (high-speed mail sorting, currency sorting, document scanning, etc.), in activities ranging from food processing to astronomy (space telescopes), and in technological fields ranging from robotics to cartography and medical imaging.

CCDs have superior dynamic range, lower FPN, and higher sensitivity to light than CMOS image sensors. However, CMOS sensors need ten times less power than CCDs and their support chips, work with a single power supply, integrate support circuitry in the same chip as the sensor, and allow random pixel access, which permits us, for example, to obtain high frame rates for limited windows of interest. There are CMOS linear arrays with 128 to 2048 pixels, area arrays with up to 2048 × 2048 pixels each 7.5 μm × 7.5 μm, and speed from 9.3 frames/s to 102 frames/s for still images. CMOS image sensors permit better infrared vision than current CCDs, which have reduced infrared sensitivity because of their antiblooming overflow structure. Because of their lower cost, CMOS image sensors are replacing CCDs in applications that do not demand high resolution or low background noise, such as those in consumer products.

9.4 FIBER-OPTIC SENSORS

The development of fiber-optic technology in the telecommunications field has led to a spilling over of knowledge for the development of fiber-optic sensors, which were almost unknown until 1977. Currently, each September issue of *Measurements & Control* lists types and manufacturers of fiber-optic sensors.

9.4.1 Fiber-Optic Basics

Fiber optics are thin transparent strands of glass or plastic that are enclosed by material of a lower index of refraction ($n_2 < n_1$) and that transmit light throughout their length by internal reflections (Figure 9.16). Plastic or stainless steel flexible coating around the fiber supplies mechanical protection. By the law of refraction, a light ray impinging at the fiber entrance is refracted so that the refracted ray lies in the same plane defined by the incident ray and the normal of the front surface but on the other side of both the surface and the normal. If the respective refractive indices are n_0 and n_1, Snell's law states that

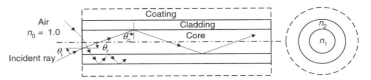

Figure 9.16 Incident rays inside the acceptance angle for a fiber optic ($n_1 > n_2$) are totally reflected and guided to the output. Incident rays outside that angle are lost into the cladding.

$$n_0 \sin \theta_i = n_1 \sin \theta_r \qquad (9.30)$$

$n_0 = 1.0$ in clean air. When the refracted ray reaches the core–cladding boundary, it refracts again. But if the incidence angle is such that $\sin \theta_c > n_2/n_1$, the sine of the angle of the ray refracted into the cladding should be larger than 1, which is impossible. Therefore, the ray reflects back into the core with a reflected angle equal to the incidence angle, so the same happens when reaching the boundary again and the ray is guided along the fiber. It follows that light rays entering the fiber with an angle smaller than α such that

$$n_0 \sin \alpha = n_1 \sin(90° - \theta_c) = n_1 \cos \theta_c = \sqrt{n_1^2 - n_2^2} \qquad (9.31)$$

are totally reflected and guided to the fiber output where they exit with an angle approximately equal to the entry angle. θ_c is termed *critical angle*, 2α is the *acceptance angle*, and $n_0 \sin \alpha$ is the *numerical aperture* (NA). Light rays outside the acceptance angle are lost into the cladding by absorption. Typical NAs for step-index fibers range from 0.2 to 0.5 (α from 11.5° to 30°). NA and the matching between the area illuminated by the light source and the core area determine the *coupling efficiency*.

Graded-index fibers have a core with nonuniform refractive index profile. The profile causes the light to travel on wavelike tracks. Outer rays have a longer path length, which is compensated by the higher speed of light in the outer core region. The acceptance angle depends on the distance from the core center: It is maximal at the center and minimal at the core–cladding boundary. NA is about 0.2.

Fiber optics are made from either plastic or glass. Plastic fiber optics are single strands of fiber material (0.25 mm to 1.5 mm in diameter) that can fit into extremely tight areas. They can easily route (visible) light to the sensing positions but absorb infrared light. Glass fiber optics are made of a single strand or a bundle of very thin (50 μm) fused silica strands. Coherent bundles have each fiber carefully lined up from one end to the other so that an image at one end can be viewed at the other. Fibers in random bundles are not lined up and cannot transmit images.

If the core diameter is very thin and n_1 is only slightly larger than n_2, the acceptance angle is very small and only axial rays are transmitted. Then we have a single mode or monomode fiber. Its core has a typical diameter of 5 μm to 12 μm, and its cladding has a diameter of about 125 μm. Larger fibers have multimode dispersion, meaning that because the path length depends on the entry angle, the exit time is also different for each ray, which introduces a phase shift and limits bandwidth.

9.4.2 Fiber-Optic Sensor Technologies and Applications

Fiber-optic sensors can sense multiple physical and chemical measurands with increased sensitivity as compared to other measurement methods, are versatile in their possible geometry, are lightweight, compact, immune to harsh environments such as intense electromagnetic fields, high temperatures, or corrosive mediums, and are intrinsically safe in explosive environments. Furthermore, they suit distributed sensing. Coiled plastic fiber optics can be mounted on reciprocating mechanisms. Plastic fiber optics, however, do not tolerate high temperatures and are sensitive to many chemicals. In contrast, glass fiber optics withstand extreme temperatures but break when sharply bent or after continuous flexing.

A measurement system based on a fiber-optic sensor consists of a light source (LED or laser), an optical fiber, and a photodetector. LEDs (infrared or visible) are highly reliable and easy to integrate in the system but have low coupling efficiency, spectral width about 40 nm, large chromatic dispersion, and bandwidth limited to about 200 MHz. Lasers have high coupling efficiency and a spectral width of about 3 nm, which results in high bandwidth and low chromatic dispersion, but are more expensive, less reliable, sensitive to overload currents, and need cooling and power stabilization. Common photodetectors in fiber-optic sensors are phototransistors, $p-i-n$ photodiodes, and avalanche photodiodes.

Fiber-optic sensors can be extrinsic or intrinsic [16]. In extrinsic, hybrid, or external modulation sensors, the fiber carries light that is modified (modulated) by the measurand in an external element. Several extrinsic sensors rely on intensity modulation because of light transmission or reflection (as in photoelectric sensors for presence detection, Figure 9.10). Figure 9.17a shows an extrinsic

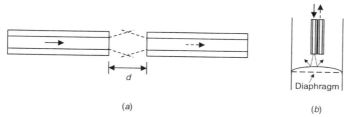

(a)   (b)

Figure 9.17 Extrinsic fiber-optics sensors based on intensity modulation because of (a) light transmission and (b) light reflection (from a diaphragm).

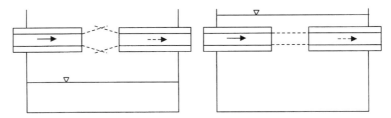

Figure 9.18   Level sensor based on light coupling through a liquid dielectric.

sensor based on the variation in the mutual coupling between two fibers depending on their relative position and NA. Position can depend, for example, on the pressure applied to a diaphragm bonded to one of the fibers or on the acceleration or vibration sensed by an inertial mass. Figure 9.17*b* shows a pressure sensor based on optical reflection. The light emerging from one fiber reflects from a curved diaphragm to another fiber. The intensity received depends on the diaphragm's curvature, which depends itself on the difference in pressure between both sides. Placing reflective areas on a rule or disk yields a position encoder.

Extrinsic sensors based on intensity modulation can also rely on the variation in refractive index. Figure 9.18 shows a liquid level sensor based on the increased light coupling from the emitter fiber to the receiver fiber when a (transparent) liquid fills the space between fibers' ends. This sensor suits flammable transparent liquids because it is intrinsically safe.

Intensity-modulation fiber-optic sensors are limited by interference that introduces variable losses and is unrelated to the measurand. Problems occur with mechanical creep and misalignment of light sources and detectors, connectors, splices, macrobending, and microbending losses when the critical angle θ_c ensuring totally internal reflection is exceeded, and so on. Sensors using dual wavelengths circumvent these problems by using a wavelength as reference that bypasses the sensing region. These sensors are used to control the processing of chemicals that absorb specific bands of infrared radiation, for in vivo blood-gas measurements, and also for air pollution monitoring.

Spectrally based fiber-optic sensors rely on the modulation in wavelength of a light beam by the measurand. For example, blackbody temperature sensors use a blackbody cavity at the end of an optical fiber. Because the light spectrum emitted by the blackbody depends on its temperature (Figure 6.23), a series of detectors combined with narrow band filters can determine the profile of emitted light and, hence, the blackbody temperature. The sensor performance improves above 200 °C and remains unaffected by strong electromagnetic fields.

Some extrinsic sensors for qualitative and quantitative identification of chemical samples use fluorescence. Fluorescence is the absorption of light of one wavelength (usually ultraviolet) by one molecule followed by the emission of light at a longer wavelength (ultraviolet or visible). Fluorescence induction in

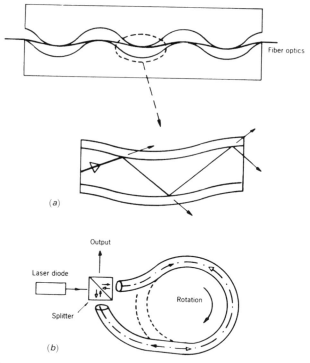

(a)

(b)

Figure 9.19 Sensors based on an intrinsic modification of light propagated inside an optical fiber: (a) intensity modulation due to microdeformations; (b) temporal phase shift due to fiber rotation (Sagnac effect).

a sample containing the substance to be analyzed is a classic technique in analytical chemistry. It is suited to optical fiber application because optical fibers can lead light to a remote region where it excites fluorescent radiation in the substance of interest. The same fibers lead the radiation back to the detector. Chemicals are identified through temporal information from their decay lifetimes, as well as through the excitation and emission wavelengths information. Fiber-optic fluorescence sensors have been applied to nuclear reactors, to underground water contamination measurement, and to intravascular sensors in medicine.

In intrinsic fiber-optic sensors, the measurand affects the characteristics of the optical fiber, either directly or through micromechanical bending. The changes induced can be in radiation intensity or phase. Some distributed sensors use optical time-domain reflectometry (OTDR).

Figure 9.19a illustrates the principle of operation for a system based on the losses resulting when the measurand (force, pressure, etc.) produces microdeformations in the fiber. Usually, in a normal straight fiber the total internal reflection results in very small transmission losses. But when external force causes the fiber to deform, the curvature of the core–cladding boundary changes and the incidence angle is smaller than the critical angle, so that there is only a partial reflection, causing losses that increase for larger deformations.

Another intrinsic sensor based on transmission losses consists of a fiber whose core and envelope materials are so chosen that their refraction indices approach each other when their temperature decreases. This results in increasing losses through the envelope. This method can be applied to distributed temperature sensing in a wide region having a fiber underneath—for example, to detect leaks in cryogenic liquid storage.

Sensors based on radiation phase variation use monomode fibers and a monochromatic laser source. The change produced by the measurand is detected by interferometry. Some interferometer-based sensors split the light beam and compare the beam coming from a monomode fiber where it has been modified by the measurand, with that coming from a reference fiber whose path is unaffected by the measurand. The measurand induces changes in the envelope, with contraction or expansions being the most frequent, thus resulting in changes in refractive index and in core fiber dimensions and therefore producing a spatial phase shift with respect to the reference fiber. This method can detect 0.025 μm displacements.

Fiber-optic gyroscopes are based on Sagnac's interferometer (Figure 9.19b) discovered in 1913 [17]. They consist of a single rotating fiber, inside which two light beams coming from the same source propagate in opposite directions. At the output they are combined again and lead to a photodetector. The fiber loop rotation makes the traveled path longer for the beam propagating in the same direction as the rotation, and shorter for the beam traveling in direction opposite to that of the rotation. The result is a short time delay that produces an amplitude modulation of the combined beam, which is proportional to the rotating speed. By arranging a loop long enough (1 km in a 10 cm diameter loop), the resulting modulation can be made significant and it is converted into an electric signal by the photodetector.

OTDR sensors use a laser diode that sends a train of light pulses through the fiber [18]. An optically perfect fiber of infinite length would not reflect any optical energy back into the launch port. Any anomaly, such as a change in refraction index, returns an echo pulse because it scatters or reflects back part of each incident light pulse. The farther the distance to the entry port, the greater the time lag between the launched pulse and the received echo. This method permits the detection of $\pm 1\,^{\circ}\text{C}$ temperature changes in a 10 km range with 1 m spatial resolution. It is applied to detect hot spots in underground power transmission lines, power transformers, and buried cables, as well as to detect leaks in pipes.

9.5   ULTRASONIC-BASED SENSORS

Ultrasound is a mechanical radiation with a frequency above the human hearing range (about 20 kHz). As for any radiation, when ultrasound strikes an

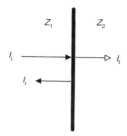

Figure 9.20   A plane wave (intensity I_i) with normal incidence on a boundary is partially reflected (I_r) and partially transmitted (I_t).

object, part is reflected, part is transmitted, and part is absorbed (Figure 9.20). In addition, when the radiation source moves relative to the reflector, there is a shift in received frequency (Doppler effect). All these properties of radiation-object interaction have been applied to the measurement of several physical quantities using ultrasound.

The penetration power for ultrasound permits noninvasive applications; that is, there is no need to install hardware where the changes to be detected occur. Noninvasive measurements are of interest in explosive and radioactive environments, in medical applications and also to prevent contamination of the medium where measurements are performed. Noninvasive sensors are generally easier to install and maintain than invasive sensors.

9.5.1   Fundamentals of Ultrasonic-Based Sensors

When a deformation is produced at a point inside an elastic medium, the deformation does not remain restricted to that point; rather it propagates to neighboring points. When the deformation is due to a vibratory movement, this movement is characterized by its frequency f, amplitude a, and atomic instantaneous velocity v. The average "net" velocity for atoms is zero.

The velocity for the perturbation to propagate from one point to another, or *wave velocity*, depends on the medium but not on the frequency. For gases and liquids, that velocity is given by

$$c^2 = \frac{K_m}{\rho} \tag{9.32}$$

where K_m is the bulk modulus of elasticity and ρ is the density. Because both parameters change with temperature, c is also temperature-dependent.

For a solid, the velocity of longitudinal waves is

$$c^2 = \frac{E(1-v)}{\rho(1+v)(1-2v)} \tag{9.33}$$

where E is Young's modulus and v is Poisson's ratio. For air, $c \approx 330$ m/s; for water, $c \approx 1500$ m/s; for steel, $c \approx 5900$ m/s; for aluminum, $c \approx 6320$ m/s. The velocity of transverse shear waves is

$$c_T^2 = \frac{G}{\rho} = \frac{E}{2\rho(1 + 2v)} \tag{9.34}$$

where G is shear modulus ($v = E/2G - 1$).

As a result of the perturbation, the pressure at a given point is not constant but changes with respect to an average value. The difference between instantaneous and average pressures is named *acoustic pressure p*. The quotient between p and v, both being considered as complex quantities (modulus and phase; we assume that the problem is analyzed in the frequency domain), is called *acoustic impedance*:

$$Z = \frac{p}{v} \tag{9.35}$$

When there are no losses in the propagation medium, p and v are in phase, so Z is real. It can be shown that its value is

$$Z = \rho c \tag{9.36}$$

Z is a characteristic parameter for each medium. For air $Z \approx 4.3 \times 10^{-4}$ Pa·s·m^{-1}; for water, $Z \approx 1.5$ Pa·s·m^{-1}; for steel, $Z \approx 45$ Pa·s·m^{-1}; for aluminum, $Z \approx 17$ Pa·s·m^{-1}.

Radiation intensity is defined as its power per unit area:

$$I = pv = \frac{p^2}{Z} \tag{9.37}$$

When the radiation propagates in a homogeneous medium, it attenuates exponentially with the distance according to

$$I = I_0 e^{-\alpha x} \tag{9.38}$$

where I_0 is the incident intensity, α is an attenuation coefficient which depends on the medium and the frequency (increases with frequency), and x is the distance traveled in the medium.

If the medium is not homogeneous, the acoustic impedance changes from region to region and the radiation is not only absorbed but also reflected. For a plane wave traveling in a direction perpendicular to a plane surface separating two mediums whose respective acoustic impedances are Z_1 and Z_2 (Figure 9.20), intensity transmission and reflection coefficients are [19]

$$R = \frac{I_r}{I_i} = \left(\frac{Z_1 - Z_2}{Z_1 + Z_2}\right)^2 \tag{9.39}$$

$$T = \frac{I_t}{I_i} = \frac{4Z_1 Z_2}{(Z_1 + Z_2)^2} \tag{9.40}$$

where I_i, I_r, and I_t are, respectively, the incident, reflected, and transmitted intensities. Note that $R + T = 1$, as expected because at normal incidence the power densities on each side must be equal. From (9.39) we deduce that the reflection increases when the difference in impedance between both mediums increases. This hinders noninvasive measurements in gases because of the high contrast in Z between gases and vessel walls.

The Doppler effect, discovered by C. Doppler in 1843, is the change in frequency undergone by a radiation (be it mechanical or electromagnetic) when it is reflected by an object that is moving with respect to the radiation transmitter. If the reflector moves with velocity v, the change in frequency is

$$f_e - f_r = 2f_e \frac{v}{c} \cos \alpha \tag{9.41}$$

where f_e is the emitted frequency, f_r is the received frequency, and α is the relative angle between reflector velocity and sound propagation directions.

9.5.2 Ultrasonic-Based Sensing Methods and Applications

Ultrasonic-based sensors are usually based on their transit time, attenuation, or velocity. Most of them use piezoelectric ceramics or polymer transducers as generators and detectors. They must work at a temperature below the respective Curie point. The resolution is limited by the wavelength, which is inversely proportional to the frequency ($\lambda = c/f$). But high frequencies have larger attenuation than low frequencies. Spurious vibrations from industrial noise may interfere with the receiver. Using systems with narrow sound beams reduces interference from background noise. Maximal power is radiated axially (i.e., perpendicularly) to the transducer face, so the sensitivity and range depend on the relative position between transducer and reflector.

Transit time ultrasonic sensors emit an ultrasonic burst and measure the elapsed time between transmission and reception (echo ranging). The elapsed time multiplied by the speed of sound equals twice the distance to the impedance discontinuity that returns the echo. The reflecting surface must be parallel to the sensing face—that is, perpendicular to the direction of sensing. Because c increases with temperature, we must measure ambient temperature to calculate the current c. The frequency is selected according to the range and reflector surface. Greater ranges need low frequencies because of the increased attenuation—for example, 23 kHz for a 30 m range and 40 kHz for a 12 m range. Smooth, nonporous surfaces reflect sound better than rough or porous surfaces.

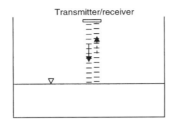

Figure 9.21 Ultrasonic level sensors determine the distance from the transducer to an impedance discontinuity by measuring the elapsed time between the emission and reception of a signal burst.

Poor reflectors return stronger echoes at low frequencies [20]. Nonflat surfaces return low-level echos [21]. The minimal distance between transducer and reflector depends on the time required for the transducer to extinguish echoes (ringing) resulting from the transmitted burst. Ringing lasts longer for lower frequencies. Evaluating a series of pulses rather than a single pulse reduces interference from background noise. Because air turbulence may blow off echoes, ultrasonic sensors are not suitable for outdoor use.

Transit time sensors yield highly linear and accurate outputs when measuring distance. Long-range sensors are required for level monitoring in large tanks or bins. Accurate distance measurement is required for position and fill-level control applications, thickness measurement, and ultrasound imaging. Instant cameras use low-cost echo ranging. Nondestructive testing also relies on sound reflections from impedance discontinuities. For liquid or solid level measurement, the transducer is usually placed above the tank and transmits a cone-shaped beam pattern (Figure 9.21). The acoustic impedance of air and liquids or solids (including powders) is so different that most of the energy propagating in one of them is reflected when reaching the interface. Dielectric constant, conductivity, color, viscosity, and specific gravity do not affect accuracy, but vapors, fumes, and dust do. Placing the sensor inside a stilling well reduces effects of surface agitation in liquids. Foamy or fluffy surfaces absorb energy, hence reducing effectiveness. The sensor can alternatively be placed at the bottom of the tank so that the sound propagates in the liquid or solid, but then the transducer and cabling must be leakproof.

Object detection and proximity switches can rely on the disruption of the sound beam as the object passes between the transmitter and the receiver, similar to beam-through photoelectric sensing (Figure 9.10a).

Sensors based on ultrasonic attenuation have been applied to air or foam bubble detectors in plastic, glass, or metal tubing—for example, to prevent air embolism in patients.

Some of the most widely used sensors based on ultrasonic velocity are ultrasonic Doppler flowmeters. Any substance having acoustic impedance different from that of the flowing fluid can act as reflector and shift the fre-

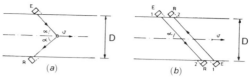

Figure 9.22 Principle for ultrasonic flowmeters: (*a*) Based on the Doppler effect; (*b*) based on transit time (E = emitter; R = receiver).

quency of an ultrasonic signal according to (9.41) (Figure 9.22*a*)—for example, trapped air or gas bubbles or suspended solids in liquids. Best reflection occurs when the reflector size is larger than 10% of the acoustic wavelength in the fluid. For a commonly used frequency of 1 MHz, the wavelength in water is $\lambda = c/f = (1500$ m/s$)/(10^6$ Hz$) = 1.5$ mm. For higher frequencies, the attenuation of the radiation in the medium would be excessive. Because the fluid velocity changes along the wave path, the receiver signal consists of a band of frequencies that must be further processed to determine flow rate.

For clean liquids (without any reflectors) and gases, flowmeters are available based on the variation in transit time between transmitter and detector, depending on whether the radiation propagates in the same direction as the flow or in the opposite direction (contrapropagating transmission). In Figure 9.22*b*, if transducer 1 acts as emitter and transducer 2 as receiver, the time for the radiation to reach the receiver is

$$t_{12} = \frac{D/\sin \alpha}{c + v \cos \alpha} \tag{9.42}$$

If transducers interchange their functions or if another transducer pair is used, then

$$t_{21} = \frac{D/\sin \alpha}{c - v \cos \alpha} \tag{9.43}$$

Flow rate is proportional to $(t_{12} - t_{21})/(t_{12}t_{21})$. Time differences to be measured are very small.

An alternative method based on the same principle uses a sing-around circuit where the transmitter emits a pulse when an associated receiver detects the radiation pulse previously transmitted. By using two transmitter–receiver pairs, one transmitting in the same direction as the flow and another one transmitting in the opposite direction, or by using a pair of reciprocal transducers, the difference in pulse repetition frequency is

$$f_1 - f_2 = \frac{2v \sin \alpha \cos \alpha}{D} \tag{9.44}$$

The velocity profile inside the tube or vessel returns a spectrum of frequencies. Further processing integrates the flow profile to yield the volumetric fluid flow.

Ultrasonic flowmeters neither produce any pressure loss, nor contaminate the fluid. They cause no wear and can measure hot and cold fluids and slurries. They are routinely used in industry to measure liquids, gases, and two-phase or multiphase fluids. They are the method of choice to estimate arterial blood flow rate noninvasively. In open channels, streams, or rivers, ultrasonic flowmeters measure the fluid level over a specific weir (Section 1.7.3). Each October issue of *Measurements & Control* lists manufacturers of ultrasonic flowmeters.

The dependence of sound velocity on temperature has been applied to measure the temperature of water in oceans and gases in chimneys.

9.6   BIOSENSORS

Biosensors are devices that incorporate biological sensing elements. They provide a specific and sensitive response to chemical species, particularly molecules of biological relevance. Figure 9.23 shows their simplified structure. There is an external membrane that must be permeable to the target analyte but exclude other substances. The biological element inside the biosensor interacts with the analyte and yields a response that is detected by a common sensor. Hence, the biological element behaves as primary sensor. It may convert the analyte to a different chemical species through a biochemical reaction, release a chemical product in response to the analyte stimulus, or change its electrical, mechanical, or optical properties. The permeability of the internal membrane near the output sensor, if present, may be different from that of the external membrane. The output sensor may be a conventional electrochemical sensor (Section 6.5), a ChemFET (Section 9.2), a piezoelectric or SAW sensor (Section 8.2.2), an optical sensor, a thermal sensor, or other [22].

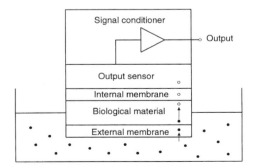

Figure 9.23   Basic structure of a biosensor. The biological element reacts to the analyte (black dots) that diffuses through the semipermeable external membrane and yields a product (circles) or undergoes a change that is detected by a common (output) sensor.

Biosensors often drift because proteins or other substances adsorbed on the membrane or the sensor surface poison the biological element. Consequently, biosensors need periodic calibration. Lifetime may depend on the total number of measurements made and on the magnitude of the analyte concentrations measured and their temperature. High concentrations and temperatures tend to reduce the sensitivity of the biological element. In order to preserve bioactive properties, it may be necessary to keep the sensor refrigerated between measurements. Temperature also affects the diffusion coefficient, gas and chemical solubilities of the liquids and solid materials used to build the sensor, and the rate of reactions. Hence, it may be necessary to operate the sensor under isothermal conditions. Biosensors to be used in human beings must be biocompatible.

Some biological materials used in sensors are enzymes, antibodies, cells, and tissue fragments. Enzymes are proteins produced by living cells that catalyze biochemical reactions in them. In the presence of a specific enzyme, a substrate is irreversibly converted into a product through the reversible formation of an intermediate enzyme–substrate complex. The products of those reactions are detected by the output sensor.

Biosensors keep the enzyme active by immobilizing it so that its diffusivity is either zero or else much less than the diffusivities for the substrate and product. There are several enzyme immobilization methods: physical entrapment, microencapsulation, adsorption, covalent cross-linking, and covalent binding. Physical entrapment methods keep the enzyme in place by viscous solutions (gels such as agarose, gelatin, and polyacrylamide) adjacent to the biosensor elements and by membranes that prevent them from diffusing away. Common membranes are cellophane, cellulose acetate or nitrate, and polyurethane. External membranes must be permeable to the analyte. Microencapsulation (for example, inside liposomes) or adsorption on glass beads or carbon particles permits the inclusion of the enzyme inside the membrane, or within viscous gels close to the output sensor. Most membranes directly adsorb enzymes, but the attachment may not be strong enough. If the membrane first adsorbs other proteins, such as albumin, enzyme attachment is stronger. Cross-linking agents such as glyoxal or glutaraldehyde significantly increase enzyme attachment, but they can interfere with enzyme activity. Some enzymes such as glucose oxidase link with covalent bonds to the polymer membrane of electrochemical sensors, which permits their direct "connection."

Enzyme stability depends on chemical and thermal environmental parameters, particularly the pH. Enzyme activity is inhibited by chemical species such as heavy metals and temperatures above $50\,°C$. Biosensors using enzymes can detect glucose and other sugars (fructose, maltose, lactose, and sucrose)—which can be decomposed by other enzymes to yield glucose—by sensing the resulting O_2 or H_2O_2 concentration or the pH; urea, by sensing NH_4^+, CO_2, or pH; and alcohols (ethanol, methanol), by sensing O_2 or H_2O_2.

Antibodies are Y-shaped proteins produced by specialized cells after stimulation by an antigen (a protein or carbohydrate substance) that act specifically against the antigen in an immune response. Monoclonal antibodies recognize and bind to a specific antigen at receptor sites at the tip of each arm. Polyclonal antibodies recognize and bind to a group of closely related antigens. Because the antigen–antibody binding does not yield any product molecule, the output sensor must detect small mass changes or variations in electrical or optical properties resulting from the binding [23]. Cloning and culturing hybrid cells permits the production of antibodies for a broad range of analytes. Their response is more specific than that of biometabolic sensors (based on reactions catalyzed by enzymes). However, some immunological reactions are irreversible, so each biosensor permits a single measurement (or immunoassay). Biosensors based on immunological reactions are termed immunosensors and bioaffinity sensors. They rely on the immobilization of either an antigen or, more often, an antibody on their external membrane.

The use of cells or subcellular elements in biosensors without altering their natural biological and biochemical functions is not as widespread as that of biosensors based on enzymes or immunosensors. The use of algae, bacteria, fungi, and yeasts immobilized on a porous membrane has been researched to detect complex mixtures of biological substances that would be difficult to detect by an array of specific enzymes. Microbial biosensors are less sensitive to inhibition by solutes and suboptimal pH and temperature than enzyme sensors, have longer lifetime, and are cheaper because there is no need to isolate an active enzyme [24]. Some difficulties to overcome are the need to keep those microorganisms alive by controlling their environment, supplying the needed nutrients (including oxygen), and keeping toxic substances away. They also have longer time response and need more time to return to the baseline after use than enzyme sensors. The output sensor detects O_2 consumption, the production of CO_2 and other metabolites, or changes in pH. A major application area is wastewater monitoring.

Because of their sensitivity, specificity, fast response, and small mass sample needed to work, biosensors may replace bioassays performed on animals and may also replace in vitro analysis. They are currently applied to quickly measure biological and chemical substances in medical tests (such as point-of-care blood analysis), food and drug testing, process control in chemical and pharmaceutical industries, environmental monitoring, and chemical warfare surveillance.

9.7   PROBLEMS

9.1   The thermometer in Figure P9.1 relies on the temperature coefficient of a silicon diode supplied by a constant current, about -2 mV/°C. The LM385-2.5 yields a 2.5 V reference voltage.

   a.   Determine the amplifier gain to obtain a 10 mV/°C output sensitivity.

   b.   The switch permits the selection of an output sensitivity of 10 mV/°F. What should then be the gain?

   c.   Determine R_1, R_3, and R_4 to obtain 100 µA through the voltage reference ICs.

d. Determine R_2 to bias the diode at 250 µA.

e. Determine the error because of op amp heating when supplied at ±15 V and their ambient temperature is $25\,°C$.

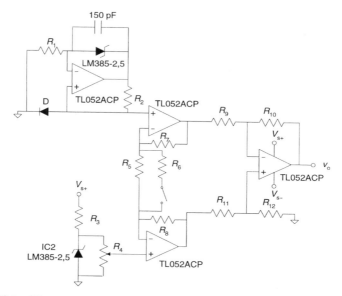

Figure P9.1   Thermometer based on the temperature coefficient of a silicon diode.

9.2   Figure P9.2 shows a thermometer based on two matched transistors, where op amps are assumed to be ideal. Design the circuit in order to have 1 mV/K output.

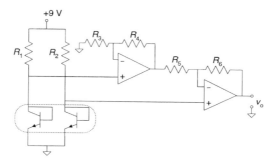

Figure P9.2   Thermometer based on two matched transistors.

9.3   Figure P9.3 shows a cold-junction compensation circuit for a type J thermocouple. The AD590 has 1 μA/K sensitivity. The output voltage is
measured with a very high impedance voltmeter. Determine the condition
to be fulfilled by V_b, V_p, R_a, and R_b, and design the complete circuit.

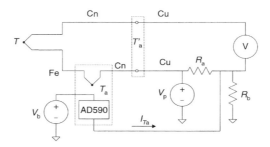

Figure P9.3   Cold-junction compensation circuit for a J-type thermocouple.

9.4   A temperature in the range from $-50\,°C$ to $+250\,°C$ is measured using a
J-type thermocouple. The reference junction is held at ambient temperature, which ranges from $10\,°C$ to $50\,°C$. In order to compensate for the
voltage generated by the reference junction, the circuit in Figure P9.4
is suggested. It is based on a temperature-to-current converter (AD592)
whose sensitivity is 1 μA/K. Calculate R and R_g in order to obtain an
output sensitivity of 10 mV/°C and a zero output at $0\,°C$, regardless of
ambient temperature. Next, consider op amp offset voltage (150 μV) and
bias currents (12 nA). Calculate the output error they would produce if
they were not compensated for. Discuss a method to cancel this error and
its implications.

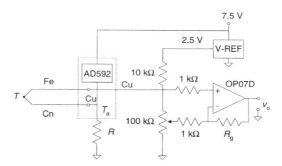

Figure P9.4   Cold-junction compensation circuit for a J-type thermocouple using a
single-ended amplifier.

9.5 The thermocouple circuit in Figure P9.5 includes reference junction compensation based on the LM35, whose output voltage is 10.0 mV/°C from 0 °C to 100 °C, with a ±0.25 °C accuracy. The temperature sensor output is connected to the reference terminal at the instrumentation amplifier. The thermocouple is type J, and we assume that the amplifier is ideal. Calculate the gain in order to obtain an output voltage independent of ambient temperature.

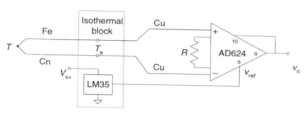

Figure P9.5 Cold-junction compensation circuit for a J-type thermocouple using an instrumentation amplifier.

9.6 A given photodiode has an equivalent output resistance of 20 GΩ shunted by 10 pF. In order to obtain an output voltage, we use a circuit like that in Figure E9.3a with $R = 100$ kΩ and the OPA121 op amp. Calculate C to obtain a flat frequency response and determine the -3 dB bandwidth.

9.7 Determine the equation for the output voltage in the circuit in Figure P9.7 and discuss the effects of op-amp offset voltage and bias currents.

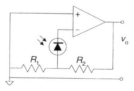

Figure P9.7 Signal conditioner for a photodiode in the photovoltaic mode.

9.8 The infrared phototransistor in Figure P9.8 yields a dark current $I_D = 3$ μA, a minimal $I_{OL(min)} = 2$ mA and a maximal collector–emitter voltage $V_{OL(max)} = 0.8$ V under total illumination, and a current $I_{OH} = I_D + (1 - n/100)I_{OL}$ when partially illuminated, where n is the percentage of blocked light. TTL gates from the series 7400 have the following parameters: $V_{IH(min)} = 2$ V, $V_{IL(max)} = 0.8$ V, $I_{IH(max)} = 40$ μA, $I_{IL(max)} = -1.6$ mA. The power supply voltage is $5(1 \pm 0.05)$ V. Design R to ensure

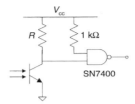

Figure P9.8   Digital interface for a phototransistor.

a digital output 0 when the incoming light is blocked by 95% or more, and a digital output 1 when the phototransistor is totally illuminated, regardless of the power supply voltage.

REFERENCES

[1] S. M. Sze (ed.). *Semiconductor Sensors.* New York: John Wiley & Sons, 1994.

[2] Y. J. Wong and W. E. Ott. *Function Circuits Design and Application.* New York: McGraw-Hill, 1976.

[3] M. P. Timko. A two-terminal IC temperature transducer. *IEEE J. Solid State Circuits*, **11**, 1976, 784–788.

[4] T. Nakamura and K. Maenaka. Integrated magnetic sensors. *Sensors and Actuators*, **A21**–**A23**, 1990, 762–769.

[5] A. Chapell (ed.). *Optoelectronics Theory and Practice.* Bedford, UK: Texas Instruments, 1976.

[6] J. Graeme. *Photodiode Amplifiers, Op Amp Solutions.* New York: McGraw-Hill, 1996.

[7] F. Daghighian. Optical position sensing with duo-lateral photoeffect diodes. *Sensors*, **11**(11), November 1994, 31–39.

[8] R. H. Garwood (ed.). *Handbook of Photoelectric Sensing.* Minneapolis, MN: Banner Engineering, 1993.

[9] P. R. Wiederhold. Chilled mirror hygrometry. *Sensors*, **14**(5), May 1997, 40–45, 71–72.

[10] G. F. Knoll. *Radiation Detectors*, 3rd ed. New York: John Wiley & Sons, 1999.

[11] P. Bergveld. The impact of MOSFET-based sensors. *Sensors and Actuators*, **8**, 1985, 109–127.

[12] P. Bergveld, J. Hendrikse, and W. Olthuis. Theory and application of the material work function for chemical sensors based on the field effect principle. *Meas. Sci. Technol.*, **9**, 1998, 1801–1808.

[13] R. Melen and D. Buss (eds.). *Charge-Coupled Devices: Technology and Applications.* New York: IEEE Press, 1977.

[14] D. K. Schroder. *Advanced MOS Devices.* Reading, MA: Addison-Wesley, 1986.

[15] H. Fieldler and K. Knupper. Market overview: Charge-coupled devices. Chapter 7

in H. Baltes, W. Gopel, and J. Hesse (eds.), *Sensors Update*, Vol. 1. New York: VCH Publishers (John Wiley & Sons), 1996.

[16] E. Udd. Fiber optic smart structures. *Proc. IEEE*, **84**, 1996, 884–894.

[17] K. Böhm and R. Rodloff. Optical rotation sensors. Chapter 17 in: E. Wagner, R. Dänliker, and K. Spenner (eds.), *Optical Sensors*, Vol. 6 of *Sensors, A Comprehensive Survey*, W. Göpel, J. Hesse, J. N. Zemel (eds.). New York: VCH Publishers (John Wiley & Sons), 1992.

[18] A. J. Rogers. Optical-fiber sensors. Chapter 15 in: E. Wagner, R. Dänliker, and K. Spenner (eds.), *Optical Sensors*, Vol. 6 of *Sensors, A Comprehensive Survey*, W. Göpel, J. Hesse, and J. N. Zemel (eds.). New York: VCH Publishers (John Wiley & Sons), 1992.

[19] L. C. Lynnworth. Ultrasonic nonresonant sensors. Chapter 8 in: H. H. Bau, N. F. de Rooij, and B. Kloeck (eds.), *Mechanical Sensors*, Vol. 7 of *Sensors, A Comprehensive Survey*, W. Göpel, J. Hesse, and J. N. Zemel (eds.). New York: VCH Publishers (John Wiley & Sons), 1994

[20] D. P. Massa. Choosing an ultrasonic sensor for proximity or distance measurement. Part 1: Acoustic considerations. *Sensors*, **16**(2), February 1999, 34–37.

[21] D. P. Massa. Choosing an ultrasonic sensor for proximity or distance measurement. Part 2: Optimizing sensor selection. *Sensors*, **16**(3), March 1999, 28–42.

[22] D. G. Buerk. *Biosensors Theory and Applications*. Lancaster, PA: Technomic, 1993.

[23] G. Gauglitz. Opto-Chemical and Opto-Immuno Sensors. Chapter 1 in: H. Baltes, W. Gopel, and J. Hesse (eds.), *Sensors Update*, Vol. 1. New York: VCH Publishers (John Wiley & Sons), 1996.

[24] I. Karube and K. Nakanishi. Microbial biosensors for process and environmental control. *IEEE Eng. Med. Biol. Magazine*, **13**(3), 1994, 364–374.

APPENDIX

SOLUTIONS TO THE PROBLEMS

This appendix gives the solutions for all the problems posed at the end of each chapter. For many of them there are certainly several valid solutions, and therefore the ones shown here are not the only acceptable options. In order to lead those having difficulties toward a solution, in some cases the solutions are partially discussed and the particular assumptions made are described. This will also aid in the judgment of the validity of the solution proposed.

CHAPTER 1

1.1 First case: For readings lower than 1/5 full scale, the first sensor is more accurate. Second case: For readings larger than 60% of full scale of the first sensor, the second sensor is more accurate.

1.2 $100\,°C \pm 1\,°C$ is correct. All other expressions are incorrect.

1.3 1%, 3%.

1.4 68%.

1.5 $[\hat{x}_n - 2.576\sigma/\sqrt{n}, \hat{x}_n + 2.576\sigma/\sqrt{n}]$.

1.6 $t = 0.52$ ms. The sensor should have a very small mass.

1.7 The transfer function from input power to oven temperature is $R/(1 + j\omega RC)$, where $R = \Delta T/P = 13.33\,°C/W$ is the thermal resistance and $C = 19.5$ J/K is the thermal capacity. $\tau = RC = 260$ ms.

1.8 $M_p = 0.45\,g$ for $\zeta = 0.7$. $M_p = 2.5\,g$ for $\zeta = 0.4$. t_p 0.4 ms for $\zeta = 0.7$.

1.9   10%.

1.10 $\zeta > 0.59$.

1.11   The resulting system is second-order low-pass with sensitivity $= 1/2pg$. $\omega_n = (2g/L)^{1/2}$, $\zeta = R/(2p)(2gL)^{1/2}$, where g is the acceleration of gravity, L is the total length for the liquid column, R is its coefficient of friction with the tube walls, and p is its density.

1.12 $f_n = 14.6$ Hz, $0.45 < \zeta < 0.55$, 9 Hz $< f_r < 11$ Hz, and $0.08 < \varepsilon(f_r) < 0.235$.

1.13   0.028 Hz.

1.14 $x = 1.26$ mm, $s_M = 882$ MPa.

1.15   In a vibrating system, $x = k \sin \omega t$, $dx/dt(\text{max}) = k\omega$, $d^2x/dt^2(\text{max}) = k\omega^2$. To determine the acceleration we can either measure distance and frequency or velocity and frequency $(a = v\omega)$. In the first method, the error of the optical displacement meter can be negligible and the error in ω has double weight. However, in the second method the error of the linear velocity meter will probably exceed that of the frequency meter. Therefore, the first method is more convenient.

1.16   **a.** Relative error $= 0.04\%$.
   b. Number of bits >11.28 (12).
   c. The accelerometer yields an output voltage $v_o = k_n\omega^2 r \cos \alpha + k_t\omega^2 r \sin \alpha$, where k_n is the nominal sensitivity and k_t is the transverse sensitivity. The calculated sensitivity will be $k = v_o/\omega^2 r$. The relative error will be $\varepsilon = (k - k_n)/k_n$. The precision required for the angular positioning system is about $3.5°$.

CHAPTER 2

2.1   If the power dissipation capability is constant all along the potentiometer, the critical point is the one next to the upper end because all the current flowing through the potentiometer and the measuring device when just below the upper end circulates through the zone between the upper end and the wiper. Maximal supply voltage, 42.6 V.

2.2   **a.** Maximal supply voltage $= 7.2$ V.
   b. Output $= 81$ μV.
   c. Calibration resistor $= 184.5$ kΩ.

2.3 $S = 3.09$ Ω/K, $\alpha_{100} = 3.82 \times 10^{-3}$ Ω/Ω/K, $R_{100} = 809$ Ω, and $R_{101} = 812$ Ω.

2.4 $S = 119\ \Omega/K$ and $\alpha = 0.04\ \Omega/\Omega/K$.

2.5   Because we supply a constant current, self-heating will be maximal at $0\,°C$ (maximal resistance, 13,640 Ω). We need $I < 0.8$ mA.

2.6   Because the parallel combination of the NTC thermistor and shunting resistor does not have a constant temperature coefficient, we cannot obtain a perfect compensation. We have to conform with an approximation: Force the inflection point for the parallel combination to fall in the center of the compensation range (20 °C), and at this point set the derivative for the total resistance (tachometer included) with respect to the temperature to zero. Result: $R_{20} = 909\ \Omega$ and $R = 639.5\ \Omega$.

2.7   Because a perfect compensation in the operating range is not possible, we select to have an exact compensation in the middle of the operating range (30 °C). For the thermistor we find $B = 3444.6$ K and $R_{30} = 1636.4\ \Omega$. Then we need $R = 1907\ \Omega$.

2.8   We can write three equations with three unknown variables. The result is $R_s = 17.8$ kΩ, $R_p = 27.13$ kΩ, and $R_G = 16.43$ kΩ.

2.9   If sensor sensitivity is $K_0[1 + \alpha(T - T_0)]$, ideally we would be interested in an amplifier gain of $G_0/[1 + \alpha(T - T_0)]$, and this leads to

$$\frac{R_6}{R_5}\frac{R_e}{R_1} - \frac{R_6}{R_4} = \frac{G_0}{1 + \alpha(T - T_0)}$$

where $R_e = R_T \| R_2 + R_3$. Because this condition cannot be fulfilled because R_e is nonlinear, we must impose other criteria. For example, we might require that at 20 °C the slopes for sensor and amplifier be equal in absolute value but with opposite sign, and impose the value for the gain at that temperature. Then, resistors must fulfill

$$\frac{R_6 R_2^2}{R_5 R_1}\frac{B}{T_0^2}\frac{R_0}{(R_2 + R_0)^2} = \alpha G_0$$

$$G_0 = \frac{R_6}{R_5}\frac{1}{R_1}\left(\frac{R_2 R_0}{R_2 + R_0} + R_3\right) - \frac{R_6}{R_4}$$

If in addition the thermistor heat dissipation is considered, since it is supplied by a current that does not depend on the thermistor resistance, it can be shown that its dissipation is maximal when its resistance equals R_2. Therefore, R_2 must be chosen equal to the minimum for the thermistor.

2.10   Using the three-point linearization method yields $R = 2531.4\ \Omega$.

CHAPTER 3

3.1 We need $I_0 = 10$ mA, which requires $R_1/R_2 = 50$. Because the quantization interval is 48 nV, we have $\Delta R = 4.8$ $\mu\Omega$ and $\Delta T = 12$ μK.

3.2 $R = 2.5$ kΩ. An output voltage independent of wire resistance (R_w) requires $R_1 \gg R_w$. FSO $= 5$ V requires $R_3 = R_2 \times (5 \text{ V})/[(1 \text{ mA})(197.8 \ \Omega)] = 24.3 R_2$.

3.3 The three-point linearization method yields $R = 2212.4 \ \Omega$. If the transistor base current is negligible, $R_b > 4568 \ \Omega$. To obtain the required output voltage we need $R_3 = 2.85 R_1$ and $R_1 = 9.06 R_2$. For example, $R_2 = 1$ kΩ, $R_1 = 9.09$ kΩ, and $R_3 = 26.1$ kΩ (standard values) approximately fulfill those conditions. For the TLV2262 we have $V_{io} = 2500 \ \mu$V and $\Delta V_{io}/\Delta T = 2 \ \mu$V/$^\circ$C, and for the D package we have $\delta(= 1/\theta) = 5.8$ mW/$^\circ$C. The actual temperature will be $40.3 \,^\circ$C, so $V_{io} = 2536 \ \mu$V, which implies a $1.3 \,^\circ$C error.

3.4 $R_r = 3320 \ \Omega$, $D(0) = 1233$, $D(600) = 4083$, and $\Delta T = 0.27 \,^\circ$C (at $600 \,^\circ$C).

3.5 Because we have three conditions but seven resistors to select, we can pose four additional conditions. For example, $R_1 = 10$ kΩ to limit the thermistor current to 0.5 mA, and $R_2 = R_5$, $R_1 = R_6$, and $R_7 = R_0$ ($= 29,490 \ \Omega$) for symmetry. If $R_2 = 50$ kΩ, $R_3 = 13$ kΩ. R_4 matches the resistance seen from the op amp inputs. Then $R_4 = R_3 \| [(R_5 + R_6 \| R_7)/2] = 8.7$ kΩ.

3.6 $R_p > 100$ kΩ, $V_{ref} = 45.5$ mV, $R_2 = 212.3 R_1$. If $R_1 = 100 \ \Omega$, $R_2 = 21$ kΩ.

3.7 The minimal error is better calculated by means of a calculator/computer program. When load resistance is more than ten times the nominal resistance for the potentiometer, the error is about $0.15/k$, $k = R_m/R_T$ (load/potentiometer). We would choose the 600 Ω unit, with a maximal supply of 51 V. Sensitivity $= 141.6$ mV/$^\circ$.

3.8 By calling $c = R/R_T$ (series resistance/potentiometer resistance), the maximal error is obtained when $\alpha = (1 - c)/2$ and its value is $(c + 1)/(4k + c + 1)$.

3.9 The absolute error referred to the FSO for the circuit with split power supplies is $e = \alpha(1 - \alpha)(2\alpha - 1)[2\alpha(1 - \alpha) + 2k]$, whereas the error in Figure 3.7a is $e = \alpha(1 - \alpha)(2\alpha - 1)/[2\alpha(1 - \alpha) + k]$, hence larger because its denominator is smaller.

3.10 **a.** The maximal relative error is produced when x is maximum and its value is 4.76%.

     b. The resistor adjacent to the sensor on the other bridge arm must be

1000 Ω; the other two resistors must equal each other and be larger than 8900 Ω.

c. The maximal sensor power dissipation in this bridge is obtained when its resistance is maximum. The maximal acceptable voltage is 47.7 V. Sensitivity = 43 mV/N.

d. When $k = 1$, maximal sensitivity is 25 mV/N. It is lower than in the previous case because the maximal supply is now 10 V.

3.11 a. $G = 20$

b. The bridge common mode signal (5 V) yields a differential mode signal and also a common mode signal at the amplifier input. The differential signal is amplified by the differential mode gain and the common mode signal by the common mode gain. When $R_x = 1100$ Ω, the two bridge arms have a 24 Ω imbalance and the signal is 238 mV. Common mode gain is 0.0006. The relative error, assuming that the common mode input resistance R is very large, is $\varepsilon \approx 0.00063 + 504/R$. If R were infinite, there would still be the error due to the common mode gain, and therefore the final error would be low, but not zero.

3.12 The fractional resistance increment is 90×10^{-6}, the output voltage 900 μV, and the maximal error because of tolerance 3214 kg/cm^2.

3.13 Because a full sensor bridge always has balanced output impedance, the effective CMRR equals that for the differential amplifier. For a 12 V supply, the bridge common mode voltage is 6 V and the amplifier output 3 mV. The bridge output for a 1% strain gage change is 1.2 V. Hence, FSO = 1.203 mV.

3.14 We must choose the value for the resistors and the point where the bridge must be balanced, according to the desired linearity. The supply voltage for the bridge must be chosen to have a high sensitivity, but avoiding self-heating. Near the zero output, it will be impossible to achieve a relative error lower than 1% because there is always a given absolute error due to sensor self-heating. We must set a limit: for example, that it does not exceed 10% of the nonlinearity error. If the bridge is balanced at 0 °C and the theoretical sensitivity is that at 0 °C, then $R_0 = 900$ Ω and the ratio of bridge resistances must be higher than 43.4. Thus the maximal dissipation is produced when the sensor reaches its maximal value. The supply voltage must not exceed about 25 V.

3.15 The output is nonlinear. Because there are two resistors to be chosen, we can force two conditions: for example, a point in the transfer characteristic and its slope. That point can be chosen, for example, so that in the center of the temperature range the output voltage is also at the center of its range. The output voltage range can reach neither +12 V nor 0 V because the op amp does not allow it. Nevertheless, because the real output curve remains below the theoretical output, we can try a $+2$ V to $+12$ V range and then verify whether the output values will be possible or not. By so doing, we have $R = 3829\ \Omega$ and $R_2 = 246\ \Omega$. At $-10\,°\text{C}$ the output would be 0.38 V, and at $+50\,°\text{C}$ it would be 11.25 V. If the op amp does not allow these values at its output, then the range should be changed: for example, from 2.5 V to 11.5 V.

3.16 The output voltage is not proportional to the temperature. If as design criteria we take that at $0\,°\text{C}$ the output be 0 V and at $+40\,°\text{C}$ be $+10$ V, then we have two conditions. R_1 is limited by the maximal current on the PTC thermistor. V_{cc} can be chosen at 10 V.

    a. If in addition we require that the ratio between resistances at the balance be 1, we have $R_1 = R_2 = 8.35\ \text{k}\Omega$, $R = 28552\ \Omega$, $R_3 = 1723\ \Omega$.

    b. If we take $v_o = 0.25T$ as the theoretical response, the maximal error results at about $19\,°\text{C}$, which coincides with the temperature where the actual sensitivity is the same as the theoretical sensitivity.

3.18 $V_r < 2.65$ V. The fractional change in resistance for a $-100\ \text{kg/cm}^2$ load is 93×10^{-6}. If $2R_1 \gg 350\ \Omega$, then $R_2/R_1 = 86$. Because the offset voltage of A1 and A2 only contribute to the bridge excitation voltage, the output zero error is due to the offset of A3 and will be 8.7 mV, which is equivalent to $87\ \text{kg/cm}^2$ applied load.

3.19 The strain is $-140\ \mu\varepsilon$ and we need $G = 2721$. The output zero error (OZE) is the sum of op amp offset voltages times the gain. The output op amp reaches $45.5\,°\text{C}$, so its offset voltage is $150\ \mu\text{V} + (45.5\,°\text{C} - 25\,°\text{C}) \times (1.8\ \mu\text{V}/°\text{C}) = 187\ \mu\text{V}$. The other op amp must supply 10 mA to the bridge, so it reaches $60\,°\text{C}$ and its offset voltage is $213\ \mu\text{V}$. For $G = 1000$, we have $\text{OZE} = 402$ mV, equivalent to $40\ \text{kg/cm}^2$.

3.20 $G = 125$. The matching condition is $R_3 R_5 = R_4(R_6 + R_7)$. For example, $R_3 = R_4 = R_5 = 10\ \text{k}\Omega$, $R_6 = 8.2\ \text{k}\Omega$, and $R_7 = 2\ \text{k}\Omega$ (trimmer). To obtain $G = 125$ we need $R_2/R_1 = 61.5$. If $R_1 = 1\ \text{k}\Omega$, $R_2 = 61.5\ \text{k}\Omega$.

3.21 If the resistors in the output amplifiers are matched so that $R_2/R_1 = R_3/R_4 = k$, the output voltage is $v_o = 4(R_5/R_4)V_r x/(1 - x^2)$. Because we excite with constant voltage, each strain gage dissipates maximal power when it has its lowest resistance, which limits $V_r < 9$ V. For a 10 V output when $x = 0.02$, we need $R_5/R_4 = 13.88$. The linearity error is about 0.04 %. Resistor tolerance yields the maximal error when R_2 and R_4 are maximal while R_1, R_3, and R_5 are minimal. Instead of FSO $= 10$ V we would obtain FSO $= 5$ V.

3.22 To obtain a -10 V output from a 50 mV input we need $G = -200$. For example, $R_1 = 1\ \text{k}\Omega$ and $R_2 = 200\ \text{k}\Omega$. Because the OPA27 has a typical dc open-loop gain $A_{d0} = 10^6$, instead of gain -200 we will obtain -199.96. Therefore, instead of FSO $= 10$ V we will obtain 9.998 V—

that is, 2 mV error. $V_{io} = 25\ \mu V$, $I_n = 40$ nA, and PSRR $= 20\ \mu V/V$ for the op amp yield an additional 19 mV error. In the isolation amplifier, $V_{io} = 70$ mV, $e_G = 50$ mV, $e_{nlG} = 0.6$ mV, IMRR (140 dB), and PSRR $= 4$ mV/V contribute 127 mV to the output error. The overall error is 149 mV—that is, about 0.015·FSO.

CHAPTER 4

4.1 Static friction makes the system nonlinear because of hysteresis.
   a. If when the inclination increases $\mu > \tan\theta$, the core displaces by $x = Mg(\sin\theta - \mu\cos\theta)/K$. When the inclination decreases, we have $x = Mg(\sin\theta + \mu\cos\theta)/K$. Null hysteresis requires $\mu = 0$. The output voltage would then be $v_o = 2.45\sin\theta$.
   b. The phase shift is about $+90°$. It can be corrected by means of a phase lagging network at the LVDT secondary, although it will attenuate the amplitude somewhat.

4.2 In order to know the phase shift at a given frequency when there is a loading effect, we must know the resistance and inductance for the primary and secondary. Secondary parameters do not affect the specified phase shift for the LVDT (assumed in open-circuit condition). But they can be deduced from the output impedance because it is specified at two different frequencies. Nevertheless, it is not granted that the parameters for the secondaries are constant with frequency. However, since there are many data for the primary, from their analysis we conclude that impedance does not significantly vary at these frequencies. We obtain for the primary 69.5 Ω and 50 mH, for the secondary windings 968 Ω and 660 mH in total, and $R_2 \approx 101$ kΩ. The phase shift to be corrected is $+29°$. It can be corrected by placing 226 nF in parallel with the meter.

4.3 From the impedance data for the primary we obtain $R_1 = 1141$ Ω and $L_1 = 0.26$ H. The sensitivity is 0.279 Ω·s/(m·rad). When exciting with 12 V at 20 kHz the FSO increases to 637 mV.

4.4 If $R = |X_C|$ at the working frequency, in addition to the relative angle between stator and rotor, there is a fixed $-45°$ phase shift.

4.5 If we first consider a single rotor loop, an excitation current $i_e = I_e\sin(\omega t + \phi)$ yields $B_x = k_e i_e$, which is linked by the loop to yield $v_r = -d\Phi/dt = -A_r d(B_x\cos 2\pi nt)/dt$. If the loop has resistance R, then v_r yields a current $i_r = v_r/R$, which creates $B_y = k_r i_r \sin 2\pi nt$. In the detection winding, B_y induces $v_o = N_o A_o dB_y/dt$, which can be written

$$v_o = K\{A\sin[(\omega + 4\pi n)t + \phi] + B\sin[(\omega + 4\pi n)t + \phi] - \omega\pi n\sin(\omega t + \phi)\}$$

where $K = N_o A_o k_r A_r k_e I_e / R$, $A = [(2\pi n)^2 + \omega^2/2 + 3\omega\pi n]$, and $B = [(2\pi n)^2 + \omega^2/2 - 3\omega\pi n]$. If the rotor has N loops separated an angle $\theta_j = 2\pi/N$ ($1 \le j \le N$), the relative angle between each loop and the x axis is $\alpha_j = 2\pi nt + \theta_j$. The output voltage is then the sum of the voltage induced by each loop. Each of those voltages includes one term that does not depend on θ_j, along with one term multiplied by A and another multiplied by B that depend on θ_j. The sum of those N phasors with phase shift θ_j will cancel out the terms with A and B, thus leading to (4.77).

4.6   The criterion to be imposed is that the derivative of the output voltage with respect to the temperature be zero. We then have

$$\frac{1}{R_3 + R_4} \left(\frac{R_4}{R_5} - \frac{R_2}{R_1} \frac{R_3}{R_5} + 1 \right) = \frac{\beta}{\alpha - \beta} \frac{1}{R_0}$$

A design option is $R = R_2 = R_3 = R_4 = 1$ kΩ, and choose R_5 low enough so the voltage applied to the sensor is not too small. Selecting $R_5 = 100$ Ω yields $R_1 = 882$ Ω.

CHAPTER 5

5.1   The circuit displays a high-pass frequency response with a dc gain of $1 + (R_2 + R_4)/R_1$, along with a high-frequency gain of $1 + R_2(R_4/R_3 + R_4/R_2 + 1)/R_1 + R_4/R_3$. If $R_1 = R_2 = R_4$, the dc gain is 3. A gain of 1000 at 1 kHz requires $R_4/R_3 = 498.5$. If $R_3 = 1000$ Ω, the other resistors must be 498.5 kΩ. In order for C not to affect the circuit at 1 kHz, we can choose a corner frequency of 100 Hz, thus leading to $C = 1.6$ μF.

5.2   Circuit inspection (or analysis) shows that if $R_2/R_1 = R_3/R_4 = k$, the first stage is a pure integrator, and that the second stage is a differential differentiator. If the ratio between resistances at the other input integrator is also k and we select $R_5 = R_1$, $R_9 = R_{10} = R$, and $C_a = C_b$, the output is

$$v_o = -v_e(k + 1) \frac{R}{R_1} \frac{C_a}{\epsilon A} 2x$$

which is frequency-independent.

5.3   The output will be independent of frequency when the resistances of R_1 and R_2 are respectively larger than the impedances of C_1 and C_2 at the working frequency. If $C_1 = C_2 = C$ and $R_4/R_3 = R_6/R_5 = k$, the output voltage is

$$v_d = k v_e \frac{C_x - C_y}{C} = k v_e \frac{\epsilon A}{C} \frac{2x}{d^2 - 2x}$$

The slew rate (9 V/μs) limits the excitation frequency to less than 143 kHz. For an FSO = 10 V when $d = 1$ cm and $x = 1$ mm and 100 kHz excitation, if $k = 1$, we need $C = 8.85$ pF and $R \gg 179$ kΩ—for example, $R = 2$ MΩ. For R_3, R_4, R_5, and R_6 we select 10 kΩ.

5.4   The output from the bridge op amp is

$$v_b = v_e \frac{R_4 - R_2 Z_3 / Z_1}{Z_3 + R_4} = v_e \frac{R_4 - R_2 Z_3 Y_1}{Z_3 + R_4}$$

If $Y_1 = Y_{10} + \Delta Y_1$, the condition for balance is $R_4 = R_2 Z_3 Y_{10}$, which leads to $R_3 = 1/(\omega^2 R_{10} C_{10} C_3)$ and $C_3 = (R_2/R_4)[C_{10} + 1/\omega^2 R_{10} C_{10})]$. By taking the derivative of v_b with respect to Y_1 we can approximate

$$\Delta v_b = v_e \frac{-R_2 Z_3}{R_4 + Z_3} \Delta Y_1 = v_e \frac{-R_2}{R_4 Y_3 + 1} \Delta Y_1 = v_a (G + jB)$$

where G is the sensor conductance and B its susceptance ($\Delta Y_1 = G + jB$). If we call $\theta + \pi$ the phase shift of v_a with respect to the excitation voltage, the output of the comparator will have amplitude V_{sat} and phase shift θ. Then, $V_1 = -V_{sat} G$ and $V_2 = V_{sat} B$.

5.5   Circuits with op amps A2 and A3 are constant current sources whose intensity is controlled by two equal resistors and that drive the sensor. A3 compensates for the drop in voltage at the temperature where a null output is desired. The desired relations are $R_{10} = R_{11} = R_5 = R_6$; $R_8 = R_9 = R_3 = R_7$; $R_0/R_3 = R_2/R_1$.

5.6   The bridge output voltage is

$$v_1 = -v_e \frac{\Delta R}{2(2R + \Delta R)} = -v_e \frac{\alpha_{100}(T - 100)}{2[2 + \alpha_{100}(T - 100)]}$$

OA1 and OA2 make a composite noninverting amplifier that amplifies v_1 with gain $G = 1 + R_2/R_1$ in the passband. OA3 and the switch implement a switched-gain demodulator. OA4 is part of a low-pass filter that yields the average of the demodulated signal—that is, $2/\pi$ times the peak value of the demodulated signal. Therefore,

$$\frac{V_s}{V_e} = \frac{2}{\pi} \left(1 + \frac{R_2}{R_1} \right) \frac{\alpha_{100}(T - 100)}{2[2 + \alpha_{100}(T - 100)]}$$

From $\alpha_0 = 0.004/K$ and $R_0 = 100$ Ω we obtain $R_{100} = 140$ Ω and $\alpha_{100} = 0.00286/K$. Then, at 100.01 °C,

$$\frac{V_s}{V_e} = \frac{2}{\pi} \left(1 + \frac{R_2}{R_1} \right) (7.4 \times 10^{-6})$$

The maximal current through the sensor is 0.845 mA, which limits the rms excitation voltage to 237 mV, hence the peak value to 335 mV. To obtain a 10 mV output for a 0.01 °C increment we need $G = 6567$. C_1 and C_2 limit the gain at, respectively, low and high frequencies, but should not attenuate the carrier frequency. If $R_2 = 10$ MΩ, we need $R_1 = 1523$ Ω, and we can select $C_1 = 1.5$ μF and $C_2 = 1.5$ pF. In the demodulator, we select $R_3 = R_4 = 1$ kΩ, so $R_5 = 500$ Ω.

5.7   If $C = 1$ nF, we need $R = 7958$ Ω for the Wien bridge to oscillate at 20 kHz. The LVDT yields 637 mV (peak) at full scale. Hence, we need $G = 18.8$, which can be distributed between the two stages for the instrumentation amplifier—for example, 5 and 3.77. If $R_1 = R_3 = 10$ kΩ, we need $R_2 = 20$ kΩ and $R_4 = 37.7$ kΩ. The minimal slew rate is 1.5 V/μs.

CHAPTER 6

6.1   By assuming that thermocouple sensitivity is constant over the temperature range for reference junctions, the only condition required is $2T_1 = T_2$. This technique simulates a 0 °C reference temperature without using melting ice.

6.2   **a.** If we measure the voltage difference between both thermocouples (voltmeter between both "iron" terminals, "constantan" terminals being connected together), according to thermocouple laws, ambient temperature does not influence the voltage. For a temperature difference of 80 °C, the voltmeter reading will be 4.535 mV.

   b. It has no influence (first law).

   c. We need a bridge output equal but opposite in sign to thermocouple voltage in the ambient temperature range and $R = R_T(0 \,°C)$, because 0 °C is the reference temperature for thermocouples. Because the bridge output is not proportional to the temperature, the other criterion can be, for example, to have an exact compensation at the center of the compensation range (20 °C), or that the bridge sensitivity at 20 °C be equal to the average thermocouple sensitivity. The first criterion yields $k \approx 104$; the second, $k \approx 103$. If $k = 103.5$, the error at 30 °C is 0.05 °C.

6.3   **a.** $v_o = (E_T - E_{T_a})\left(1 + \dfrac{R_2}{R_1} + \dfrac{R_2}{R_4}\right) - (10 \text{ V})\dfrac{R_2}{R_4} + IR_1$

   b. $G = 1778$.

   c. $R_2 = 405R_3 = 7.38R_4$. If $R_3 = 100$ Ω, then $R_2 = 40.5$ kΩ and $R_4 = 5.5$ kΩ.

6.4   **a.** $v_o = E_T(A/B) - E_{T_a}(A/B) + v_c$.

   b. $v_c = I_{T_a}T_a R_b \dfrac{R_2}{R_1 + R_2} - V_z \dfrac{R_3 + \alpha R_p}{R_3 + R_4 + R_p} \dfrac{R_2}{R_1 + R_2}$.

c. $R_1 = 240R_2$. If $R_2 = 100\ \Omega$, then $R_1 = 24\ k\Omega$.

d. $R = 8100\ \Omega$, $R_4 + (1 - \alpha)R_p = R_3 + \alpha R_p$. If $R_3 = 100\ k\Omega$, then $R_4 = 150\ k\Omega$ and $R_p = 10\ k\Omega$.

6.5 a. $v_o = G\left[V_r\left(\dfrac{R_{T_a}}{R + R_{T_a}} - \dfrac{R_3}{R_1 + R_3}\right) + S_K(T - T_a)\right]$.

b. $R_0 R_1 = R_2 R_3$.

c. If $R_2 \gg R_0$ and $R_1 \gg R_3$, we need $R_2 = 95625\ \Omega$. Also, $R_1 = 95625\ \Omega$ and $R_3 = 100\ \Omega$.

d. $G = 261$.

e. Because the output sensitivity is 10 mV/°C, we need $V_{io} < 3.8\ \mu V$, which is very restrictive. Hence, we would need to null out the actual offset voltage.

6.6 To obtain a null output at $0\,°C$ we need

$$I_{T_a} - E_{T_a}\left(\frac{1}{R_1} + \frac{1}{R_2} + \frac{1}{R_3}\right) = \frac{V_r}{R_1}$$

To obtain the desired sensitivity we need

$$10\ \text{mV/°C} = S_J\left[1 + R_2\left(\frac{1}{R_1} + \frac{1}{R_3}\right)\right]$$

where $S_J = 52.146\ \mu V/°C$ is obtained from thermocouple tables. These equations lead to $R_1 = 36{,}630\ \Omega$ and, if $R_2 = 10\ k\Omega$, $R_3 = 51.8\ \Omega$.

6.7 The input impedance for the meter is so high that the sensor can be considered to be open-circuited. The output voltage is that obtained in static conditions but considering the voltage divider effect between sensor and meter impedances. Thus the 9 mV at sensor output terminals reduces to about 1.5 mV. The peak-to-peak deformation is about 2.2 pm.

6.8 a. 40 kV/cm.

b. 8889 V/cm.

c. 1076 Hz.

d. 47.5 nF.

e. 83.3 V/cm.

CHAPTER 7

7.1 The data sheet for the OP77GP yields the following maximal values: $V_{io} = 150\ \mu V$, $\Delta V_{io}/\Delta T = 1.2\ \mu V/°C$, $I_b = 6\ nA$, $\Delta I_b/\Delta T = 60\ pA/°C$, $I_{io} = 4.5\ nA$, $\Delta I_{io}/\Delta T = 85\ pA/°C$, $P_d = 75\ mW$. It also lists the following minimal values: $A_{d0} = 2 \times 10^6$, PSRR $= 5\,\mu V/V$ at dc, and PSRR ≈ 75 dB at 120 Hz. For the "P" package, $\theta_{ja} = 103\,°C/W$. From these data, the actual operating temperature is about $8\,°C$ above the ambient temperature (i.e., $48\,°C$); and IZE $= 175\,\mu V$, mostly contributed by V_{io} and its drift because the estimated input currents are about 10.7 nA (I_n) and 4.1 nA (I_p). If the power supply variations are very slow, the equivalent input error is negligible, but if they are 120 Hz ripples, the corresponding input error is about $27\,\mu V$. The gain error because of the finite A_{d0} is 5×10^{-4}, hence negligible, but resistor tolerance yields a 2% uncertainty in the gain. Because resistor drift is very small and currents through resistors are small, most drift will be IZE drift due to that of V_{io}.

7.2 **a.** To compensate bias currents we need $R_1 = R_2 \| R_3$. R_2 includes the resistance seen from potentiometer wiper to the left. To make sure that this resistance remains low, we should place a resistor $R_0 \ll R_2$ between the potentiometer wiper and ground and a series resistor R_s between the wiper and R_0 to avoid loading the potentiometer. Because R_1 is very large and a voltage gain of 400 needs $R_3 = 399R_2$, R_2 should be similar to R_1, which would result in a very large R_3. Alternatively, we can place $R_b = R_1$ in series with the op amp inverting input and select $R_3 = 3.97$ MΩ, $R_2 = 10$ kΩ, $R_a = 10$ kΩ, $R_s = 10$ kΩ, and $R_0 = 100$ Ω.

   b. The actual output is $V_o = I_s Z_e A_d [1 + A_d R_2/(R_2 + R_3)]$, where Z_e is the parallel of R_1, sensor impedance, and op amp common mode input impedance (basically, a capacitance C_c). At 10 kHz, $A_d \approx 1/(j400)$, so the circuit gain is $200\sqrt{2}$ instead of 400.

7.3 **a.** $G = 16$. If $R_1 = 1$ kΩ, we need $R_2 = 15$ kΩ.

   b. $R = 212$ MΩ, $C_2 = 663$ pF.

   c. $R_0 = 212$ MΩ, $C_0 > 1$ nF (to reduce noise bandwidth for R_0).

7.4 **a.** $G = 10.6$.

   b. If $R_4 \| R_5 \| R_6 \ll R_4$, a corner frequency of 20 Hz needs $R_4 = 53$ MΩ. To have an output voltage centered at midrange of the supply voltage, we need

$$\frac{1}{2} = \frac{R_5 \| R_6}{R_5} + R_3 \left(\frac{R_5 \| R_6}{R_5} \frac{1}{R_1 \| R_2} - \frac{1}{R_1} \right)$$

   If in addition we impose $R_5 = R_6$ and $R_1 = R_2$, to obtain the desired gain we need $10.6 = 1 + 2R_3/R_1$. If $R_1 = 20$ kΩ $= R_2$, R_3 must be 96 kΩ. To reduce the number of different resistor values we can select $R_5 = R_6 = 20$ kΩ.

   c. For example, $R_7 = R_9 = 1$ MΩ and $R_8 = 10$ kΩ.

7.5 **a.** The output voltage is $v_o = -i[R_2 + R(1 + R_2/R_1)]$. If $R = 10$ MΩ, we need $R_2/R_1 = 9$—for example, $R_1 = 10$ kΩ and $R_2 = 90$ kΩ.

    b. V_{io} and $I_n \times R_s$ ($R_s = 10$ GΩ), and their respective drifts, are both amplified by ten to yield OZE = 5.15 V with about 1 V/°C drift.

    c. $Z_i \approx R(1 + R_2/R_1)/A_d$, where A_d is the op amp open-loop gain.

7.6 $RC = R_1 C_1$, and $R_1 \ll R$ and $C_1 \gg C$.

7.7 **a.** $S_v = 31$ μV/Pa.

    b. $C = 2$ nF.

    c. $R > 24.2$ MΩ.

    d. R_1 limits the passband to frequencies below 9824 Hz.

7.8 The conditions to be fulfilled are $C_0 = G_{min} \times (1$ nF$)$ and $C_0 = G_{max} \times (10$ pF$)$. If $G_{min} = 1$, we need $C_0 = 1$ nF and $G_{max} = 100$. For example, $R_1 = R_3 = R_4 = 10$ kΩ and $R_2 = 100\ \Omega + 9900\ \Omega$. Then, $R_0 \gg 1.5$ GΩ and $R \ll 150$ kΩ.

7.9 Under ideal conditions, 10 pF in the charge amplifier would be enough, and the frequency response would be flat. But a bias resistor R_0 must be added; and sensor and cable resistance and capacitance, along with the op amp finite gain A_d, must be considered. If $R_0 = 16$ GΩ, the corner frequency would be 1 Hz. Therefore, either R_0 should be higher or the external amplifier should have a corner frequency well below 1 Hz. This means that the op amp should have a very low bias current, and that bias current would flow in part through the sensor and cable leakage resistance.

7.10 **a.** $C_0 = 100$ pF.

    b. The upper corner frequency limits R to less than 16 kΩ. Bias currents limit R_1 to less than 10 GΩ. If $R = 1$ kΩ and $R_1 = 1$ GΩ, the lower corner frequency requires $C_1 \gg 208$ pF. We can select $C_1 = 100$ nF, so R_1 could be smaller.

7.11 The hot spot temperature is 100 °C and the drift after 1000 h for a 95% confidence level is 1%. Since 5 years is about 43,800 h, the actual drift can be estimated as 6.6 times the drift for 1000 h. Therefore, the 95% confidence interval is [93.4 kΩ $< R <$ 106.6 kΩ].

7.12 The bridge gives a differential output voltage proportional to the strain and with balanced 60 Ω output resistance. The maximal supply voltage for the bridge is 6 V. Because the bridge output signal is very small, the instrumentation amplifier must have a very high gain—for example 1000—so the noise from its (internal) input stage predominates over that of the output stage. Because the bridge output resistance is very small, the contribution of noise currents is negligible. Noise voltage for the AD624 is specified as a peak-to-peak value from 0.1 Hz to 10 Hz and as rms spectral density at 1 kHz. For a gain of 1000, the noise spectral density from 10 Hz to 1 kHz can be assumed constant. Because noise in different frequency bands is random and uncorrelated, the total noise is obtained by adding noise power. If the factor for converting rms values to peak-to-peak values is 6.6, the total noise is 130 nV. The resolution determined by this noise is about 2200 Pa.

7.13   We select $R_1 = 1$ kΩ, $R_2 = 200$ kΩ, $C_2 = 8.2$ nF, $R_3 = 330$ kΩ, $C_3 = 3$ μF, $R_4 = 230$ Ω, $C_4 = 680$ nF, and $R_s = 1$ kΩ. R_2 and C_2 determine $f_2 = 97$ Hz, R_3 and C_3 determine $f_3 = 0.16$ Hz, and R_4 and C_4 determine $f_4 = 1018$ Hz. Thus we have a passband amplifier from 0.16 Hz to 97 Hz, with an additional low-pass first-order filter. The passband for each noise source, however, is not the same as for the signal. For simplicity, we assume that noise bandwidth equals filter bandwidth. The bandwidth for the thermal noise of R_s is $B = 97$ Hz $- 0.16$ Hz ≈ 97 Hz, and the noise voltage (RTI) is 40 nV. The bandwidth for the thermal noise from R_1 is limited to 97 Hz. We assume that, in addition, the storage oscilloscope will set a minimal frequency about 0.1 Hz. Hence the noise bandwidth is about 97 Hz and the noise voltage (RTI) is 40 nV. Similarly, the noise bandwidth for the thermal noise of R_2 is about 97 Hz and the noise voltage (RTO) is 594 nV. The noise bandwidth for the thermal noise of R_3 is 0.16 Hz $- 0.1$ Hz $= 0.06$ Hz, and the noise voltage (RTI) is 18 nV. The noise bandwidth for the op amp noise voltage is 97 Hz $- 0.1$ Hz ≈ 97 Hz. The maximal peak-to-peak op amp voltage noise from 0.1 Hz to 10 Hz is 0.65 μV, and the maximal noise voltage spectral density is 20 nV/$\sqrt{\text{Hz}}$ at 10 Hz and 13.5 nV/$\sqrt{\text{Hz}}$ at 100 Hz. Dividing by 6.6 to convert peak-to-peak values into rms values, and using the data at 10 Hz for a worst-case prediction, the noise voltage (RTI) is 220 nV. Op amp current noise is also separately specified from 0.1 Hz to 10 Hz (35 pA, peak-to-peak) and at 10 Hz (0.9 pA/$\sqrt{\text{Hz}}$) and 100 Hz (0.27 pA/$\sqrt{\text{Hz}}$). Using the same approach as above for the op amp voltage noise, the noise current at the inverting input yields 2.1 μV (RTO), and the noise current at the noninverting input yields negligible contribution because of its reduced noise bandwidth (0.16 Hz to 0.1 Hz). The noise bandwidth for the thermal noise of R_4 is about 1018 to 0.1 Hz, and the noise voltage (RTO) is 62 nV. Because the noise gain for the circuit is 201, the contribution from the op amp voltage noise dominates and the rms input noise voltage is 220 nV, equivalent to 1.5 μV (peak-to-peak).

7.14   The equivalent circuit for noise analysis is that in Figure 7.21. Because $C = 2$ pF and $C_s = 20$ pF, and the op amp has $f_T = 1.8$ MHz, the circuit bandwidth is about 180 kHz. Therefore, the noise bandwidth is determined by the external filter from 10 Hz to 10 kHz. The output power spectral density is

$$e_{no}^2 = \left(1 + \frac{R}{R_s}\right)^2 e_n^2 + \left(\frac{R}{R_s}\right)^2 e_{ts}^2 + e_t^2 + i_n^2 R^2$$

and the output noise power will be the result of integrating e_{no}^2 over the noise bandwidth. For the TLC2201B, $i_n = 0.6$ fA/$\sqrt{\text{Hz}}$, and $e_n = 35$ nV/$\sqrt{\text{Hz}}$ at 10 Hz and $e_n = 15$ nV/$\sqrt{\text{Hz}}$ at 1 kHz. From these data we determine $f_{ce} = 13$ Hz. The thermal noise is $e_{ts} = 1.3$ μV for R_s and $e_t = 0.4$ μV for R. By integrating e_{no}^2 we obtain $E_{no} = 40$ μV, contributed mostly by R.

CHAPTER 8

8.1 Below 60,000 r/min (1000 Hz), measuring period rather than frequency yields better resolution.

8.2 28.248 MHz and 10 s.

8.3 $\Delta T = 11\,°C$ and the corresponding correction factor is 6.6 με.

8.4 In Figure 8.28, a convenient place for a sensor is R_2 because it is grounded. The output frequency will be $f(-40) = 0.333f(0) = 333$ Hz, $f(85) = 4.275f(0) = 4275$ Hz. If $C_1 = C_2 = C$, in order for $f(0) = 1$ kHz we need $R_1 C^2 = 1.84 \times 10^{-12}$. If $R_1 = 10$ kΩ, we need $C = 13.6$ nF. To ensure oscillation, we can select $R_3 = R_1$, $R_4' = 20$ kΩ, and $R_4'' = 210$ kΩ.

8.5 The oscillation period is $2t_1 = 2R_1 C_1 \ln(R_2/R_T) = 2R_1 C_1(\ln R_2 - \ln R_0 - \Delta T \ln b)$, where $b = 1.028$. If $R_2 = R_0 = (7989\ \Omega)$, we need $R_1 C_1 = 0.9$ ms. If $R_1 = 10$ kΩ, we need $C_1 = 90$ nF. For simplicity, $R = 10$ kΩ.

8.6 When the astable output is high, the signal connected to the up input has frequency $f_r/2$ and the signal connected to the down input is at low level. When the astable output is low, the signal connected to the up input is at low level and the signal connected to the down input has frequency f_r. The counter output will therefore be

$$N = t_H \times (f_r/2) - t_L \times f_r$$
$$= 0.693 \times 2R_0 C(f_r/2) - 0.693 \times R_0(1 - x)C$$

where t_H and t_L are, respectively, the times to charge and discharge C, and $x = G \times \varepsilon$ is the fractional change of resistance ($G = 2$ is the gage factor). In order to have $N = 1$ when $\varepsilon = 1$ με we need $1 = 0.693 \times (120\ \Omega) \times C \times f_r \times 2$. If $f_r = 1$ MHz, we need $C = 6$ nF.

8.7 The compensation circuit must provide an adequate average supply voltage whose thermal drift counterbalances the oscillator drift. The counter output is

$$N = \frac{f_c}{2 \times f_r/100} = \frac{f_c}{0.559} \frac{100}{2}(C_H + C_G)$$

where the 2 factor considers that the gate is open only during half a period of the variable oscillator. The sensor capacity is $C_H = C_0\{1 + a[\text{RH} + b(T - 20\,°\text{C})] + P(\text{RH})\}$, where $b = 0.05\ \text{RH/K}$ and $P(\text{RH})$ is a polynomial describing nonlinearity. Gate capacitance depends on supply voltage according to $C_G = C_{G0} + c(V_{cc} - 6\ \text{V})$. When analyzing the dependence of C_G on V_{cc}, we have

$$C_G = \frac{0.559N}{50Rf_r} - C_H(50\%, 20\,°\text{C})$$

From the measurements at 6 V and 9 V we respectively deduce $C_{G0} = 3.6\ \text{pF}$ and $C_{G9} = 1.6\ \text{pF}$. Therefore, $c = -0.67\ \text{pF/V}$. The supply voltage for the oscillator is $V_{cc} = G \times T - V_0$, where $G = (1\ \mu\text{A/K}) \times R_2$, and $V_0 = (273\ \mu\text{A}) \times R_2 - (V_c R_2)/R_1$. $C = C_H + C_G$ will not depend on the temperature when $C_0 \times a \times b + c \times G = 0$ and $C_0 \times a \times b \times (-20\,°\text{C}) + C_{G0} + c(V_0 - 6\ \text{V}) = 0$. From the first equation we obtain $G = 11.25\ \text{mV/K}$ and, hence, $R_2 = 11,250\ \Omega$. From the second equation we obtain $V_0 = 375\ \text{mV}$, and, if $V_r = 1.235\ \text{V}$ (AD589), we need $R_1 = 5,153\ \Omega$. $R_3 = R_1 \| R_2$.

8.8 The maximal counting time to prevent overflowing is 127.5 µs. Because the maximal sensor resistance is 2956 Ω, we need $C < 22.7\ \text{nF}$. If we select $C = 20\ \text{nF}$ (standard), the number of counts will be 224 at 500 °C and 210 at 450 °C. Because of the uncertainty in each counting process, the maximal uncertainty for the difference is 2 counts, which implies a 7 °C uncertainty. The minimal high output voltage is 0.7 V below the supply voltage, in which case the time to the threshold would be 263 µs and the counter would overflow.

8.9 We need 11 bit resolution in the measurement of the period, which lasts for 49.2 ms in the worst case (246 cycles of the slowest frequency).

8.10 We need 10 bit resolution in the measurement of the period and must count 173 cycles of the input signal.

8.11 **a.** The RH dynamic range is 40, $C_{min} = 108.8\ \text{pF}$, and $C_{max} = 140.8\ \text{pF}$. If the frequency of the relaxation oscillator is $f = k_1/C$, we need to count a number k of cycles large enough to have the desired dynamic range $(k > 1.5k_1\ \mu\text{s/pF})$ but small enough to count for less than 10 ms $[k \times (140.8\ \text{pF}/k_1) < 10\ \text{ms}]$. Therefore, $1.5\ \mu\text{s/pF} < k/k_1 < 71\ \mu\text{s/pF}$. If we select $k = 10k_1\ \mu\text{s/pF}$, since the internal counters of the 8051 have 16 bits, k should be less than 255, and therefore $k_1 = 25.5\ \text{pF/µs}$. When $\text{RH} = 43\%$ the relaxation oscillator would oscillate at 209 kHz, which is a bit too high. Selecting $k = 50k_1$ µs/pF would yield $k_1 = 5.1$ pF/µs and $f \approx 42$ kHz at 43% RH.

b. At RH $= 10\%$, $f = 112$ kHz, the calculated change in capacitance would be -14.8 pF and the result would be RH $= 6\%$. Hence, $e = 4\%$. At RH $= 90\%$, $f = 86.6$ kHz, the calculated change in capacitance would be 16.3 pF and the result would be RH $= 83.7\%$. Hence, $e = 6.3\%$.

8.12 a. 41 MHz

b. $0.36°$ and 12.5 µs.

CHAPTER 9

9.1 a. $G = 5$.

b. $G = 9$.

c. $R_1 = 5$ kΩ. If $R_4 = 100$ kΩ, we need $R_3 = 100$ kΩ.

d. $R_2 = 10$ kΩ.

e. The maximal quiescent current for the TL052 is 5.6 mA and the "P" package has $\theta_{ja} = 125$ °C/W. Hence, the actual temperature for op amps is 46 °C. If the instrumentation amplifier concentrates its gain in the first stage, we have OZE $= G(V_{io1} - V_{io2}) + 2V_{io3}$. For a worst-case condition, we suppose $V_{io1}(25$ °C$) = 0.8$ mV with maximal drift $(25$ µV/°$)$, $V_{io2}(25$ °C$) = -0.8$ mV with typical drift $(6$ µV/°C$)$, and $V_{io3} = V_{io1}$. Then, OZE $= 12.6$ mV, which implies about 1.3 °C error.

9.2 If $R_1 = 2R_2$, the gain for the instrumentation amplifier must be 16.74. Therefore, $R_6/R_5 = R_3/R_4 = 16.74 - 1$. We can select $R_5 = R_4 = 1$ kΩ, and $R_6 = R_3 = 15.8$ kΩ, all with a 0.1% tolerance for best CMRR (72 dB minimum with this tolerance and differential gain). $R_1 = 100$ kΩ yields acceptable current levels.

9.3 The design conditions are $V_p = (273$ K$) \times R_b \times (1$ µA/K$)$ and $R_a \| R_b = 52$ Ω. V_b can be freely chosen inside the range accepted by the AD590. R_b should be less than, say, 10 kΩ to avoid any voltmeter loading effect. V_p must be small so the current on R_a is small. If we choose $V_p = 1.35$ V (mercury cell), we need $R_b = 4945$ Ω and $R_a = 52.55$ Ω.

9.4 We must consider that the circuit for reference voltage adjustment at the op amp has a finite equivalent output resistance that depends on the adjustment level. We obtain $R = 52$ Ω and $R_g = 303$ kΩ. The voltage to be compensated by adjusting the potentiometer is 14.272 mV; therefore the resistance from the wiper to ground will be 628 Ω. Offset voltage and bias currents yield a 29.4 mV output. This can be compensated by the potentiometer, but a modified gain will result, thus requiring us to readjust R_g.

9.5 $G = 185.2$.

9.6 The parallel capacitance of the photodiode connected to the OPA121 is about 14 pF, and for the OPA121, $f_T = 2$ MHz. The transimpedance is a second-order low-pass function. For a flat frequency response we impose $\zeta = 1/\sqrt{2}$, which requires $C = 5$ pF. When $\zeta = 1/\sqrt{2}$ the -3 dB bandwidth equals the natural frequency, here 409 kHz.

9.7 $v_o = (1 + R_2/R_1)(kT/q)\ln(1 + i_p/I_s)$, where i_p is the photocurrent and I_s is the reverse saturation current. The offset voltage appears at the output multiplied by $1 + R_2/R_1$, and the bias current adds to i_p.

9.8 The design equations are $R_{min} = (V_{cc,max} - V_{IL})/(I_{OH} + I_{IL})$ and $R_{max} = (V_{cc,min} - V_{IH})/(I_{OH} + I_{IH})$, which lead to $11{,}125\ \Omega < R < 19{,}200\ \Omega$. For example, $R = 15$ kΩ.

INDEX

571